MC 68HC11   An Introduction

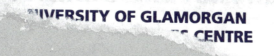

Software and Hardware Interfacing

MC 68HC11   An Introduction
Software and Hardware Interfacing

Han-Way Huang
Mankato State University

West Publishing Company
Minneapolis/St. Paul,
New York, Los Angeles,
San Francisco

Copyediting: Pamela S. McMurry
Proofreading: Chris Thillen
Composition: G & S Typesetters, Inc.
Artwork: G & S Typesetters, Inc.
Cover Image: Courtesy of Motorola

WEST'S COMMITMENT TO THE ENVIRONMENT

In 1906, West Publishing Company began recycling materials left over from the production of books. This began a tradition of efficient and responsible use of resources. Today, up to 95 percent of our legal books and 70 percent of our college and school texts are printed on recycled, acid-free stock. West also recycles nearly 22 million pounds of scrap paper annually—the equivalent of 181,717 trees. Since the 1960s, West has devised ways to capture and recycle waste inks, solvents, oils, and vapors created in the printing process. We also recycle plastics of all kinds, wood, glass, corrugated cardboard, and batteries, and have eliminated the use of Styrofoam book packaging. We at West are proud of the longevity and the scope of our commitment to the environment.

Production, Prepress, Printing and Binding by West Publishing Company.

 PRINTED ON 10% POST CONSUMER RECYCLED PAPER   Printed with Printwise
Environmentally Advanced Water Washable Ink

British Library Cataloguing-in-Publication Data. A catalogue record for this book is available from the British Library.

COPYRIGHT ©1996   by WEST PUBLISHING COMPANY
                   610 Opperman Drive
                   P.O. Box 64526
                   St. Paul, MN 55164-0526

Library of Congress Cataloging-in-Publication Data

Huang, Han-Way.
    MC 68HC11 an introduction : software and hardware interfacing /
Han-Way Huang.
        p.   cm.
    Includes index.
    ISBN 0-314-06735-3 (soft : alk. paper)
    1. Digital control systems.   2. Electronic controllers.
    I. Title.
    TJ223.M53H95   1996
    629.8'9—dc20   96-4815
                                                    CIP

CONTENTS

PREFACE

Recent advances in electronic semiconductor technology have resulted in the development of highly integrated microprocessors and microcontrollers. The invention of microprocessors revolutionized the electronics industry. Microprocessors are used not only in many desktop personal computers but also in many more embedded applications such as fax and copying machines, laser printers, and communication controllers. However, a microprocessor needs external memories, peripheral interface chips and other logic circuits to build a complete system, and these requirements complicate the design of microprocessor-based products. A microcontroller incorporates a small amount of memory, timers, and other peripheral functions in one chip (along with the CPU), and these extra resources are adequate for many embedded applications. Microcontrollers not only simplify the design of many embedded products but also have the advantage of smaller size and lower power consumption.

This book is intended for the engineer, technologist or technician who is interested in learning how to use a microcontroller, in particular, the Motorola 68HC11, in the design of an instrument or some other device. This book has four parts. It begins with the programming model of the 68HC11 and explores the CPU registers, addressing modes, instructions, and the instruction execution process. The second part introduces assembly language programming using the 68HC11 instruction set. Single-and multi-precision arithmetic, looping, delay creation, branching, logical operations, and software and hardware development tools for the 68HC11 are all covered. The third part discusses data structure manipulations, subroutine calls, and input/output using the 68HC11 EVB library functions. Finally, the book covers all the I/O functions, including interrupts and resets, external memory expansion, the 68HC11 I/O ports and parallel I/O, timer functions, asynchronous and synchronous serial I/O interfacing, and A/D conversion. Both the programming and the hardware-interfacing issues are explored in detail. Software and hardware development go hand in hand throughout the book.

Microcontroller expansion methods are examined for the recourse they provide when on-chip resources turn out to be insufficient for the task at hand. Timing and electrical loads are thoroughly examined. A wide variety of I/O devices are explored, and their interactions with the 68HC11 are examined in detail.

This book is intended to be used in a one-semester or two-quarter course at the introductory level for students with or without programming background. The reader is, however, assumed to have some knowledge of number systems. An extensive appendix on number systems is provided for those who are not familiar with this subject. Because this book also provides many more advanced examples, it can be used by senior-level students or practicing engineers who want to learn the 68HC11.

Organization of the Book

Chapter 1 gives a brief overview of the hardware and software of a computer system; it introduces different types of memory technologies and presents the programming model and the instruction execution cycles of the 68HC11 microcontroller. Chapter 2 introduces assembly language syntax and assembler directives and explores the implementation of single- and multi-precision arithmetic, logic operations, delay times, and program loops using appropriate 68HC11 instructions. Software and hardware development tools are also reviewed in this chapter. Chapter 3 describes different types of data structures and their operations and covers the issues related to subroutine calls, including parameter passing, returning, results, and allocation of local variables. Many examples using the 68HC11 EVB library routines are included at the end of this chapter.

Chapter 4 introduces the setup of the 68HC11 operation mode and the DRAM and SRAM technologies and explores the memory system design issues, including memory space assignment, decoding methods, and the decoder design. It presents the conventions used in timing diagrams and the 68HC11 read and write bus cycle timing diagrams. An extensive example explains the design process for adding an external SRAM chip to the 68HC11. Many diagrams are included to explain the timing verification process.

Chapter 5 introduces interrupts and reset handling and explores the details of the 68HC11 interrupt and reset mechanisms. Chapter 6 presents the basic input/output concepts and examines the 68HC11 I/O ports. Examples demonstrate the interfacing of input/output devices such as DIP switches, LEDs, seven-segment displays, D/A converters, keyboards, and printers to the 68HC11.

Chapter 7 discusses the functions and applications of the 68HC11 timer system, including the input-capture, output-compare, real-time interrupt, and pulse accumulator functions. Numerous examples demonstrate measurement of frequencies and pulse widths, generation of delays and digital waveforms, generation of periodical interrupts, capture of arrival times, and so on.

Chapter 8 discusses asynchronous serial communication and the serial communication interface (SCI) of the 68HC11. Examples illustrate hardware interfacing, timing analysis, and programming both the SCI and the Motorola 6850 chip. Chapter 9 presents the 68HC11 serial peripheral interface (SPI). This versatile interface is a protocol that enables any peripheral device that conforms to it to interface to the 68HC11. A table listing major components from several manufacturers that conform to this protocol is provided. Chapter 10 deals with the 68HC11 A/D converter. This chapter explains not only the operation of the 68HC11 A/D converter but also analog signal conditioning and interpretation and processing of results.

Numerous exercises and lab assignments at the end of each chapter enhance the study of each topic. Solutions to the exercises and transparencies on disks are available for instructors.

Acknowledgments

This book would not have been possible without the help of a number of people, and I would like to express my gratitude to all of them. I would like to thank Jay Farnam, Chad Friedler, Tony Plutino, and others at Motorola who have supported my teaching over the years. I would also like to thank my EE and EET students, who motivated me to write this book. Many thanks to my editor Christopher Conty of West Publishing for his enthusiastic support during the preparation of the book. I also appreciate the outstanding work of the production staff at West Publishing, especially Emily Autumn and Tamborah Moore. I would also like to express my thanks for the many useful comments and suggestions of my colleagues who reviewed this text during the course of its development, especially Roger A. Mussell and William Rainey, Harper College; Paul D. Johnson, Grand Valley State University; Elvin Stepp, University of Cincinnati; David G. Delker, Kansas State University; John Blankenship, DeVry Institute of Technology; Gene S. Bruner, Oregon Institute of Technology; Mohammed Rachedine, Hampton University; Paul Butler, Ocean County College; Hossein Heydari, Penn State University at Harrisburg; and Charles Swain, Rochester Institute of Technology. Finally, I am grateful to my wife, Su-Jane, and my sons, Craig and Derek, for their encouragement, tolerance, and support during the entire preparation of this book.

Han-Way Huang
December 1995

1

INTRODUCTION

TO

MOTOROLA

68HC11

1.1 Objectives

After you have completed this chapter, you should be able to:

- define or explain the following terms: computer, processor, microprocessor, microcontroller, hardware, software, cross assembler, RAM, SRAM, ROM, EPROM, EEPROM, byte, nibble, bus, KB, MB, mnemonic, opcode, and operand
- explain the differences between the immediate, direct, extended, indexed, relative, and inherent addressing modes
- disassemble machine code into mnemonic assembly language instructions
- explain the 68HC11 instruction execution cycles

1.2 What is a Computer?

A computer is made up of hardware and software. The computer hardware consists of four main components: (1) a processor, which serves as the computer's "brain," (2) an input unit, through which programs and data can be entered into the computer, (3) an output unit, on which computational results can be displayed, and (4) memory, in which the computer software programs and data are stored. Figure 1.1 shows a simple block diagram of a computer. The processor communicates with memory and input/output (I/O) devices through a set of signal lines referred to as a *bus.* The common bus actually consists of three buses: a *data* bus, an *address* bus, and a *control* bus.

1.2.1 The Processor

The processor, which is also called the central processing unit (CPU), can be further divided into three major parts:

Registers. A register is a storage location in the CPU. It is used to hold data and/or a memory address during execution of an instruction. Access to data in registers is faster than access to data in memory. Registers play an essential role in the efficient execution of programs. The number of registers varies greatly from computer to computer.

Arithmetic logic unit. The arithmetic logic unit (ALU) is the computer's numerical calculator and logical operation evaluator. The ALU receives data from main memory and/or registers, performs a computation, and, if necessary, writes the result back to main memory or registers.

Control unit. The *control unit* contains the hardware instruction logic. The control unit decodes and monitors the execution of instructions. The control unit also acts as an arbiter as various portions of the computer system compete for the resources of the CPU. The activities of the CPU are synchronized by the system clock. All CPU activities are

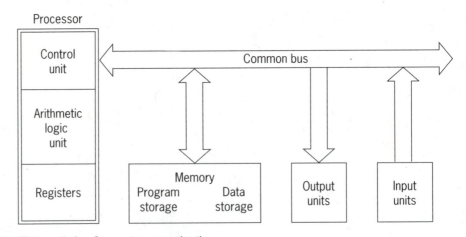

Figure 1.1   Computer organization

measured by *clock cycles*. The clock rates of modern microprocessors can be as high as several hundred MHz, where

1 MHz = 1 million ticks (or cycles) per second

(for example, the Alpha 21164 microprocessor from DEC runs at 300 MHz). The period of a 1 MHz clock signal is 1 microsecond (10^{-6} second). The control unit also maintains a register called the *program counter* (PC), which controls the memory address of the next instruction to be executed. During the execution of an instruction, the presence of overflow, an addition carry, a subtraction borrow, and so forth, are flagged by the system and stored in another register called a *status register*. The resulting flags are then used by the programmer for program control and decision making.

WHAT IS A MICROPROCESSOR?

The processor in a very large computer is built from a number of integrated circuits. A *microprocessor* is a processor packaged in a single integrated circuit. A *microcomputer* is a computer that uses a microprocessor as its CPU. Early microcomputers were quite simple and slow. However, many of today's desktop microcomputers have become very sophisticated and are even faster than many large computers manufactured only a few years ago.

Microprocessors come in 4-bit, 8-bit, 16-bit, 32-bit, and 64-bit models. The number of bits refers to the number of binary digits that the microprocessor can manipulate in one operation. A 4-bit microprocessor, for example, is capable of manipulating 4 bits of information in one operation. Four-bit microprocessors are used for the electronic control of relatively simple machines. Some pocket calculators, for example, contain 4-bit microprocessors.

Many 32-bit and 64-bit microprocessors also contain on-chip memory to enhance their performance. Because microprocessors and input/output devices have different characteristics and speeds, peripheral chips are required to interface I/O devices to the microprocessor. For example, the integrated circuit M6821 is often used to interface a parallel device such as a printer or seven-segment display to the Motorola M6800 8-bit microprocessor.

Microprocessors have been widely used since their invention. It is not exaggerating to say that the invention of microprocessors has revolutionized the electronics industry. However, the following limitations of microprocessors led to the invention of microcontrollers:

■ A microprocessor requires external memory to execute programs.

■ A microprocessor cannot directly interface to I/O devices; peripheral chips are needed.

■ Glue logic (such as address decoders and buffers) is needed to interconnect external memory and peripheral interface chips to the microprocessor.

Because of these limitations, a microprocessor-based design cannot be made as small as might be desirable. The invention of microcontrollers not only eliminated most of these problems but also simplified the hardware design of microprocessor-based products.

WHAT IS A MICROCONTROLLER?

A *microcontroller* is a computer implemented on a single very large scale integration (VLSI) chip. A microcontroller contains everything contained in a microprocessor along with one or more of the following components:

memory

a timer

an analog-to-digital converter

a digital-to-analog converter

a direct memory access (DMA) controller

a parallel I/O interface (often called a parallel port)

a serial I/O interface

memory component interface circuitry

The Motorola 68HC11 is an 8-bit microcontroller family developed in 1986. The 68HC11 microcontroller family has more than 50 members, and the number is still increasing. The microcontrollers in this family differ mainly in the size of their on-chip memories and in their I/O capabilities. The characteristics of different memory technologies will be discussed shortly. As shown in Figure 1.2, the 68HC11A8 has the following features:

- on-chip static random access memory (SRAM)
- on-chip electrically erasable, programmable read-only memory (EEPROM)
- on-chip read-only memory (ROM)
- four-stage programmable prescaler
- three input capture functions (IC_i, $i = 1, \ldots, 3$)
- five output compare functions (OC_i, $i = 1, \ldots, 5$)
- an 8-bit pulse accumulator circuit
- a serial communication interface (SCI)
- an 8-channel, 8-bit analog-to-digital converter
- a serial peripheral interface (SPI)
- a real-time interrupt (RTI) circuit
- a computer operating properly (COP) watchdog system

APPLICATIONS OF MICROCONTROLLERS

Since their introduction, microcontrollers have been used in every application that we can imagine. They are used as controllers for displays, printers, keyboards, modems, charge card phones, and home appliances such as refrigerators, washing machines, and microwave ovens. They are also used to control the operation of automobile engines and machines in factories. Today, most homes have one or more microcontroller-controlled appliances.

1.2.2 Memory

Memory is where software programs and data are stored. A computer may contain semiconductor, magnetic, and/or optical memory. Only semiconduc-

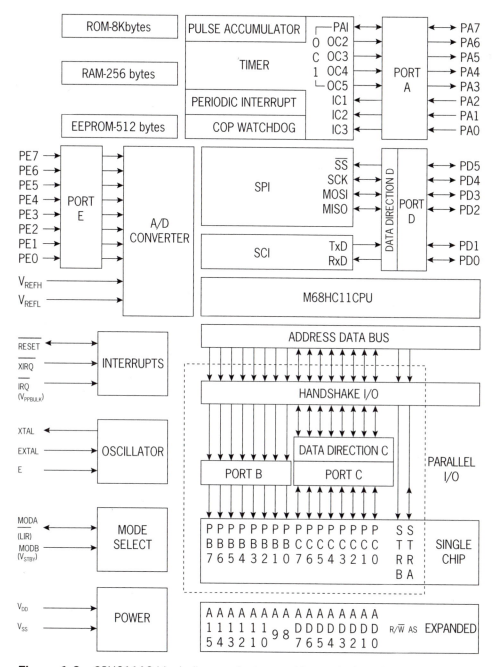

Figure 1.2 68HC11A8 block diagram (redrawn with permission of Motorola)

tor memory will be discussed in this book. Semiconductor memory can be further classified into two major types: *random-access memory* (RAM) and *read-only memory* (ROM).

RANDOM-ACCESS MEMORY
Random-access memory is *volatile* in the sense that it cannot retain data in the absence of power. RAM is also called *read/write memory* because it

allows the processor to read from and write into it. The microprocessor can temporarily store or write data into RAM, and it can later read that data back. Reading memory is nondestructive to the contents of the memory location. Writing memory is destructive. When the microprocessor writes data to memory, the old data is written over and destroyed.

There are two types of RAM technology: static RAM (SRAM) and dynamic RAM (DRAM). Static RAM uses from four to six transistors to store one bit of information. As long as power is stable, the information stored in the SRAM will not be degraded. Dynamic RAM uses one transistor and one capacitor to store one bit of information. The information is stored in the capacitor in the form of electric charge. The charge stored in the capacitor will leak away over time, so periodic refresh operation is needed to maintain the contents of DRAM.

RAM is mainly used to store *dynamic* programs or data. A computer user often wants to run different programs on the same computer, and these programs usually operate on different sets of data. The programs and data must therefore be loaded into RAM from the hard disk or other secondary storage, and for this reason they are called dynamic.

READ-ONLY MEMORY

ROM is nonvolatile. If power is removed from ROM and then reapplied, the original data will still be there. However, as its name implies, ROM data can only be read. If the processor attempts to write data to a ROM location, ROM will not accept the data, and the data in the addressed ROM memory location will not be changed.

Mask-programmed read-only memory (MROM) is a type of ROM that is programmed as it is manufactured. The semiconductor manufacturer places binary data in the memory according to the request of the customer. To be cost-effective, many thousands of MROM memory units, each containing a copy of the same data (or program), must be sold.

Programmable read-only memory (PROM) is a type of read-only memory that can be programmed in the field (often by the end user) using a device called a PROM programmer or a PROM "burner." Once a PROM has been programmed, its contents cannot be changed.

Erasable programmable read-only memory (EPROM) is a type of programmable read-only memory that can be erased by subjecting it to strong ultraviolet light. It can then be reprogrammed. A quartz window on top of the EPROM integrated circuit permits ultraviolet light to be shone directly on the silicon chip inside. Once the chip is programmed, the window can be covered with dark tape to prevent gradual erasure of the data. EPROM is often used in prototype computers, where the software may be revised many times until it is perfected. EPROM does not allow erasure of the contents of an individual location. The only way to make changes is to erase the entire EPROM chip and reprogram it.

Electrically erasable programmable read-only memory, EEPROM, is a type of nonvolatile memory that can be erased by electrical signals and reprogrammed. EEPROM allows each individual location to be erased and reprogrammed.

Flash memory was invented to incorporate the advantages and avoid the drawbacks of the EPROM and EEPROM technologies. Flash memory can be erased and reprogrammed in the system without using a dedicated programmer. It achieves the density of EPROM, but it does not require a window for erasure. Like EEPROM, flash memory can be programmed and erased electrically. However, it does not allow individual locations to be erased—the user can only erase the whole chip. Today, the monitor programs (also called the basic I/O systems (BIOS)) of many high-performance PCs are stored in flash memory.

1.3 The Computer's Software

A computer is useful because it can execute programs. Programs are known as *software*. A program is a set of instructions that the computer hardware can execute. The program is stored in the computer's memory in the form of binary numbers called *machine instructions*. It is difficult and not productive to program a computer in machine instructions, so *assembly language* was invented to simplify the programming job. An *assembly program* consists of assembly instructions. An assembly instruction is the mnemonic representation of a machine instruction. For example, in the 68HC11

> ABA stands for "add the contents of accumulator B to accumulator A."
> The corresponding machine instruction is 00011011.

> DECA stands for "decrement the contents of accumulator A by 1."
> The corresponding machine instruction is 01001010.

A *text editor* is used to develop a program using a computer. A text editor allows the user to type, modify, and save the program source code in a text file. The assembly program that the programmer enters is called the *source program* or *source code*. A software program called an *assembler* is then invoked to translate the program written in assembly language into machine instructions. The output of the assembly process is called *object code*. It is a common practice to use a *cross assembler* to assemble assembly programs. A cross assembler is an assembler that runs on one computer but generates machine instructions that will be executed by another computer that has a different instruction set. In contrast, an assembler runs on a computer and generates machine instructions to be executed by machines that have the same instruction set. The Motorola freeware as11 is a cross assembler that runs on an IBM PC or Apple Macintosh and generates machine code that can be downloaded into a 68HC11-based computer for execution.

There are several drawbacks to programming in assembly language:

■ The programmer must be very familiar with the hardware organization of the computer in which the program is to be executed.

■ A program (especially a long one) written in assembly language is extremely difficult to understand for anyone other than the author of the program. Even the author of the program can lose track of the program logic if the program is very long.

■ Programming productivity is not satisfactory for large programming projects.

For these reasons, high-level languages such as FORTRAN, PASCAL, C, and C++ were invented to avoid the problems of assembly language programming. A program written in a high-level language is also called a *source pro-gram*, and it requires a software program called a *compiler* to translate it into machine instructions. A compiler compiles a program into *object code*. Just as there are cross assemblers, there are *cross compilers* that run on one computer but translate programs into machine instructions to be executed on a com-puter with a different instruction set.

High-level languages are not perfect either. One of the major problems with high-level languages is that the machine code compiled from a program in a high-level language cannot run as fast as its equivalent in the assembly language. For this reason, many time-critical programs are still written in as-sembly language. Only assembly language will be used in this text.

1.4 The 68HC11 Registers

The 68HC11 microcontroller has many registers. These registers can be classified into two categories: CPU registers and I/O registers. CPU registers are used solely to perform general-purpose operations such as arithmetic, logic, and program flow control. I/O registers are mainly used to control the opera-tions of I/O subsystems and record the status of I/O operations, etcetera. I/O registers are treated as memory locations when they are accessed. CPU regis-ters do not occupy the 68HC11 memory space.

The CPU registers of the 68HC11 are shown in Figure 1.3 and are listed be-low. Some of the registers are 8-bit and some are 16-bit.

> *General-purpose accumulators A and B*. Both A and B are 8-bit reg-isters. Most arithmetic functions are performed on these two registers.

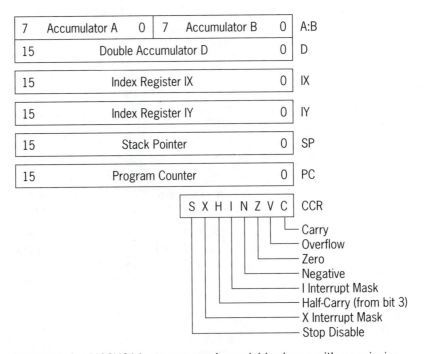

Figure 1.3   M68HC11 programmer's model (redrawn with permission of Motorola)

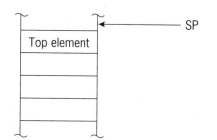

Figure 1.4 68HC11 Stack Structure

These two accumulators can also be cascaded to form a single 16-bit accumulator that is referred to as the D accumulator.

Index registers IX (or X) and IY (or Y). These two registers are used mainly in addressing memory operands. However, they are also used in several other operations.

Stack pointer (SP). A stack is a first-in-last-out data structure. The 68HC11 has a 16-bit stack pointer that initially points to the location above the top element of the stack, as shown in Figure 1.4. The stack will be discussed in Chapter 3.

Program counter (PC). The address of the next instruction to be executed is specified by the 16-bit program counter. The 68HC11 fetches the instruction one byte at a time and increments the PC by 1 after fetching each instruction byte. After the execution of an instruction, the PC is incremented by the number of bytes of the executed instruction.

Condition code register (CCR). This 8-bit register is used to keep track of the program execution status and control the program execution. The contents of the CCR are shown in Figure 1.3. The function of each condition code bit will be explained in later sections and chapters.

All of these registers are available to the programmer.

1.5 Memory Addressing

Memory consists of a sequence of directly addressable "locations." In this book a location will be referred to as an *information unit.* This term is deliberately generic to emphasize that the contents of a location can be data, instructions, the status of peripheral devices, and so on. An information unit has two components: its *address* and its *contents.*

Each location in memory has an address that must be supplied before its contents can be accessed. The CPU communicates with memory by first identifying the location's address and then passing this address on the address bus. The data are transferred between memory and the CPU along the data bus (see Figure 1.5). The number of bits that can be transferred on the data bus at one time is called the *data bus width* of the processor.

The 68HC11 has an 8-bit data bus and can access only one memory byte at a time. The 68HC11 has an address bus of 16 signal lines and can address up to

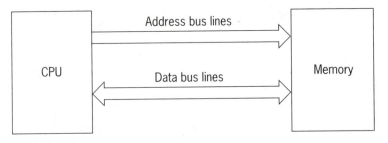

Figure 1.5   Transferring data between CPU and memory

2^{16} (65536) different memory locations. The accessible memory addresses are in the range from 0000_{16} to $FFFF_{16}$. The size of memory is measured in bytes. Each byte has 8 bits. A 4-bit quantity is called a *nibble*. To simplify the quantification of memory, the unit *kilobyte* (KB) is often used. K is given by the following formula:

$$K = 1024 = 2^{10}$$

Using powers of 2, the size of memory can be computed in decimal and hexadecimal format. For example,

$$64KB = 2^6 \times 2^{10} \text{ bytes} = 2^{16} \text{ bytes} = 65536 \text{ bytes}$$

Another frequently used unit is *megabyte* (MB), which is given by the following formula:

$$M = K^2 = 1024 \times 1024 = 1048576$$

1.6 The 68HC11 Addressing Modes

Addressing modes are used to specify the operands needed in an instruction. Six addressing modes are provided in the 68HC11: *immediate, direct, extended, indexed* (with either of two 16-bit index registers and an 8-bit offset), *inherent,* and *relative.* All except the immediate and inherent modes can be used to specify a memory operand.

Each of the addressing modes (except immediate and inherent) results in an internally generated, double-byte value referred to as the *effective address.* This value appears on the address bus during the external memory reference portion of the instruction.

The following paragraphs describe each of the addressing modes. In these descriptions, the effective address is used to indicate the memory address from which the argument is fetched, at which the argument is to be stored, or from which execution is to proceed.

Address and data values are represented in binary format inside the computer. However, a large binary number is not easy for a human being to deal with, so decimal and hexadecimal formats are often used instead. Octal numbers are also used in some cross assemblers. In this text, we will use a notation that adds a prefix to a number to indicate the base used in the number repre-

Base	Prefix
binary	%
octal	@
decimal	(nothing)*
hexadecimal	$

*Note: Some assemblers use &.

Table 1.1 ■ Prefix for number representation

sentation. The prefixes for binary, octal, decimal, and hexadecimal numbers are given in Table 1.1. This method is also used in Motorola microcontroller and microprocessor manuals.

1.6.1 Immediate (IMM)

In the immediate addressing mode, the actual argument is contained in the byte or bytes immediately following the instruction opcode. The number of bytes matches the size of the register. In assembly language syntax, an immediate value is preceded by a # character. The following instructions illustrate the immediate addressing mode:

    LDAA #22

loads the decimal value 22 into the accumulator A.

    ADDA #@32

adds the octal value 32 to the accumulator A.

    LDAB #$17

loads the hexadecimal value 17 into the accumulator B.

    LDX #$1000

loads the hexadecimal value 1000 into the index register X, where the upper byte of X receives the value $10 and the lower byte of X receives the value $00.

1.6.2 Direct Mode (DIR)

In the direct addressing mode, the least significant byte of the effective address of the instruction appears in the byte following the opcode. The high-order byte of the effective address is assumed to be $00 and is not included in the instruction. This limits use of the direct addressing mode to operands in the $0000–$00FF area of the memory. Instructions using the direct mode execute one clock cycle faster than their counterparts using the extended mode.

The following instructions illustrate the direct addressing mode:

    ADDA $00

adds the value stored at the memory location with the effective address $0000 to the accumulator A.

    SUBA $20

subtracts the value stored at the memory location with the effective address $0020 (hexadecimal) from the accumulator A.

    LDD $10

loads the contents of the memory locations at $0010 and $0011 into double accumulator D, where the contents of the memory location at $0010 are loaded into accumulator A and those of the memory location at $0011 are loaded into accumulator B.

1.6.3 Extended Mode (EXT)

In the extended addressing mode, the effective address of the operand appears explicitly in the two bytes following the opcode:

    LDAA $1003

loads the 8-bit value stored at the memory location with effective address $1003 into accumulator A.

    LDX $1000

loads the 16-bit value stored at the memory locations with the effective addresses $1000 and $1001 (hexadecimal) into the index register X. The byte at $1000 will be loaded into the upper byte of X and the byte at $1001 will be loaded into the lower byte of X.

    ADDD $1030

adds the 16-bit value stored at the memory locations with the effective addresses $1030 and $1031 to the double accumulator D.

1.6.4 Indexed Mode (INDX, INDY)

In the indexed addressing mode, one of the index registers (X or Y) is used in calculating the effective address. Thus the effective address is variable and depends on the current contents of the index register X (or Y) and a fixed, 8-bit, unsigned offset contained in the instruction. Because the offset byte is unsigned, only positive offsets in the range from 0 to 255 can be represented. If no offset is specified or desired, the machine code will contain $00 in the offset byte. For example,

    ADDA 10,X

adds the value stored at the memory location pointed to by the sum of 10 and the contents of the index register X to the accumulator A.

Each of the following instructions subtracts the value stored at the memory location pointed to by the contents of the index register X from the accumulator A:

    SUBA 0,X
    SUBA ,X
    SUBA X

Please note that the third format is not acceptable to the Motorola as11 freeware assembler.

You probably wonder why the index addressing mode is useful. Programs often need to change the address part of an instruction as the program runs. The index addressing mode requires part of the address to be placed in the index register in the microprocessor. The contents of the index register can be changed by the program, thus changing the effective address of the instruction while the offset byte of the instruction remains unchanged. Using the index addressing mode also shortens the instruction by one byte if the effective address is higher than the hexadecimal value $FF.

1.6.5 Inherent Mode (INH)

In the inherent addressing mode, everything needed to execute the instruction is encoded in the opcode. The operands are CPU registers and thus are not fetched from memory. These instructions are usually one or two bytes.

    ABA

adds the contents of accumulator B to accumulator A.

    INCB

increments the value of accumulator B by 1.

    INX

increments the value of the index register X by 1.

1.6.6 Relative Mode (REL)

The relative addressing mode is used only for branch instructions. Branch instructions, other than the branching versions of bit-manipulation instructions, generate two machine-code bytes, one for the opcode and one for the *branch offset*. The branch offset is the distance relative to the first byte of the instruction immediately following the branch instruction. The branch offset has a range of −128 to +127 bytes. When the branch is taken, the branch offset is added to the program counter to form the effective branch address. The source program specifies the destination of any branch instruction by its absolute address, given as either a numerical value or a symbol or expression that can be numerically evaluated by the assembler.

In Figure 1.6, the 68HC11 will branch to execute the instruction DECB if the Z bit in the CCR register is 1 when the instruction BEQ $e164 is executed.

Address	Opcode	Operand	
$e100	BEQ	$e164	
	ADDA	#10	
	⋮		$64 bytes
$e164	DECB		

Figure 1.6 Example of the relative addressing mode. The opcode byte of the instruction DECB is $64 bytes away from the opcode byte of the instruction ADDA #10

Address	Opcode	Operand
	BEQ	there
	ADDA	#10
	⋮	
there	DECB	

Figure 1.7   Using a label to specify the branch target

A better way to specify the branch target is to use a symbolic label. Figure 1.7 is an improvement to the example in Figure 1.6.

1.7 A Sample of 68HC11 Instructions

Most 68HC11 instructions consist of one to two bytes of opcode and zero to three bytes of operand information. The opcode specifies the operation to be performed. A 68HC11 instruction can have from zero to three operands. One of the operands is used both as a source and as the destination of the operation. The operand information is represented by one of the addressing modes.

1.7.1 The LOAD Instruction

LOAD is the generic name of a group of instructions that place a value or copy the contents of a memory location (or memory locations) into a register. Most 68HC11 arithmetic and logical instructions include a register as one of the operands. Before a meaningful operation can be performed, a value must be placed in the register. The LOAD instruction places or copies a value from a memory location into a register. The 68HC11 has LOAD instructions to load values into accumulator A, accumulator B, double accumulator D, the stack pointer, index register X, and index register Y.

For example, the following instruction will load the decimal value 10 into accumulator A:

    LDAA #10

where, the # character indicates that the value that follows (that is, 10) is to be placed into the accumulator A.

The following instruction will copy the contents of the memory location at $1000 into accumulator A:

    LDAA $1000

A more extensive sample of LOAD instructions is given in Table 1.2.

When two consecutive memory bytes are loaded into a 16-bit register, the contents of the memory location at the lower address is considered more significant and is loaded into the upper half of the register, while the contents of the memory location at the higher address is considered less significant and is loaded into the lower half of the register. When the contents of a 16-bit register are saved in the memory, the upper byte of the register is saved at the lower address while the lower byte is saved at the higher address.

Instruction	Meaning	Addressing mode
LDAA #10	Place the decimal value 10 (hexadecimal A) into accumulator A.	immediate
LDAA $1000	Copy the contents of the memory location at $1000 into accumulator A.	extended
LDAB #10	Place the decimal value 10 (hexadecimal A) into accumulator B.	immediate
LDAB $1000	Copy the contents of the memory location at $1000 into accumulator B.	extended
LDD #10	Place the decimal value 10 (hexadecimal 000A) into double accumulator D.	immediate
LDD $1000	Copy the contents of the memory locations at $1000 and $1001 into double accumulator D.	extended
LDS #255	Place the decimal value 255 (hexadecimal $FF) into the stack pointer SP.	immediate
LDS $1000	Copy the contents of the memory location at $1000 into the upper eight bits of register SP, and then copy the contents of the memory location at $1001 into the lower eight bits of register SP.	extended
LDX #$1000	Place the hexadecimal value $1000 into index register X.	immediate
LDX $1000	Copy the contents of the memory location at $1000 into the upper eight bits of register X, and then copy the contents of the memory location at $1001 into the lower eight bits of register X.	extended
LDY #1000	Place the decimal value 1000 (hexadecimal 3E8) into index register Y.	immediate
LDY 1000	Copy the contents of the memory location at 1000 (decimal) into the upper eight bits of register Y, and then copy the contents of the memory location at 1001 into the lower eight bits of register Y.	extended

Table 1.2 ■ A sample of LOAD instructions

Example 1.1

Write an instruction to place the decimal value 1023 (hexadecimal 3FF) into the stack pointer SP.

Solution: The following instruction will place the decimal value 1023 into the stack pointer SP:

LDS #1023

The binary representation of the decimal value 1023 is 1111111111_2. After the execution of this instruction, the contents of the 16-bit register SP become:

15	14	13	12	11	10	9	8	7	6	5	4	3	2	1	0
0	0	0	0	0	0	1	1	1	1	1	1	1	1	1	1

The numbers above the register SP stand for bit positions. The most significant bit is at the leftmost position (bit 15), and the least significant bit is at the rightmost position (bit 0).

In order to unify the representation, we will use the following notations throughout this book:

[*reg*]: refers to the contents of the register reg. reg can be any one of the following:

A, B, D, X, Y, SP, or PC.

[*addr*]: refers to the contents of memory location at address *addr.*

mem [*addr*]: refers to the memory location at addr.

Example 1.2

Write an instruction to load the contents of the memory locations at $0000 and $0001 into double accumulator D. Initially, the contents of D and the memory locations at $0000 and $0001 are $1010, $20, and $30, respectively. Show the new values in these registers after the execution of the instruction.

Solution: To load two consecutive memory bytes into the double accumulator, we need to specify only the address of the most significant byte, that is, $0000 in this example. The instruction is:

LDD $0000

The contents of D and memory locations $0000 and $0001 before and after execution of the instruction are as follows:

Before execution of LDD $0000	After execution of LDD $0000
[D] = $1010 [$0000] = $20 [$0001] = $30	[D] = $2030 [$0000] = $20 [$0001] = $30

1.7.2 The ADD Instruction

ADD is the generic name of a group of instructions that perform the addition operation. The ADD instruction is one of the most important arithmetic instructions in the 68HC11. The ADD instruction can have either two or three operands. In a three-operand ADD instruction, the C bit of the condition code register is always included as one of the source operands. Three-operand ADD instructions are used mainly in multiple-precision arithmetic, which will be discussed in Chapter 2. The ADD instruction has the following constraints:

■ The ADD instruction can specify at most one memory location as a source operand.

■ The memory operand can be used only as a source operand.

■ The destination operand must be a register (it can be A, B, X, Y, or D).

■ The register specified as the destination operand must also be used as a source operand.

For example,

ADDA #20

adds the decimal value 20 (hexadecimal $14) to the contents of accumulator A and places the result in accumulator A.

ADDA $40

adds the contents of the memory location at $40 to the contents of the accumulator A and places the result in accumulator A.

ADCA $00

adds the carry bit (in the CCR register) and the contents of the memory location at $00 to accumulator A and places the result in accumulator A.

More examples of ADD instructions are given in Table 1.3.

Instruction	Meaning	Addressing mode
ABA	Add accumulator B to accumulator A and store the result in A.	inherent
ABX	Add the 8-bit accumulator B to the 16-bit register X and store the sum in X.	inherent
ABY	Add the 8-bit accumulator B to the 16-bit register Y and store the sum in Y.	inherent
ADCA #12	Add the decimal value 12 and the C flag in the CCR register to accumulator A and store the sum in accumulator A.	immediate
ADCA $20	Add the contents of the memory location at $20 and the C flag in the CCR register to accumulator A and store the sum in accumulator A.	direct
ADCB #12	Add the decimal value 12 and the C flag in the CCR register to accumulator B and store the sum in accumulator B.	immediate
ADCB $20	Add the contents of the memory location at $20 and the C flag in the CCR register to accumulator B and store the sum in accumulator B.	direct
ADDA #12	Add the decimal value 12 to accumulator A and store the sum in accumulator A.	immediate
ADDA $20	Add the value of the memory location at $20 to accumulator A and store the sum in accumulator A.	direct
ADDB #12	Add the decimal value 12 to accumulator B and store the sum in accumulator B.	immediate
ADDB $20	Add the value of the memory location at $20 to accumulator B and store the sum in accumulator B.	direct
ADDD #0012	Add the value 12 to double accumulator D and store the sum in double accumulator D.	immediate
ADDD $0020	Add the 16-bit value stored at memory locations $20 and $21 to double accumulator D and store the sum in double accumulator D. The value at $20 is the upper 8 bits, while the value at location $21 is the lower 8 bits.	direct

Table 1.3 ■ A sample of ADD instructions

The ADD instructions that specify one of the index registers as the destination are mainly used in address calculation, not for general-purpose 16-bit addition. The instruction ABX adds the contents of accumulator B to the lower byte of index register X. If there is a carry out, it will be added to the upper byte of index register X. The instruction ABY is similar except that the destination is index register Y.

Example 1.3

Write an instruction sequence to add the contents of the memory locations at $10 and $20 and leave the sum in the accumulator A.

Solution: This problem can be solved by loading the contents of one of the memory locations into accumulator A and then adding the contents of the other memory location into accumulator A, as is done by the following instructions:

```
LDAA $10
ADDA $20
```

The first instruction loads the contents of the memory location at $10 into accumulator A. The second instruction then adds the contents of the memory location at $20 to accumulator A.

1.7.3 The SUB Instruction

SUB is the generic name of a group of instructions that perform the subtraction operation. Like the ADD instruction, the SUB instruction can have either two or three operands. The three-operand SUB instruction includes the C bit of the CCR register as one of the source operands. Three-operand SUB instructions are used mainly in multiple-precision arithmetic. The SUB instruction has the following constraints:

- The SUB instruction can specify at most one memory location as a source operand.
- The memory operand can be used only as a source operand.
- The destination operand must be an accumulator (it can be either A, B, or D).
- The register specified as the destination operand is also used as a source operand.

For example,

```
SUBA #10
```

subtracts the decimal value 10 from accumulator A and leaves the difference in accumulator A.

```
SUBB $10
```

subtracts the contents of the memory location at $10 from accumulator B and leaves the difference in accumulator B.

Example 1.4

> Write an instruction sequence to subtract the value of the memory location at $00 from that of the memory location at $30 and leave the result in accumulator B.
>
> **Solution:** We need to load the contents of the memory location at $30 into accumulator B and then subtract the contents of the memory location at $00 directly from accumulator B. The appropriate instructions are:
>
> ```
> LDAB  $30
> SUBB  $00
> ```

Further examples of SUB instructions are given in Table 1.4.

Instruction	Meaning	Addressing mode
SBA	Subtract the value of accumulator B from accumulator A and store the difference in accumulator A.	inherent
SBCA #10	Subtract the decimal value 10 and the C bit of the CCR register from accumulator A and store the difference in accumulator A.	immediate
SBCA $20	Subtract the contents of the memory location at $20 and the C flag of the CCR register from accumulator A and store the difference in A.	direct
SBCB #10	Subtract the decimal value 10 and the C bit of the CCR register from accumulator B and store the difference in accumulator B.	immediate
SBCB $20	Subtract the contents of the memory location at $20 and the C flag of the CCR register from accumulator B and store the difference in accumulator B.	direct
SUBA #10	Subtract the decimal value 10 from accumulator A and store the difference in accumulator A.	immediate
SUBA $20	Subtract the contents of the memory location at $20 from accumulator A and store the difference in accumulator A.	direct
SUBB #10	Subtract the decimal value 10 from accumulator B and store the difference in accumulator B.	immediate
SUBB $20	Subtract the contents of the memory location at $20 from accumulator B and store the difference in accumulator B.	direct
SUBD #$0010	Subtract the hexadecimal value $0010 from double accumulator D and store the difference in double accumulator D.	immediate
SUBD $0020	Subtract the 16-bit value stored at memory locations $0020 and $0021 from double accumulator D and store the difference in double accumulator D. The contents at location $0020 is the upper 8 bits, while the contents at location $0021 is the lower 8 bits.	extended

Table 1.4 ■ A sample of SUB instructions

1.7.4 The STORE Instruction

STORE is the generic name of a group of instructions that store the contents of a register into a memory location or memory locations. The 68HC11 has six STORE instructions. The STORE instruction allows the contents of accumulator A, accumulator B, double accumulator D, the stack pointer SP, index register X, or index register Y to be stored at one or two memory locations. The destination must be a memory location.

For example,

    STAA $10

stores the contents of accumulator A in the memory location at $10.

    STAB $10

stores the contents of accumulator B in the memory location at $10.

    STD $10

stores the upper and lower 8 bits of double accumulator D in the memory locations at $10 and $11, respectively.

    STX $2000

stores the upper and lower 8 bits of index register X in the memory locations at $2000 and $2001.

Example 1.5

Write an instruction sequence to add the contents of the memory locations at $00 and $01 and then store the sum in the memory location at $10.

Solution: This problem can be solved in three steps:

> **Step 1**
> Load the contents of the memory location at $00 into accumulator A.
> **Step 2**
> Add the contents of the memory location at $01 to accumulator A.
> **Step 3**
> Store the contents of accumulator A in the memory location at $10.

The appropriate instructions are:

```
LDAA  $00    ;load the contents of the memory location at $00 into A
ADDA  $01    ;add the contents of the memory location at $01 to A
STAA  $10    ;store the sum in A at the memory location at $10
```

Example 1.6

Write an instruction sequence to swap the contents of the memory locations at $00 and $10.

Solution: To swap, we need to load the contents of the memory locations at $00 and $10 into accumulators A and B, respectively, and then store the contents of A and B in the memory locations at $10 and $00, respectively. The following instructions would be used:

```
LDAA  $00    ; load the contents of $00 into A
LDAB  $10    ; load the contents of $10 into B
STAA  $10    ; store A into $10
STAB  $00    ; store B into $00
```

1.8 The 68HC11 Machine Code

We have learned that each 68HC11 instruction consists of 1 to 2 bytes of opcode and 0 to 3 bytes of operand information. In this section, we will look at the machine codes of a sample of 68HC11 instructions, the disassembly of machine instructions, and instruction execution timing.

1.8.1 A Machine-Code Sequence

The basic assembly language or machine code instructions can be sequenced to perform calculations, as we have seen in previous examples. Consider the following high-level language statements:

```
I := 29
L := I + M
```

Assume that the variables I, L, and M refer to memory locations $00, $01, $02, respectively. The first statement assigns the value 29 to variable I, and the second statement assigns the sum of variables I and M to variable L. The high-level language statements translate to the following equivalent assembly language and machine instructions:

Assembly instructions	Machine instructions (in hexadecimal format)
LDAA #29	86 1D
STAA $00	97 00
ADDA $02	9B 02
STAA $01	97 01

Note that the decimal number 29 is equivalent to hexadecimal $1D. Assume that these four instructions are stored in consecutive memory locations starting at $C000. Then the contents of the memory locations from $C000 to $C007 are as follows:

Address	Machine code
$C000	86
$C001	1D

(continued)

Address	Machine code
$C002	97
$C003	00
$C004	9B
$C005	02
$C006	97
$C007	01

If the memory location at $02 contains $20, then the memory location at $01 is assigned the value $1D + $20 = $3D. Figure 1.8 shows the changes in the values stored in the memory locations and in accumulator A when the instructions are executed. (These values are represented in hexadecimal format.)

1.8.2 Decoding Machine-Language Instructions

The process of decoding (disassembling) a machine-language instruction is more difficult than assembling it. The opcode is the first one or two bytes of an instruction. By decomposing its bit pattern, the assembly instruction mnemonic and the addressing mode of the operand can be identified.

■

Example 1.7

A segment of program machine code contains the following opcode and addressing information:

96 30 8B 07 97 30 96 31 8B 08 97 31

Using the machine opcodes and corresponding assembly instructions in Table 1.5, decode the given machine code into assembly instructions.

Address	Before program execution	After program execution
$00	???	1D
$01	???	3D
$02	20	20
A	???	3D

Figure 1.8 Changes in the contents of memory locations and accumulator A after program execution

Machine code	Assembly instruction format
01	NOP
86	LDAA IMM
96	LDAA DIR
C6	LDAB IMM
D6	LDAB DIR
CC	LDD IMM
DC	LDD DIR
8B	ADDA IMM
9B	ADDA DIR
CB	ADDB IMM
DB	ADDB DIR
C3	ADDD IMM
D3	ADDD DIR
97	STAA DIR
D7	STAB DIR
DD	STD DIR

Note:
1. IMM is a one-byte immediate value for instructions that involve A or B, and it is a two-byte immediate value for instructions that involve D.
2. DIR stands for a one-byte direct address between 00 and FF.

Table 1.5 ■ Machine opcodes and their corresponding assembly instructions (all numbers are in hexadecimal format)

Solution: The process of decoding the machine language instruction begins with the opcode byte 96.

a. The opcode byte 96 corresponds to the following LOAD instruction format:

LDAA DIR

To complete the decoding of this instruction, the byte that immediately follows 96 (that is, 30) should be included. Therefore, the machine code of the first instruction is 96 30. The corresponding assembly instruction is LDAA $30.

b. The opcode of the second instruction is 8B, which corresponds to the following ADD instruction format:

ADDA IMM

To decode this instruction completely, the byte that immediately follows 8B (that is, 07) should be included. The machine code of the second instruction is thus 8B 07. The corresponding assembly instruction is ADDA #07.

c. The opcode of the third instruction is 97, which corresponds to the following STORE instruction format:

STAA DIR

Including the byte that immediately follows 97 (that is, 30), we see that the machine code of the third instruction is 97 30. The corresponding assembly instruction is STAA $30.

Continuing in this manner, we can decode the remaining machine code bytes into the following assembly instructions:

```
LDAA  $31
ADDA  #08
STAA  $31
```

A program that can disassemble machine code into assembly instructions is called a *disassembler*. A disassembler can be used to translate the machine code in ROM into assembly instructions.

1.8.3 The Instruction Execution Cycle

In order to execute a program, the microprocessor or microcontroller must access memory to fetch instructions or operands. The process of accessing a memory location is called a *read cycle,* the process of storing a value in a memory location is called a *write cycle,* and the process of executing an instruction is called an *instruction execution cycle.*

When executing an instruction, the 68HC11 performs a combination of the following operations:

- ■ It performs a read cycle (or a sequence of read cycles) to fetch instruction opcode byte(s) and addressing information.
- ■ It performs the read cycle(s) required to fetch the memory operand(s) (optional).
- ■ It performs the operation specified by the opcode.
- ■ It writes back the result to either a register or a memory location (optional).

We will illustrate the instruction execution cycle using the LOAD, ADD, and STORE instructions shown below. The details of the data transfer on the buses are included to illustrate the read/write cycles. Assume the program counter PC is set at $C000, the starting address for the machine instructions. The contents of the memory locations at $2000 and $3000 are $19 and $37, respectively.

Assembly language instructions	Memory location address	Machine code
LDAA $2000	$C000	B6 20 00
ADDA $3000	$C003	BB 30 00
STAA $2000	$C006	B7 20 00

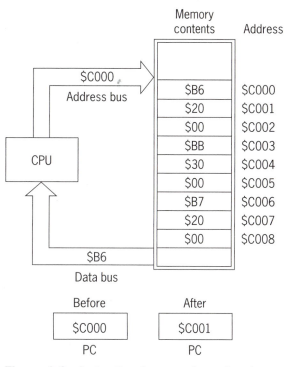

Figure 1.9   Instruction 1—opcode read cycle

Instruction 1 LDAA $2000

Execution of this instruction involves the following steps:

Step 1
The value in PC ($C000) is placed on the address bus with a request to "read" the contents of that location.

Step 2
The eight-bit value at the location $C000 is the instruction opcode byte $B6. This value is placed on the data bus by the memory hardware and returned to the processor, where the control unit begins interpretation of the instruction. On the read cycle, the control unit causes the PC to be incremented by 1, so it now points to location $C001. Figure 1.9 shows the opcode read cycle.

Step 3
The control unit recognizes that the LOAD instruction requires a two-byte value for the operand address. This is found in the two bytes immediately following the opcode byte (at locations $C001 and $C002). Therefore two read cycles are executed, and the value of the PC is incremented by 2. The PC has a final value of $C003, and the address $2000 is stored in an internal register (invisible to the programmer) inside the CPU. Figure 1.10 shows the address byte read cycles.

Step 4
The actual execution of the LOAD instruction requires an additional read cycle. The address $2000 is put on the address bus with a read

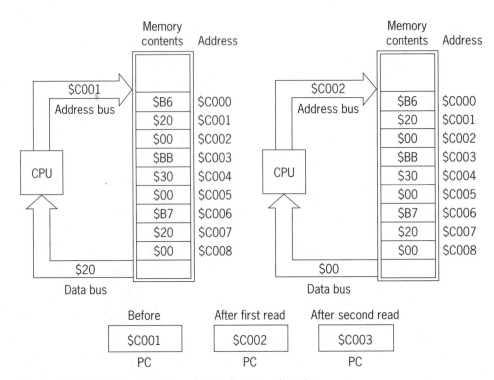

Figure 1.10   Instruction 1—address byte read cycles

request. The contents of memory location $2000 are placed on the data bus and stored in accumulator A, as shown in Figure 1.11.

Instruction 2 ADDA $3000

The PC initially has the value $C003. Three read cycles are required to fetch the second instruction from memory. The execution cycle for this instruction involves the following steps:

Step 1
Fetch the opcode byte at location $C003. At the end of this read cycle, the PC is incremented to $C004 and the opcode byte $BB has been fetched. The control unit recognizes that this version of the ADD instruction requires two more read cycles to fetch the extended address. These two read cycles are performed in the following two steps.

Step 2
Fetch the upper extended address byte ($30) from the memory location at $C004. The PC is then incremented to $C005.

Step 3
Fetch the lower extended address byte ($00) from the memory location at $C005. The PC is then incremented to $C006.

Step 4
Execution of this instruction requires an additional read cycle to read in the operand at location $3000. The control unit places the value

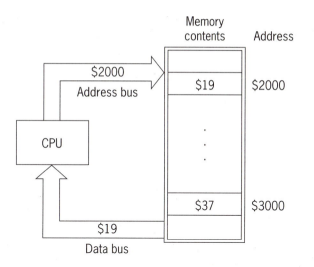

Figure 1.11 Instruction 1—execution read cycle

$3000 on the address bus to fetch the contents of memory location $3000.

Step 5

The returned value $37 is added to accumulator A. The accumulator now has the value $50 ($19 + $37 = $50).

Instruction 3 STAA $2000

As was the case for the previous two instructions, three read cycles are required to fetch this instruction from memory. The PC initially has the value $C006. The execution cycle for this instruction involves the following steps:

Step 1

Fetch the opcode byte at the location $C006. At the end of the read cycle, the PC is incremented to $C007 and the opcode byte $B7 has been fetched. The control unit recognizes that this version of the STORE instruction requires two more read cycles to fetch the extended address. These two read cycles are performed in the following steps.

Step 2

Fetch the upper extended address byte ($20) from the memory location at $C007. The PC is then incremented to $C008.

Step 3

Fetch the lower extended address byte ($00) from the memory location at $C008. The PC is then incremented to $C009.

Step 4

The purpose of this instruction is to store the contents of accumulator A in memory, so the control unit places the extended address $2000 on the address bus, and the value in accumulator A ($50) is written into the memory location at $2000. Figure 1.12 shows the execution write cycle.

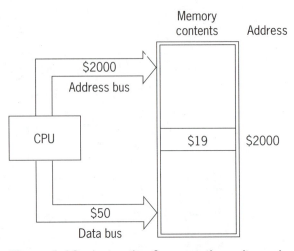

Figure 1.12 Instruction 3—execution write cycle

1.8.4 Instruction Timing

The 68HC11 Reference Manual lists the CPU execution time for all instructions. The 68HC11 internal operations and external read/write cycles are controlled by the E clock signal. The E clock signal is derived by dividing the frequency of an external crystal oscillator by 4.

1.9 Summary

This chapter started with a brief discussion of the basic structure of a computer system and then progressed to the study of the 68HC11 instructions and addressing modes. The methods of mnemonic instruction assembly, machine code disassembly, and the instruction execution process were also covered in this chapter.

1.10 Glossary

Accumulator   A register in a computer that contains an operand to be used in an arithmetic operation.

ALU (arithmetic logic unit)   The part of the processor in which all arithmetic and logical operations are performed.

Assembly instruction   A mnemonic representation of a machine instruction.

Assembler   A program that converts a program in assembly language into machine instructions so that it can be executed by a computer.

Bus   A set of signal lines through which the processor of a computer communicates with memory and I/O devices.

Central processing unit (CPU)   The combination of the registers, the ALU, and the control unit.

Computer   A computer consists of hardware and software. The hardware includes four major parts: the central processing unit, the memory unit, the input unit, and the output unit. Software is a sequence of instructions that control the operations of the hardware.

Control unit   The part of the processor that decodes and monitors the execution of instructions. It arbitrates the use of computer resources and makes sure that all computer operations are performed in proper order.

Cross assembler   An assembler that runs on one computer but generates machine instructions that will be executed by another computer that has a different instruction set.

EPROM (erasable programmable read-only memory)   A type of read-only memory that can be erased by subjecting it to strong ultraviolet light. It can be reprogrammed using an EPROM programmer. A quartz window on top of the EPROM chip allows light to be shone directly on the silicon chip inside.

EEPROM (electrically erasable programmable read-only memory)   EEPROM can be erased and reprogrammed using electrical signals. EEPROM allows each individual location inside the chip to be erased and reprogrammed.

Machine instruction   A set of binary digits that tells the computer what operation to perform.

Masked ROM (MROM)   A type of ROM that is programmed when it is fabricated.

Microprocessor   A CPU packaged in a single integrated circuit.

Microcontroller   A computer system implemented on a single, very large scale integrated circuit. A microcontroller contains everything that is in a microprocessor and may contain memories, an I/O device interface, a timer circuit, an A/D converter, etcetera.

Memory   Storage for software and information.

Nibble   A group of four-bit information.

Object code   The sequence of machine instructions that results from the process of assembling and/or compiling a source program.

Program   A set of instructions that the computer hardware can execute.

Program counter (PC)   A register that keeps track of the address of the next instruction to be executed.

PROM (programmable read-only memory)   A type of ROM that allows the end user to program it once and only once using a device called a PROM programmer.

RAM (random-access memory)   RAM technologies allow read and write access to every location inside the memory chip. Furthermore, read and write access take the same amount of time for any location within the RAM chip.

Register   A storage location in the CPU. It is used to hold data and/or a memory address during the execution of an instruction.

ROM (read-only memory)   A type of memory that is nonvolatile in the sense that when power is removed from ROM and then reapplied, the original data are still there. ROM data can only be read—not written—during normal computer operation.

Source code   A program written in either assembly language or a high-level language; also called a source program.

Status register   A register located in the CPU. It keeps track of the status of instruction execution by noting the presence of carries, zeros, negatives, overflows, etcetera.

1.11 Exercises

E1.1 What is a processor? What sections of a computer make up a processor?

E1.2 What makes a microprocessor different from the processors used in large computers?

E1.3 What makes a microcontroller different from the microprocessors used in personal computers?

E1.4 How many bits of data are stored in each memory location of a micro-controller trainer built around the 68HC11 microcontroller?

E1.5 How many different memory locations can the 68HC11 microcontroller address?

E1.6 Why must every computer have some nonvolatile memory?

E1.7 What are the differences between MROM, PROM, EPROM, and EEPROM? For what type of application is each most suitable?

E1.8 What is the difference between source code and object code?

E1.9 What register is used to keep track of the address of the next instruction to be executed?

E1.10 Convert 24K into decimal representation.

E1.11 Write instructions to swap the contents of accumulators A and B. *Hint:* Use a memory location (for example, $00) as a swap buffer.

E1.12 Write an instruction sequence to subtract 4 from memory locations $00 to $02.

E1.13 Write an instruction sequence that translates the following high-level language statements into 68HC11 assembly instructions.

```
I := 10
J := 20
K := J - I
```

Assume variables I, J, and K are located at $01, $04, and $10, respectively.

E1.14 Translate the following assembly instructions into machine instructions using Table 1.5.

```
LDAA #$20
ADDA $10
LDAB #$90
STD $00
```

E1.15 Translate the following assembly instructions into machine instructions using Table 1.5.

```
LDAA #00
LDAB $01
STD $60
LDAB $03
STD $62
```

E1.16 Disassemble the following machine code into 68HC11 assembly instructions using Table 1.5.

```
96 40 8B F8 97 40 96 41 8B FA 97 41
```

E1.17 Disassemble the following machine code into 68HC11 assembly instructions using Table 1.5.

```
DC 00 8B FE CB FD DD 00
```

E1.18 Determine the number of read and write cycles performed during the execution of the following instructions:

    a. LDX #$1000

    b. LDS #$FF

    c. SUBA $2000

    d. ADDD 0,Y

E1.19 Determine the memory contents at $00, $01, and $02 after execution of the following instruction sequence, given that [$00] = $12, [$01] = $34, and [$02] = $56.

```
LDAA $00
LDAB $01
STAB $00
LDAB $02
STAB $01
STAA $02
```

E1.20 Determine the contents of the memory locations at $00, $01, and $02 after execution of the following instruction sequence, given that [$00] = $11, [$01] = 22, and [$02] = 33.

```
LDAA $02
ADDA $01
STAA $01
ADDA $00
STAA $00
```

E1.21 Refer to the 68HC11 Reference Manual to find the execution time (in E clock cycles) of the following instructions.

    a. LDAA #$00

    b. STAA $1000

    c. LDAB 10,Y

    d. ADDA 9,X

    e. STD $2000

E1.22 Assume that the E clock signal frequency is 2 MHz. You are given the following instructions:

```
LDD #$0000
LDX #$1000
LDY #$2000
```

Find

    a. the period of the E clock signal in seconds.

    b. the total execution time (in E clock cycles and in seconds) of these three instructions.

    c. the number of times that these three instructions must be executed in order to create a delay of one second.

2

68HC11

ASSEMBLY

PROGRAMMING

2.1 Objectives

After completing this chapter, you should be able to:

- use assembler directives to allocate memory blocks, define constants, etcetera
- write assembly programs to perform simple arithmetic operations
- write program loops to perform repetitive operations
- use a flowchart to describe program flow
- communicate to evaluation boards such as EVB, EVBU, and EVM using a PC
- enter commands to display and modify the registers and memory locations of the evaluation board
- enter and download programs onto the evaluation board for execution
- create delays of any length using program loops

2.2 Introduction

Assembly language programming is a method of creating instructions that are the symbolic equivalent of machine code. The syntax of each instruction is structured to allow direct translation to machine code.

This chapter begins a formal study of Motorola 68HC11 assembly language programming. Although Chapter 1 included several basic instructions, this chapter introduces the formal rules of program structure, specification of variables and data types, and the syntax rules for program statements. The development in this text will follow the standards set out in the M68HC11 Reference manual. The opcode mnemonics, register names, and directives are fairly standard. However, particular assemblers may permit considerable variation in symbol names (such as use of uppercase and lowercase letters), operand expressions, delimiters, and so forth. This text will note some of the typical variations, and the specific ones permitted by the assembler on your system will be listed in your manual.

The rules for the Motorola portable cross assembler (PASM) will be followed in this chapter. Of all cross assemblers, the Motorola freeware as11.exe is especially popular in universities. Notes on the use of this cross assembler will be made at the appropriate places. All programs in this text are compatible with this freeware.

Evaluation boards are very useful for learning to use the microcontroller and for product development. Three popular evaluation boards will be introduced in this chapter: the 68HC11EVB, the 68HC11EVBU, and the 68HC11EVM. All are designed and manufactured by Motorola. Because software tools are indispensable for the software development process, the development tools available for the 68HC11EVM, EVB, and EVBU will also be discussed.

2.3 Assembly Language Program Structure

Let us begin by examining a C program and its equivalent assembly language code. Many of the details of the assembly language code will not be clear on a first reading. However, it will serve as an introduction to the general structure of an assembly language program and some of the specific rules of syntax. The following C program assigns values to two variables and performs a simple arithmetic computation:

```
main ()
{
    int i, j, k;          ;i, j, & k are integer variables
    i = 75;               ;assign 75 to i
    j = 10;               ;assign 10 to j
    k = i + j − 6;
}
```

The equivalent assembly language code is shown below. It illustrates the structure of an assembly language program and includes directives, comments, and symbolic addresses. Line numbers are added for reference.

```
(1)*Data storage declaration
(2)         ORG  $00
(3) i       RMB  1              ;variable i
(4) j       RMB  1              ;variable j
(5) k       RMB  1              ;variable k
(6)*Program instruction section
(7) start   ORG  $C000          ;starting address of programs
(8)         LDAA #75            ;initialize i to 75
(9)         STAA i              ;    "
(10)        LDAA #10            ;initialize j to 10
(11)        STAA j              ;    "
(12)        ADDA i              ;compute i + j
(13)        SUBA #6             ;compute i + j − 6
(14)        STAA k              ;store i + j − 6 at k
(15)        END
```

Before we examine the structure of assembly language programs, you should note the following key concepts in this program:

- Lines (1) and (6) are full-line comments, while lines (3)–(5), (7), (9), and (11)–(14) contain in-line comments that explain the function (or operation) of the corresponding assembler directive (or instruction). Comments are used extensively in assembly language programs to provide documentation.

- Lines (2) and (7) introduce the directive ORG. A directive is an instruction that tells the assembler to perform a support function such as reserving memory. ORG sets up the location counter, and it must be followed by instructions or some other directive such as RMB or FCB. The functions of RMB and FCB will be explained in section 2.4.

- Lines (3)–(5) allocate data space. The symbolic addresses i, j, and k are given as labels. The directive RMB (reserve memory byte) 1 tells the assembler to reserve one byte of data storage. The value of this byte is not initialized.

- Lines (3)–(5) and (7) begin with a label. A label is a symbolic name for a memory address.

- Lines (9), (11), (12), and (14) use the symbolic addresses i, j, i, and k.

- Line (15) is the END directive. It signals the assembler to stop assembling instructions and interpreting directives. It must be the last statement of every 68HC11 assembly program.

2.3.1 A Global View of a 68HC11 Assembly Program

An assembly language program is divided into five sections that contain the main program components. In some cases these sections can be mixed to provide better algorithm design.

- Assembler Directives: These instructions are supplied to the assembler by the user; they define data and symbols, set assembler and linking conditions, and specify output format. Assembler directives do not produce machine code.

- Assembly Language Instructions: These instructions are 68HC11 instructions; some are defined with labels.

- Data Storage Directives: These directives allocate data storage locations containing initialized or uninitialized data.

- The END Directive: This is the last statement in a 68HC11 assembly language source code, and it causes termination of the assembly process. The END directive is ignored by the as11 assembler.

- Comments: There are two types of comments in an assembly program. The first type is used to explain the function of a single instruction or directive. The second type explains the function of a group of instructions or directives or a whole routine. Adding comments makes a program more readable. The rules for forming comments will be explained shortly.

2.3.2 A Local View of a 68HC11 Assembly Program

Each line of a 68HC11 assembly program, excluding certain special constructs, is comprised of four distinct fields. Some of the fields may be empty. The order of these fields is:

1. Label
2. Operation
3. Operand(s)
4. Comment

THE LABEL FIELD

Labels are symbols defined by the user to identify memory locations in the program or data areas of the assembly module. For most instructions and directives, the label is optional. The rules for forming a label are:

- A label name must begin with a letter (A–Z, a–z), and the letter can be followed by letters, digits (0–9), or special symbols. Some assemblers permit special symbols (such as a period, a dollar sign, or an underscore), some permit only uppercase or lowercase letters, and others will internally convert all letters to either uppercase or lowercase and thus will not distinguish between them.

- Most assemblers restrict the number of characters in a label name (usually to the most significant eight characters), but others allow many more.

- A label can be defined in one of two ways: either (1) the name starts in column 1 and is separated from the rest of the line by a space character, or (2) the name starts in any column and terminates with a colon (:). The Motorola freeware requires the label to start in the first column.

The Motorola freeware allows the following characters to be included in a label:

[a–z][A–Z]_.[0–9]$

where . and _ count as nondigits and the $ counts as a digit to avoid confusion with hexadecimal constants. All characters of a label are significant, and upper- and lowercase characters are distinct. The maximum number of characters in a label is currently set at 15.

Example 2.1 ■ Valid and invalid labels

The following instructions contain valid labels:

  a. START   ADDA #10   ;label begins in column 1
  b. PRINT: JSR HEXOUT   ;label is terminated by a colon (:)
               JMP START   ;instruction references label "START"

The following instructions contain invalid labels:

  c. GO OUT   ADDA $00   ;a blank is included in the label
  d. LOOP   ADDB $10   ;the label begins in column 2, so a colon is required

---

THE OPERATION FIELD

This field contains the mnemonic names for machine instructions and assembler directives. If a label is present, the opcode or directive must be separated from the label field by at least one space. If there is no label, the operation field must be at least one space from the left margin.

Example 2.2 ■ Examples of operation fields

          ADDA #$20   ;ADDA is the instruction mnemonic
  ZERO   EQU 0   ;equate directive EQU occupies the operation field

---

THE OPERAND FIELD

If an operand field is present, it follows the operation field and is separated from the operation field by at least one space. The operand field may contain operands for instructions or arguments for assembler directives. The following examples include operand fields:

          LDAB 0,X   ;0,X is the operand field
  LOOP: BNE LOOP   ;LOOP is the operand field

THE COMMENT FIELD

The comment field is optional and is added mainly for documentation. The comment field is ignored by the assembler. A comment may be inserted in one of three ways:

  1. If an asterisk (*) is the first printable character in the line, a comment may be inserted at the beginning of a line. In this case the entire line is a comment—an instruction or directive preceded by an asterisk will not be recognized. The asterisk generally does not have

to be in the first column; however, the Motorola freeware requires it to be in the first column.

2. A comment may follow the operation and operand fields of an assembly instruction or directive; in this case it is preceded by at least one space.

3. If the first non-white-space character in a line is an exclamation point (!), the rest of the line is a comment. However, the Motorola freeware does not recognize comments starting with !.

A comment cannot occur on a line containing only the label field, since the first word of the comment would be interpreted as an opcode. In this text we will use ; as the first character of a comment after the operand field. The first two types of comments are illustrated in the following examples:

   a.   ADDA #2   ;add 2 to accumulator A
   b. ABC: an invalid comment—no opcode is present
   c. *The whole line is a comment

Example 2.3

Identify the four fields in the following source statement:

LOOP   ADDA #40   ;add 40 to accumulator A

Solution: The four fields of the source statement are as follows:

   a. LOOP is a label.
   b. ADDA is an instruction mnemonic.
   c. #40 is the operand.
   d. ;add 40 to accumulator A is a comment.

2.4 A Sample of Assembler Directives

Assembler directives look just like instructions in an assembly language program, but they tell the assembler to do something other than create the machine code for an instruction. The available directives vary with the assembler. Interested readers should refer to the appropriate assembler user's manual. In this section, the most often used assembler directives will be discussed. Most of them are supported by the Motorola freeware. Statements enclosed in brackets [ ] are optional.

The END statement is the assembler directive that is used to end a program to be processed by an assembler. The program has the following form:

(your program)
END

The END directive indicates the logical end of the source program. Any statement following the END directive is ignored. A warning message will be printed if the END directive is not included in the source code; however, the program will still be assembled correctly if there is no other error in the program. The Motorola freeware assembler ignores the END directive.

When writing an assembly program, the programmer often needs to allocate storage in one way or another. To *allocate* storage means to find room for a variable or a program in memory. The assembler uses a *location counter* to keep track of the place to put data or a program during the assembling process. The ORG directive is used to set the value of the location counter, thus telling the assembler where to put the next byte it generates after the ORG directive. The syntax of this directive is as follows:

```
ORG <expression>
```

For example, the sequence

```
ORG $C000
LDAB #$FF
```

will put the opcode byte for the instruction LDAB #$FF at location $C000.

There are several directives that allow the programmer to reserve an area of memory. The *reserve memory byte* directive RMB reserves a block of memory whose size is specified by the number that follows the directive. The syntax of this directive is as follows:

```
[<label>]   RMB <expression> [<comment>]
```

For example, the statement

```
BUFFER   RMB 100
```

allocates 100 (decimal) bytes for some data and lets the programmer refer to it using the label BUFFER. The bytes in this area can be accessed by using the label for the RMB directive with an offset. For example, to load the first byte into accumulator A, use LDAA BUFFER; to load the tenth byte into accumulator A, use LDAA BUFFER+9; and so on. The programmer can also use an ORG directive to tell the assembler where to reserve a block of memory. For example, the sequence

```
         ORG $100
BUFFER   RMB 100
```

will reserve a block of 100 bytes starting at address $100.

Sometimes it is desirable to give initial values to the reserved memory area. There are five directives that can give initial values to the reserved memory block: BSZ, FCB, FDB, FCC, and DCB.

The *block storage of zeros* directive BSZ causes the assembler to allocate a block of bytes and assign each byte the initial value of zero. The number of bytes allocated is given by the <expression> in the operand field. The standard format of this directive is

```
[<label>]   BSZ <expression> [<comment>]
```

where <expression> specifies the number of bytes to allocate for storage. The expression must not contain undefined or forward references and must be an absolute expression.

For example,

```
LINELEN   EQU 80
BUFFER    BSZ LINELEN     ;allocate storage for a CRT display buffer of 80 characters
```

The *form constant byte* directive FCB will put a byte in memory for each argument of the directive. The standard format of this directive is as follows:

```
[<label>]    FCB [<expression>][,<expression>, . . . ,<expression>][<comment>]
```

For example, the statement

```
ABC    FCB $11,$22,$33
```

will initialize three bytes in memory to

```
$11
$22
$33
```

and will tell the assembler that ABC is the symbolic address of the first byte, whose initial value is $11. The programmer can also force these bytes to a particular address by adding the ORG directive. For example, the sequence

```
       ORG $100
ABC    FCB $11,$22,$33
```

will initialize the contents of memory locations at $100, $101, and $102 to $11, $22, and $33, respectively.

The *form double byte* directive FDB will initialize two consecutive bytes for each argument. The format of this directive is as follows:

```
[<label>]    FDB [<expression>][,<expression>, . . . ,<expression>][<comment>]
```

For example, the directive

```
ABC    FDB $11,$22,$33
```

will initialize six consecutive bytes in memory to

```
$00
$11
$00
$22
$00
$33
```

and will tell the assembler that ABC is the symbolic address of the first byte in this area, whose value is $00. The FDB is especially useful for putting addresses in memory so that they can be picked up into an index register. Suppose that OCHND is the label of the first instruction of a program and that it has the value of $D000. Then the directive

```
FDB OCHND
```

will generate the following two bytes in memory:

```
$D0
$00
```

The *form constant character* directive FCC will generate the code bytes for the letters in the arguments of the directive, using the ASCII code. All letters to be coded and stored are enclosed in double quotes. The format of this directive is as follows:

    [label]   FCC "<string>" [<comment>]

For example, the directive

    ALPHA   FCC "DEF"

will generate the following values in memory:

    $44
    $45
    $46

and will let the assembler know that the label ALPHA refers to the address of the first letter, which is stored as the byte $44. A character string to be output is often stored in memory using this directive.

The *define constant block* directive DCB will reserve an area of memory and initialize each byte to the same constant value. The format of this directive is as follows:

    [label]   DCB <length>,<value>

For example, the directive

    SPACE   DCB 80,$20

will generate a line of 80 space characters and tell the assembler that the label SPACE refers to the first space character. The value $20 is the ASCII code of the space character.

The DCB directive is not supported by the Motorola freeware assembler. Instead, the freeware assembler includes the directive FILL, which serves the same purpose as the DCB directive. The syntax of the FILL directive is

    [<label>]   FILL <value>,<length>

For example, the following directive

    ones   FILL 1,40

will force the freeware assembler to fill each of the 40 memory locations starting from the label *ones* with a 1.

A program will be more readable and maintainable if symbolic names are used instead of numbers in most situations—this can be achieved by using the *equate* directive. The equate directive EQU allows the programmer to use a symbolic name in place of a number. The syntax of the equate directive is as follows:

    <label>   EQU <expression> [<comment>]

For example, the directive

    ROM   EQU $E000

tells the assembler that the value $E000 is to be substituted wherever ROM appears in the program.

2.5 Flowcharts

After writing programs for a while, the reader will learn the importance of program documentation. Adding comments to a program is one method of program documentation. Another form of program documentation is the flowchart, which describes the logical flow of a program.

Figure 2.1 shows the flowchart symbols used in this book. The terminal symbol is used at the beginning and end of each program. When it is used at the beginning of a program, the word Start is written inside it. When it is used at the end of a program, it contains the word Stop.

The *process box* tells what must be done at this point in the program execution. The operation specified by the process box could be to shift accumulator A to the right one place, decrement the index register by 1, etcetera.

The *input/output box* is used to represent data that either enters or exits the computer.

The *decision box* contains a question that can be answered either yes or no. A decision box has two exits, also marked yes and no. The computer will take one action if the answer is yes and will take a different action if the answer is no.

The *on-page connector* indicates that the flowchart continues elsewhere on the same page. The place where it is continued will have the same label as the on-page connector. The *off-page connector* indicates that the flowchart continues on another page. To find where the flowchart continues, you need to look at the following pages of the flowchart to find the matching off-page connector.

Normal flow on a flowchart is from top to bottom and from left to right. Any line that does not follow this normal flow should have an arrowhead on it.

2.6 Problem-Solving with a Computer

To use a computer to solve a problem, a sequence of appropriate steps must be followed. These steps are best illustrated by the flowchart in Figure 2.2. Using a computer to solve a problem is often an iterative process. The

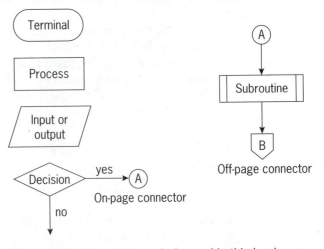

Figure 2.1   Flowchart symbols used in this book

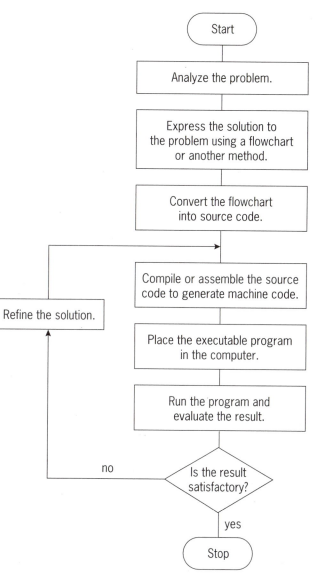

Figure 2.2   The procedure for using a computer to solve a problem

problem solver may need to refine the solution several times before a satisfactory solution is obtained.

2.7 Writing Programs to Do Arithmetic

In this section, we will use small programs that perform simple calculations to demonstrate how a program is written.

Example 2.4

Write a program to add the values of three memory locations ($00, $01, and $02) and save the sum at $03.

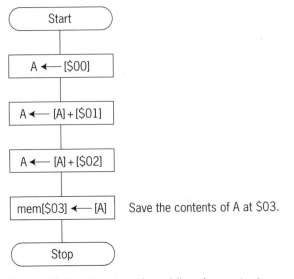

Figure 2.3   Flowchart for adding the contents
of three memory locations

Solution: The problem can be solved by the following four steps:

> **Step 1**
> Load the contents of the memory location at $00 into accumulator A.
> **Step 2**
> Add the contents of the memory location at $01 to accumulator A.
> **Step 3**
> Add the contents of the memory location at $02 to accumulator A.
> **Step 4**
> Save the contents of accumulator A at $03.

These steps are shown in the flowchart in Figure 2.3. The program is as follows:

```
LDAA  $00    ;load the contents of the memory location at $00 into A
ADDA  $01    ;add the contents of the memory location at $01 to A
ADDA  $02    ;add the contents of the memory location at $02 to A
STAA  $03    ;save the sum at $03
END
```

Example 2.5

>   Write a program to subtract 6 from three 8-bit numbers stored at $00, $01,
> and $02.

Solution: In the 68HC11, a memory location cannot be the destination of an
ADD or a SUB instruction. Therefore, three steps must be followed to add or
subtract a number to or from a memory location:

> **Step 1**
> Load the memory contents into an accumulator.

Step 2
Add (subtract) the number to (from) the accumulator.
Step 3
Store the contents of the accumulator back to the memory location.

The program is as follows:

```
LDAA  $00      ;load the first number into A
SUBA  #06      ;subtract 6 from the first number
STAA  $00      ;store the decremented value at $00
LDAA  $01      ;load the second number into A
SUBA  #06      ;subtract 6 from the second number
STAA  $01      ;store the decremented value at $01
LDAA  $02      ;load the third number into A
SUBA  #06      ;subtract 6 from the third number
STAA  $02      ;store the decremented value at $02
END
```

2.7.1 The Carry/Borrow Flag

So far we have been working with one-byte hexadecimal numbers because these are the largest numbers that can fit into accumulators A and B. However, programs can also be written to add and subtract numbers that contain two or more bytes. Arithmetic performed in an 8-bit microprocessor on numbers that are larger than one byte is called *multiprecision arithmetic.* Multiprecision arithmetic makes use of the condition code register (CCR).

Bit 0 of the CCR register is the carry (C) flag. It can be thought of as a temporary ninth bit that is appended to any 8-bit register. The C flag enables the programmer to write programs to add and subtract hexadecimal numbers that are larger than one byte. For example, consider the following two instructions:

```
LDAA  #$93
ADDA  #$8B
```

These two instructions add the numbers $93 and $8B.

```
   $93
 + $8B
 ─────
  $11E
```

The result is $11E, a 9-bit number, which is too large to fit into the 8-bit accumulator A. When the 68HC11 executes these two instructions, the lower 8 bits of the answer, $1E, are placed in accumulator A. This part of the answer is called the *sum.* The ninth bit is called a *carry.* A carry of 1 following an addition instruction sets the C flag of the CCR register to 1. A carry of 0 following an addition instruction will clear the C flag to 0. For example, execution of the following two instructions

```
LDAB  #$23
ADDB  #$44
```

will clear the C flag to 0 because the carry resulting from this addition is 0. In summary,

- If the addition produces a carry of binary 1, the carry flag is set to 1.
- If the addition produces a carry of binary 0, the carry flag is reset to 0.

2.7.2 Multiprecision Addition

Multiprecision addition is the addition of numbers that are larger than one byte in an 8-bit computer. The numbers $24E8 and $B234 are both two-byte numbers. To find their sum using the 68HC11 microcontroller, we have to write a program that uses multiprecision addition. If we add these two numbers on paper we get:

```
      1 ←————————— carry from the low byte to high byte
   $24E8
 + $B234
   ─────
   $D71C
```

Starting from the right,

Step 1
Add $8 and $4, which gives the sum $C.

Step 2
Add $E and $3, which gives a result of $11, or a sum of $1 and a carry of $1. The carry is written above the column of numbers on the immediate left. This transfers the carry from the low byte of the number to its high byte.

Step 3
Add $4 and $2 and the carry $1, which gives a sum of $7.

Step 4
Add $2 and $B, which gives a sum of $D.

The important point is that the carry had to be transferred from the low byte of the sum to its high byte. The following program can add these two-byte numbers and store the high byte of the sum at $00 and the low byte at $01:

```
LDAA  #$E8    ;load the low byte of the first number into A
ADDA  #$34    ;add the low byte of the second number to A
STAA  $01     ;save the low byte of the sum
LDAA  #$24    ;load the low byte of the second number into A
ADCA  #$B2    ;add the high byte of the second number and the carry to A
STAA  $00     ;save the high byte of the sum
END
```

Note that the LOAD and STORE instructions do not affect the value of the C flag (otherwise, the program would not work).

Example 2.6

Write a program to add the numbers $123456 and $CDEF66. Use the accumulator B. Save the low byte of the sum at $02, the middle byte at $01, and the high byte at $00.

Solution: The addition of these two 3-byte numbers is illustrated in the following diagram:

	12	34	56
+	CD	EF	66

memory locations	$00	$01	$02
	E0	23	BC

The addition should start from the least significant byte, then proceed to the second significant byte, and then to the most significant byte. The program is as follows:

```
LDAB  #$56    ;load the low byte of the first number into B
ADDB  #$66    ;add the low byte of the second number to B
STAB  $02     ;save the low byte of the sum at $02
LDAB  #$34    ;load the middle byte of the first number into B
ADCB  #$EF    ;add the middle byte of the second number and carry to B
STAB  $01     ;save the middle byte of the sum at $01
LDAB  #$12    ;load the high byte of the first number into B
ADCB  #$CD    ;add the high byte of the second number and carry to B
STAB  $00     ;save the upper byte of the sum at $00
END
```

2.7.3 Subtraction and the C Flag

The C flag also enables the 68HC11 to borrow from the high byte to the low byte during a multiprecision subtraction. Consider the following subtraction problem:

$$\begin{array}{r} \$2D \\ -\$89 \\ \hline \end{array}$$

We are attempting to subtract a larger number from a smaller number. Subtracting $9 from $D is not a problem:

$$\begin{array}{r} \$2D \\ -\$89 \\ \hline 4 \end{array}$$

Now we need to subtract $8 from $2. To do this, we need to borrow from somewhere. The 68HC11 borrows from the C flag, thus setting the C flag. When we borrow from the next higher digit of a hexadecimal number, the borrow has a value of decimal 16. After the borrow from the C flag, the problem can be completed:

$$\begin{array}{r} \$2D \\ -\$89 \\ \hline \$A4 \end{array}$$

When the 68HC11 executes a subtract instruction, it always borrows from the C flag. The borrow is either a 1 or a 0. The C flag operates as follows during a subtraction:

- If the 68HC11 borrows a 1 from the C flag during a subtraction, the C flag is set to 1.
- If the 68HC11 borrows a 0 from the C flag during a subtraction, the C flag is set to 0.

2.7.4 Multiprecision Subtraction

Multiprecision subtraction is the subtraction of numbers that are larger than one byte. To subtract the hexadecimal number $45D9 from $562A, the 68HC11 has to perform multiprecision subtraction:

```
 $562A
-$45D9
```

Like multiprecision addition, multiprecision subtraction is performed one byte at a time, beginning with the low byte. The following two instructions can be used to subtract the low byte:

```
LDAB #$2A
SUBB #$D9
```

Because a larger number is being subtracted from a smaller one, a 1 is borrowed from the C flag, causing it to be set to 1. The contents of accumulator B should be saved before the high bytes are subtracted. Let's save the low-byte result at $01:

```
STAB $01
```

When the high bytes are subtracted, the borrow of 1 has to be subtracted from the high-byte result. In other words, we need a "subtract with borrow" instruction. There is such an instruction, but it is called *subtract with carry*. It comes in two versions, a SBCA instruction for accumulator A and a SBCB instruction for accumulator B. The instructions to subtract the high bytes are:

```
LDAB #$56
SBCB #$45
```

We also need to save the high byte of the result at $00 with the following instruction:

```
STAB $00
```

The complete program with comments is as follows:

```
LDAB #$2A    ;load the low byte of the minuend into B
SUBB #$D9    ;subtract the low byte of the subtrahend from B
STAB $01     ;save the low byte of the difference
LDAB #$56    ;load the high byte of the minuend into B
SBCB #$45    ;subtract the high byte of the subtrahend and C bit from B
STAB $00     ;save the high byte of the difference
END
```

Example 2.7

Write a program to subtract the hexadecimal number $568DCA from $783FCB. Use accumulator B. Save the low byte of the difference at $02, the middle byte at $01, and the high byte at $00.

Solution: The subtraction is illustrated in the following diagram:

78	3F	CB
56	8D	CA

	$00	$01	$02
memory locations	21	B2	01

Multiprecision subtraction proceeds from low byte to high byte:

```
LDAB  #$CB    ;load the low byte of the minuend into B
SUBB  #$CA    ;subtract the low byte of the subtrahend from B
STAB  $02     ;save the low byte of the difference
LDAB  #$3F    ;load the middle byte of the minuend into B
SBCB  #$8D    ;subtract the middle byte of the subtrahend and C bit from B
STAB  $01     ;save the middle byte of the difference
LDAB  #$78    ;load the high byte of the minuend into B
SBCB  #$56    ;subtract the high byte of the subtrahend and C bit from B
STAB  $00     ;save the high byte of the difference
END
```

2.7.5 Binary-Coded Decimal Addition

Although virtually all digital systems are binary in the sense that all signals within the systems can take on only two values, some nevertheless perform arithmetic in the decimal system. In some circumstances, the identity of decimal numbers is retained to the extent that each decimal digit is individually represented by a binary code. There are ten decimal digits, so four binary bits are required for each code element. The most obvious choice is to use binary numbers 0000 through 1001 to represent the decimal digits 0 through 9. This form of representation is known as the *BCD* (binary-coded decimal) *representation* or *code.* The binary numbers 1010 through 1111 do not represent decimal digits, and they are therefore illegal BCD codes.

Most input to and output from computer systems is in the decimal system, because this is the most convenient system for human users. On input, the decimal numbers are converted into some binary form for processing, and this conversion is reversed on output. In a "straight binary" computer, a decimal number is converted into its binary equivalent; for example, the decimal number 46 is converted to 101110. In a computer using the BCD system, 46 would be converted into 0100 0110. If the BCD format is used, it must be preserved during arithmetic processing.

The principal advantage of the BCD system is the simplicity of input/output conversion; its principal disadvantage is the complexity of arithmetic processing. The choice (between binary and BCD) depends on the type of problems the system will be handling.

The 68HC11 microcontroller can add only binary numbers—not decimal numbers. The following program appears to cause the 68HC11 to add the decimal numbers 12 + 34 and store the sum in the memory location at $0000:

```
LDAB  #$12
ADDB  #$34
STAB  $00
END
```

The program performs the following addition:

$$
\begin{array}{r}
\$12 \\
+\$34 \\
\hline
\$46
\end{array}
$$

When the 68HC11 executes this program, it adds the numbers according to the rules of hexadecimal addition and produces the sum $46. This is the correct BCD answer, because the result represents the decimal sum of 12 + 34. In this example, the 68HC11 gives the appearance of performing decimal addition. However, a problem occurs when the 68HC11 adds two BCD digits and gets a sum greater than 9. Then the sum is incorrect in the decimal number system, as the following three examples illustrate.

$$
\begin{array}{r}
\$12 \\
+\$08 \\
\hline
\$1A
\end{array}
\qquad
\begin{array}{r}
\$25 \\
+\$37 \\
\hline
\$5C
\end{array}
\qquad
\begin{array}{r}
\$29 \\
+\$49 \\
\hline
\$72
\end{array}
$$

The answers to the first two problems are obviously erroneous in the decimal number system because the hexadecimal digits A and C are not between 0 and 9. The answer to the third example appears to contain valid BCD digits, but in the decimal system 29 plus 49 equals 78, not 72; this example involves a carry from the low nibble to the high nibble.

In summary, a sum in the BCD format is incorrect if it is greater than $9 or if there is a carry to the next higher nibble. Incorrect BCD sums can be adjusted by adding $6 to them. To correct the examples,

1. Add $6 to every sum digit greater than 9.
2. Add $6 to every sum digit that had a carry of 1 to the next higher digit.

Here are the problems with their sums adjusted:

$$
\begin{array}{r}
\$12 \\
+\$08 \\
\hline
\$1A \\
+\$\ 6 \\
\hline
\$20
\end{array}
\qquad
\begin{array}{r}
\$25 \\
+\$37 \\
\hline
\$5C \\
+\$\ 6 \\
\hline
\$62
\end{array}
\qquad
\begin{array}{r}
\$29 \\
+\$49 \\
\hline
\$72 \\
+\$\ 6 \\
\hline
\$78
\end{array}
$$

The fifth bit of the condition code register is the *half-carry,* or H flag. A carry from the low nibble to the high nibble during addition is a half-carry. A half-carry of 1 during addition sets the H flag to 1, and a half-carry of 0 during addition clears it to 0. If there is a carry from the high nibble during addition, the C flag is set to 1, which indicates that the high nibble is incorrect. A $6 must be added to the high nibble to adjust it to the correct BCD sum.

Fortunately, the programmer doesn't have to write instructions to detect illegal BCD sums following a BCD addition. The 68HC11 provides a *decimal adjust accumulator A* instruction, DAA, which takes care of all these detailed detection and correction operations. The DAA instruction monitors the sums of BCD additions and the C and H flags and automatically adds $6 to any nibble that requires it. The rules for using the DAA instruction are:

- The DAA instruction can be used only for BCD addition. It does not work for subtraction or hexadecimal arithmetic.
- The DAA instruction must be used immediately after one of the three instructions that leave their sum in accumulator A. (These three instructions are ADDA, ADCA, and ABA).
- The numbers added must be legal BCD numbers to begin with.

Example 2.8

Write a program to add the BCD numbers stored at memory locations $00 and $01 and save the sum at $02.

Solution:

```
LDAA  $00      ;load the first BCD number into A
ADDA  $01      ;perform the addition
DAA            ;decimal adjust the sum in A
STAA  $02      ;save the sum
END
```

Execution of the DAA instruction may set the C flag, allowing the programmer to implement multiprecision BCD addition.

2.7.6 Multiplication and Division

The 68HC11 provides one multiply and two divide instructions. The MUL instruction multiplies the 8-bit unsigned binary value in accumulator A by the 8-bit unsigned binary value in accumulator B to obtain a 16-bit unsigned result in double accumulator D. The upper byte of the product is in accumulator A, and the lower byte of the product is in accumulator B.

Example 2.9

Multiply $45 by $24 and store the result at $00 and $01.

Solution: Before a multiplication is performed, the operands must be loaded into accumulators A and B, as is done in the following program:

```
LDAA #$45    ;load the multiplicand into A
LDAB #$24    ;load the multiplier into B
MUL          ;perform multiplication
STD  $00     ;save the result
END
```

The unsigned multiply procedure also allows multiprecision operations. In multiprecision multiplication, the multiplier and the multiplicand must be broken down into 8-bit chunks, and several 8-bit by 8-bit multiplications must be performed. Assume we want to multiply a 16-bit hexadecimal number M by another 16-bit hexadecimal number N. To illustrate the procedure, we will break M and N down as follows:

$$M = M_H M_L$$
$$N = N_H N_L$$

where M_H and N_H are the upper 8 bits of M and N, respectively, and M_L and N_L are the lower 8 bits. Four 8-bit by 8-bit multiplications are performed, and then the partial products are added together as shown in Figure 2.4. The procedure to add these four partial products is as follows:

Step 1
Allocate four bytes to hold the sum. Assume these four bytes are located at P, P + 1, P + 2, and P + 3.

Step 2
Generate the partial product $M_L N_L$ and save it at locations P + 2 and P + 3.

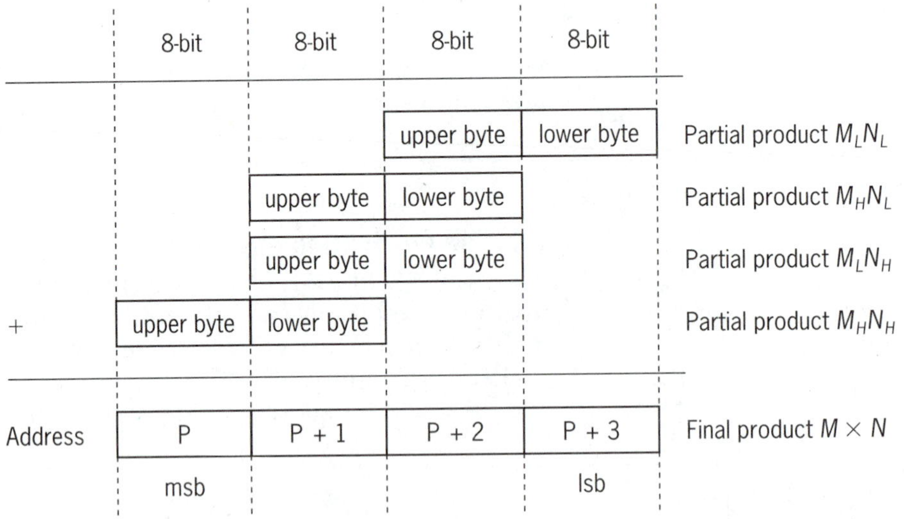

Note: msb stands for most significant byte and lsb for least significant byte.

Figure 2.4   Sixteen-bit by 16-bit multiplication

Step 3
Generate the partial product $M_H N_H$ and save it at locations P and P + 1.
Step 4
Generate the partial product $M_H N_L$ (in D).
Step 5
Add the contents of the accumulator D to memory locations P + 1 and P + 2. The C flag may be set to 1 after this addition.
Step 6
Add the C flag to memory location P.
Step 7
Generate the partial product $M_L N_H$ (in D).
Step 8
Add the contents of accumulator D to memory locations P + 1 and P + 2. The C flag may be set to 1 after this addition.
Step 9
Add the C flag to memory location P.

Example 2.10

Write a program to multiply $5678 by $1234 and store the result at memory locations $0000 to $0003. Use the method illustrated in Figure 2.4.

Solution: The highest, the next-to-highest, the next-to-lowest, and the lowest 8 bits of the product are to be stored at $0000, $0001, $0002, and $0003, respectively. The following program is a direct translation of the previous multiprecision multiplication algorithm:

```
          LDAA  #$78    ;load M_L into A
          LDAB  #$34    ;load N_L into B
          MUL           ;multiply M_L by N_L
          STD   $02     ;store the partial product M_L N_L at $02 and $03
          LDAA  #$56    ;load M_H into A
          LDAB  #$12    ;load N_H into B
          MUL           ;multiply M_H by N_H
          STD   $00     ;store the partial product M_H N_H at $00 and $01
          LDAA  #$56    ;load M_H into A
          LDAB  #$34    ;load N_L into B
          MUL           ;generate partial product M_H N_L
* The following instructions add M_H N_L to memory locations at $01 and $02
          ADDD  $01
          STD   $01
* The following three instructions add the C flag to memory location at $00
          LDAA  $00
          ADCA  #00
          STAA  $00

          LDAA  #$78    ;load M_L into A
          LDAB  #$12    ;load N_H into B
          MUL           ;generate the partial product M_L N_H
```

```
* The following instructions add M_L N_H to memory location $01 and $02
        ADDD  $01
        STD   $01
* The following three instructions add the C flag to memory location $00
        LDAA  $00
        ADCA  #00
        STAA  $00
        END
```

The integer divide instruction IDIV performs an unsigned integer divide of the 16-bit numerator in the D accumulator by the 16-bit denominator in index register X and sets the condition code register accordingly. The quotient is placed in X, and the remainder is placed in D. Depending on the base (radix) of the number, radix point can be called binary point, decimal point, etc. The radix point is assumed to be in the same place for both the numerator and the denominator. The radix point is to the right of bit zero for the quotient. In the case of division by zero, the quotient is set to $FFFF, and the remainder is indeterminate.

Example 2.11

Divide $4567 by $1234 and leave the quotient in X and the remainder in D.

Solution:

```
LDD  #$4567    ;load the dividend into D
LDX  #$1234    ;load the divisor into X
IDIV           ;perform the division
END
```

The fractional divide instruction FDIV performs an unsigned fractional divide of the 16-bit numerator in the D accumulator by the 16-bit denominator in index register X and sets condition code register accordingly. The quotient is placed in X, and the remainder is placed in D. The radix point is assumed to be in the same place for both the numerator and the denominator. The radix is to the left of bit 15 for the quotient. The numerator is assumed to be less than the denominator. In the case of division by zero or overflow (i.e., the denominator is less than or equal to the numerator), the quotient is set to $FFFF and the remainder is indeterminate.

Example 2.12

Divide the fractional number $.2222 by $.4444 using the FDIV instruction.

Solution: Before the fractional division can be performed, we need to load these two numbers into D and X, respectively.

```
LDD  #$2222    ;place dividend in D
LDX  #$4444    ;place divisor in X
FDIV
END
```

Since most arithmetic operations can be performed only on accumulators, the contents of accumulator D and index register X must be swapped so that further operations can be performed on the quotient. The 68HC11 provides two swap instructions:

- The XGDX instruction exchanges the contents of accumulator D and index register X.
- The XGDY instruction exchanges the contents of accumulator D and index register Y.

The following example illustrates the application of instructions IDIV and XGDX.

Example 2.13

Write a program to convert the 16-bit binary number stored at $00–$01 to BCD format and store the result at $02–$06. Each BCD digit is stored in one byte.

Solution: A binary number can be converted to BCD format (decimal number) using repeated division by 10. The largest 16-bit binary number corresponds to decimal 65536 which has five decimal digits. The first division by 10 computes the least significant digit and should be saved at memory location $06, the second division by 10 computes the second least significant digit, and so on. The program is as follows:

```
LDD  $00     ;place the 16-bit binary number in D
LDX  #10
IDIV         ;compute the least significant digit
STAB $06     ;save the least significant BCD digit
XGDX         ;place the quotient in D to prepare for the next division
LDX  #10
IDIV         ;compute the second most significant digit
STAB $05     ;save the second least significant BCD digit
XGDX         ;place the quotient in D to prepare for the next division
LDX  #10
IDIV         ;compute the middle BCD digit
STAB $04     ;save the middle BCD digit
XGDX         ;place the quotient in D to prepare for the next division
LDX  #10
IDIV         ;compute the second most and most significant digits
STAB $03     ;save the second most significant BCD digit
XGDX         ;swap the most significant BCD digit to B
STAB $02     ;save the most significant BCD digit
END
```

2.8 Program Loops

A computer is very good at performing repetitive operations. A programmer can write a program loop to tell the computer to repeat the same series of instructions many times. A *finite loop* is a program loop that will be executed only a finite number of times, while an *endless loop* is one in which the computer stays forever.

We will consider four major variants of the looping mechanism:

■ DO statement S FOREVER
This is an endless loop in which statement S is repeated forever. In some applications, the programmer might add the statement IF C then EXIT to leave the endless loop.

■ FOR I = n1 to n2 DO S or FOR I = n2 down to n1 DO S
Here, the variable I is the *loop counter*, which keeps track of the number of remaining times statement S is to be executed. The loop counter can be incremented (in the first case) or decremented (in the second case). Statement S is repeated $n2 - n1 + 1$ times. The value $n2$ is assumed to be no smaller than $n1$. If there is concern that the relationship $n1 \leq n2$ may not hold, then it must be checked at the beginning of the loop. Five steps are required to implement a FOR loop:

Step 1
Initialize the loop counter.
Step 2
Perform the specified operations and increment (or decrement) the loop counter.
Step 3
Compare the counter with the limit.
Step 4
Exit if the value of the loop counter is greater (or less) than the upper (lower) limit.
Step 5
Return to step 2.

■ WHILE C DO S
Whenever a WHILE construct is executed, the logical expression C is evaluated first. If it yields a true value, statement S is carried out. If it does not yield a true value, statement S will not be executed. The action of a WHILE construct is illustrated in Figure 2.5.

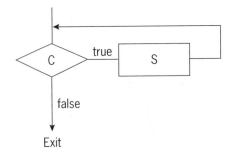

Exit

Figure 2.5   The WHILE . . . DO construct

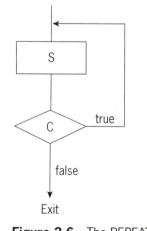

Figure 2.6 The REPEAT . . . UNTIL construct

■ REPEAT S UNTIL C
Statement S is first executed, then the logical expression C is evaluated. If C is false, the next statement will be executed. Otherwise, statement S will be executed again. The action of this construct is illustrated in Figure 2.6. Statement S will be executed at least once.

To implement a finite loop in 68HC11 assembly language, the programmer must use one of the conditional branch instructions. When executing the conditional branch instruction, the 68HC11 decides whether the branch should be taken by checking the values of condition flags.

2.8.1 The Condition Code Register

The contents of the condition code register are shown in Figure 2.7. The shaded characters are condition flags that reflect the status of an operation. The meanings of these condition flags are as follows:

■ C, the carry flag
Whenever a carry is generated as the result of an operation, this flag will be set to 1. Otherwise, it will be set to 0.

■ V, the overflow flag
Whenever the result of a two's complement arithmetic operation is out of range, this flag will be set to 1. Otherwise, it will be set to 0. The V flag is set to 1 when the carry from the most significant bit and the second most significant bit differ as the result of an arithmetic operation.

■ Z, the zero flag
Whenever the result of an operation is zero, this flag will be set to 1. Otherwise, it will be set to 0.

Figure 2.7 Condition code register

■ N, the negative flag
Whenever the most significant bit of the result of an operation is 1, this flag will be set to 1. Otherwise, it will be set to 0. This flag indicates that the result of an operation is negative.

■ H, the half-carry flag
Whenever there is a carry from the lower four bits to the upper four bits as the result of an operation, this flag will be set to 1. Otherwise, it will be set to 0.

2.8.2 Conditional Branch Instructions

A conditional branch instruction is required to implement a finite loop. The format of a conditional branch instruction is:

[<label>]   Bcc rel   [<comment>]

where

cc is one of the condition codes listed in Table 2.1.

rel is the distance of the branch from the first byte of the instruction that follows the Bcc instruction. The programmer normally uses a label instead of a value to specify the branch target.

The 68HC11 also has an unconditional branch instruction that is often used in a program loop. The format of this instruction is:

[<label>]   BRA rel   [<comment>]

As you can see in Table 2.1, some of the comparisons are unsigned while others are signed. Unsigned comparisons should be applied to quantities that are never negative. Memory address is an example of non-negative quantity.

Condition code	Meaning
CC	carry clear
CS	carry set
EQ	equal to 0
GE	greater than or equal to 0 (signed comparison)
GT	greater than 0 (signed comparison)
HI	higher (unsigned comparison)
HS	higher or same (unsigned comparison)
LE	less than or equal to 0 (signed comparison)
LO	lower (unsigned comparison)
LS	lower or same (unsigned comparison)
LT	less than 0 (signed comparison)
MI	minus (signed comparison)
NE	not equal to 0
PL	plus (signed comparison)
VC	overflow bit clear
VS	overflow bit set

Table 2.1 ■ Branch condition codes

LOOP: instruction X

.

.

.

Bcc LOOP

where cc is one of the condition codes

Figure 2.8   Program loop format

Signed comparisons should be applied to quantities that can be both positive and negative. Voltages and temperatures are examples of signed quantities.

The format for using one of the conditional branch instructions to implement a loop is shown in Figure 2.8. There is no conditional branch instruction that checks the H flag to determine whether a branch should be taken. Most conditional branch instructions check only one condition flag, as shown in the following list:

- C flag   BCC—branch if C = 0
  BCS—branch if C = 1
  BLO—branch if C = 1
  BHS—branch if C = 0
- Z flag   BEQ—branch if Z = 1
  BNE—branch if Z = 0
- N flag   BPL—branch if N = 0
  BMI—branch if N = 1
- V flag   BVS—branch if V = 1
  BVC—branch if V = 0

The following conditional branch instructions check more than one condition flag to determine whether the branch should be taken:

- BGE: branch if $(N \oplus V) = 0$
- BGT: branch if $(Z + (N \oplus V)) = 0$
- BHI: branch if $(C + Z) = 0$
- BLE: branch if $(Z + (N \oplus V)) = 1$
- BLS: branch if $(C + Z) = 1$
- BLT: branch if $(N \oplus V) = 1$

where $\oplus$ stands for the exclusive-or operation.

■

Example 2.14

For which of the following conditional branch instructions will the branch be taken after the execution of the instruction CMPA #10? Accumulator A contains 5.

a. BCC target
b. BPL target
c. BNE target

> **d.** BCS target
> **e.** BMI target
> **f.** BHS target

Solution: The 68HC11 subtracts 10 from 5 when executing the instruction CMPA #10. The result is negative. After execution of the instruction, the N and C flags are set to 1, while the V and Z flags are cleared to 0. Therefore, instructions (c), (d), (e) will cause the branch to be taken.

2.8.3 Decrementing and Incrementing Instructions

A program loop requires a counter, often called the *loop count*, to keep track of the number of times the loop remains to be executed. The loop count can be either decremented or incremented by 1. The following instructions decrement a register or a memory location:

- DECA: subtract one from the contents of accumulator A.
- DECB: subtract one from the contents of accumulator B.
- DEC opr: subtract one from the contents of the memory location opr, which must be specified in either the extended or the index addressing mode.
- DES: subtract one from the stack pointer SP.
- DEX: subtract one from index register X.
- DEY: subtract one from index register Y.

These instructions increment a register or memory location:

- INCA: add one to the contents of accumulator A.
- INCB: add one to the contents of accumulator B.
- INC opr: add one to the contents of the memory location opr, which must be specified in either the extended or the index addressing mode.
- INS: increment the value of the stack pointer by 1.
- INX: increment the value of index register X by 1.
- INY: increment the value of index register Y by 1.

Example 2.15

Write a program to compute $1 + 2 + 3 + \ldots + N$ (say $N = 20$) and save the sum at $00.

Solution: The procedure for computing the sum of the integers from 1 to 20 is illustrated in Figure 2.9. We will use accumulator B for the loop count *i* and the

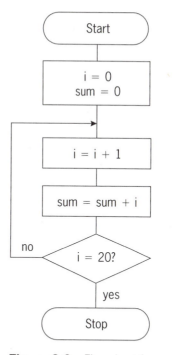

Figure 2.9 Flowchart for computing
$1 + 2 + \ldots + 20$

accumulator A for the variable *sum*. The assembly program that implements this algorithm is as follows:

```
N          EQU   20        ;loop count limit

           LDAB  #0        ;initialize the loop count to 0.
  * accumulator A is used to accumulate the sum
           LDAA  #00       ;initialize the sum to 0
again      INCB            ;increment the loop count
           ABA             ;add i to sum
           CMPB  #20       ;compare the loop count with the upper limit
           BNE   again     ;branch if the loop count is smaller than 20
           STAA  $00       ;save the sum
           END
```

2.8.4 Compare Instructions

The 68HC11 has a group of compare instructions that are dedicated to setting the condition flags. Compare instructions and conditional branch instructions are often used together to implement a program loop. Compare instructions are executed to change the condition flags of the condition code register, which are then checked by conditional branch instructions to determine

Instruction format	Function
[<label>] CBA [<comment>]	compare A to B
[<label>] CMPA opr [<comment>]	compare A to a memory location or value
[<label>] CMPB opr [<comment>]	compare B to a memory location or value
[<label>] CPD opr [<comment>]	compare D to a memory location or value
[<label>] TST opr [<comment>]	test a memory location for negative or zero
[<label>] TSTA [<comment>]	test A for negative or zero
[<label>] TSTB [<comment>]	test B for negative or zero
[<label>] CPX opr [<comment>]	compare X to a memory location or value
[<label>] CPY opr [<comment>]	compare Y to a memory location or value

opr is specified in one of the following addressing modes: EXT, INDX, INDY, IMM (not applicable to TST opr), or DIR (not applicable to TST opr).

Table 2.2 ■ 68HC11 compare instructions

whether a branch should be taken. The compare instructions are listed in Table 2.2. The functions of these compare instructions are as follows:

- CBA: subtracts the value in accumulator B from accumulator A and sets the condition flags accordingly; the contents of both A and B are unchanged.

- CMPA: subtracts the value of a memory location (or immediate value) from accumulator A and sets the condition flags accordingly; the contents of A and the memory location are not modified.

- CMPB: subtracts the value of a memory location (or immediate value) from accumulator B and sets the condition flags accordingly; the contents of B and the memory location are not modified.

- CPD: subtracts the 16-bit value stored at the specified memory locations (or just the immediate value) from accumulator D and sets the condition flags accordingly; the contents of D and the memory location are not modified.

- TST: subtracts 0 from the value stored at the specified memory location and sets the condition flags accordingly without modifying the contents of the memory location.

- TSTA: subtracts 0 from accumulator A and sets the condition flags accordingly without modifying the value in A.

- TSTB: subtracts 0 from accumulator B and sets the condition flags accordingly without modifying the value in B.

- CPX: subtracts the specified value or the value of the specified memory location from X and sets the condition flags accordingly without modifying the value in X.

- CPY: subtracts the specified value or the value of the specified memory location from Y and sets the condition flags accordingly without modifying the value in Y.

Example 2.16

Write a program to determine the largest of the twenty 8-bit numbers stored at $00–$13. Save the largest number at $20.

Solution: The procedure for finding the largest element of an array is illustrated in Figure 2.10.

Accumulator A will be used to hold the temporary largest element. Accumulator B will be used as the loop counter, and index register X will be used as the pointer to the array element. The program should stop when the loop counter B is equal to the array count minus one. The program that implements this algorithm is as follows:

```
N          EQU   $13

           LDAB  #1
           LDAA  $00      ;set [$00] as the temporary max of the array
           LDX   #$01     ;points X to memory location $01
loop       CMPA  0,X      ;compare A to the next element
           BHS   chkend   ;do we need to update the temporary array max?
           LDAA  0,X      ;update the temporary array max
```

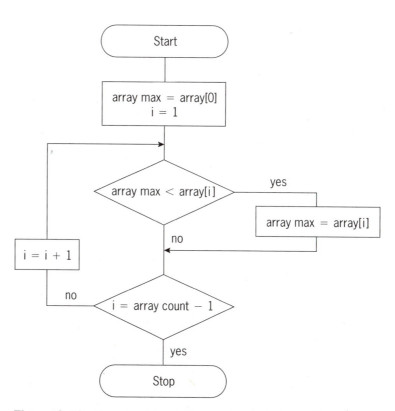

Figure 2.10 Flowchart for finding the largest element in an array

```
chkend    CMPB  #N          ;compare the loop index with the loop limit
          BEQ   exit        ;have we checked the whole array yet?
          INX               ;not yet, and move the array pointer
          INCB              ;increment the loop counter
          BRA   loop
exit      STAA  $20         ;save the array max
          END
```

■

2.8.5 Special Conditional Branch Instructions

The 68HC11 provides two special conditional branch instructions for implementing program loops. The syntax of the first special conditional branch instruction is

[<label>]   BRCLR (opr)(msk)(rel)   [<comment>]

where

opr specifies the memory location to be checked and must be specified using either the direct or the index addressing mode.

msk is an 8-bit mask that specifies the bits of the specified memory location to be checked. The bits of the memory byte to be checked correspond to those bit positions that are ones in the mask.

rel is the branch offset and is specified in the relative mode.

This instruction tells the 68HC11 to perform the logical AND of the memory location specified by opr and the mask supplied in the instruction and then to branch if the result is zero (only if all bits corresponding to ones in the mask byte are zeros in the tested byte).

For example, for the instruction sequence

```
          LDX #$1000
HERE      BRCLR $30,X %10000000 HERE
          LDAA $31,X
```

the 68HC11 will continue to execute the second instruction if the most significant bit of the memory location at $1030 is 0. Otherwise, the instruction LDAA $31,X will be executed.

The syntax of the second special conditional branch instruction is

[<label>]   BRSET (opr)(msk)(rel)   [<comment>]

where

opr specifies the memory location to be checked and must be specified using either the direct or the index addressing mode.

msk is an 8-bit mask that specifies the bits of the specified memory location to be checked. The bits of the memory byte to be checked correspond to those bit positions that are ones in the mask.

rel is the branch offset and is specified in the relative mode.

This instruction tells the 68HC11 to perform the logical AND of the specified memory location inverted and the mask supplied in the instruction and then to branch if the result is zero (only if all bits corresponding to ones in the mask byte are ones in the tested byte).

For example, for the instruction sequence

```
LOOP     ADDA #01
         . . . . . . .
         BRSET $20 %11110000 LOOP
         LDAB $01
```

the branch will be taken if the most significant four bits of the memory locations at $20 are all ones.

___

Example 2.17

Write a program to compute the sum of the odd numbers in an array with twenty 8-bit elements. The starting address of the array is $00. Save the sum at $20 and $21.

Solution: The address of the last element of the given array is $13. The sum may be larger than 8 bits, and it will be stored in memory locations $20 and $21. The least significant bit of an odd number is a 1. The algorithm for solving this problem is as follows:

> **Step 1**
> Initialize the sum to 0. Use the index register X to step through the array, and initialize it to 0.
> **Step 2**
> Check the least significant bit of the value of the memory location pointed to by X. If it is a one, then add it to the sum. Because the sum and the array element are of different lengths, addition proceeds in two steps. In the first step, the array element is added to the lower byte of the sum. In the second step, the carry resulting from the first step is added to the upper byte.
> **Step 3**
> Compare [X] to $13. If [X] equals $13, then exit. Otherwise, increment X by 1 and go to step 2.

The program is as follows:

```
N        EQU   $13
SUM      EQU   $20

         LDAA  #$00
         STAA  SUM          ;initialize the sum to 0
         STAA  SUM+1        ;      "
         LDX   #$00         ;point X to the first number
loop     BRCLR 0,X $01 chkend    ;is it an odd number?
         LDD   SUM          ;load the sum into accumulator D
         ADDB  0,X          ;add the odd number to the lower byte of the sum
         ADCA  #00          ;add the carry from the lower byte to the upper byte
  *                         ;of the sum
         STD   SUM          ;update the sum
chkend   CPX   #N           ;compare X to the array limit
```

```
            BHS    exit        ;are we done?
            INX                ;check the next number
            BRA    loop        ;go back to the beginning of the loop
    exit    END
```

The third instruction of this program uses the expression SUM+1, which stands for the address that is one byte higher than SUM (i.e., the upper byte of the sum). Most assemblers support this expression.

2.8.6 Instructions for Variable Initialization

When writing a program, the programmer often needs to initialize a variable to zero. The 68HC11 has three instructions for this purpose. They are:

```
[<label>]    CLR  opr    [<comment>]
```

where opr stands for a memory location specified using either the extended or the index addressing mode. The memory location is initialized to zero after execution of this instruction.

```
[<label>]    CLRA    [<comment>]
```

Accumulator A is set to zero by this instruction.

```
[<label>]    CLRB    [<comment>]
```

Accumulator B is set to zero by this instruction.

2.9 Shift and Rotate Instructions

Shift and *Rotate* instructions are useful for bit field operations, and they can be used to speed up the integer multiply and divide operations if one of the operands is a power of 2. A shift/rotate instruction shifts/rotates the operand by one bit. The 68HC11 has shift instructions that can operate on accumulators A, B, and D or a memory location. A memory operand (specified as opr in the following paragraphs) must be specified using the extended or the index addressing mode. Both left and right shifts/rotates are available. The format and function of each shift/rotate instruction are described in the following paragraphs.

The following arithmetic shift left instructions operate on 8-bit operands:

```
[<label>]    ASL  opr    [<comment>]  ;the memory location opr is the operand
[<label>]    ASLA        [<comment>]  ;accumulator A is the operand
[<label>]    ASLB        [<comment>]  ;accumulator B is the operand
```

The operation of these three instructions is illustrated in the following diagram:

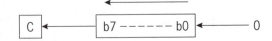

where C is the carry flag. The bit 0 is vacated and is filled with a 0.

The arithmetic shift left instruction can also operate on the D accumulator:

[<label>]   ASLD   [<comment>]

The operation of this instruction is as follows:

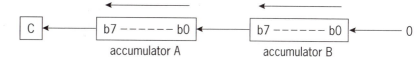

The following arithmetic shift right instructions operate on 8-bit operands:

[<label>]   ASR opr   [<comment>] ;the memory location opr is the operand
[<label>]   ASRA   [<comment>] ;accumulator A is the operand
[<label>]   ASRB   [<comment>] ;accumulator B is the operand

The operation of these three instructions is shown in the following diagram:

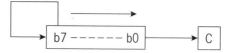

The sign of the operand is maintained because the sign bit is duplicated after the shift right operation. This feature can be used to divide a signed number by a power of 2.

The following logical shift left instructions operate on 8-bit operands:

[<label>]   LSL opr   [<comment>] ;the memory location opr is the operand
[<label>]   LSLA   [<comment>] ;accumulator A is the operand
[<label>]   LSLB   [<comment>] ;accumulator B is the operand

The operation of these three instructions is as follows:

The 68HC11 also has a logical shift left instruction that operates on D:

[<label>]   LSLD   [<comment>]

The operation of this instruction is:

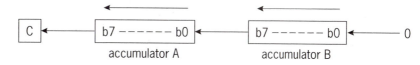

The 68HC11 also has three logical shift right instructions that operate on 8-bit operands:

[<label>]   LSR opr   [<comment>] ;the memory location opr is the operand
[<label>]   LSRA   [<comment>] ;accumulator A is the operand
[<label>]   LSRB   [<comment>] ;accumulator B is the operand

The operation of these three instructions is as follows:

The 68HC11 also has a logical shift right instruction that operates on D:

[<label>]   LSRD   [<comment>]

The operation of this instruction is as follows:

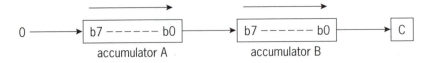

The 68HC11's rotate instructions operate only on 9-bit operands because they include the C flag in the rotate operation. Only one place is rotated at a time.

The following are *rotate left* instructions:

[<label>]   ROL opr   [<comment>] ;the memory location opr is the operand
[<label>]   ROLA   [<comment>] ;accumulator A is the operand
[<label>]   ROLB   [<comment>] ;accumulator B is the operand

The operation of these three instructions is:

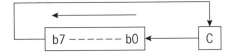

The following are *rotate right* instructions:

[<label>]   ROR opr   [<comment>] ;the memory location opr is the operand
[<label>]   RORA   [<comment>] ;accumulator A is the operand
[<label>]   RORB   [<comment>] ;accumulator B is the operand

The operation of these three instructions is:

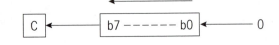

Example 2.18

Compute the new values of accumulator A and the C flag after execution of the instruction ASLA. Assume that the original value in A is $74 and the C flag is 1.

Solution: The operation of this instruction is shown in Figure 2.11.

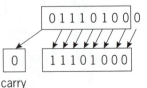

carry

Figure 2.11   Operation of the ASLA instruction

The result is as follows:

Original value	New value
[A] = 01110100 C = 1	[A] = 11101000 C = 0

Example 2.19

Compute the new values of the memory location at $00 and the carry flag after execution of the instruction ASR $00. Assume that the original value of the memory location at $00 is $F6 and that the C flag is 1.

Solution: The operation of this instruction is shown in Figure 2.12.

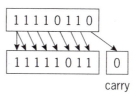

carry

Figure 2.12   Operation of the instruction ASR $00

The result is:

Before ASR $00	After ASR $00
[$00] = 11110110 C = 1	[$00] = 11111011 C = 0

Example 2.20

If the memory location at $00 contains $F6 and the C flag is 1, what are the new values of this memory location and the carry flag after execution of the instruction LSR $00?

Solution: The operation of this instruction is illustrated in Figure 2.13.

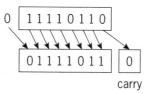

carry

Figure 2.13   Operation of the instruction LSR $00

The result is:

Before LSR $00	After LSR $00
[$00] = 11110110 C = 1	[$00] = 01111011 C = 0

Example 2.21

Compute the new values of accumulator B and the C flag after execution of the instruction ROLB. Assume the original value of B is $BE and C = 1.

Solution: The operation of this instruction is illustrated in Figure 2.14.

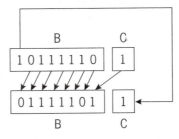

Figure 2.14   Operation of the instruction ROLB

The result is:

Before ROLB	After ROLB
[B] = 10111110 C = 1	[B] = 01111101 C = 1

Example 2.22

Compute the new values of accumulator B and the C flag after execution of the instruction RORB. Assume the original value of B is $BE and C = 1.

Solution: This instruction is illustrated in Figure 2.15.

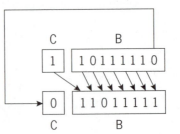
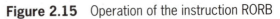

Figure 2.15   Operation of the instruction RORB

The result is:

Before RORB	After RORB
[B] = 10111110 C = 1	[B] = 11011111 C = 0

Example 2.23

Write a program to count the number of ones contained in the 16-bit number stored at $00 and $01 and save the result at $02.

Solution: The logical shift right instruction is available for the double accumulator D. We can load the 16-bit number into D and shift it to the right 16 times or until D becomes zero. Initialize the number of ones to zero. With each shift right operation, check the carry flag. If the carry flag is a one, increment the number of ones.

The following program implements this idea:

```
          ORG  $C000
          LDAA #$00       ;clear A to 0
          STAA $02        ;initialize the count to 0
          LDD  $00        ;load 2 bytes into D
loop      LSRD            ;logical shift right D
          BCC  testzero   ;branch if the carry bit is 0
          INC  $02        ;increment the number of 1s
testzero
          CPD  #00        ;compare D with 0
          BNE  loop       ;if [D] = 0 then stop
          END
```

Occasionally the programmers may need to shift a number larger than 16 bits, but the 68HC11 does not have an instruction that does this. This problem can be dealt with as follows: Assume that the number has k bytes (or $8k$ bits) and that the most significant byte is located at loc. The remaining $k - 1$ bytes are located at loc + 1, loc + 2, . . . , loc + $k - 1$, as shown in the following diagram:

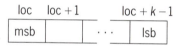

The shift-one-bit-to-the-right operation can be illustrated by

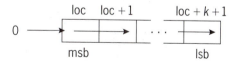

As this diagram shows,

■ Bit 7 of each byte will receive bit 0 of the byte on its immediate left, with the exception of the most significant byte, which will receive a zero.

■ Each byte will be shifted to the right by one bit. Bit 0 of the least significant byte will be shifted out and lost.

This operation can therefore be implemented as follows:

Step 1
Shift the byte at loc to the right one place (using the LSR loc instruction).
Step 2
Rotate the byte at loc + 1 to the right one place (using the ROR <opr> instruction).
Step 3
Repeat step 2 for the remaining bytes.

By repeating this procedure, the given number can be shifted to the right as many bits as desired.

Example 2.24

Write a program to shift the 32-bit number stored at $20–$23 to the right four places.

Solution: The most significant, second-most significant, second-least significant, and the least significant bytes of this 32-bit number are stored at $20, $21, $22, and $23, respectively. The following instruction sequence is a direct translation of the procedure described above.

```
        LDAB #4        ;set up the loop count
        LDX  #$0020
again   LSR  0,X
        ROR  1,X
        ROR  2,X
        ROR  3,X
        DECB
        BNE  again
```

2.10 Program Execution Time

Instruction execution in the 68HC11 is controlled by the E clock signal. The E clock signal is derived from an external crystal oscillator. A 68HC11 instruction may take from 2 to 41 E clock cycles to execute. The execution time of each 68HC11 instruction can be found in Appendix B. Knowledge of the execution time of each instruction and the clock period allows a delay of any length to be created. The execution times of a sample of instructions are shown in Table 2.3.

Instruction	Execution time (unit is E clock cycle)
BNE <rel>	3
DECB	2
DEX	3
LDAB (imme)	2
LDX (imme)	3
NOP	2

Table 2.3 ■ Execution times of a few instructions

Example 2.25

Write a program to create a delay of 100 ms. Assume the frequency of the E clock signal is 2 MHz.

Solution: A delay requires the execution of a program loop or loops. Let x be the execution time of the program loop, and let T be the desired delay. Then the loop count N can be calculated from the following equation:

$$N = T \div x$$

For example, the execution time of the following four instructions is 10 E clock cycles (5 μs):

```
again    NOP           ;2 cycles
         NOP           ;2 cycles
         DEX           ;3 cycles
         BNE  again    ;3 cycles
```

To create a 100-ms delay, this loop must be executed 20000 times:

$$100 \text{ ms} \div 5 \ \mu s = 0.1 \text{ sec} \div 0.000005 \text{ sec} = 20000$$

A delay of 100 ms is created by the following program loop:

```
         LDX  #20000    ;use X as the loop count
again    NOP
         NOP
         DEX
         BNE  again     ;not done yet
         END
```

The actual execution time of this program segment is three E clock cycles more than 100 ms because the instruction LDX #20000 takes three E clock cycles to execute. This instruction is the overhead for setting up the delay loop.

A longer delay can be created by using nested program loops.

Example 2.26

Write a program to create a delay of 10 seconds.

Solution: We need to write a two-layer loop that includes the loop in the previous program and executes it for the following number of times:

10 s ÷ 100 ms = 100

The program is as follows:

```
(1)          LDAB  #100      ;use B as the outer loop count
(2)outer     LDX   #20000    ;use X as the inner loop count
(3)inner     NOP
(4)          NOP
(5)          DEX             ;decrement the inner loop count
(6)          BNE   inner     ;branch back to the inner loop
(7)          DECB            ;decrement the outer loop count
(8)          BNE   outer     ;branch back to the outer loop
(9)          END
```

This program also has overhead that makes the delay slightly over 10 seconds. As explained earlier, it takes 5 μs to execute instructions (3) to (6). These four instructions are repeated 2000000 times to create a delay of 10 seconds. However, as you have learned, program loops repeat a sequence of instructions, while certain other instructions are needed to set up the program loops. Those instructions are often not included in the calculation of execution time just for simplicity—they are the overhead of creating program loops. The overhead in this example is created by the instructions listed in Table 2.4. Therefore, the total overhead is 401 μs. To create an exact delay, these instructions must be taken into account when setting up the program loop. More accurate delays can be created by using timer functions, which will be studied in Chapter 7.

Instruction	Execution time (in E clock cycles)	Times of execution	Execution time
LDX #20000	3	100	150 μs
LDAB #100	2	1	1 μs
DECB	2	100	100 μs
BNE OUTER	3	100	150 μs

Table 2.4   ■   Execution time of sample instructions

2.11 68HC11 Development Tools

There are numerous development tools for microcontroller-based products. These tools can be classified into two categories: software tools and hardware tools. Software tools include text editors, communication (terminal emulation) programs, cross assemblers, cross compilers, simulators, and integrated software. Hardware tools include evaluation boards, oscilloscopes, logic analyzers, in-circuit emulators, etcetera. In this text, we will discuss the software tools and three evaluation boards made by Motorola.

2.11.1 Software Tools

TEXT EDITORS

The text editor is a software program that allows the user to enter and edit program files. Many readers will probably use the Microsoft EDIT editor that comes with the Microsoft DOS (disk operating system) and/or the Windows Notepad program. Listed in Table 2.5 are the common editing functions available in the Microsoft EDIT text editor. Although it is not a sophisticated text editor, the Microsoft EDIT editor is adequate for assembly programming. To invoke the EDIT editor, simply type **EDIT** <**filename**> at the DOS command prompt.

In an editing session, the user can invoke the commands listed in Table 2.5 by simultaneously pressing the alt key and the first character of the category.

Category	Command	Function
File	new	allows the user to enter and edit a new file
	open	allows the user to edit an existing file
	save	saves the file being edited
	save as	saves the current file in different name
	exit	leave the editor
Edit	cut	allows the user to cut (delete) a block of text into a buffer
	copy	allows the user to copy a block of text
	paste	allows the user to paste the text in the buffer to the cursor position
	clear	deletes the character at the cursor position (rarely used)
Search	find	allows the user to search the specified word
	repeat last find	allows the user to repeat the previous search
	change	allows the user to search for a word and replace it with another word
Option	display	allows the user to set the monitor screen foreground and background colors, display scroll bars, and set tab positions
	path	allows the user to select a different location (path) for the editor help file (edit.hlp)

Table 2.5 ■ Microsoft DOS EDIT editing functions

For example, to open a file, press alt and **f** simultaneously to display the commands in the file category and then press **o**. The EDIT editor will then display a *dialog box* that asks the user to enter the name of the file to be opened. To invoke commands in the edit category, press alt and **e** simultaneously, followed by the character (**t** for cut, **p** for paste, **c** for copy, or **e** for clear) representing the edit command. The other commands can be invoked in the same way. Before invoking the Cut and Copy commands, the user must specify a block of text to be cut and copied by pressing the shift key and an arrow key simultaneously. There are four arrow keys: right, left, up, and down. The right and down arrow keys are used more often.

Another editor that the reader may want to use is the Notepad program in the Microsoft Windows system. To use the Notepad program, follow this procedure:

Step 1
Start the Windows program (if it is not running yet) by typing **win** at the DOS command prompt.

Step 2
Open the Program Manager Window (if it is not opened in your setup) by double-clicking on the Program Manager icon. Skip this step if this window is already opened.

Step 3
Open the Accessories window by double-clicking on the Accessories icon within the Program Manager Window.

Step 4
Open the Notepad window by double-clicking on the Notepad icon within the Accessories window.

Step 5
Start to enter and/or edit a program or any other document.

The Notepad program supports commands in four categories: file, edit, search, and help. The use of most commands is fairly straightforward, and the mouse can be used to invoke all commands. The cut-and-paste operation is easier to use than the one on the EDIT editor.

The main reason for using the EDIT editor and the Notepad program is that they come with the computer system without extra charge. For a software developer who mainly programs in high-level languages, these two editors are probably not adequate. He or she would probably want to use a sophisticated editor that provides functions such as automatic keyword completion, automatic indentation, parenthesis matching, and syntax checking. These functions can increase the productivity of a high-level language programmer significantly.

CROSS ASSEMBLERS AND COMPILERS

Cross assemblers and cross compilers generate the executable code that is placed in the ROM, EPROM, or EEPROM of a 68HC11-based product. Since cross assemblers and compilers have considerable variations in many areas, we encourage readers to refer to the vendors' reference manuals.

The Motorola freeware cross assembler as11 is used in many universities and colleges simply because it is free and easy to use. The Motorola freeware assembles 68HC11 assembly programs into executable code in the S-record format. To invoke the freeware, enter the following command at the DOS command prompt:

    as11 filename.asm

where *asm* is the file name extension. After this command is entered and if there is no syntax error, the user should see a new file, *filename.s19*, generated in the current working directory.

TERMINAL EMULATION PROGRAMS

A terminal emulation program enables a PC to act like a terminal so that it can talk to another computer. A PC user can run a terminal emulation program such as the Microsoft Windows Terminal program, Kermit, or Procomm to communicate to the 68HC11 single-board computer.

The procedure for invoking the Terminal program is as follows:

Step 1
Start Windows by typing **win** at the DOS command prompt.
Step 2
Open the Program Manager window (if it is not opened in your setup) by double-clicking on the Program Manager icon.
Step 3
Open the Accessories window by double-clicking on the Accessories icon.
Step 4
Start the Terminal program by double-clicking on the Terminal icon. You should then see the Terminal program window (Figure 2.16).

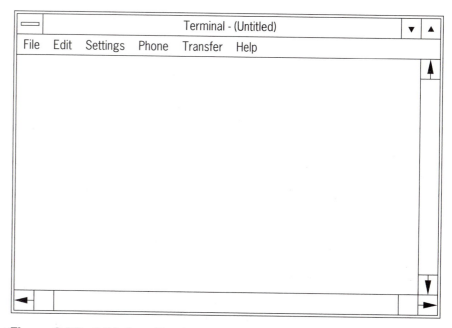

Figure 2.16 A Windows Terminal program session

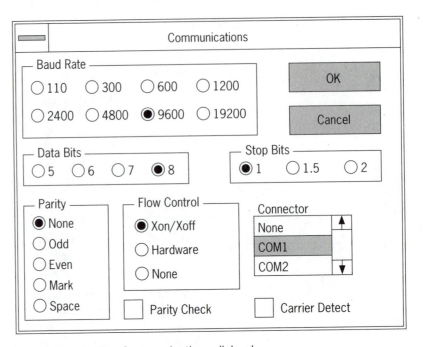

Figure 2.17   The Communications dialog box

In order to communicate with the 68HC11 single-board computer, the first thing to do is to set up the communication parameters. This can be done by clicking on Settings to bring out the Settings menu. There are 11 items under the Settings menu—select the item Communications. The Communications dialog box (see Figure 2.17) will then appear. It allows the user to select the following parameters:

■ baud rate (select 9600)

■ data bits (select 8)

■ number of stop bits (select 1)

■ parity (select none)

■ flow-control (select Xon/Xoff)

■ connector (communication port) click the communication port connected to the single-board computer (can be from 1 to 4)

■ parity check (don't select)

■ carrier detect (don't select)

It is a good idea to save the communication settings so that the same settings can be used in the future. To do this:

Step 1
Select File in the Terminal window. The pull-down menu under File will appear.

Step 2
Select the item Save or Save As. This will bring out a dialog box that asks the user to specify the file name. The user can specify both the file name and the directory in which to save the communication settings.

To use the saved communication settings, do the following steps:

Step 1
Select File in the Terminal window. The pull-down menu will appear.

Step 2
Select Open. This will bring out a dialog box that asks the user to specify the file to be opened. Suppose the name of the file containing the communication settings is *com.trm*—click on the file com.trm, and the saved communication settings will be used in the present session. Click on the OK button to close the dialog box.

To exit the Windows Terminal program, select the command Exit in the File category. The procedure for downloading a program to the 68HC11EVB or 68HC11EVM computer will be illustrated shortly in the section on evaluation boards.

For a user who is not running Microsoft Windows, the freeware Kermit is probably a good choice because it offers many excellent features. Before running Kermit to communicate to an evaluation board such as the 68HC11EVB, 68HC11EVBU, or 68HC11EVM, several parameters must be set up properly. The most important parameters include character length, parity, baud rate, and communication port. A common setting is to choose an 8-bit character length and no parity. The baud rate of 9600 is standard on Motorola evaluation boards, and the communication port can be either 1 or 2, depending on the PC communication port being used. The user can set up these parameters by editing the file *MSKERMIT.INI*, which is located in the directory where the Kermit program is installed. Make sure that the following lines are included in *MSKERMIT.INI*:

    set speed 9600

    set parity none

    set port x (x is 1 or 2 depending on the communication port being used)

MSKERMIT.INI is a batch file that will be executed when Kermit is started. To invoke the Kermit program, simply type **kermit** then press Enter. If version 3.11 of the Kermit program is used, the following message should appear on your monitor screen:

```
IBM-PC MS-DOS kermit: 3.11    7 Sept 1991 patch level 0
Copyright (C) Trustees of Columbia University 1982, 1991.
Type ? or HELP for help
Smile!
MS-Kermit>
```

The last line of the message is the Kermit prompt for the user to enter commands. A summary of the Kermit commands that can be used to communicate with the 68HC11 single-board computer is in Table 2.6.

Category	Command	Function
Local file management	DIR	list files
	CD	change directory
	DELETE	delete files
	RUN	execute a DOS command
	TYPE	display a file
	SPACE	show disk space
Kermit program management	EXIT	from Kermit, return to DOS
	QUIT	same as EXIT
	TAKE	execute Kermit commands from file
	CLS	clear screen
	PUSH	enter DOS, EXIT returns to Kermit
	Ctrl-C	interrupt a command
Communication settings	SET PORT, SET SPEED	
	SET PARITY	
	SET FLOW-CONTROL	
	SHOW COMMUNICATIONS	

Table 2.6 ■ Functional summary of Kermit commands

Communication parameters can also be set interactively, i.e., when Kermit is running, as shown in the following example.

Example 2.27

Set up the following communication parameters after Kermit is invoked:

communication port = 2
baud rate = 9600
parity = none

Solution: After the user sees the Kermit prompt, the following commands should be entered one at a time:

set port 2 (or com2)
set speed 9600
set parity none

The session will look like this (commands typed by the user are in boldface):

```
MS-Kermit>set port 2
MS-Kermit>set speed 9600
MS-Kermit>set parity none
```

The user can view all communication parameters by entering the command **show communications.** To exit Kermit, simply enter the command **exit** or **quit** followed by Enter.

To communicate to the evaluation board, type **c**(onnect) and press the Enter key at the Kermit prompt as follows:

MS-Kermit>**c**

The PC will then be in terminal emulation mode, and the most important communication parameters, such as baud rate, communication port, parity, and terminal type, will be displayed on the bottom line of the screen. All Kermit commands can be abbreviated to the shortest unique string, as illustrated in the previous example. The procedure for communicating and downloading executable programs to the evaluation board will be discussed in the section on Buffalo commands. Readers who are interested in the details of Kermit commands should refer to the book *Using MS-DOS Kermit* by Christine M. Gianone (Digital Press, 1992).

THE 68HC11 SIMULATOR

A simulator allows the user to execute microcontroller programs without having the actual hardware. It uses PC memory to represent microcontroller registers and memory locations and to interpret each microcontroller instruction, and it saves execution results in the PC memory. The simulator also allows the user to set the contents of memory locations and registers before the simulation run starts. However, software simulators have limitations—the most severe one is that they cannot effectively simulate I/O behavior. Motorola has a freeware simulator (sim11) for the 68HC11 that can be downloaded from the bulletin board at (512) 891-3733. The following ftp sites also have the Motorola freeware software:

1. ftp.ee.ualberta.ca
2. ernie.uvic.ca
3. cherupakha.media.mit.edu
4. ftp.luth.edu

INTEGRATED DEVELOPMENT TOOLS

Ideally, integrated development software would provide an environment that combines a text editor, a cross assembler and/or compiler, a simulator, a communication program, and a front-end interface to an evaluation board so that the user could enter the assembly (or high-level) program, assemble and debug the program, download the executable code to the evaluation board, and execute the program all in the same environment, without needing to exit any program.

Dr. Noel Anderson developed a front-end interface (PDE6811) for the EVB when he taught in the Electrical Engineering Department at North Dakota

State University. The PDE6811 allows the user to enter and assemble his or her assembly program and then download the executable code to either the EVB or the simulator (SIM6811). The PDE6811 user interface has the appearance of a Borland compiler environment, which has separate windows for pull-down menus, messages, the editor, and instruction set references. The SIM6811 simulator features pull-down menus, register displays, memory displays, breakpoints, instruction disassembly, and bus activity traces.

Motorola has included a demo version of P&E Microcomputer System's integrated cross assembler, IASM11, and the front-end user interface EVM11 in their 68HC11EVB and 68HC11EVM products. The IASM11 includes a text editor, a cross assembler, and a communication program that allows users to download the executable code onto the evaluation board. The front-end user interface EVM11 works with the 68HC11EVM and has separate windows for displaying the contents of CPU registers and memory locations, the instruction being executed, breakpoints, messages, and command menus. The EVM user can see the program execution results easily with this front-end interface.

2.11.2 68HC11 Evaluation Boards

Motorola has developed several evaluation boards to facilitate the prototyping and evaluation of 68HC11-based designs. These single-board computers have a resident *monitor* program that can, in response to the user's request, provide the following services:

- display the contents of registers and/or memory locations
- modify the contents of registers and/or memory locations
- set program execution breakpoints where program execution will be stopped
- trace program execution
- download the user's program from a PC or other computer
- allow the user to enter short programs
- perform other functions and services

The monitor program takes control of the computer when the power is first turned on. It will initialize the I/O registers and the contents of some memory locations, and it may perform some power-on testing. It may also provide many other functions and services that can be invoked by the user program.

The 68HC11EVB and the 68HC11EVM are the earliest evaluation board designs. The newer 68HC11EVBU provides a wire-wrap area where the user can add other logic. These three boards provide a one-line assembler/disassembler which allows the user to enter short programs onto the board for direct execution. The on-board monitor program provided by each board allows the user to download programs into the board. The 68HC11EVM board also allows the user to program the on-chip EEPROM of other 68HC11 chips. The monitor program resident on the 68HC11EVB and 68HC11EVBU boards is called Buffalo.

In the following sections we will refer to these boards as EVB, EVBU, and EVM.

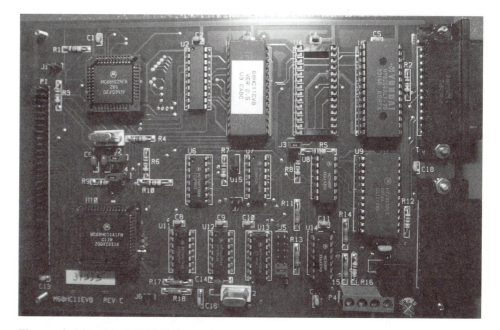

Figure 2.18 68HC11EVB Computer

THE 68HC11EVB

Figure 2.18 is a photograph of the EVB. The major chips on this board are shown in the block diagram in Figure 2.19. The details of the communication ports will be discussed in Chapter 8.

The microcontroller unit (MCU) used in the EVB computer is a 68HC11A1. An 8-MHz crystal oscillator is used to generate the 2-MHz E clock signal to control the operation of the 68HC11A1. The MCU is configured to operate in the expanded mode, which makes ports B and C unavailable to the user. Port B and C pins can be restored and made available to the user, however, by adding the port replacement unit (PRU). The EVB is controlled by the Buffalo monitor program, which resides in the 8 KB of EPROM. The user RAM is 8 KB, ranging from $C000 to $DFFF, and it can be expanded to 16 KB. The EVB provides two connectors to communicate to a host computer and a terminal. The user can interact with the EVB via the terminal connector.

The idea of providing a host port can be traced back to the 68000-based ECB computer designed by Motorola in 1979. In those days, the price of a PC was very high. Most companies used either a mainframe or a minicomputer as a host to run development software. The ECB computer had a terminal port and a host port. Design engineers used a terminal to talk to the ECB computer through the terminal port, while the host port was connected to the host computer. When the engineer needed to edit or assemble the program, he or she set the ECB computer to the *transparent mode*, in which characters coming in from the terminal port were sent directly to the host port by the ECB computer and the characters coming in from the host port were sent directly to the terminal port. In this way the design engineer could log in the host computer to do all the development work. When he or she needed to evaluate the program on the ECB computer, he or she would set the ECB to normal mode and download the program to the ECB for execution.

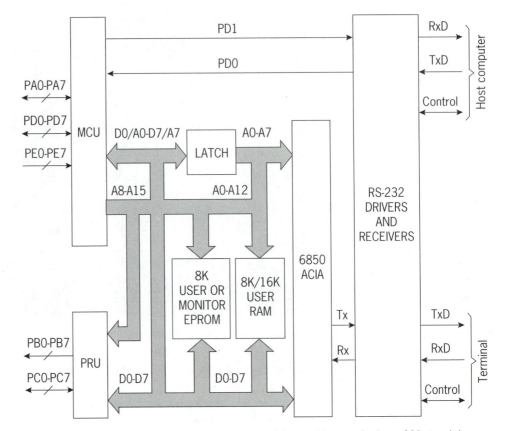

Figure 2.19   68HC11EVB block diagram (redrawn with permission of Motorola)

Today most people rarely use the host port because PCs are so inexpensive that there is no need to use a terminal and run the cross assembler or compiler on a mini- or mainframe computer.

EVB configuration   It is very useful to know the memory space allocation of a computer. The memory space allocation is often described by a map called the *memory map*. By examining the memory map, the user can identify the memory blocks occupied by different devices and registers and also figure out quickly what memory blocks are still available for allocation.

The EVB memory map has a single-map design. User RAM resides at different address locations from the MCU ROM, as shown in Figure 2.20. User RAM is an area where the user can write and test programs. The internal RAM is 256 bytes. However, the area from $0040 to $00FF is used by the Buffalo monitor, so the user is restricted to the area from $0000 to $003F. Small programs that should be retained when power is turned off can be placed in the 512-byte EEPROM.

The Buffalo monitor provides a self-contained operating environment. The user can interact with the monitor through predefined commands entered from a terminal. The commands supported by the Buffalo monitor are listed in Table 2.7.

Using Buffalo Commands   Before using the EVB, the user must either invoke the Windows Terminal program or run a terminal program such as Kermit. If

Internal Ram (MCU reserved)	$0000 $00FF
Not used	$0100 $0FFF
PRU + REG. Decode	$1000 $17FF
Not used	$1800 $3FFF
Flip-flop decode	$4000 $5FFF
Optional 8K RAM	$6000 $7FFF
Not used	$8000 $97FF
Terminal ACIA	$9800 $9FFF
Not used	$A000 $B5FF
EEPROM	$B600 $B7FF
Not used	$B800 $BFFF
User RAM	$C000 $DFFF
Monitor EPROM	$E000 $FFFF

Figure 2.20 EVB memory map diagram

the Windows Terminal program is used, the user can simply press the Enter key to communicate to the EVB (after first setting up the communication parameters properly or opening the file of communication settings). The following message will appear on the screen:

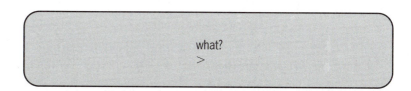

Command	Description
ASM [<address>]	assembler/disassembler
BF <address1><address2><data>	block-fill memory with data
BR [−] [<address>] . . .	breakpoint set
BULK	bulk-erase EEPROM
BULKALL	bulk-erase EEPROM + CONFIG register
CALL [<address>]	execute subroutine
G [<address>]	execute program
HELP	display monitor commands
LOAD <host download command>	download (S-records) via host port
LOAD <T>	download (S-records) via terminal port
MD [<address1>[<address2>]]	dump memory to terminal
MM [<address>]	memory modify
MOVE <address1><address2>[<destination>]	move memory to a new location
P	proceed/continue from breakpoint
RM [p,y,x,a,b,c,s]	register modify
T [<n>]	trace $1-$FF instructions
TM	enter transparent mode
VERIFY <host download command>	compare memory to download data via host port
VERIFY <T>	compare memory to download data via terminal port

where <> enclose syntactical variable
    [ ] enclose optional fields
    [ ] . . . enclose repeated optional fields
All numbers are entered in hexadecimal format

Table 2.7 ■ EVB monitor program commands

where > is the Buffalo monitor prompt. If Kermit is used, then the user should type **c** to the right of the Kermit prompt and press the Enter key to establish the connection. He or she will then see the message

on the screen.

The command

    BF <addr1><addr2><data>

allows the user to fill the block of memory locations from *addr1* to *addr2* with the value given by *data.* The command

    MD [<addr1>[<addr2>]]

allows the user to display the contents of memory locations; either one, two, or no addresses can be specified in this command. If no address is specified, the monitor will display the memory locations whose addresses were given in the previous MD command. If only one address is specified, the monitor will display the contents of nine rows of memory locations, starting from the address specified in the command. Each row contains 16 memory locations in hexadecimal format.

(In the following examples, bold letters indicate commands or responses entered by the user, while normal letters represent messages displayed by the Buffalo monitor program. The Enter key is required for all commands but will not be shown.)

Example 2.28

Use Buffalo monitor commands to fill the memory locations from 00 to $3F of the EVB with the value $FF and display them on the monitor screen.

Solution: At the Buffalo monitor prompt enter the command BF 00 3F FF. Wait until the next monitor prompt appears, and then enter the command MD 00 3F. You should see the following on your screen:

```
>BF 00 3F FF
>MD 00 3F
0000 FF FF FF FF FF FF FF FF FF FF FF FF FF FF FF FF
0010 FF FF FF FF FF FF FF FF FF FF FF FF FF FF FF FF
0020 FF FF FF FF FF FF FF FF FF FF FF FF FF FF FF FF
0030 FF FF FF FF FF FF FF FF FF FF FF FF FF FF FF FF
```

The command MM [<address>] allows the user to modify the contents of the specified memory location. The Buffalo monitor will display the current contents of that memory location, and the user should enter the new value to its right. This command allows you to modify the contents of one location at a time. The monitor prompt will reappear when the command is completed.

Example 2.29

Set the value of the memory location at $00 to $00 and display the contents of the memory locations from $00 to $0F on the screen.

Solution: Enter the command MM 00 at the Buffalo monitor prompt. The monitor will display the message 0000 FF on the screen, as shown below:

```
>MM 00
0000 FF_
```

The cursor (underscore character) will be blinking to the right of the second line. Type the new value (00) and press the Enter key at the cursor:

```
                              0000 FF 00
```

You can then verify the new value of the memory location $0000 by entering the command MD 00 0F. The whole session should look like this:

```
>MM 00
0000 FF 00
>MD 00 0F
0000 00 FF FF FF FF FF FF FF FF FF FF FF FF FF FF FF
```

Sometimes the user needs to set the values of CPU registers before running the program. The command

RM [p,x,y,a,b,c,s]

allows the user to display and modify the contents of all CPU registers. The letters inside the brackets stand for program counter PC, index register X, index register Y, accumulator A, accumulator B, condition code register CCR, and stack pointer SP, respectively. If the user does not specify the register in the command, then PC is selected by default. If the user changes his or her mind and decides not to modify the register, he or she can press the Enter key.

Example 2.30

Use a Buffalo command to set the value of the stack pointer to $3F.

Solution: You should use the command RM S. The Buffalo monitor will display the contents of all the CPU registers on one line and the contents of the stack pointer on the next line, and then it will wait for the new value to be entered by the user. The whole session should look like this:

```
>RM S
P-C000 Y-FFFF X-FFFF A-FF B-FF C-D0 S-004A
S-004A 3F
>
```

You can verify the contents of the stack pointer by reentering the command **RM.**

Example 2.31

Use the RM command to set the contents of Y to $0001 and the contents of B to $00.

Solution: First enter the command **RM Y.** The Buffalo command will display the contents of all the CPU registers on one line and the contents of Y on the next line, and then it will wait for you to enter the new value. Enter the new value, and press the space bar to display the next register. The whole session will look like this:

```
>RM Y
P-C000 Y-FFFF X-FFFF A-FF B-FF C-D0 S-3F
Y-FFFF 001<space bar>
X-FFFF <space bar>
A-FF <space bar>
B-FF 00 <Enter key>
```

After learning how to set and display the contents of memory locations and CPU registers, the next step is to learn how to enter and run a program on the EVB. Note that the user program should be entered in the area from $C000 to $DFFF. There are two ways to enter programs into the EVB:

1. Use the EVB assembler/disassembler. In this method, the command

ASM [<addr>]

is used to enter instructions directly into the SRAM of the EVB. For example, to enter the instruction ADDA #$20 at $C000, the user first types **ASM C000** to the right of the Buffalo prompt >. The Buffalo monitor will display the current instruction at $C000, then display the assembler prompt (the > character indented four columns from the leftmost column) and wait for the user to enter the new instruction. The instruction entered by the user will be assembled into machine code and displayed in the following line. The procedure can be continued as long as the user has more instructions to be entered. The whole process is something like this:

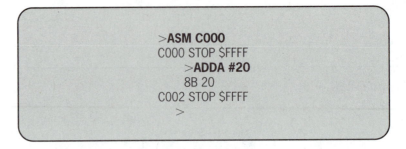

```
>ASM C000
C000 STOP $FFFF
      >ADDA #20
      8B 20
C002 STOP $FFFF
      >
```

When the complete program has been entered, the user can press the Control and **A** keys simultaneously to exit the assembler; the Buffalo monitor prompt will reappear.

2. Download from a PC. The following steps will download a program into the EVB computer:

Step 1
Use a text editor (such as the EDIT editor) to enter the program as a text file.

Step 2
Invoke the cross assembler to assemble the program.

Step 3
Invoke the linker to link the object code files created in step 2 (this step is not necessary when using the Motorola freeware).

Step 4
Invoke the software that converts the executable code into the S-record format (not needed if the Motorola freeware is used). S-record is a common data format defined by Motorola so that program codes generated by assemblers and compilers from different vendors can be loaded into the same hardware for execution.

Step 5
Use the command LOAD T to download the executable code onto the EVB. This step requires a specific procedure dictated by the terminal emulation program. If you are using the Windows Terminal program, select the command Send text file . . . under the category Transfers. A dialog box will appear and will ask you to specify the file name to be downloaded. Enter the file name with full path and use the mouse to click on OK. After the file is downloaded, the Buffalo command prompt will reappear.

If you are using Kermit, do the following:

1. Press the Control and] keys simultaneously, and then type **c** to return to the Kermit prompt, MS-Kermit. Another way to return to the Kermit prompt is to press the alt and **x** keys simultaneously.

2. Enter **type filename.s19>com1** and press Enter (assuming communication port com1 is being used).

3. Press the Enter key one or a few times, and the Buffalo monitor prompt will reappear.

After the program has been entered into the RAM, the user can begin to run it. To start executing the resident program, type the following command (startaddress is the starting address of the program):

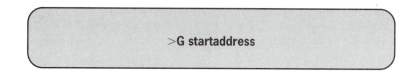

>**G startaddress**

It is desirable to return the CPU control back to the Buffalo monitor when the user program has completed execution. This can be done by adding the SWI instruction as the last instruction of the user program or by setting a breakpoint to the memory location next to the last byte of the user program. Using the SWI instruction is easier. The EVM user should set a breakpoint to return to the EVM monitor, because the SWI instruction does not return the CPU control to the EVM monitor.

If a program has bugs and cannot execute correctly, the user needs to locate the bugs and fix them. There are two ways to identify the problems:

■ Use the *breakpoint set* (BR) command to examine the program execution result at any suspected locations. Up to four breakpoints can be set at one time. The contents of the CPU registers will be displayed at the breakpoint(s). Program execution can be continued from a breakpoint by using the P (proceed) command. Breakpoints can be deleted using the command **BR −** when they are not needed.

■ Use the *trace* command (T) to monitor program execution on an instruction-by-instruction basis. The contents of the CPU registers will be displayed at the completion of each instruction. You must set the address (PC) of the instruction to be traced before issuing this command. One or several instructions can be traced at the same time.

Example 2.32

Enter the Buffalo commands to view the breakpoints, set two new breakpoints at $DF00 and $DF20, and then delete the breakpoint at $DF20.

Solution: Breakpoints can only be set in RAM. To display the current breakpoints, type the command **BR** followed by Enter. Up to four breakpoints can be set at one time. If no breakpoints have been set, the screen should look like this:

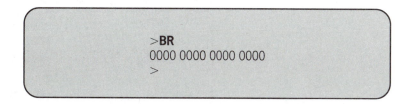

```
>BR
0000 0000 0000 0000
>
```

To set breakpoints at $DF00 and $DF20, type the command **BR DF00 DF20** then press Enter. After this command, you should see the following display on the screen:

>**BR DF00 DF20**
DF00 DF20 0000 0000

>

To delete the breakpoint at $DF20, type the command **BR −DF20** followed by Enter. You should then see the following message on the screen:

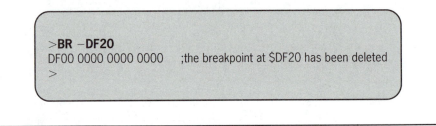

>**BR −DF20**
DF00 0000 0000 0000   ;the breakpoint at $DF20 has been deleted
>

To delete all breakpoints, type **BR −.**

Example 2.33

Use Buffalo command to enter the following instructions, starting from $C000, and trace the execution result of these instructions.

```
LDX #$1000
LDS #$3F
CLRA
STAA $10
```

Solution: You can use the direct method to enter these instructions. First use the command **ASM C000** to specify where to enter the instructions. After entering the four instructions, press the Control and **A** keys simultaneously to exit the assembler mode. The whole process will look like this:

>**ASM C000**
C000 STX $FFFF   ;this line may be different on the reader's EVB
       >**LDX #1000**
       CE 10 00

```
C003 STX $FFFF      ;this line may be different on the reader's EVB
     >LDS #3F
     8E 00 3F
C006 STX $FFFF      ;this line may be different on the reader's EVB
     >CLRA
     4F
C007 STX $FFFF      ;this line may be different on the reader's EVB
     >STAA 10
     97 10
C009 STX $FFFF
     >
     >
```

In the trace mode, the Buffalo monitor executes one instruction at a time and displays the contents of CPU registers. Before tracing the program, the user should set the program counter to the address of the first instruction to be traced. To trace these four instructions, simply type the command **T 4.** You should see the following message on the screen:

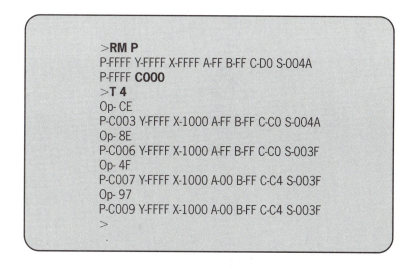

```
>RM P
P-FFFF Y-FFFF X-FFFF A-FF B-FF C-D0 S-004A
P-FFFF C000
>T 4
Op- CE
P-C003 Y-FFFF X-1000 A-FF B-FF C-C0 S-004A
Op- 8E
P-C006 Y-FFFF X-1000 A-FF B-FF C-C0 S-003F
Op- 4F
P-C007 Y-FFFF X-1000 A-00 B-FF C-C4 S-003F
Op- 97
P-C009 Y-FFFF X-1000 A-00 B-FF C-C4 S-003F
>
```

THE 68HC11EVBU

The 68HC11EVBU is designed as a low-cost tool for debugging/evaluation of 68HC11A8, E9, 711E9, 811A8, and 811E2 microcontrollers. The EVBU is the smallest 68HC11-based evaluation board designed by Motorola. The resident microcontroller unit on the EVBU is a 68HC11E9. The 68HC11E9 MCU on the EVBU is configured to operate in the single-chip mode. However, the EVBU can be reconfigured to operate in either expanded, bootstrap, or test

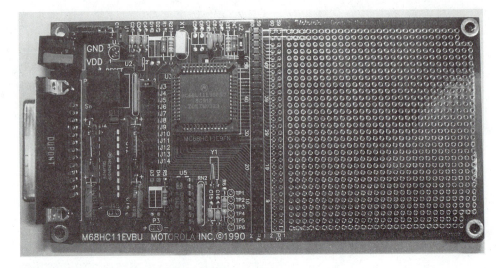

Figure 2.21   Motorola 68HC11EVBU computer

modes. A photograph and block diagram of the EVBU are shown in Figures 2.21 and 2.22, respectively.

The EVBU memory map has a single-map design reflecting the resident 68HC11E9. The 68HC11E9 has 12 KB of ROM, 512 bytes of SRAM, and 512 bytes of EEPROM on the chip. The memory map of the EVBU is shown in Figure 2.23. The on-chip SRAM locations $0048–$00FF are used by the Buffalo monitor, so approximately 325 bytes remain for the user. Only the 512 bytes of EEPROM and the 325 bytes of SRAM are available for program development. The EVBU requires only a 5-V power supply and an RS-232C compatible terminal for operation.

The operation of the EVBU is controlled by the Buffalo monitor program (a

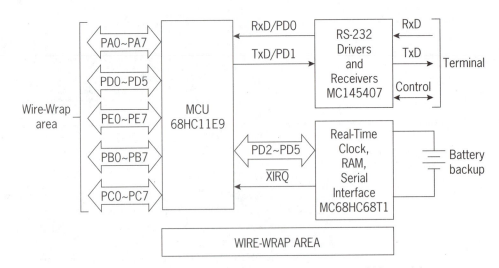

Figure 2.22   EVBU block diagram (redrawn with permission of Motorola)

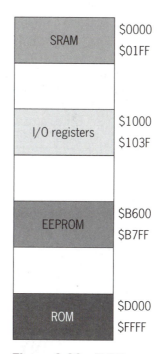

Figure 2.23 EVBU memory map

later version than that in the EVB). Therefore, in addition to the commands supported by the EVB, the EVBU supports two more commands:

    STOPAT <address>

and

    XBOOT [<address1> [<address2>]

The STOPAT command causes a user program to be executed one instruction at a time until the specified address is encountered. This command should be used only if the current PC value is known. The XBOOT command loads/transfers a block of data from address1 through address2 via the serial communication interface (SCI) to another MC68HC11 device that has been reset in the bootstrap mode. This command allows the user to program another 68HC11's on-chip EEPROM using the EVBU. However, a lot of work must be done before this command can be used. To program the on-chip EEPROM of the 68HC11, it would be easier to use the software PCbug11, which Motorola bundled with the EVBU. When used as a front-end to the evaluation board, PCbug11 requires the evaluation board to operate in the bootstrap mode and communicates to the EVBU through the SCI subsystem of the 68HC11. PCbug11 allows the user to program the on-chip EEPROM of the microcontroller on the EVBU. EVBU owners should refer to the PCbug11 user's manual for more details.

The **68HC11EVM**

The 68HC11EVM is another evaluation board developed by Motorola. In addition to allowing users to run their programs, the EVM also provides the capability to program other MCUs. Both the AX and EX series of the 68HC11 MCUs can be programmed. A photograph and a block diagram of the EVM are shown in Figures 2.24 and 2.25, respectively.

The EVM provides a tool for designing, debugging, and evaluating the 68HC11AX or 68HC11EX series MCU-based target system equipment. In either the AX or EX series MCU configuration, the EVM operates in one of two maps (monitor or user map). Both maps are 64 KB and are decoded via a field programmable gate array (FPGA). Both maps also include the 68HC11 MCU internal 256-byte RAM and 64-byte registers. The monitor program executes in the monitor map, while the user program, which is usually downloaded from a PC, executes in the user map. The monitor and user maps of the 68HC11AX series are shown in Figure 2.26. In Figure 2.26, the reader can see two 8-KB user pseudo ROMs. The first pseudo ROM occupies the address space from $C000 to $DFFFF, and the second occupies the address space from $E000 to $FFFF. User pseudo ROMs are actually external SRAMs. These RAMs are write-protected when a user program is being executed, but a user program can be downloaded to these two pseudo ROMs for execution. The first pseudo ROM was not installed by the factory in earlier EVMs.

Figure 2.24　68HC11EVM computer

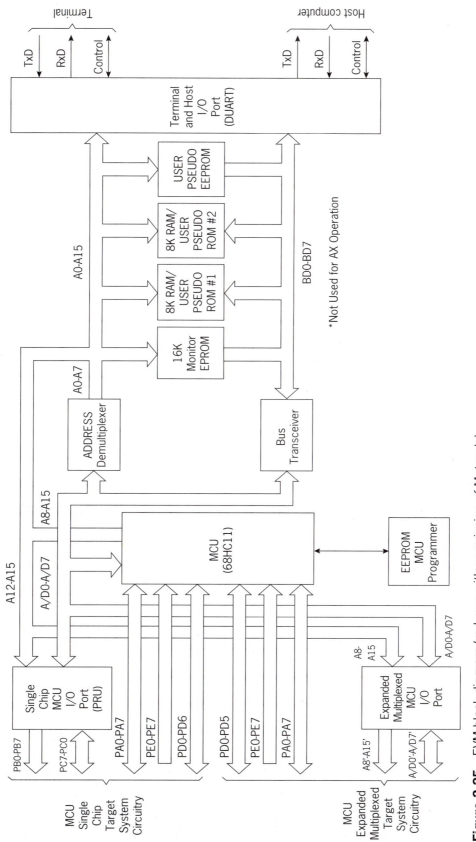

Figure 2.25 EVM block diagram (redrawn with permission of Motorola)

Monitor map

Address	Description
0000–00FF	Internal MCU 256-byte RAM
0100–0FFF	Reserved
1000–103F	Internal MCU 64-byte register block
1040–A1FF	Reserved
A200–A20F	DUART MC2681
A210–AFFF	Reserved
B400–B5FF	512-byte scratch pad RAM 6264
B600–BFFF	Reserved
C000–FFFF	16KB EVMbug Monitor ROM

User map

Address	Description
0000–00FF	Internal MCU 256-byte RAM
0100–0FFF	Unused
1000–103F	Default internal MCU 64-byte register block
1040–B5FF	User-available
B600–B7FF	512-byte MCU EEPROM or 6264 RAM (pseudo EEPROM)
B800–DFFF	User-available
E000–FFFF	8KB Pseudo ROM

Figure 2.26 EVM (AX) memory map

Command	Description
ASM <address>	assembler/disassembler (interactive)
BF <addr1> <addr2> <data>	block-fill memory with data
BR [<address>]...	breakpoint set
BULK [<103F>]	bulk-erase EEPROM or CONFIG register
CHCK <addr1> [<addr2>]	check EEPROM
COPY <addr1> [<addr2>] [<addr3>]	copy EEPROM to user map
ERASE <addr1> [<addr2>]	erase individual EEPROM byte
G [<address>]	go (execute program)
HELP	help (display commands)
LOAD <port> [=<text>]	load (S-records) from I/O port
MD <addr1> [<addr2>]	memory display
MM <address>	memory modify (interactive)
NOBR [<address>]	remove breakpoint
P [<count>]	proceed (through breakpoint)
PROG <addr1> [<addr2> [<data>]]	program EEPROM via pseudo EEPROM
RD	register display
RM	register modify (interactive)
SPEED [<baud rate>]	baud rate select for host I/O port
T [<count>]	trace
TM [<exit character>]	transparent mode
VERF <addr1> [<addr2>] [<addr3>]	verify EEPROM against user map

Table 2.8 ■ EVM monitor commands

	EVB	EVBU	EVM
Controller	68HC11A1	68HC11E9	68HC11A1 or 68HC11E1
Monitor	Buffalo	Buffalo	EVMbug
External SRAM	8/16 KB	0 KB	8/16 KB (Pseudo ROM)[1]
Internal SRAM (on MCU)	256 bytes	512 bytes	256 bytes (A1) or 512 bytes (E1)
Number of serial ports	2 (one terminal and one host port)	1	2 (one terminal and one host port)
Power requirement	+5V, +12V, & −12V	+5V only	+5V, +12V, & −12V
Wire-wrap area	no	yes	no
EEPROM programming capability	no	yes[2]	yes
Real-time clock chip (68HC68T1)	no	yes	no
Operating frequency	8 MHz crystal oscillator 2 MHz E clock	8 MHz crystal oscillator 2 MHz E clock	8 MHz crystal oscillator 2 MHz E clock
MCU Support	A0, A1, A8, E0, E1, E9, & 811E2	A8, E9, 711E9, 811A8, & 811E2	A0, A1, A8, E0, E1, E9, & 811E2

1. This SRAM is read-only for user programs and hence is called pseudo ROM.
2. The EVBU requires the use of PCbug11 software, which lets the user modify and assemble code into EEPROM as if it were RAM. PCbug11 also allows the programming of EPROM (e.g., 711E9) parts on EVBU.

Table 2.9 ■ A comparison of EVB, EVBU, and EVM features

The monitor commands of EVM are similar to those of EVB and are listed in Table 2.8.

A comparison of the features of these three evaluation boards is given in Table 2.9.

2.12 Summary

An assembly program consists of five major parts: assembler directives, assembly language instructions, data storage directives, the END directive, and comments. Comments and flowcharts provide program documentation. Using a computer to solve a problem should follow a specific procedure. In this chapter, the assembler directives and 68HC11 instructions are discussed in detail, category by category. The 68HC11 can perform only 8-bit arithmetic. Numbers longer than 8-bit must be manipulated using multiprecision arithmetic. Examples show the techniques of dealing with multiprecision arithmetic. Program loops require the use of branch instructions and the setting of condition codes using compare or test instructions. Shift and rotate instructions are useful for bit field manipulations and for performing multiplications and divisions by a power of 2. A delay can easily be created by using program loops. Software and hardware tools for developing 68HC11-based products are discussed. The features and commands of the Motorola EVB, EVBU, and EVM single-board computers are also summarized in this chapter.

2.13 Glossary

Assembler directive   A command to the assembler for defining data and symbols, setting assembler and linking conditions, and specifying output format. Assembler directives do not produce machine code.

Binary-coded decimal (BCD)   A coding method that uses four binary digits to represent a decimal digit. The binary codes 0000_2–1001_2 correspond to the decimal digits 0–9.

Breakpoint   A memory location in a program where the user program execution will be stopped and the monitor program will take over the CPU control and display the contents of CPU registers.

Comment   A statement that explains the function of a single instruction or directive or a group of instructions or directives. Comments make a program more readable.

Flowchart   A diagrammatic method of representing the logical flow of a program.

Integrated development tool   A tool that provides a development environment that combines the text editor, assembler and/or compiler, simulator, and communication software in one package so that the user does not need to exit any program during the development process.

Label field   The first field in an assembly program line; used by the programmer to identify memory locations or data areas in the program.

Multiprecision arithmetic   Arithmetic performed on numbers that are larger than the word length of a computer.

Program loop   A sequence of instructions that will be executed repeatedly for a finite or infinite number of times.

Microprocessor simulator   A program that runs on a computer and allows the user to run the programs (in the microprocessor's instruction set) without having the actual hardware. The simulator allows the user to set the contents of registers and memory locations and to examine the program execution result(s).

Terminal emulation program   A program that allows the user to use a computer as a terminal to communicate with another computer. File transfer is one of the most common operations supported by this type of program.

Text editor   A program that allows the user to create and edit text files.

2.14 Exercises

E2.1. Identify the four fields of the following instructions:

```
a.        LDAA  #$20     ;initialize A to $20
b. LOOP   BNE   NEXT     ;branch if not equal
```

E2.2. Find the valid and invalid labels in the following assembly statements, and explain why the invalid labels are not valid.

column 1

↓

```
a. ABC      CLRA        ;initialize A to 0
b. LP       LDAB  0,X   ;get the next element
c. TOO:     ABA
d. HI+LO    STAA  0,X
e. NO_T:    INX
f. LO_HI    CPX  #$20
```

E2.3. Write assembler directives to reserve 100 bytes from $00 to $63 and initialize them to 0.

E2.4. Write assembler directives to build a table to hold the ASCII codes of the capital letters A–Z. The table should be stored in memory locations $00 to $19.

E2.5. Write assembler directives to store the following message in memory locations starting from $00: Welcome to computer engineering program!

E2.6. Write a program to compute the sum of $DA03 and $934A and save the result at $00 and $01.

E2.7. Write a program to subtract the three-byte number stored at $00–$02 from the three-byte number stored at $03–$05 and save the difference at $10–$12.

E2.8. Write a program to compute the average of an array of ten 8-bit numbers. The array is stored at $F0–$F9. Save the result at $00.

E2.9. Find the values of accumulators A and B after the execution of the MUL instruction if they originally contain the values

    a. $33 and $80, respectively

    b. $7C and $55, respectively

E2.10. Which of the conditional branch instructions in the following list will cause the branch to be taken if the condition flags $N = C = 1$ and $Z = V = 0$.

    a. BCC target   b. BNE target

    c. BGE target   d. BLS target

    e. BMI target   f. BCS target

    g. BLT target

E2.11. Write a program to swap the last element of an array with the first element, the next-to-last element with the second element, etcetera. Assume that the array has $20 eight-bit numbers and that the array starts at $00.

E2.12. Write a program to compute the sum of the positive numbers of an array with $30 eight-bit numbers. Store the sum at $00–$01. The array starts at $10.

E2.13. Generate the machine code for the following conditional branch instructions, if each of the instructions occurs at location $C100. Let ALPHA be the address $C090, and let BETA be the address $C150.

    a. BNE   ALPHA

    b. BGT   BETA

E2.14. Find the contents of the CCR register after execution of each of the following instructions, given that [A] = $50 and the condition flags are N = 0, C = 1, Z = 1, and V = 0.

    a. TSTA   b. CMPA #$60

    c. ADDA #$67   d. SUBA #$70

    e. LSLA   f. RORA

E2.15. Find the contents of the CCR register after execution of each of the following instructions, given that [B] = $00, and the condition flags are N = 0, C = 0, Z = 1, and V = 0.

    a. TSTB   b. ADDB #$30

    c. SUBB #$7F   d. LSLB

    e. ROLB   f. ADDB #$CF

E2.16. Determine the branch instruction and the offset relative to the PC from the following machine code.

    a. 2B 80   b. 2D 20

    c. 25 E0   d. 2F F0

E2.17. Write a program to compute the average of the square of each element of an array with 20 eight-bit numbers. The array is stored at $00–$13. The result should be saved at $21–$22.

E2.18. Write a program to compute the product of two 3-byte numbers stored at $00–$02 and $03–$05 and save the result at $10–$15.

E2.19. Write a program to compute the sum of the array elements that are a multiple of 4, given an array of 40 eight-bit numbers. The array starts at $00.

E2.20. Determine the number of times the following loop will be executed.

```
          LDAA #%10000000
   LOOP   LSRA
          STAA $00
          ADDA $00
          BMI LOOP
            .
            .
            .
```

E2.21. Write a small program to shift the 24-bit number stored at $10–$12 three places to the left.

2.15 Lab Exercises and Assignments

L2.1. Turn on the PC and run a terminal program such as Kermit to connect to the evaluation board. Then perform the following operations:

    a. Enter a command to set the contents of the memory locations from $00 to $3F to 00.

    b. Enter a command to display the contents of the memory locations from $00 to $3F to verify the previous operation.

L2.2. Enter monitor commands to display the breakpoints, set new breakpoints at $D000 and $D100, and then delete all breakpoints.

L2.3. Enter monitor commands to set the contents of accumulators A and B to $00 and $01, respectively.

L2.4. Invoke the one-line assembler/disassembler to enter the following instructions to the evaluation boards, starting from address $C000, and trace through the program:

```
LDAA #12
LDAB #08
MUL
STD $10
SWI
```

L2.5. Use a text editor to enter the following assembly program as a file with the file name *learn1.asm:*

```
            ORG  $00
            FCB  1,2,3,4,5,6,7,8,9,10,11,12,13,14,15,16,17,18,19,20
            ORG  $C000
            LDY  #$14
            LDX  #$00
            LDAA #0
            STAA $30
AGAIN       LDAA 0,X
            LSRA
            BCC  CHEND
            LDAA $30
            ADDA 0,X
            STAA $30
CHEND       INX
            DEY
            BNE  AGAIN
            SWI
            END
```

After entering the program, do the following:

    a. Assemble the program.

    b. Download the S-record file (file name *learn1.s19*) to the evaluation board.

    c. Display the contents of the memory locations from $00 to $3F.

    d. Execute the program.

    e. Reexamine the contents of the memory location at $30.

Note: this program adds all odd numbers in the given array and stores the sum at $30.

L2.6. Write a program to compute the sum of the integers from 1 to 100 and store the sum at $10 and $11.

L2.7. Write a program to divide each element of an array by 4. The array has 20 elements and is stored at $00–$13. To test the program, define an array using the directive FCB.

L2.8. Write a program to determine how many elements in an array are divisible by 8. The array has 20 eight-bit elements and is stored at $00–$13. Store the result at $20. To test the program, define an array using the directive FCB.

3

DATA

STRUCTURES

AND

SUBROUTINE

CALLS

3.1 Objectives

After completing this chapter, you should be able to:

- access stack elements and manipulate the stack data structure
- manipulate array, matrix, and string data structures
- write subroutines
- make subroutine calls
- invoke EVB I/O routines to input data from the keyboard and output messages to the PC monitor screen

3.2 Introduction

The main function of a computer is to manipulate information. In order to manipulate information efficiently, we need to study data structures to learn how information is organized and how elements of particular structures can be manipulated. The data structures most commonly used in application programs are:

Strings. A string is a sequence of characters.

Arrays. An array is an ordered set of elements of the same type. The elements of the array are arranged so that there is a zeroth, first, second, third, and so forth. An array may be one-, two-, or multidimensional. A vector is a one-dimensional array. A matrix is a two-dimensional array.

Stacks. A stack is a data structure with a top and a bottom. Elements can be added or removed only at the top of a stack. A stack is a last-in-first-out (LIFO) data structure.

Queues. A queue is a data structure in which elements can be added at only one end and removed only from the other end. The end to which new elements can be added is called the *tail* of the queue, and the end from which elements can be removed is called the *head* of the queue. The queue is a first-in-first-out (FIFO) data structure.

Dequeues. A dequeue is a data structure in which elements can be added and removed at both ends.

Trees. A tree is a finite, nonempty set of elements in which one element is called the *root* and the remaining elements are partitioned into $m (\geq 0)$ disjoint subsets, each of which is itself a tree. Each element in a tree is called a *node* of the tree.

Graphs. A graph consists of a set of nodes and a set of *arcs* (or *edges*). Each arc in a graph is specified by a pair of nodes.

Linked lists. A linked list is a data structure consisting of linked nodes. Each node consists of two fields, an information field and a next address field. The information field holds the actual element on the list, while the next address field contains the address of the next node in the list.

Common operations applied to data structures include adding and deleting elements, traversing and searching a data structure, etcetera. A complete study of data structures and their associated operations is beyond the scope of this book. This chapter will discuss only strings, arrays, and the stack data structure.

3.3 The Stack

Conceptually, a stack is a list of data items whose elements can be accessed from only one end. A stack data structure has a top and a bottom. The operation that adds a new item to the top is called *push*. The top element can be removed by performing an operation called *pull* or *pop*. Physically, a stack is

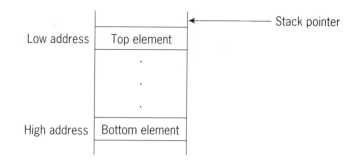

Figure 3.1 Diagram of a stack

a reserved area in main memory where programs perform only push and pull operations. The structure of a stack is shown in Figure 3.1. A stack has a stack pointer that points to the top element or to the memory location above the top element.

The 68HC11 has a 16-bit stack pointer (SP) to facilitate the implementation of the stack data structure. The stack pointer points to the memory location one byte above the top element of the stack. By convention, the stack grows from high addresses toward lower addresses. An area in main memory is allocated for use as the stack area. If the stack grows into addresses lower than the stack buffer, a stack overflow error occurs; if the stack is pulled too many times, a stack underflow occurs. The on-chip SRAM of the 68HC11 is often used to implement the stack.

The various push and pull instructions available for the 68HC11 are listed in Table 3.1. A push instruction writes data from the source to the stack and then decrements the SP. There are four push instructions: PSHA, PSHB, PSHX, and PSHY.

Mnemonic	Function
PSHA	push A onto the stack
PSHB	push B onto the stack
PSHX	push X onto the stack
PSHY	push Y onto the stack
PULA	pull A from the stack
PULB	pull B from the stack
PULX	pull X from the stack
PULY	pull Y from the stack

Table 3.1 ■ The 68HC11 push and pull instructions

Example 3.1

Suppose that

[A] = \$33
[SP] = \$00FF
[B] = \$20

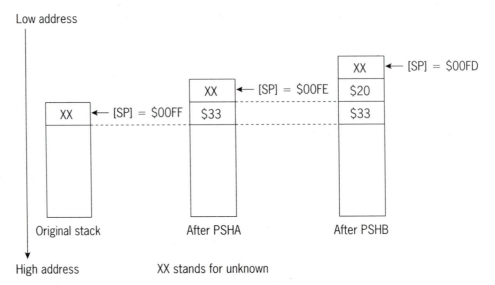

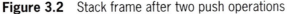

Figure 3.2   Stack frame after two push operations

What will be the contents of the top byte of the stack before and after the execution of PSHA? What will be the contents of the top two bytes of the stack if PSHB is also executed?

Solution: The contents of the stack before and after the execution of PSHA and PSHB are shown in Figure 3.2.

■

Pull instructions are the inverse of pushes. First, the stack pointer is incremented. Then the designated register is loaded from the stack at the address contained in the stack pointer. There are also four pull instructions: PULA, PULB, PULX, and PULY. Some microprocessors have pop instructions, which are equivalent to pull instructions.

Before the stack can be used, the stack pointer must be initialized using the LDS instruction. All modes except the inherent addressing mode can be used with the LDS instruction. When necessary, the stack pointer can be saved in memory by using the STS instruction. The memory location to hold the stack pointer can be specified in the direct, extended, or index addressing mode. The on-chip SRAM area of the 68HC11 that is not used by the monitor program can be used as the stack area. For example the user can initialize the stack pointers of the EVB, EVBU, and EVM to $3F, $1FF, and $FF, respectively.

The 68HC11 has instructions that facilitate access to variables in the stack, since the stack may contain variables that the user program will need to access during its execution. By making index register X or Y point to the top element of the stack, the program can access the in-stack variables using the index addressing mode. The 68HC11 instructions are:

■ TSX: The contents of the stack pointer plus one are loaded into index register X.

■ TSY: The contents of the stack pointer plus one are loaded into index register Y.

Instruction	Operation
DES	decrement the stack pointer by 1
INS	increment the stack pointer by 1
LDS	load the contents of a memory location or immediate value into SP
STS	store the contents of the stack pointer in a memory location
TSX	load the contents of the stack pointer plus one into X
TSY	load the contents of the stack pointer plus one into Y
TXS	load the contents of index register X minus one into SP
TYS	load the contents of index register Y minus one into SP

Table 3.2 ■ Instructions for the stack pointer

Additional instructions for the stack pointer are listed in Table 3.2. Suppose the top three bytes (from low address to high address) of the stack are $11, $22, $33. Then after execution of the TSX instruction, the index register points to $11, as illustrated in Figure 3.3. It is now fairly straightforward to access the three bytes in the stack. For example, the middle byte (value $22) can be loaded into A by using the instruction LDAA 1,X. This technique will be used extensively throughout this chapter.

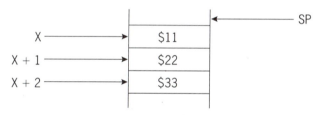

Figure 3.3 The top three bytes of a stack

Example 3.2

Write an instruction sequence to load the top element of a stack into accumulator A and the 9th element from the top of the stack into accumulator B.

Solution: We need to load the stack pointer into an index register so that the index addressing mode can be used to access the stack elements. The following instruction sequence does this and accesses the first and ninth stack elements:

```
TSX        ; index register X points to the top element of the stack
LDAA 0,X   ; access the top element of the stack
LDAB 9,X   ; access the 9th element of the stack
```

3.4 Indexable Data Structures

Vectors and matrices are indexable data structures. A *vector* is a sequence of elements in which each element is associated with an index i that can be used to access it. Conceptually, a vector is a one-dimensional data structure. To make address calculation easy, the first element is usually associated with the index 0 and each successive element with the next integer, but you can change the index origin of the array to 1 if you are willing to modify the routines. The elements of a vector are considered numbers of the same precision.

A vector can be defined by the assembler directive FCB (or FDB, depending on the length of the element). Suppose that the vector VEC has elements 1, 2, 3, 4, 5, and 6. It can be defined as follows:

```
VEC      FCB  1,2,3,4,5,6
```

The first element, which is 1, is referred to as VEC(0). To access the ith element, use the following instructions:

```
LDAB  #i      ;load the index i into B
LDX   #VEC    ;load the vector base address into X
ABX           ;compute the address of VEC(i)
LDAA  0,X     ;A contains VEC(i)
```

The array, which was used in Chapter 2, is also a vector data structure.

Example 3.3 ■ Sequential search

Ten 16-bit numbers are stored at locations starting from $10. A search key is stored at $00 and $01. The sequential search program searches for the first occurrence of the key in the list of numbers. The address of the first matched element will be stored at locations $02–$03. If the key is not found, −1 will be stored at those locations.

Solution: The flowchart for the sequential search is shown in Figure 3.4. To implement the sequential search algorithm described in the flowchart, we will

- use double accumulator D to hold the search key
- use index register Y as the loop count
- use index register X to point to the array element so that the index addressing mode can be used to reference the array elements
- allocate two bytes to hold either the address of the matched element or −1 (if no match)

The program is as follows:

```
N         EQU 10            ;array count
notfound  EQU −1

          ORG $00
key       FDB xxxx          ;the key to be used for searching
result    RMB 2             ;memory locations to store the address

          ORG $10
array     FDB $20,$40,$202,$1,$200,$10,$22,$21,$300,$101

          ORG $C000
          LDD #notfound     ;store −1 to result and result + 1
```

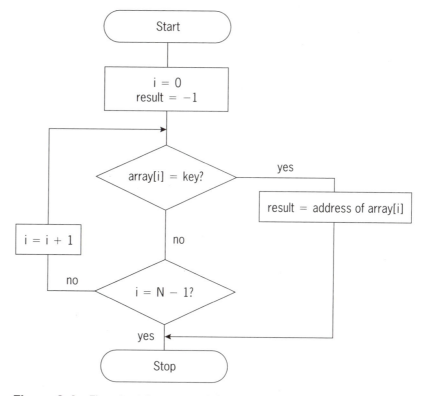

Figure 3.4 Flowchart for sequential search

```
         STD  result       ;    "
         LDD  key           ;get the key
         LDX  #array        ;get the starting address of the array
         LDY  #0            ;initialize the loop count
loop     CPD  0,X           ;compare with the key
         BEQ  found         ;found it ?
         INX                ;increment the pointer by 2
         INX                ;    "
         INY                ;increment the loop count
         CPY  #N−1
         BNE  loop          ;are we done?
         JMP  stop          ;stop the search
found    STX  result        ;store the address of the first occurrence
stop     END                ;
```

A *matrix* is a vector whose elements are vectors of the same length. We normally think of a matrix as a two-dimensional pattern, as in

$$\text{MAT} = \begin{matrix} 1 & 2 & 3 & 4 & 5 \\ 6 & 7 & 8 & 9 & 10 \\ 11 & 12 & 13 & 14 & 15 \\ 16 & 17 & 18 & 19 & 20 \\ 21 & 22 & 23 & 24 & 25 \end{matrix}$$

A matrix can be stored in row major order, as shown in the following example:

```
MAT1    FCB    1,2,3,4,5
        FCB    6,7,8,9,10
        FCB    11,12,13,14,15
        FCB    16,17,18,19,20
        FCB    21,22,23,24,25
```

or it can be stored in column major order, which is created as follows:

```
MAT2    FCB    1,6,11,16,21
        FCB    2,7,12,17,22
        FCB    3,8,13,18,23
        FCB    4,9,14,19,24
        FCB    5,10,15,20,25
```

In the following discussion, matrices are assumed to have N rows and N columns, and the notation (i,j)th is used to refer to the matrix element in the ith row and the jth column. For convenience, the first element in a row or a column is associated with the index 0. The address of the matrix element in the ith row and jth column can be computed by using a polynomial equation that depends on which order is used. For example, in a row major order matrix MAT1 where each element is one byte, the address of the (i,j)th element is

$$\text{address of the } (i,j)\text{th element} = (i \times N) + j + \text{address of MAT1}(0,0) \tag{3-1}$$

For a similar matrix MAT2 defined in column major order, the address of the (i,j)th element is

$$\text{address of the } (i,j)\text{th element} = (j \times N) + i + \text{address of MAT2}(0,0) \tag{3-2}$$

For example, suppose memory locations i and j contain the row and column indices, respectively. The following instruction sequence computes the address of the element MAT1(i,j) and leaves the address in index register X:

```
LDAA  i         ;place row index in A
LDAB  #N        ;put N in B
MUL             ;compute i × N
ADDB  j         ;compute i × N + j
ADCA  #0        ;    "
ADDD  #MAT1     ;compute MAT(0,0) + i × N + j
XGDX            ;place the result in X
```

The following program computes the sum of two matrices MA and MB and creates a new matrix MC. Each element of the new matrix is the sum of the corresponding elements of MA and MB.

Example 3.4 ■ Matrix sum program

Write a program to compute the sum of two 8×8 matrices. The names and starting addresses of these two matrices are MA and MB and that of the resulting matrix is MC. These three matrices are stored in memory in row major order. Each element of the matrix is one byte.

Solution: To solve this problem, we must add together all the corresponding elements of two matrices. The addresses of the matrix elements can be calculated using the equations given above. Assume that matrix MA is stored in memory locations starting at $00 and that matrix B is stored in memory locations immediately after MA. The following program simply steps through every pair of (i,j)th elements, adds them together, and stores the result in the corresponding element of matrix MC:

```
N        EQU  $8
         ORG  $00
MA       FCB  . . .        ;matrix MA starts from here and is defined by directive
*        . . .             ;FCBs
MB       FCB  . . .        ;matrix MB starts from here and is defined by directive
*        . . .             ;FCBs

MC       RMB  64           ;reserve 64 bytes to hold the sum
i        RMB  1            ;row index i
j        RMB  1            ;column index j
BUF      RMB  1            ;buffer to hold one element
DISP     RMB  2            ;memory locations to hold the value of N × i + j

         ORG  $C000        ;starting address of the program
         LDAA #N           ;start from the last row
         STAA i            ;initialize index i

OUT_LP   DEC  i            ;compute the next row index
         LDAB #N           ;start from the last column back to 0th column
         STAB j
```

* The following 10 instructions fetch MA(i,j) and save it at BUF

```
IN_LP    DEC  j            ;compute the correct j value
         LDAA i            ;place row index in A
         LDAB #N           ;put N into B
         MUL               ;multiply i by N
         ADDB j            ;compute i × N + j
         ADCA #0           ;     "
         STD  DISP         ;save the value of i × N + j
         ADDD #MA          ;compute MA(0,0) + i × N + j
         XGDX              ;place the result in X
         LDAA 0,X          ;load MA(i,j) into accumulator A
         STAA BUF          ;save MA(i,j)
```

* The following 4 instructions fetch MB(i,j) into A

```
         LDD  DISP         ;load the (i,j)th matrix element displacement
         ADDD #MB          ;compute the address of MB(i,j)
         XGDX              ;exchange D with X so that X points to MB(i,j)
         LDAA 0,X          ;load MB(i,j) into accumulator A

         ADDA BUF          ;add MA(i,j) and MB(i,j)
         STAA BUF          ;save MA(i,j) + MB(i,j) temporarily
```

```
* The following 3 instructions compute the address of MC(i,j) and leave it in X
        LDD    DISP
        ADDD   #MC
        XGDX

        LDAA   BUF       ;get the sum
        STAA   0,X       ;save the sum in MC(i,j)
        TST    j         ;is this the end of a column?
        BNE    IN_LP     ;if not, branch back to the start of the inner loop
        TST    i         ;is this the end of all the computation?
        BNE    OUT_LP    ;if not, go back to compute the next row
        END
```

In mathematics, the *transpose* M^T of a matrix M is the matrix obtained by writing the rows of M as the columns of M^T. To obtain the transpose of a matrix, the (i,j)th element must be swapped with the (j,i)th element for each i and j from 0 to $N - 1$.

Example 3.5 ■ Matrix Transpose Program

Write a program to transpose an $N \times N$ matrix.

Solution: An $N \times N$ matrix can be divided into three parts, as shown in Figure 3.5. The elements on the diagonal do not need to be swapped. When swapping the matrix elements, the row index i runs from 0 to $N - 2$. For each i, the index j runs from $i + 1$ to $N - 1$. The procedure for transposing a matrix is illustrated in the flowchart in Figure 3.6.

The following assembly program computes the transpose of a given matrix.

```
N        EQU   8        ;dimension of the matrix
ILIMIT   EQU   N-2      ;upper limit of i
JLIMIT   EQU   N-1      ;upper limit of j
         ORG   $00
```

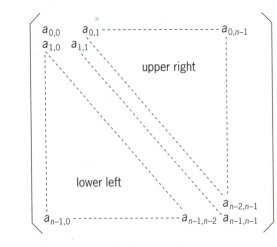

Figure 3.5 Matrix breakdown

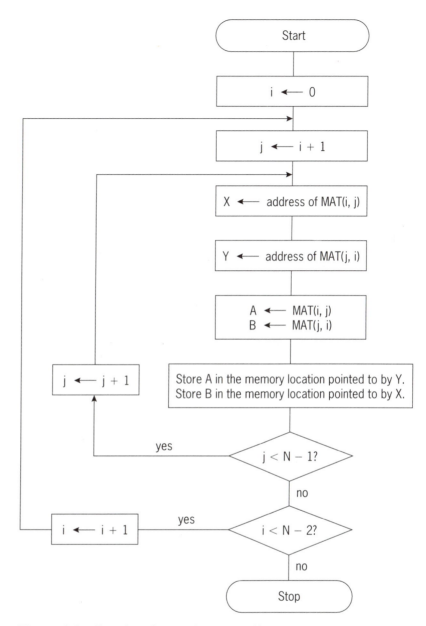

Figure 3.6 Flowchart for matrix transposition

```
i       RMB   1              ;row index
j       RMB   1              ;column index
MAT     FCB   01,02,03,04,05,06,07,08        ;matrix elements
        FCB   09,10,11,12,13,14,15,16        ;    "
        FCB   17,18,19,20,21,22,23,24        ;    "
        FCB   25,26,27,28,29,30,31,32        ;    "
        FCB   33,34,35,36,37,38,39,40        ;    "
        FCB   41,42,43,44,45,46,47,48        ;    "
        FCB   49,50,51,52,53,54,55,56        ;    "
        FCB   57,58,59,60,61,62,63,64        ;    "
```

```
              ORG   $C000      ;starting address of the program
              CLR   i          ;initialize i to 0
LOOP1         LDAA  i
              INCA
              STAA  j          ;initialize j to i+1
```

* The following 7 instructions compute the address of element MAT(i,j) and
* leave it in X

```
LOOP2         LDAA  i          ;place the row index in A
              LDAB  #N         ;put N into B
              MUL              ;compute i × N
              ADDB  j          ;compute i × N + j
              ADCA  #0         ;      "
              ADDD  #MAT       ;compute MAT(0,0) + i × N + j
              XGDX             ;place the result in X
```

* The following 7 instructions compute the address of element MAT(j,i) and leave it in Y

```
              LDAA  j          ;place the row index in A
              LDAB  #N         ;put N into B
              MUL              ;multiply j by N
              ADDB  i          ;compute j × N + i
              ADCA  #0         ;      "
              ADDD  #MAT       ;compute MAT(0,0) + j × N + i
              XGDY             ;place the result in X
```

* The following 4 instructions swap MAT(i,j) with MAT(j,i)

```
              LDAA  0,X
              LDAB  0,Y
              STAA  0,Y
              STAB  0,X

              LDAB  j
              INC   j
              CMPB  #JLIMIT    ;is j = N − 1?
              BNE   LOOP2
              LDAA  i
              INC i
              CMPA  #ILIMIT    ;is i = N − 2?
              BNE   LOOP1
              END
```

3.5 Strings

A string is a sequence of characters terminated by a null (ASCII code 0) or other special character such as EOT (ASCII code $04). In this book, we will use the null character to terminate strings. Common operations applied to strings include concatenation, character and word counting, string matching, etcetera.

Example 3.6 ■ Append one string to the end of another string

Given two strings, *string1* and *string2*, write a program to append *string2* to the end of the *string1*.

Solution: This problem seems easy. However, it raises the issue of whether a string can be arbitrarily expanded. To simplify the programming job, we will create a new string to hold the result, and we will assume that the new string has enough space to hold the concatenation of two strings. The process of concatenating the two strings is as follows:

Step 1
Copy the first string to the new string.
Step 2
Copy the second string at the end of the new string.

The program is:

```
            ORG   $00
string1     FCC   "xxxxxxxx"    ;first string
            FCB   0             ;terminate the first string
string2     FCC   "yyyyyyyyy"   ;second string
            FCB   0             ;terminate the second string
newstrng    RMB   80

            ORG   $C000
            LDX   #string1
            LDY   #newstrng
copy1ststr  LDAB  0,X
            BEQ   copy2ndstr    ;is this the end of the first string?
            STAB  0,Y           ;store the character in the new string
            INX                 ;move to the next character
            INY                 ;      "
            BRA   copy1ststr
copy2ndstr  LDX   #string2      ;prepare to copy the second string
loop2       LDAB  0,X           ;get the next character
            BEQ   exit          ;is this the end of the second string?
            STAB  0,Y
            INX
            INY
            BRA   loop2
exit        STAB  0,Y           ;terminate the new string with the Null character
            END
```

Example 3.7 ■ String character and word count

Write a program to count the number of characters and words contained in a given string. Words are separated by one or more white spaces—space, tab, carriage return, and line-feed characters are considered white spaces. A word may consist of only one character. White spaces and the null character should be included in the character count.

Solution: Let the character count, word count, and current character be represented by *char_cnt*, *wd_cnt*, and *curr_char*, respectively. The starting address of the string to be processed is at string_X. The pointer (*char_ptr*) to the current character is in X. An empty string consists of a null character.

The logic of this program is shown in the flowchart in Figure 3.7. The flowchart is based on the following plan:

- Every character that the program encounters should cause the character count to be incremented by 1.

- The first non-white-space character to the right of one or more white spaces is the beginning of a new word and should increment the word count by 1.

The program corresponding to the flowchart is:

```
tab        EQU    $09          ;ASCII code for horizontal tab
sp         EQU    $20          ;ASCII code for SPACE
CR         EQU    $0D          ;ASCII code for carriage return
LF         EQU    $0A          ;ASCII code for line feed

           ORG    $00
char_cnt   RMB    1
wd_cnt     RMB    1
string_X   FCC    "xxxxxxxxxxxxxxxxxxx"    ;the string to be processed
           FCB    0            ;the null character to terminate the previous string

           ORG    $C000
           LDX    #string_X
           CLR    char_cnt     ;initialize character count to 0
           CLR    wd_cnt       ;initialize word count to 0
loop       LDAB   0,X          ;read the current character
           INC    char_cnt
           INX                 ;move to next character
           TSTB
           BEQ    exit         ;is this a null character?
```

* The following eight instructions skip the spaces between words

```
           CMPB   #sp
           BEQ    loop         ;skip the space character
           CMPB   #tab
           BEQ    loop         ;skip the tab character
           CMPB   #CR
           BEQ    loop         ;skip the carriage return character
           CMPB   #LF
           BEQ    loop         ;skip the line feed character
```

* A non-space character is the beginning of a word

```
           INC    wd_cnt       ;a non-space character is the beginning of a new word
wd_loop    LDAB   0,X
           INX                 ;move the character pointer
           INC    char_cnt
```

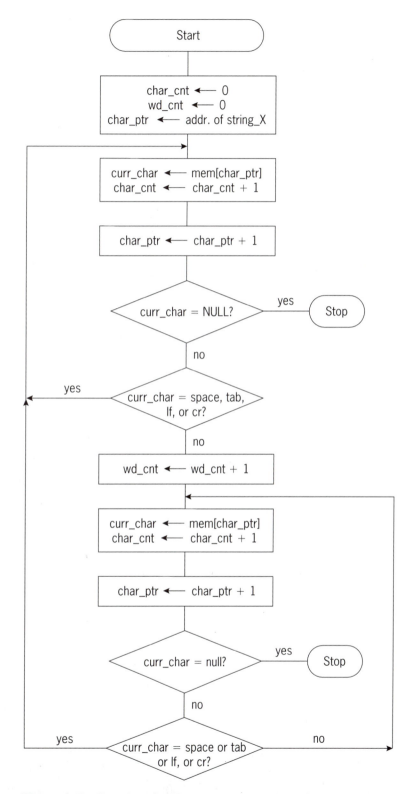

Figure 3.7 Flowchart for character count and word count program

```
* The following ten instructions check the end of a word

          TSTB
          BEQ   exit          ;is this a null character?
          CMPB  #sp
          BEQ   loop
          CMPB  #tab
          BEQ   loop
          CMPB  #LF
          BEQ   loop
          CMPB  #CR
          BEQ   loop

* a non-space character is part of a word

          BRA   wd_loop
exit      END
```

Example 3.8 ■ Word Searching

Given a string and a word, write a program to determine whether the word occurs in the given string.

Solution: Assume the string and the word are stored at *string_X* and *word_X*, respectively. In this program the basic plan is to identify the beginning of each new word and compare the new word with the given word to see if they are equal. If they are equal, then stop. Otherwise, continue until the end of the string. The memory location *search* will be set to 1 if the word is found; otherwise, it will be cleared to 0. The given word must be matched character by character, and the comparison must continue one character beyond the last non-white-space character of the given word if the word occurs in the string. There are three possible outcomes:

Case 1
The word is not the last word in the string. Comparison of the last characters will yield "not equal."

Case 2
The word is the last word in the given string, but there are one or a few white spaces between the last word and the null character. Comparison of the last characters will again give the result "not equal."

Case 3
The word is the last word in the given string, and it is followed by the null character. In this case, the comparison result for the last character is "equal."

The logic of the program is shown in the flowchart in Figure 3.8. The program is as follows:

```
tab      EQU   $09          ;ASCII code of horizontal tab
sp       EQU   $20          ;ASCII code of space
CR       EQU   $0D          ;ASCII code of carriage return
LF       EQU   $0A          ;ASCII code of line feed
NULL     EQU   $00          ;ASCII code of null
```

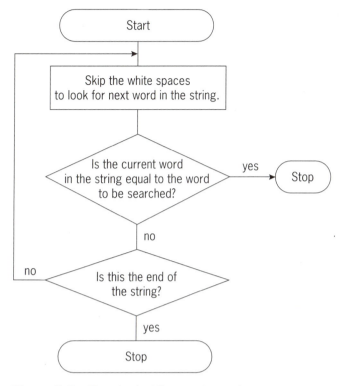

Figure 3.8 Flowchart of the word search program

```
          ORG    $0000
search    RMB    1
string_X  FCC    "xxxxxxxxxxxxxxxxxxxx"  ;a string terminated by null
          FCB    0
word_X    FCC    "yyyyy"                 ;a word terminated by null
          FCB    0


          ORG    $C000
          CLR    search                 ;initialize the search flag to 0
          LDX    #string_X              ;place the starting address of the string in X
loop      LDAB   0,X
          INX                           ;move the string pointer
```

* The following ten instructions skip the white spaces to search for the next
* word in the string

```
          TSTB
          BEQ    exit                   ;is this the end of the string?
          CMPB   #sp                    ;is the current character of the string a space?
          BEQ    loop                   ;if yes, check the next character of the string
          CMPB   #tab                   ;is the current character of the string a tab?
          BEQ    loop                   ;if yes, check the next character of the string
          CMPB   #CR                    ;is the current character of the string a CR?
          BEQ    loop                   ;if yes, check the next character of the string
          CMPB   #LF                    ;is the current character of the string a LF?
          BEQ    loop                   ;if yes, check the next character of the string
```

* The occurrence of the first non-white character indicates the beginning of
* a word, and the comparison should be started

```
            LDY   #word_X      ;place the starting address of the word in Y
            LDAA  0,Y          ;place the current character of word_X in A
            INY                ;move the word pointer
next_ch     CBA                ;compare characters in A and B
            BNE   end_of_wd    ;check the next word
            CMPA  #NULL        ;is this the end of a word?
            BEQ   matched      ;if yes, the word is found in the string
            LDAA  0,Y          ;get the next character of word_X
            LDAB  0,X          ;get the next character of string_X
            INX                ;move the string pointer
            INY                ;move the word pointer
            BRA   next_ch      ;check the next character
```

* The following ten instructions check to see if the end of the given word
* is reached

```
end_of_wd
            CMPA  #NULL
            BNE   next_wd      ;if not the end of the given word, then not matched
            CMPB  #CR
            BEQ   matched
            CMPB  #LF
            BEQ   matched
            CMPB  #tab
            BEQ   matched
            CMPB  #sp
            BEQ   matched
```

* The following twelve instructions skip the unmatched word in the string

```
next_wd     LDAB  0,X          ;get the next character in the string
            BEQ   exit         ;stop if this is the end of the string
            INX
            CMPB  #CR
            BEQ   jmp_loop     ;the label loop is too far away to use a conditional
 *                            ;branch
            CMPB  #LF
            BEQ   jmp_loop
            CMPB  #tab
            BEQ   jmp_loop
            CMPB  #sp
            BEQ   jmp_loop
            BRA   next_wd

jmp_loop    JMP   loop
matched     LDAB  #1           ;set the search flag to one
            STAB  search       ;     "
exit        END
```

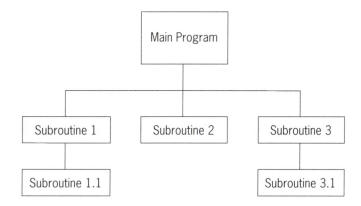

Figure 3.9 A structured program

3.6 Subroutines

Good program design is based on the concept of modularity—the partitioning of programming code into subroutines. A main module contains the logical structure of the algorithm, while smaller program units execute many of the details.

The principles of program design in high-level languages apply even more to the design of assembly language programs. Begin with a simple main program whose steps clearly outline the logical flow of the algorithm, and then assign the execution details to subroutines. Of course, subroutines may themselves call other subroutines. Figure 3.9 shows an example of the partitioning of a structured program code.

3.7 Subroutine Calls and Returns

A subroutine is a sequence of instructions stored in memory at a specified address. The subroutine can be called from various places in the program. When a subroutine is called, the address of the next instruction is saved and program control passes to the called subroutine, which then executes its instructions. The subroutine terminates with a return instruction directing program control back to the instruction following the one that called the subroutine. Figure 3.10 illustrates this process.

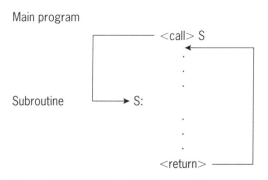

Figure 3.10 Subroutine processing

The situations in which a programmer would use a subroutine include the following:

■ There are several places in a program that require the same computation to be performed with different parameters. The program can be shortened by using subroutine calls.

■ The program is complex. If the programmer can divide a big program into smaller ones and solve each smaller problem in a subroutine, the problem will be easier to manage and understand, and it will also be easier to maintain.

The 68HC11 provides two instructions for calling subroutines:

```
[<label>]    BSR  <rel>    [<comment>]
[<label>]    JSR  <opr>    [<comment>]
```

where

<rel> is the offset to the subroutine.

<opr> is the address of the subroutine and is specified in the DIR, EXT,or INDexed addressing mode.

In terms of program control, these instructions are similar to the Bcc (branch on condition) and JMP (jump) instructions and have the same syntax. The BSR and JSR instructions, however, provide for automatically saving the program counter on the stack. The address of the first byte of the instruction that follows the BSR or JSR instruction is saved in the stack, as shown in Figure 3.11. Upon return from the subroutine, the saved return address is pulled from the stack and placed in the program counter. Program execution then continues.

The BSR instruction can use only the relative addressing mode to specify the subroutine address. The range of the branch can be from −128 to 127 bytes. The JSR instruction can use the direct, extended, or index addressing modes to specify the subroutine address; thus the subroutine can be as far as 64 KB away from the JSR instruction.

The last instruction of a subroutine is the RTS instruction, which causes the top two bytes of the stack to be pulled and loaded into the PC. Program execution will then continue from the new PC value.

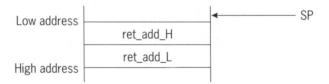

ret_add_H: upper byte of the return address
ret_add_L: lower byte of the return address

Figure 3.11   The top two bytes of the stack after the execution of BSR or JSR

3.8 Issues in Subroutine Calls

In the following discussion we will refer to the program or routine that makes the subroutine call as the *caller* and the subroutine called by other routines as the *callee*. We say that a subroutine is *entered* when it is being executed. There are three important issues to consider in subroutine calls:

- *Parameter passing.* The programmer often wants the subroutine to perform computations using parameters passed to it. There are four methods for passing parameters to a subroutine:

  1. *Use registers.* In this method, the parameters are placed in registers before a subroutine call is made. This method is most convenient when there are only a few parameters to be passed to the callee; it is also the easiest method and has the lowest overhead in accessing the passed parameters.

  2. *Use program memory.* The parameters to be passed follow the BSR or JSR instruction. This method involves more overhead than the first method, and accessing the passed parameters is more complicated. The callee will need to adjust the return address saved in the stack.

  3. *Use the stack.* The parameters are pushed onto the stack before a subroutine call is made. This method is easier than the second method but slightly more complicated than the first. Either the caller or the callee must clean up the stack space used for passing parameters after the computation is completed.

  4. *Use global memory.* Global memory is accessible to both the caller and the callee. The caller simply places the parameters in global memory, and the callee will be able to access them.

- *Returning results.* The result of a computation can be returned by three methods:

  1. *Use registers.* This method is most convenient when there are only a few bytes of values to be returned. When this method is used, the caller may need to save registers before making the subroutine call.

     In the 68HC11, registers A, B, X, and Y can be used to hold arguments or results. The carry bit in the condition code register can be used to pass a one-bit result that can be used in instructions like BCC or BCS. If there are only a few arguments, using the registers is the best method.

  2. *Use the stack.* The caller creates a hole in the stack by decrementing the stack pointer before making a subroutine call. The callee saves the computational result in the hole before returning to the caller.

  3. *Use global memory.* Global memory is accessible to both the caller and the callee. The callee simply saves the result in the designated locations.

■ *Allocation of local variables.* In addition to the parameters passed to it, a subroutine may need memory locations to hold loop indices and working buffers that hold temporary variables and results. These variables are useful only during execution of the subroutine and are called *local variables* because they are local to the subroutine. Local variables are often allocated in the stack so that they are not accessible to the caller.

There are two methods for allocating local variables:

1. Use as many DES instructions as needed if fewer than six bytes are needed. This approach takes less time if no more than five bytes are needed for local variables.

2. Use the following instruction sequence if more than five bytes are needed:

```
TSX
XGDX
SUBD  #N      ;allocate N bytes
XGDX
TXS           ;move the stack pointer up by N bytes
```

It takes 18 E clock cycles to allocate six bytes in the stack using the DES instruction, while the instruction sequence for the second method takes only 16 E clock cycles for any number of bytes. The TXS instruction decrements the value in X by 1 and places it in the stack pointer. A similar operation can be performed on index register Y using the TYS instruction.

The space allocated to local variables must be deallocated before the subroutine returns to the caller. There are two corresponding methods for deallocating local variables:

1. Use as many INS instructions as needed if fewer than six bytes are to be deallocated. This approach takes less time if no more than five bytes are to be deallocated.

2. Use the following instruction sequence if there are six or more bytes to be deallocated.

```
TSX
XGDX
ADDD  #N
XGDX
TXS           ;move down the stack pointer by N bytes
```

It takes 18 E clock cycles to deallocate six bytes from the stack using the INS instruction, while it takes only 16 E clock cycles to deallocate any number of bytes using this instruction sequence.

3.9 The Stack Frame

The stack is used heavily during a subroutine call: the caller may pass parameters using the stack, and the callee may need to save registers and allocate

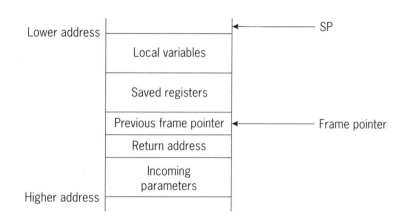

Figure 3.12   A stack frame

local variables in the stack. The region in the stack that holds incoming parameters, the subroutine return address, local variables, and saved registers is referred to as the *stack frame*. Some microprocessors dedicate a register to the management of the stack frame—this register is called the *frame pointer*. (The 68HC11, however, does not have a register dedicated to the function of the frame pointer.) Since the stack frame is created during a subroutine call, it is also called the *activation record* of the subroutine. The stack frame exists as long as the subroutine is not exited. An example of a stack frame is shown in Figure 3.12.

The frame pointer primarily facilitates access to local variables and incoming parameters. A negative offset relative to the frame pointer is used to access a local variable, and a positive offset relative to the frame pointer is used to access an incoming parameter. Since the 68HC11 does not have a dedicated frame pointer, we will not use the term frame pointer in the following discussion. The user can use either the TSX (or TSY) instruction so that index register X (or Y) points to the top byte of the stack. The index addressing mode can then be used to access all the values in the stack frame. The stack frame of a 68HC11 subroutine is shown in Figure 3.13.

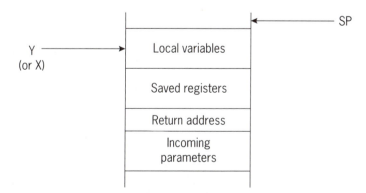

Figure 3.13   Example of a 68HC11 stack frame

Example 3.9

Draw the stack frame for the following program segment after the ninth instruction of the subroutine MAT_REV has been executed.

```
            LDAA  #N
            PSHA
            LDX   #MAT
            PSHX
            BSR   MAT_REV
              .
              .
              .
MAT_REV
(1)         PSHB
(2)         PSHA
(3)         PSHX
(4)         PSHY
(5)         TSX
(6)         XGDX
(7)         SUBD #6
(8)         XGDX
(9)         TXS
            ────
```

Solution: Two parameters are passed by the caller. These two parameters occupied three bytes. The callee allocates six bytes for local variables. Two 8-bit and two 16-bit registers are saved in the stack. The stack frame is shown in Figure 3.14.

When a subroutine pushes registers onto the stack, it must restore them by pulling their old values from the stack. Since the stack is a first-in-last-out data structure, the register that was last pushed onto the stack must be the first to be pulled. For example, if a subroutine has the following push instructions at its entry point:

```
    PSHA
    PSHB
    PSHX
    PSHY
```

then it must have the following pull instructions before it returns:

```
    PULY
    PULX
    PULB
    PULA
```

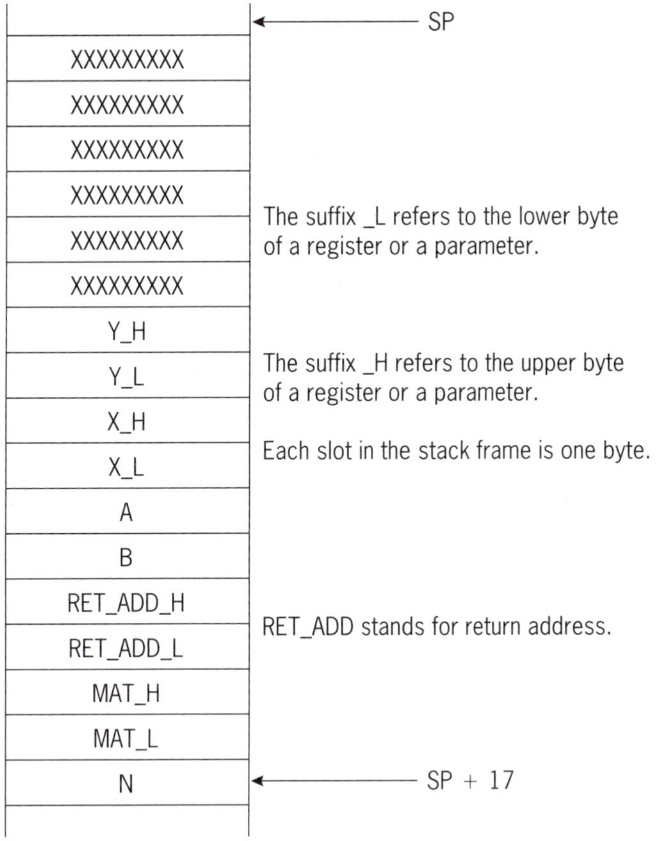

	← SP
XXXXXXXX	
XXXXXXXX	
XXXXXXXX	
XXXXXXXX	
XXXXXXXX	The suffix _L refers to the lower byte of a register or a parameter.
XXXXXXXX	
Y_H	
Y_L	The suffix _H refers to the upper byte of a register or a parameter.
X_H	
X_L	Each slot in the stack frame is one byte.
A	
B	
RET_ADD_H	
RET_ADD_L	RET_ADD stands for return address.
MAT_H	
MAT_L	
N	← SP + 17

Figure 3.14 Stack frame for Example 3.9

3.10 Examples of Subroutine Calls

The examples in this section illustrate passing parameters, returning results, and allocation of local variables.

3.10.1 Using Registers to Pass Parameters

In the following example, registers are used to pass parameters and to return the result. The subroutine does not allocate local variables in the stack.

Example 3.10 ■ Average of an Array

Write a subroutine to compute the average of an array with N 8-bit elements, and write an instruction sequence to call the subroutine. The array starts at ARRAY. Use registers to pass parameters and return the average in the B accumulator.

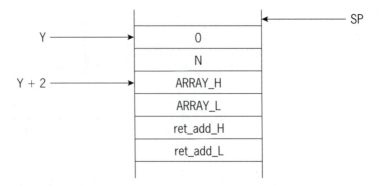

The suffix _H refers to the upper byte of a parameter.
The suffix _L refers to the lower byte of a parameter.
ret_add is the return address.

Figure 3.15   Another example of stack frame

Solution: The caller passes the array count N and the array base address ARRAY in registers A and X, respectively, and would use the following three instructions to make the subroutine call:

```
LDX  #ARRAY
LDAA #N
BSR  AVERAGE    ;call the subroutine "average"
        .
        .
        .
```

Registers X and A are used during the computation of the array average, so both X and A must be saved on the stack. X is pushed onto the stack first, followed by A. The array average is computed by dividing the array sum by the array count. The sum must be loaded into the D accumulator and the array count N must be loaded into index register X before the division is performed. The array count is 8-bit, but the 68HC11 does not have an instruction to load an 8-bit value into X. Therefore, for convenience, a zero is pushed onto the stack above the memory location that holds N so that the LDX 0,Y instruction can be used to load the value N into X. Index register Y will be used to point to the top byte of the stack to facilitate access to all the parameters in the stack. The stack frame is shown in Figure 3.15.

The logic of the program is shown in the flowchart in Figure 3.16. To convert the flowchart into the 68HC11 assembly instructions, we will

■ use index register X as the array pointer
■ use index register Y as the loop count
■ use accumulator D to accumulate the sum

The following subroutine computes the array average:

```
* This routine returns the average of an array in accumulator B

average  PSHX
         PSHA
```

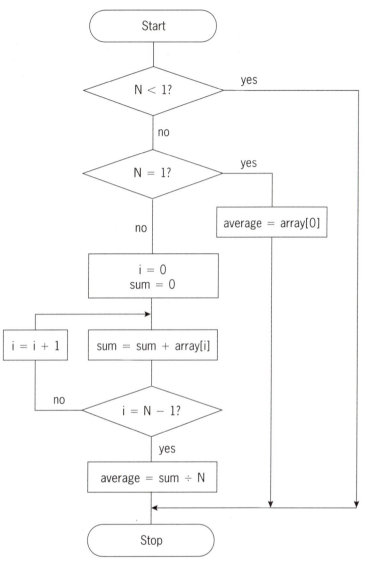

Figure 3.16 Flowchart for computing the average of an array

```
        CLRA
        PSHA              ;place a 0 on the top of the stack
        TSY               ;Y points to the top byte of the stack
        LDAB  1,Y         ;place N in Y
        CMPB  #1          ;check if N < 1
        BLO   exit        ;is the array empty?
        BHI   do          ;does the array have more than one element?
        LDAB  0,X         ;get the single element and return
        BRA   exit        ;        "
do      XGDY              ;exchange D with Y and place N in Y
        CLRA              ;initialize the array sum to 0
        CLRB              ;        "
again   ADDB  0,X         ;add an element to the sum
        ADCA  #0          ;add carry to the upper 8-bit of the sum
```

```
              INX                ;move the array pointer
              DEY                ;decrement the loop count
              BNE    again       ;is this the end of the loop?
              TSY                ;point Y to the top of the stack
              LDX    0,Y         ;load N into X
              IDIV               ;compute the array average
              XGDX               ;exchange X and D so that D contains the quotient
       *                         ;in which A contains 0 and B contains the quotient
       *                         ;the quotient is returned in B
exit          PULA               ;clean up the stack
              PULA               ;         "
              PULX               ;         "
              RTS
```

■

3.10.2 Using the Stack to Pass Parameters

The caller would use the following instruction sequence to push parameters onto the stack and call the subroutine SUB1:

```
LDAA  #N      ;N is the argument to sub1
PSHA          ;push the argument of sub1 onto the stack
BSR  SUB1
```

If the callee does not save any registers, it could use the following instruction sequence to retrieve the argument from the stack:

```
TSX           ;transfer the stack pointer to X
LDAA  2,X     ;get the argument
```

In the following example, parameters are passed in the stack and the result is returned in the memory location specified by the caller.

■

Example 3.11

Write a subroutine to find the maximum element of an array, and write an instruction sequence to call this subroutine. The following parameters are passed in the stack:

- *array:* the starting address of the given array
- *arcnt:* the array count
- *amax:* address to hold the maximum element of the array

Solution: The caller would include the following instruction sequence to pass parameters, make the subroutine call, and restore the stack:

```
       .
       .
       .
LDX   #array
PSHX
LDAA  #arcnt
```

```
PSHA
LDX    #amax
PSHX
BSR    MAX
TSX                 ;clean up the stack
LDAB #5             ;       "
ABX                 ;       "
TXS                 ;       "
 .
 .
 .
```

The logic of the program is illustrated in the flowchart in Figure 3.17.

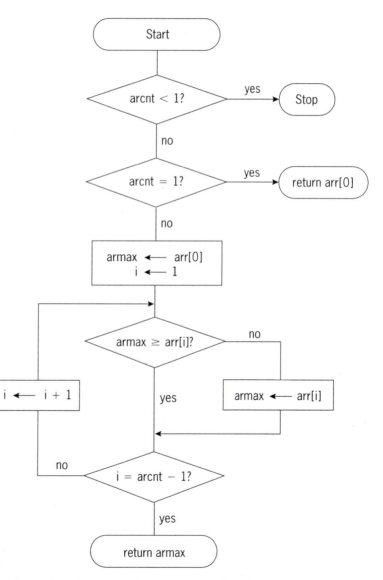

Figure 3.17 Flowchart for array max subroutine

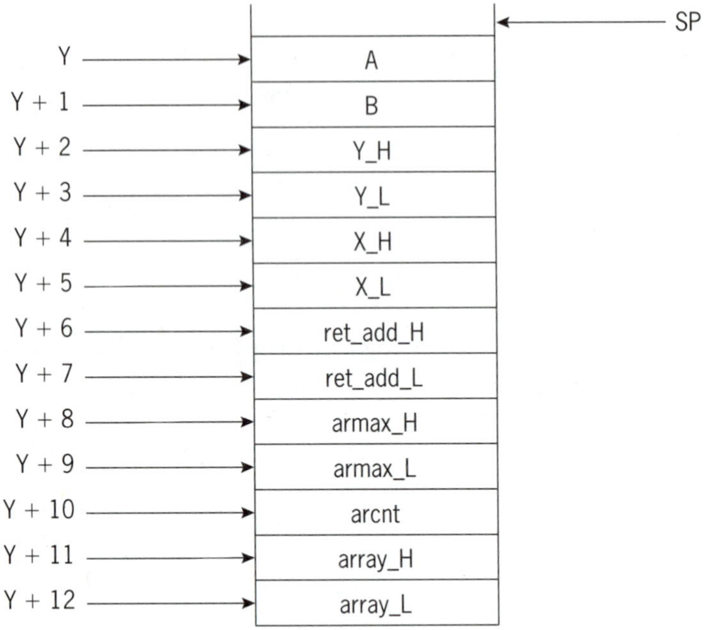

Each slot in the stack is one byte.
The suffix _H specifies the upper byte of a parameter.
The suffix _L specifies the lower byte of a parameter.

Figure 3.18   Stack frame for the array max subroutine

The registers that will be used in the subroutine must be saved. The stack frame and the offset of each parameter from the top element of the stack are shown in Figure 3.18.

The subroutine is as follows:

```
array    EQU    11          ;array base address offset from the top of the stack
arcnt    EQU    10          ;array count offset from the top of the stack
armax    EQU    8           ;array max offset

MAX      PSHX               ;save all registers
         PSHY               ;      "
         PSHB               ;      "
         PSHA               ;      "
         TSY                ;point Y to the top element of the stack
         LDX    array,Y     ;load the array base address into X
         LDAB   arcnt,Y     ;load the array count into B
         LDAA   0,X         ;assign the first element as the temporary MAX
         CMPB   #1          ;check array count
         BLO    EXIT        ;return if the array count is less than 1
         BHI    START       ;look for array max if array count is larger than 1
         BRA    DONE        ;the array has a single element
START    INX                ;set X to point to the second element of the array
         DECB               ;loop limit is arcnt − 1
AGAIN    CMPA   0,X         ;compare the next element with the MAX
         BGE    noswap      ;should MAX be updated?
```

```
              LDAA  0,X          ;update MAX
    noswap    INX                ;move to the next element
              DECB               ;decrement the loop count
              BNE   AGAIN        ;is it done?
    DONE      LDX   armax,Y      ;get the address for the array MAX
              STAA  0,X          ;save the array MAX
    EXIT      PULA               ;restore registers
              PULB               ;
              PULY               ;
              PULX               ;
              RTS
```

3.10.3 Using Program Memory to Pass Parameters

In this method, the parameters to be passed to the subroutine are placed immediately after the BSR (or JSR) instruction:

```
        .
        .
        .

    BSR  SUB
    FDB  I,J,K
    <next instruction>
```

I, J, and K are the addresses of parameters, and each occupies two bytes. After execution of the BSR instruction, the address of the location where the address of variable I is stored is pushed onto the stack, as shown in Figure 3.19. The subroutine can retrieve parameters by using this address. For example, the value of K can be fetched by using the following instructions:

```
    (1)   TSX            ;X points to the top byte of the stack
    (2)   LDX   0,X      ;get the addresses of the argument list
    (3)   LDX   4,X      ;get the address of K into X
    (4)   LDAB  0,X      ;get the value of K into B
```

Instruction 1 places the address of the top byte of the stack into X. Instruction 2 places the address of the byte after the BSR SUB instruction into X. Instruction 3 places the address of variable K into X, and the instruction 4 loads

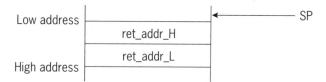

Each stack slot is one byte.
The suffix _H refers to the upper byte of a parameter.
The suffix _L refers to the lower byte of a parameter.
ret_addr actually points to the location where the address
of variable I is stored.

Figure 3.19   Stack frame after the execution of BSR

the value of K into B. The return address in the stack must be adjusted so that the callee can return to the instruction that follows the BSR instruction. If the argument list is six bytes long, the following instructions will correct the return address:

```
TSY
LDD   ret_add,Y     ;ret_add is the offset of the return address from the top of the stack
ADDD  #6
STD   ret_add,Y
```

The following example illustrates passing parameters using the program memory and allocating local variables in the stack.

Example 3.12 ■ Bubble sort

Write a subroutine to sort an array of 16-bit elements using the bubble sort method, and write an instruction sequence to call this routine. The starting address of the array is ARR, and the array count is *arcnt*. Use program memory to pass parameters to the callee.

Solution: The caller would use the following instruction sequence to call the BUBBLE subroutine:

```
        .
        .
        .
JSR  BUBBLE
FDB  ARR
FCB  ARCNT
<next instruction>
```

The key ideas of the bubble sort are:

- N is the array count
- Starting with the first element and continuing up to the next-to-last element, compare every pair of adjacent elements. Swap elements when necessary so that the largest (or smallest) element is swapped to the last place of the array in the first round. $N - 1$ comparisons are performed during this round.
- Further comparisons and swaps are not needed if the array is already sorted after a round. To avoid unnecessary operations, a flag is used to indicate whether further comparisons and swaps are still needed. This flag is set to 1 at the beginning of each round and is cleared to 0 when a swap is performed. The array is probably not in sorted order if the flag is 0 at the end of a round.
- In the second round of comparisons swap elements when necessary so that the second largest (or second smallest) element becomes the next-to-last element of the array. $N - 2$ comparisons are performed during this round.

■ Continue making similar comparisons and swaps until the array is sorted.

■ Up to $N - 1$ rounds will be needed.

The logic of the bubble sort is shown in the flowchart in Figure 3.20.

To facilitate the comparison and swapping operations, the following five local variables are required in the subroutine BUBBLE:

1. *outer:* a one-byte number to keep track of the number of rounds remaining to be performed

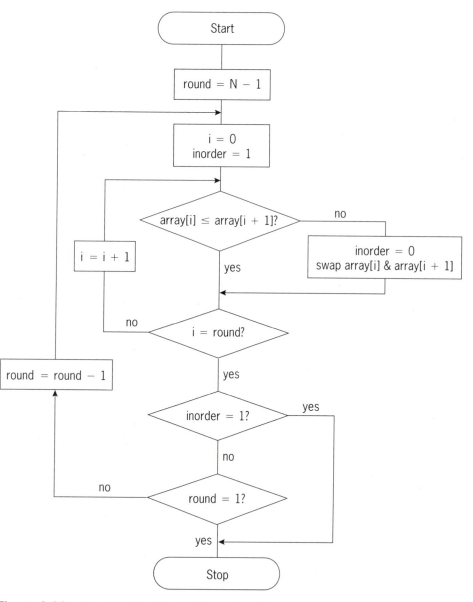

Figure 3.20 Flowchart for the bubble sort routine

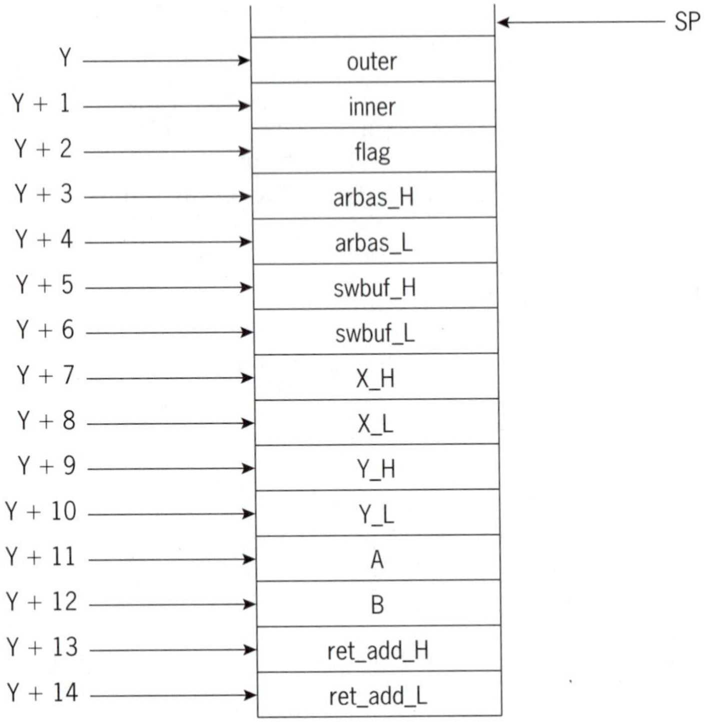

The suffix _H refers to the upper byte of a parameter.
The suffix _L refers to the lower byte of a parameter.

Figure 3.21   Stack frame after calling the bubble sort routine

2. *inner:* a one-byte number to keep track of the number of comparisons remaining to be performed in a round.

3. *flag:* a one-byte flag that indicates whether the given array is already in sorted order at the end of a compare-and-swap round

4. *arbas:* the two-byte base address of the next subarray to be sorted

5. *swbuf:* a two-byte buffer space for swapping elements

These five local variables are allocated in the stack. Figure 3.21 shows the stack frame after saving registers and allocating local variables.

The bubble sort subroutine is as follows:

```
local     EQU   7      ;number of bytes of local variables
ret_add   EQU   13     ;offset of the return address from the top of the stack
arcnt     EQU   2      ;offset of the array count from the JSR instruction
outer     EQU   0      ;offset of the local variable outer from the top of the stack
inner     EQU   1      ;offset of the local variable inner from the top of the stack
flag      EQU   2      ;offset of the local variable flag from the top of the stack
arbas     EQU   3      ;offset of the local variable arbas from the top of the stack
swbuf     EQU   5      ;offset of the local variable swbuf from the top of the stack
adjust    EQU   3      ;return address adjustment value
true      EQU   1      ;mnemonic for true
false     EQU   0      ;mnemonic for false
```

```
bubble  PSHB              ;save registers
        PSHA              ;   "
        PSHY              ;   "
        PSHX              ;   "
```

* The following five instructions allocate space (7 bytes) for local variables

```
        TSX               ;make room for local variables
        XGDX              ;   "
        SUBD  #local      ;   "
        XGDX              ;   "
        TXS               ;   "

        TSY               ;Y points to the top of the stack
```

* The following three instructions fetch parameters from the program memory

```
        LDX   ret_add,Y   ;get the pointer to the array base
        LDAB  arcnt,X     ;transfer the array count to B
        LDX   0,X         ;put the array base in X

        DECB              ;initialize the number of rounds
        STAB  outer,Y     ;   "
        STAB  inner,Y     ;initialize the inner loop count
        STX   arbas,Y     ;save the array base for the next loop
```

* The following three instructions adjust the return address

```
        LDD   ret_add,Y   ;get the incorrect return address
        ADDD  #adjust     ;correct the return address in the stack
        STD   ret_add,Y   ;   "
```

```
ploop   LDAA  #true       ;set the array in-order flag to true
        STAA  flag,Y      ;   "
        LDX   arbas,Y     ;get the array base for the next loop
        LDAA  outer,Y     ;initialize the inner loop count
        STAA  inner,Y     ;   "
cloop   LDD   0,X         ;get the first element
        CPD   2,X         ;compare to the next element
        BLE   lptest      ;is swap needed?
```

* The following 5 instructions swap the two adjacent elements

```
        STD   swbuf,Y     ;swap two elements
        LDD   2,X         ;   "
        STD   0,X         ;   "
        LDD   swbuf,Y     ;   "
        STD   2,X         ;   "

        LDAB  #false      ;reset the in-order flag
        STAB  flag,Y      ;   "
lptest  INX               ;check the next two adjacent elements
        INX               ;   "
```

```
        DEC   inner,Y         ;update the inner loop count
        BNE   cloop           ;is this the end of a round?
        TST   flag,Y          ;is the array already in order?
        BNE   done            ;    "
        DEC   outer,Y         ;decrement the number of rounds to be performed
        BNE   ploop
```

* The following five instructions deallocate space allocated to local variables

```
done    TSX                   ;deallocate local variables
        XGDX                  ;    "
        ADDD  #local          ;    "
        XGDX                  ;    "
        TXS                   ;    "


        PULX                  ;restore registers
        PULY                  ;    "
        PULA                  ;    "
        PULB                  ;    "
        RTS                   ;return to the caller
```

3.11 Input and Output Routines

Both the EVB and EVM have commands that allow the user to modify and examine the contents of memory locations. However, it would be more convenient if the executing program could read data directly from the terminal (or PC) keyboard and output results directly to the monitor screen. The EVB and EVBU not only allow the programmer to write I/O routines to perform these two operations but also provide a set of I/O routines that can be called by the user program. These I/O routines are listed in Table 3.3. The subject of writing I/O routines will be discussed in Chapter 8. In this section, the reader will learn how to call the I/O routines in Table 3.3 to read data from the keyboard and output data to the monitor screen.

The subroutines in Table 3.3 are in ROM. A jump table allows the user program to call these I/O routines by jumping to the desired entry of the table. Each entry is a three-byte jump instruction. The first byte is the opcode of the JMP instruction. The second and the third bytes are the address of the actual subroutine that performs the I/O operation. To invoke one of the I/O routines, execute a JSR instruction to the corresponding address given in Table 3.4. For example, to output the ASCII character contained in accumulator A, use the following instruction:

```
    JSR  $FFB8
```

To output a string, use the following instruction sequence:

```
    LDX  #STRING    ;STRING is the starting address of the string
    JSR  $FFC7
```

Routine Name	Function
UPCASE	converts the character in accumulator A to uppercase
WCHEK	tests the character in A and returns with the Z bit set if the character is a white space (space, comma, tab)
DCHEK	tests the character in A and returns with the Z bit set if the character is a delimiter (carriage return or whitespace)
INIT	initializes I/O device
INPUT	reads I/O device
OUTPUT	writes I/O device
OUTLHLF	converts left nibble of A to ASCII and outputs to terminal port
OUTRHLF	converts right nibble of A to ASCII and outputs to terminal port
OUTA	outputs the ASCII character in A
OUT1BYT	converts the binary byte at the address in index register X to two ASCII characters and outputs them; returns address in index register X pointing to next byte
OUT1BSP	converts the binary byte at the address in index register X to two ASCII characters and outputs them followed by a space; returns address in index register X pointing to next byte
OUT2BSP	converts two consecutive binary bytes starting at address in index register X to four ASCII characters and outputs the characters followed by a space; returns address in index register X pointing to next byte
OUTCRLF	outputs ASCII carriage return followed by a line feed
OUTSTRG	outputs string of ASCII bytes pointed to by address in index register X until character is an end-of-transmission ($04)
OUTSTRG0	same as OUTSTRG, except that leading carriage returns and line feeds are skipped
INCHAR	inputs ASCII character to A and echoes back; this routine loops until character is actually received

Table 3.3 ■ EVB I/O Routines

Address	Instruction
$FFA0	JMP UPCASE
$FFA3	JMP WCHEK
$FFA6	JMP DCHEK
$FFA9	JMP INIT
$FFAC	JMP INPUT
$FFAF	JMP OUTPUT
$FFB2	JMP OUTLHLF
$FFB5	JMP OUTRHLF
$FFB8	JMP OUTA
$FFBB	JMP OUT1BYT
$FFBE	JMP OUT1BSP
$FFC1	JMP OUT2BSP
$FFC4	JMP OUTCRLF
$FFC7	JMP OUTSTRG
$FFCA	JMP OUTSTRG0
$FFCD	JMP INCHAR

Table 3.4 ■ 68HC11EVB I/O routine jump table

Example 3.13

Write a program to be run on the EVB that

1. prompts the user to enter the current time in the order of hours, minutes, and seconds
2. reads the current time entered by the user
3. checks the time entered by the user; if the time is not valid, prompts the user to reenter
4. clears the screen and displays the current time at the center of the screen

Solution: This program will use the following message to prompt the user to enter the current time:

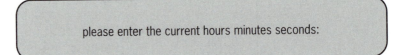

please enter the current hours minutes seconds:

The current hours, minutes, and seconds must be entered in BCD format and must be separated by spaces. If an invalid time is entered, the user will be prompted to reenter the current hours, minutes, and seconds. The instruction sequence for prompting the user to enter the current time is as follows:

```
        LDX  #T_PROMPT    ;load the address of the first character of prompt
        JSR  $FFC7        ;output the prompt
          .
          .
          .
T_PROMPT  FCC  "please enter the current hours minutes seconds:"
          FCB  $04              ;terminate the prompt
```

The program will call the INCHAR subroutine to read in the current time from the keyboard, one digit at a time, and check the validity of each digit. If the entered character is neither a white space (tab, space, or carriage return) nor a valid BCD digit, the program will prompt the user to reenter the time.

The ASCII codes of valid BCD digits range from $30 (0) to $39 (9). The value of $30 is subtracted from the entered ASCII code to convert it to the corresponding binary value. The valid hours digits are between 00 and 23. The valid minutes and seconds digits are between 00 and 59. Every two digits are stored in one byte. The upper digit is stored in the upper four bits, and the lower digit is stored in the lower four bits.

The key ideas for reading current time from the keyboard are illustrated in the flowchart in Figure 3.22. The routine that inputs the current time and does the checking is:

```
        hh     RMB   1              ; memory location to store the hours digits
        mm     RMB   1              ; memory location to store the minutes digits
        ss     RMB   1              ; memory location to store the seconds digits
```

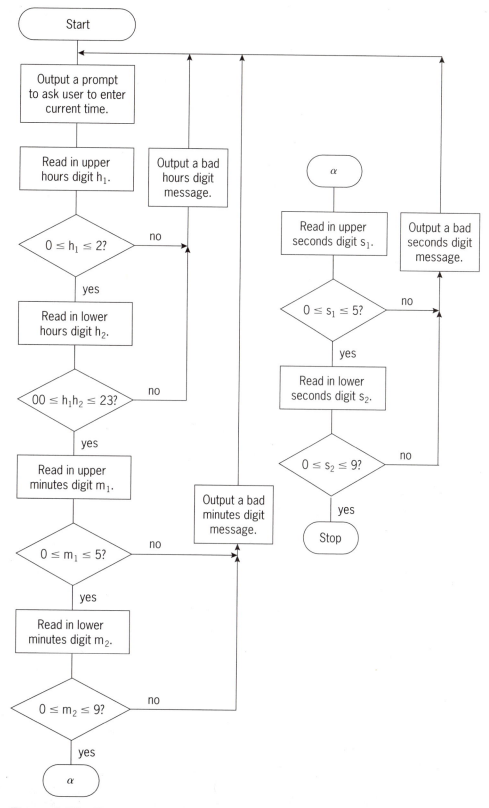

Figure 3.22 Flowchart for reading current time from keyboard

* following are BUFFALO I/O routines used in this subroutine

```
outa      EQU   $FFB8        ; output the character in A
wchek     EQU   $FFA3        ; test the character in A for white space
dchek     EQU   $FFA6        ; test the character in A for delimiter
outcrlf   EQU   $FFC4        ; output ASCII carriage return followed by line feed
outstrg   EQU   $FFC7        ; output the string pointed to by X
inchar    EQU   $FFCD        ; input a character and echo it
outlhlf   EQU   $FFB2        ; convert left nibble of A to ASCII and output
outrhlf   EQU   $FFB5        ; convert right nibble of A to ASCII and output
colon     EQU   $3A          ; ASCII code of the colon character

          ORG   $C000        ; starting address of EVB user RAM space
start     LDX   #t_prompt
          JSR   outstrg      ; prompt the user to enter current time
          JSR   outcrlf      ; output carriage return and line feed
```

* The following instruction sequence reads in hours digits and checks to make
* sure they are between 00 and 23

```
again     JSR   inchar       ; read the upper hours digit and echo it
          JSR   dchek        ; is it a delimiter character?
          BEQ   again        ; skip white spaces and delimiter characters
          CMPA  #$30         ; compare with ASCII code for 0
          BLO   badhour      ; is it an invalid hours digit?
          CMPA  #$32         ; compare with ASCII code for 2
          BHI   badhour      ; is it an invalid hours digit?
          SUBA  #$30         ; convert it to binary
          LSLA               ; shift to upper four bits
          LSLA               ;      "
          LSLA               ;      "
          LSLA               ;      "
          STAA  hh           ; save the upper hours digit

          JSR   inchar       ; read the lower hours digit and echo it
          CMPA  #$30         ; compare with the ASCII code for 0
          BLO   badhour      ; is it an invalid hours digit?
          CMPA  #$39         ; compare with the ASCII code for 9
          BHI   badhour      ; is it an invalid hours digit?
          SUBA  #$30         ; convert to binary
          LDAB  hh           ; check the upper hours digit
          CMPB  #$20         ;      "
          BLO   saveit       ; the upper hours digit is 1 or 0
```

* If the upper hours digit is 2 then compare the lower hours digit with 3

```
          CMPA  #3
          BHI   badhour      ; is it larger than 23?
saveit    ABA                ; combine lower and upper hours digits
          STAA  hh           ; save the hours digits
          JMP   minute       ; go and read minutes digits
badhour   LDX   #bhpmt       ; output the bad hour prompt
          JSR   outstrg      ;      "
          JSR   outcrlf      ; output carriage return and linefeed
          JMP   start        ; start all over again
```

* The following instruction sequence reads in minutes digits and checks to make sure they
* are between 00 and 59

```
minute    JSR   inchar      ; read the upper minutes digit and echo it
          JSR   wchek       ; is it a white space character?
          BEQ   minute      ; skip white spaces
```

* The following four instructions check to see if the ASCII code is between $30 and $35

```
          CMPA  #$30        ; compare with the ASCII code for 0
          BLO   badmin      ; is it an invalid minutes digit?
          CMPA  #$35        ; compare with the ASCII code for 5
          BHI   badmin      ; is it an invalid minutes digit?
          SUBA  #$30        ; convert it to binary
          LSLA              ; shift to upper 4 bits
          LSLA              ;     "
          LSLA              ;     "
          LSLA              ;     "
          STAA  mm          ; save the upper minutes digit
          JSR   inchar      ; read the lower minutes digit
```

* The following four instructions check to see whether the ASCII code is between $30
* and $39

```
          CMPA  #$30        ; compare with ASCII code for 0
          BLO   badmin      ; is it an invalid minutes digit
          CMPA  #$39        ; compare with ASCII code for 9
          BHI   badmin      ; is it an invalid minutes digit

          SUBA  #$30        ; convert to binary
          ADDA  mm          ; combine the upper and lower minutes digits
          STAA  mm          ;     "
          BRA   second      ; go and read second digits
badmin    LDX   #bmpmt      ; output a bad minutes digit prompt
          JSR   outstrg     ;     "
          JSR   outcrlf     ; output carriage return and linefeed characters
          JMP   start       ; start all over again
```

* The following instruction sequence reads in second digits and checks to make sure they
* are between 00 and 59

```
second    JSR   inchar      ; read the upper seconds digit and echo it
          JSR   wchek       ; is it a white space character?
          BEQ   second      ; skip white spaces
```

* The following four instructions check whether the ASCII code is between $30 and $35

```
          CMPA  #$30        ; compare with ASCII code for 0
          BLO   badsec      ; is it an invalid seconds digit?
          CMPA  #$35        ; compare with ASCII code for 5
          BHI   badsec      ; is it an invalid seconds digit?
          SUBA  #$30        ; convert it to binary
          LSLA              ; shift to upper four bits
          LSLA              ;     "
          LSLA              ;     "
          LSLA              ;     "
```

```
              STAA   ss              ; save the upper seconds digit
              JSR    inchar          ; read the lower seconds digit
```

* The following four instructions check whether the ASCII code is between $30 and $39

```
              CMPA #$30              ; compare with ASCII code for 0
              BLO    badsec          ; is it an invalid seconds digit?
              CMPA #$39              ; compare with ASCII code for 9
              BHI    badsec          ; is it an invalid seconds digit?

              SUBA #$30              ; convert it to binary
              ADDA ss                ; combine two seconds digits
              STAA   ss              ;     "
              BRA    display         ; go and display the current time
badsec        LDX    #bspmt          ; output a bad seconds prompt
              JSR    outstrg         ;     "
              JSR    outcrlf         ; output carriage return and linefeed
              JMP    start           ; start all over again

t_prompt      FCC    "please enter the current hours minutes seconds:"
              FCB    $04             ; end of prompt
bhpmt         FCC    "Bad hour digits!"
              FCB    $04             ; end of bad hour prompt
bmpmt         FCC    "Bad minute digits!"
              FCB    $04             ; end of bad minute prompt
bspmt         FCC    "Bad second digits!"
              FCB    $04             ; end of bad second prompt
```

The resolution of a monitor screen is 24 by 80 characters. To display the current time in the center of the monitor screen, the program must perform the following four operations:

1. Output a form-feed character (ASCII code $0C) to clear the screen and move the cursor to the upper left corner of the screen.
2. Output 11 pairs of carriage return and line feed to move the cursor to the first column of the center line of the screen.
3. Output 36 space characters to move the cursor to the center of the screen.
4. Output the current time (hh:mm:ss) on the screen.

* The following two instructions clear the screen and move the cursor to the upper left
* corner of the screen

```
display       LDAA #$0C              ; output a form feed character
              JSR    outa            ;     "
```

* The following instruction sequence outputs 11 CR/LF pairs and moves the cursor to the
* first column of the line in the center of the screen

```
              LDAB #11               ; initialize the count for outputting 11 CR/LF pairs
here          JSR    outcrlf
              DECB
              BNE    here
              LDX    #spaces
              JSR    outstrg
```

* The following four instructions output hours digits

```
            LDAA  hh
            JSR   outlhlf       ; output the upper hours digit
            LDAA  hh
            JSR   outrhlf       ; output the lower hours digit

            LDAA  #colon        ; output the colon character
            JSR   outa          ;    "
```

* The following four instructions output minutes digits

```
            LDAA  mm
            JSR   outlhlf       ; output the upper minutes digit
            LDAA  mm
            JSR   outrhlf       ; output the lower minutes digit

            LDAA  #colon        ; output a colon character
            JSR   outa          ;    "
```

* The following four instructions output seconds digits

```
            LDAA  ss
            JSR   outlhlf       ; output the upper seconds digit
            LDAA  ss
            JSR   outrhlf       ; output the lower seconds digit

            SWI                 ; return to Buffalo monitor

spaces      FILL  $20,36        ; 36 space characters
            END
```

3.12 Summary

This chapter explored methods of processing data structures, subroutine call issues, and the I/O routines available in the EVB board. Extensive examples illustrated the concepts introduced here. Both arrays and matrices were discussed. Storage methods and address calculation for individual elements in an array or a matrix are among the most important issues related to the array and matrix processing. Two of the most common operations on arrays—sorting and searching—were illustrated in this chapter.

A programmer needs to use good programming style, especially when working on a large project. Good programming style is based on the concept of modularity—partitioning code into subroutines. Three important issues related to the subroutine call—parameter passing, result returning, and local variable allocation—were studied in detail in this chapter. Input/output is another important aspect of the development of any microprocessor-based product. With good I/O as feedback, problems encountered in product development can be quickly identified. Good I/O capability is also important in the regular use of microcontroller-based products. Use of the I/O routines provided by the EVB computer was illustrated in this chapter.

3.13 Glossary

Activation record   Another term for stack frame.

Array   An ordered set of elements of the same type. The elements of the array are arranged so that there is a zeroth, first, second, third, and so forth. An array may be one-, two-, or multi-dimensional.

Dequeue   A data structure in which elements can be added and removed at both ends.

Frame pointer   A pointer used to facilitate access to parameters in a stack frame.

Global memory   Memory that is available for all programs in a computer system.

Graph   A data structure that consists of a set of nodes and a set of arcs (or edges). Each arc in a graph is specified by a pair of nodes.

Linked list   A data structure that consists of linked nodes. Each node consists of two fields, an information field and a next address field. The information field holds the actual element on the list, and the next address field contains the address of the next node in the list.

Local variable   Temporary variables that exist only when a subroutine is called. They are used as loop indices, working buffers, etcetera. Local variables are often allocated in the system stack.

Matrix   A two-dimensional data structure that is organized into rows and columns. The elements of a matrix are of the same length and are accessed using their row and column numbers (i,j), where i is the row number and j is the column number.

Parameter passing   The process and mechanism of sending parameters from a caller to a subroutine, where they are used in computations; parameters can be sent to a subroutine using CPU registers, the stack, program memory, or global memory.

Program memory   The memory space occupied by user programs; can be used to pass parameters to a subroutine.

Pull   The operation that removes the top element from a stack data structure.

Push   The operation that adds a new element to the top of a stack data structure.

Queue   A data structure to which elements can be added at only one end and removed only from the other end. The end to which new elements can be added is called the tail of the queue, and the end from which elements can be removed is called the head of the queue.

Return address   The address of the instruction that immediately follows the subroutine call instruction (either JSR or BSR).

Stack   A last-in-first-out data structure whose elements can be accessed only from one end. A stack structure has a top and a bottom. A new item can be added only to the top, and the stack elements can be removed only from the top.

Stack frame   A region in the stack that holds incoming parameters, the subroutine return address, local variables, saved registers, etcetera.

String   A sequence of characters.

Subroutine   A sequence of instructions that can be called from various places in the program and will return to the caller after its execution. When a subroutine is called, the return address will be saved on the stack.

Subroutine call   The process of invoking the subroutine to perform the desired operations. The 68HC11 has BSR and JSR instructions for making subroutine calls.

Transpose   An operation that converts the rows of a matrix into columns and vice versa.

Tree   A finite nonempty set of elements in which one element is called the root and the remaining elements are partitioned into m (≥ 0) disjoint subsets (branches), each of which is itself a tree. Each element in a tree is called a node of the tree. A node that does not have any branches is called a *leaf.*

Vector   A vector is a unidimensional data structure in which each element is associated with an index *i*. The elements of a vector are of the same length.

3.14 Exercises

E3.1. Write instructions to set up the top six bytes of the stack as follows:

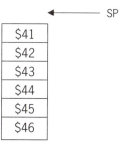

E3.2. Write instructions to load the tenth element of an array into the A accumulator. The base address of the array is ARRAY. The first element is associated with index 0.

E3.3. Write a program to find the median of the sorted array VECT with N 8-bit elements. The median of an array is the middle element of the array after the array is sorted. If n is odd, the median is arr$[\frac{n}{2}]$. If n is even, then the median is the average of elements arr$[\frac{n}{2} - 1]$ and arr$[\frac{n}{2}]$. Place the median in A. The array index starts from 0.

E3.4. Write a program to perform a binary search on a sorted array. The sorted array is located at ARR and has N 8-bit numbers. The key for searching is stored at KEY. The program leaves the address of the element that matches the key in X or leaves $FFFF in X if no element matches the key. Assume the array is sorted in ascending order. The binary search algorithm is as follows:

Step 1
Initialize *max* and *min* to $N - 1$ and 0, respectively.
Step 2
If *max* < *min*, then exit. No element matches the key.
Step 3
Let *mean* = (*min* + *max*)/2.

Step 4
If *key* equals ARR(*mean*), key is found; exit.
Step 5
If *key* < ARR(*mean*), then set *max* to *mean* − 1 and go to Step 2.
Step 6
If *key* > ARR(*mean*), then set *min* to *mean* + 1 and go to Step 2.

E3.5. Compute the addresses of the (2,3)th, (4,9)th, and (10,8)th elements of a 12 × 12 matrix of 8-bit elements. The given matrix is stored in column major order at $D000–$D08F.

E3.6. Write a program to multiply two 8 × 8 matrices, MA and MB, made up of 8-bit elements, and create an 8 × 8 matrix MC with 16-bit elements to hold the result. Assume that MA, MB, and MC are the starting addresses of the matrices and that all matrices are stored in row major order.

E3.7. Write a program to manipulate an *N* × *N* matrix so that

1. all elements on the diagonal are reset to 0

2. all elements in the upper right are divided by 2

3. all elements in the lower left are multiplied by 2.

The base address of the matrix is MA. MA is stored in memory in row major order.

E3.8. The label ARR is the starting address of an array of twenty-five 8-bit elements. Trace the following code sequence and describe what the subroutine TESTSUB does.

```
                LDX    #ARR
                LDAA   #25
                BSR    testsub
                .
                .
                .
testsub   PSHA
          DECA
          PSHX
          LDAB  0,X
          INX
again     CMPB  0,X
          BLE   next
          LDAB  0,X
next      INX
          DECA
          BNE   again
          PULX
          PULA
          RTS
```

E3.9. Write a subroutine that can generate a delay of 1 to 60 seconds and a program to call the subroutine. The program writes a 1 to the memory location at $1004 and calls the subroutine to create a delay of 1 second; it then writes a 2 to the same location and calls the subroutine to delay for 2 seconds. The same operation is repeated until the delay is 60 seconds, and then the program starts all over again. Repeat this operation forever. The delay in seconds should be passed to the subroutine in the A accumulator.

E3.10. Draw the stack frame after the execution of the eighth instruction in the following instruction sequence:

```
address   instruction

          LDAB #$10
          LDAA #$20
          PSHA
          LDX #0000
          PSHX
$C040     BSR XYZ
             .
             .
             .

(1) XYZ   PSHB
(2)       PSHA
(3)       PSHX
(4)       TSX
(5)       XGDX
(6)       SUBD #10
(7)       XGDX
(8)       TXS
             .
             .
             .
```

E3.11. Draw the stack frame and enter the value of each stack slot (if it is known) at the end of the following instruction sequence:

```
          DES
          DES
          CLRB
          LDAA  #10
          PSHA
          LDAA  #$C0
          PSHA
          LDX   #$D000
          PSHX
          BSR   SUB1
             .
             .
             .
SUB1      PSHB
          PSHY
          TSX
          XGDX
          SUBD #12
          XGDX
          TXS
             .
             .
             .
```

E3.12. Given the following instruction sequence,

```
JSR  SUB2
FDB  N,M,K,P,Q
```

where N,M,K,P, and Q are addresses of variables to be passed to the subroutine SUB2, write an instruction sequence to load the value of variable M into A as part of subroutine SUB2.

E3.13. Write a routine to concatenate two strings P and Q to form a new string R, and also write the instruction sequence to call this subroutine. P, Q, and R are terminated by the end-of-transmission (EOT) character. P, Q, and R are the starting addresses of these three strings and are passed to the subroutine in the stack. The subroutine will output the strings P, Q, and R and the following three messages on the screen:

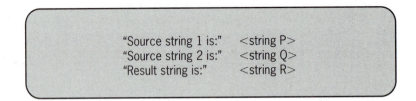

E3.14. Write a subroutine to compare two strings. The starting addresses of these two strings are passed in the stack. If the strings are equal, the subroutine returns a 1 to the caller in accumulator A. Otherwise, it returns a 0.

E3.15. Convert the character and word count program in Example 3.7 into a subroutine. The starting address of the string is passed in the stack. The caller of this subroutine will allocate two bytes in the stack for returning the character and word counts. The character count will be below the word count in the stack.

E3.16. Convert the word-matching program in Example 3.8 into a subroutine. Parameters are passed to the subroutine in the stack. The starting address of the string is pushed into the stack first, followed by the starting address of the word to be searched. If the word is found, return a 1 in A. Otherwise, return a 0 in A.

E3.17. In the following instruction sequence, *arr* is the starting address of an array with *n* 8-bit elements, and *i* and *j* are indices of two array elements.

```
address    instruction

           LDX  #$1000
           LDY  #$0100
           LDAA #10
           LDAB #$24
           JSR  swap
$c024      FDB  arr
           FCB  n,i,j
```

The subroutine swap has the following instructions at its entry point:

```
swap       PSHX
           PSHY
```

        PSHB
        PSHA
        TSX

    a. Draw the stack frame after the instruction TSX in the subroutine *swap* has been executed.

    b. Add instructions to place the array elements arr(*i*) and arr(*j*) into accumulators A and B, respectively.

3.15 Lab Exercises and Assignments

L3.1. Write a subroutine that finds and returns the maximum odd element of an array of *N* 8-bit integers. The base address of the array and the array count (the number of elements in the array) are passed to this subroutine in index register X and accumulator B, respectively. The result should be returned in accumulator B. A 0 should be returned if there is no odd element in the array. Also write a main program to test this subroutine.

L3.2. Modify the bubble sort subroutine so that parameters are passed in the stack, with the array count pushed onto the stack before the array base address. The elements of the array are 8-bit integers. Write a main program to test the new version of the bubble sort subroutine.

L3.3. Write a subroutine to convert the hexadecimal number in accumulator A to two BCD digits and return the result in A. The number to be converted is assumed to be less than 100. Also write a main program to test your subroutine. Include the following steps.

    1. Call the EVB I/O routines OUTLHLF and OUTRHLF to output the hexadecimal number in accumulator A.

    2. Call the EVB I/O routine OUTCRLF to output the carriage return and line feed pair.

    3. Call the hex-to-BCD subroutine to perform the conversion.

    4. Call the OUTLHLF and OUTRHLF routines to output the result.

    5. Call the EVB I/O routine OUTCRLF to output the carriage return and line feed pair.

    6. Test the subroutine using an array of 10 integer numbers.

L3.4. Write a subroutine to swap the first row of a matrix with the last row, the second row with the second-last row, etcetera. The matrix is *N* by *M* and is stored in row-major order. The starting address of the matrix and the matrix dimensions (*N* and *M*) are passed to this routine in the stack in this order: matrix base address, *N*, *M* (that is, the matrix base address is pushed onto the stack first, followed by *N* and *M*). The matrix elements are 8-bit. Write a main program to test this routine, including the following features:

    1. Call the EVB I/O routine OUTSTRG to output the message "The original matrix is as follows:" and call the routine OUTCRLF to move the screen cursor to the next line.

    2. Call the EVB I/O routines OUTLHLF and OUTRHLF to output the matrix elements, one row per line. Elements should be separated by one space.

3. Call the matrix swap subroutine to swap the rows of the matrix.

4. Call the I/O routine OUTSTRG to output the message "The swapped matrix is as follows:" and call the routine OUTCRLF to move the screen cursor to the next line.

5. Call the EVB I/O routines OUTLHLF and OUTRHLF to output the matrix elements, one row per line.

L3.5. Write a subroutine that can count the number of characters and words contained in a string. The string is terminated by an EOT character, and words are separated by spaces. All printable characters are legal and can be part of a word. The following parameters should be passed to this subroutine in the stack:

> P, the starting address of the string to be checked
>
> ccnt, the address of the memory location to hold the character count
>
> wcnt, the address of the memory location to hold the word count

Parameters are pushed onto the stack in the order P, ccnt, and wcnt. Each parameter occupies two bytes. The subroutine will output the string, character count, and word count to the screen before returning to the caller. Also write a short program to test this routine.

4

OPERATION
MODES
AND
MEMORY
EXPANSION

4.1 Objectives

After completing this chapter, you should be able to

- set up the 68HC11 operation mode
- make memory space assignments and design an address decoder
- add external memory to the 68HC11
- perform timing analysis for a memory system

4.2 Introduction

The 68HC11 is a family of microcontrollers developed in 1986. Members of the 68HC11 family have the same instruction set and addressing modes but differ in the size of their on-chip memories and in their I/O functions. Each member is named by appending the part number to the 68HC-prefix, for example, 68HC11A1, 68HC11E9, etcetera. The A series parts (part numbers A8, A1, and A0) are the foundation of the 68HC11 family. Table 4.1 summarizes the members of the 68HC11 family at the time of this writing.

Operation modes, on-chip memories, and external memory expansion methods will be covered in this chapter. The discussion will be based principally on the 68HC11A8.

4.3 A Summary of the 68HC11 Signal Pins

The block diagram and signal pins of the 68HC11A8 are shown in Figure 1.2. The 68HC11A8 is available in the 48-pin dual in-line package (DIP), 52-pin plastic leaded chip carrier (PLCC), and 64-pin quad flat pack (QFP) versions. The 68HC11 has five I/O ports that can interface to I/O devices directly. An I/O port consists of registers and I/O pins. Registers allow the user to set operation parameters, perform data transfer, and check the status of I/O operations. I/O pins allow direct data transfer to and from an I/O device. Most signal pins serve multiple functions. The function of each I/O port will be discussed in Chapter 6.

4.4 The 68HC11 Operation Modes

The 68HC11 has four different operation modes: expanded, single-chip, special bootstrap, and special test. The expanded mode allows the user to add off-chip memory devices and I/O peripheral chips to the microcontroller, whereas these additions are not permitted in the single-chip mode. The special test mode is mainly used during Motorola's internal production testing. The bootstrap mode allows the user to load a program into the on-chip SRAM from the serial communication interface (SCI). The loaded program can perform any function the user wants. One such application is to program the on-chip EEPROM or EPROM.

4.4.1 Single-Chip Mode

In the single-chip mode, the 68HC11 functions without external address and data buses. Port B, port C, strobe A (STRA), and strobe B (STRB) pins are available for general-purpose parallel I/O. In this mode, all software needed to control the microcontroller is contained in internal memories.

4.4.2 Expanded Mode

In the expanded mode, the 68HC11 has the capability to access a 64KB address space. This total address space includes the same on-chip memory addresses used in the single-chip mode plus external peripheral and memory devices. The expansion bus consists of ports B and C and control signals AS and

Part number	EPROM	ROM	EEPROM	RAM	Description
68HC11A8	—	8K	512	256	8-bit A/D, 38 I/O, 3 MHz mux bus
68HC11A1	—	—	512	256	8-bit A/D, 38 I/O, 3 MHz mux bus
68HC11A0	—	—	—	256	8-bit A/D, 38 I/O, 3 MHz mux bus
68HC11D3	—	4K	—	192	32 I/O, 3 MHz bus
68HC11D0	—	—	—	192	14 I/O
68HC711D3	4K	—	—	192	32 I/O, 3 MHz mux bus
68HC811E2	—	—	2K	512	8-bit A/D, 38 I/O, 4 input capture, 3 MHz mux bus
68HC711E9	12K	—	512	512	8-bit A/D, 38 I/O, 4 input capture, 3 MHz mux bus
68HC11E9	—	12K	512	512	8-bit A/D, 38 I/O, 4 input capture, 3 MHz mux bus
68HC11E1	—	—	512	512	8-bit A/D, 38 I/O, 4 input capture, 3 MHz mux bus
68HC11E0	—	—	—	512	8-bit A/D, 38 I/O, 4 input capture, 3 MHz mux bus
68HC11E20	—	20K	512	768	8-bit A/D, 38 I/O, 4 input capture, 3 MHz mux bus
68HC11F1	—	—	512	1K	8-bit A/D, 30 I/O, 4 MHz non-mux bus
68HC11G5	—	16K	—	512	10-bit A/D, 4 PWMs, 66 I/O, 2 MHz non-mux bus
68HC11G7	—	24K	—	512	10-bit A/D, 4 PWMs, 66 I/O, 2 MHz non-mux bus
68HC711G7	16K	—	—	512	10-bit A/D, 4 PWMs, 66 I/O, 2 MHz non-mux bus
68HC11J6	—	16K	—	512	54 I/O, 2 MHz non-mux bus
68HC711J6	16K	—	—	512	54 I/O, 2 MHz non-mux bus
68HC11K4	—	24K	640	768	8-bit A/D, 4 PWMs, 62 I/O, 4 MHz non-mux bus
68HC711K4	24K	—	640	768	8-bit A/D, 4 PWMs, 62 I/O, 4 MHz non-mux bus
68HC11KA4	—	24K	640	768	8-bit A/D, 4 PWMs, 51 I/O, 4 MHz non-mux bus
68HC11K3	—	24K	—	768	8-bit A/D, 4 PWMs, 62 I/O, 4 MHz non-mux bus
68HC11K1	—	—	640	768	8-bit A/D, 4 PWMs, 62 I/O, 4 MHz non-mux bus
68HC11K0	—	—	—	768	8-bit A/D, 4 PWMs, 62 I/O, 4 MHz non-mux bus
68HC11L6	—	16K	512	512	8-bit A/D, 46 I/O, 3 MHz mux bus
68HC711L6	16K	—	512	512	8-bit A/D, 46 I/O, 3 MHz mux bus
68HC11L5	—	16K	—	512	8-bit A/D, 46 I/O, 3 MHz mux bus
68HC11L1	—	—	512	512	8-bit A/D, 46 I/O, 3 MHz mux bus
68HC11L0	—	—	—	512	8-bit A/D, 46 I/O, 3 MHz mux bus
68HC11M2	—	32K	—	1.25K	8-bit A/D, 16-bit math coprocessor, 62 I/O 4 MHz non-mux bus
68HC711M2	32K	—	—	1.25K	8-bit A/D, 16-bit math coprocessor, 62 I/O 4 MHz non-mux bus
68HC11N4	—	24K	640	768	8-bit A/D, 8-bit D/A, 6 PWMs, 16-bit math coprocessor, 62 I/O, 4 MHz non-mux bus
68HC711N4	24K	—	640	768	8-bit A/D, 8-bit D/A, 6 PWMs, 16-bit math coprocessor, 62 I/O, 4 MHz non-mux bus
68HC11P2	—	32K	640	1K	8-bit A/D, 62 I/O, 4 MHz non-mux bus, 2 additional SCI
68HC711P2	32K	—	640	1K	8-bit A/D, 62 I/O, 4 MHz non-mux bus, 2 additional SCI
68HC711E20	20K	—	—	768	expanded 711E9, 20K EPROM
68HC11ED0	—	—	—	512	SPI, SCI
68HC11G0	—	—	—	512	4 PWM, 8-ch 10-bit A/D, SPI, SCI
68HC11KA0	—	—	—	768	4 PWM, 8-ch 8-bit A/D, SPI, SCI
68HC11KA1	—	—	640	768	4 PWM, 8-ch 8-bit A/D, SPI, SCI
68HC11KA2	—	32K	640	1K	4 PWM, 8-ch 8-bit A/D, SPI, SCI
68HC711KA2	32K	—	640	1K	4 PWM, 8-ch 8-bit A/D, SPI, SCI
68HC11KA3	—	24K	—	768	4 PWM, 8-ch 8-bit A/D, SPI, SCI

Table 4.1 ■ MC68HC11 family members

R/$\overline{\text{W}}$. Port B functions as the upper eight address pins (A15–A8) and port C functions as the multiplexed data and low address pins (A7/D7, ..., A0/D0). The AS signal is used by the external memory system to latch the low address signals. All bus cycles, whether internal or external, execute at the E-clock frequency.

4.4.3 Special Bootstrap Mode

The 68HC11A8 has a 192-byte bootstrap ROM where a bootloader program is located. This ROM is enabled only if the 68HC11 is reset in the special bootstrap mode; it appears as internal memory space at locations $BF40–$BFFF. The bootloader program uses the serial communication interface (SCI) to read a 256-byte program into the on-chip RAM at locations $0000–$00FF. After the character for address $00FF is received, control is automatically passed to that program at location $0000. There are almost no limitations on the functions of the programs that can be loaded and executed through the bootstrap process.

The bootloaded program is often used to perform on-chip EEPROM or EPROM programming. The executable code or data to be programmed into the EEPROM or EPROM is frequently stored in a PC as a file. During the bootloading and programming process, the PC executes a terminal program to communicate with the bootloader and the bootloaded program to send the executable code. The user should create or get a copy of the PC-resident program in order to perform the bootload process or EEPROM and EPROM programming. Readers who are interested in using this technique to program the 68HC11 on-chip EEPROM or EPROM should refer to Motorola application notes AN1010 and AN1060.

4.4.4 Special Test Mode

The special test mode, a variation of the expanded multiplexed mode, is primarily used during Motorola's internal production testing.

4.4.5 Setting the Operation Mode

The 68HC11 is in the *reset state* when the voltage level on the $\overline{\text{RESET}}$ pin is low. The operation mode is set when the 68HC11 exits the reset state. On the rising edge of the $\overline{\text{RESET}}$ signal, the voltage levels of pins MODA and MODB are latched into the HPRIO register, which sets the operation mode of the 68HC11. The voltage levels of pins MODA and MODB and the corresponding operation modes are shown in Table 4.2.

Inputs		
MODB	MODA	**Mode description**
1	0	normal single-chip
1	1	normal expanded
0	0	special bootstrap
0	1	special test
1 = high voltage, 0 = low voltage		

Table 4.2 ■ Hardware mode-select summary

4.5 Memory Technology

Memory chips are a major component of many digital systems. Thus it is very important for a digital system designer to understand the characteristics of memory chips and know how to interface them to a microprocessor and microcontroller.

4.5.1 Memory Component Types

Semiconductor memories retain the binary information stored in their memory cells for varying amounts of time, depending on the type of memory they have. The different types of memory devices are therefore categorized according to their storage characteristics as volatile, nonvolatile, dynamic, or static.

VOLATILE AND NONVOLATILE MEMORIES

Volatile memories lose their stored information when power is removed. The internal transistor circuits that store the information in memories of this type require a constant supply voltage to operate. Nonvolatile memory retains stored information even when power to the memory is removed.

ROMs AND RAMs

There are two major types of memory: read-only memory (ROM) and random-access memory (RAM). Random-access memories are also called read-write memories because read and write accesses take roughly equal amount of time. Within these two major types, additional classifications reflect the operation of their memory devices. ROMs are nonvolatile, and RAMs are normally volatile. Readers can refer to Chapter 1 for a discussion of ROMs.

DYNAMIC AND STATIC MEMORIES

Semiconductor random-access memories are further classified as dynamic or static.

Dynamic memories are memory devices that require periodic refreshing of the stored information. *Refresh* is the process of restoring binary data stored in a particular memory location. The internal circuitry of dynamic memories uses a very simple capacitive charge storage circuit in which only one transistor and one capacitor are required to store one bit of information. The simplicity of the circuit allows a very large number of memory bits to be fabricated onto the semiconductor chip, but the penalty for the simple circuitry is that the capacitive charge "leaks." The time interval over which each memory location of a DRAM chip must be refreshed at least once in order to maintain its contents is called its *refresh period*. Refresh periods typically range from a few milliseconds to tens of milliseconds for present-day high-density DRAMs.

Static memories are designed to store binary information without needing periodic refreshes and require the use of more complicated storage circuitry for each bit. Four to six transistors are needed to store one bit of information. Thus, static memories store less information per unit area of semiconductor material than do dynamic memories.

Memory chip	Power of 2	Capacity (bits)
1KBit	2^{10}	1,024
4KBit	2^{12}	4,096
16KBit	2^{14}	16,384
64KBit	2^{16}	65,536
256KBit	2^{18}	262,144
1MBit	2^{20}	1,048,576
4MBit	2^{22}	4,194,304
16MBit	2^{24}	16,777,216

Table 4.3 ■ Semiconductor memory chip capacities

4.5.2 Memory Capacity and Organization

Memory devices are often labeled with a shorthand notation, such as $1M \times 1$ or $128K \times 8$, that indicates their storage capacity and organization. It is necessary to understand this notation in order to select the appropriate size and type of memory chip.

MEMORY CAPACITY

The capacity of a memory device is the total amount of information that the device can store. The capacity of semiconductor memories is often referred to as their *memory density*. Because they are binary in nature, semiconductor memories are always manufactured in powers of 2. For example, a 64KBit memory chip has 2^{16} memory cells of storage, for a total of 65,536 bits. A 1MBit memory chip has 2^{20} or 1,048,576 total memory bits of storage. Table 4.3 lists several common memory chip capacities.

MEMORY ORGANIZATION

Information on a single memory chip can be retrieved (read) or stored (written) one bit at a time or several bits at a time. The term *memory organization* describes the number of bits that can be written or read onto or from a memory chip during one input or output operation. For example, if only one bit can be stored or retrieved at a time, the organization is "by one," written $\times 1$. If eight bits can be read or written at a time, the organization is "by eight," written $\times 8$. Another common organization is by 4, which is referred to as a *nibble-organized memory*. A memory chip that is labeled $m \times n$ has m locations and each location has n bits. For example, a $128K \times 8$ DRAM chip has 128K different locations and each location has eight bits.

Example 4.1

Using the following memory chips, how many SRAM chips will be needed to build a 512KB, 16-bit memory system for a 16-bit microprocessor? The memory is to be designed so that 16-bit data can be accessed in one read/write operation.

    a. $256K \times 1$ SRAM

    b. $256K \times 4$ SRAM

c. 256K × 8 SRAM

d. 64K × 8 SRAM

Solution:

    a. If we use × 1 organization memory chips, 16 chips will be needed to build a 16-bit-wide memory system. Sixteen 256K × 1 SRAM chips have a capacity of 512KB.

    b. With × 4 organization memory chips, 4 chips will be needed to build a 16-bit-wide memory system. Four 256K × 4 SRAM chips have a capacity of 512KB.

    c. With × 8 organization memory chips, 2 chips will be required to build a 16-bit memory system. Two 256K × 8 SRAM chips have a capacity of 512KB.

    d. Two 64K × 8 SRAM chips have a capacity of 128KB, so 8 chips will be required to build a 512KB, 16-bit memory system.

4.5.3 Memory Addressing

Every type of memory chip has external pins or connections that are called *addresses*. *Addressing* involves the application of a unique combination of high and low logic levels to select a correspondingly unique memory location.

Small capacity memories can have a unique external connection or pin for each address line. However, as memory capacities grow, it becomes impractical to provide an external pin for each address line. For example, a 1M × 1 bit memory device requires 20 address lines to allow for unique selection of each memory bit. Because external connections must be kept to a minimum so that the package used for the memory chip can be as small as possible, a technique known as *multiplexing* was invented. Multiplexing keeps the external pin count to a minimum. It is a circuit technique that allows the same external connection to be used for two or more different signals that are differentiated according to the time at which the external connection is examined. The most common way to multiplex memory chip pins is to use half the number of address lines that are necessary for unique address selection. Address signals are then divided into *row* and *column* address signals. The memory chip looks for and latches the row address during the first part of the address cycle, then it looks for and latches the column address during the second part of the cycle. Two signal pins are associated with the multiplexed address operation. The *row address strobe*, RAS, is the signal that indicates that row address logic levels are being applied, while the *column address strobe*, CAS, is the signal that indicates that column address logic levels are being applied. Address multiplexing techniques have been used to reduce the pin count and package size in dynamic memory chips for a long time because DRAMs are used in such large quantities in so many digital systems. Other types of memory chips do not use address multiplexing.

4.6 External Memory Expansion for the 68HC11

In the expanded mode, the user can add external memory to the 68HC11. The port B pins are used as the upper address signals (A15–A8), while the port

C pins become the *time-multiplexed* lower address and data signals (A7/D7–A0/D0).

Several issues need to be dealt with when adding external memory chips to the 68HC11:

- address space assignment
- address decoding
- timing considerations

4.6.1 Memory Space Assignments

The 68HC11A8 supports 64KB of memory space. Part of this 64KB space is occupied by the on-chip memories and I/O registers listed in Table 4.4.

The on-chip SRAM and I/O registers can be repositioned to any 4KB page boundary in the 64KB memory map by programming the INIT register. The INIT register is located at $1030, and its contents are:

7	6	5	4	3	2	1	0	INIT
RAM3	RAM2	RAM1	RAM0	REG3	REG2	REG1	REG0	located at $1030

value after
RESET: 0   0   0   0   0   0   0   1

Four bits of the INIT register control the SRAM address, and four control the address of the internal registers:

- RAM3–RAM0: RAM map position

  These four bits, which specify the upper hexadecimal digit of the SRAM address, control the position of the SRAM in the memory map. By changing these bits, the SRAM can be repositioned to the beginning of any 4KB page in the memory map. After reset, these four bits are zeros ($0); thus, the SRAM is initially positioned at $0000–$00FF. If these four bits are written to ones ($F), the SRAM moves to $F000–$F0FF.

- REG3–REG0: 64-byte register block position

  These four bits, which specify the upper hexadecimal digit of the address for the 64-byte block of internal registers, control the position of these registers in the memory map. These bits can be changed to reposition the register block to the beginning of any 4KB page in the memory map. After reset, these four bits are 0001 ($1); therefore, the registers are initially positioned at $1000–$103F. If these four bits are written to ones ($F), the registers move to $F000–$F03F.

Address range	Allocation
$0000–$00FF	on-chip SRAM
$1000–$103F	I/O registers
$B600–$B7FF	on-chip EEPROM
$E000–$FFFF	on-chip ROM

Table 4.4 ■ 68HC11A8 on-chip memory map

The register INIT can be programmed only within the 64 E clock cycles after reset, so such programming is often done in the *reset handling routine.* When relocating the I/O registers, the user must make sure that two different devices are not mapped to the same location. However, no physical harm will be caused if two devices are mapped to the same location. If the on-chip ROM and RAM are mapped to the same area, RAM will take priority and ROM will become inaccessible. If the I/O registers are relocated so that they conflict with the RAM and/or ROM, then the I/O registers will take priority and the RAM and/or ROM at those locations become inaccessible. Similarly, if an internal resource conflicts with an external device, no harmful conflict results—data from the external device will not be applied to the internal data bus and cannot interfere with internal read operations.

Example 4.2

Remap the on-chip SRAM to $2000–$20FF and remap the I/O registers to $3000–$303F.

Solution: To remap the SRAM and I/O registers to the specified memory spaces, a value of $23 must be programmed into the INIT register. The following instruction sequence must be placed in the reset handling routine and be executed within 64 E clock cycles after the $\overline{\text{RESET}}$ signal goes to high:

```
SRAM    EQU  $20          ;this value will remap SRAM to $2000–$20FF
IOREG   EQU  $3           ;this value will remap IO regs to $3000–$303F
REMAP   EQU  SRAM+IOREG
INIT    EQU  $1030
        LDAB #REMAP
        STAB INIT
         .
         .
         .
```

The unoccupied memory space can be assigned to external memories. It is most convenient to allocate memory space in blocks of equal size, with each block comprising contiguous memory locations. The major consideration in memory space allocation is the number of locations in the memory chips. The number of memory locations in the memory chip is a power of 2, for example, 2K, 4K, 8K, 16K, and so on. The address decoder can be simplified if all the memory chips have the same capacity.

Example 4.3

Assign the 68HC11 memory space using a block size of 4KB.

Solution: The 64KB memory space can be divided into sixteen 4KB blocks. The address ranges of the blocks are as follows:

Block number	Address range
0	$0000–$0FFF
1	$1000–$1FFF
2	$2000–$2FFF
3	$3000–$3FFF
4	$4000–$4FFF
5	$5000–$5FFF
6	$6000–$6FFF
7	$7000–$7FFF
8	$8000–$8FFF
9	$9000–$9FFF
10	$A000–$AFFF
11	$B000–$BFFF
12	$C000–$CFFF
13	$D000–$DFFF
14	$E000–$EFFF
15	$F000–$FFFF

Example 4.4

Assign the 68HC11 memory space using a block size of 8KB.

Solution: The 64KB memory space can be divided into eight 8KB blocks. The address ranges of the blocks are:

Block number	Address range
0	$0000–$1FFF
1	$2000–$3FFF
2	$4000–$5FFF
3	$6000–$7FFF
4	$8000–$9FFF
5	$A000–$BFFF
6	$C000–$DFFF
7	$E000–$FFFF

4.6.2 Address-Decoding Methods

To transfer data correctly and protect the computer from damage, only one device at a time should be allowed to drive the data bus. The address decoder selects and enables one and only one data transfer device at a time. Eight-bit microprocessors and microcontrollers can use two address-decoding schemes: full and partial address decoding. A memory component is said to be *fully decoded* when each of its addressable locations responds to only a single address

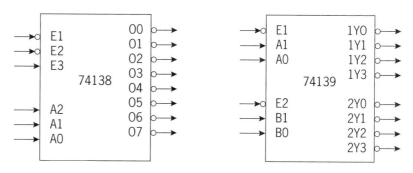

Figure 4.1 The 74138 and 74139 decoder chip pin layouts

on the system bus. A memory component is said to be *partially decoded* when each of its addressable locations responds to more than one address on the system bus.

The 74138 3-to-8 decoder chip and the 74139 2-to-4 decoder chip have been popular in address decoder design. The 74139 is a dual 2-to-4 decoder chip with two identical 2-to-4 decoders. The pin layouts of these two chips are shown in Figure 4.1. In these chips,

■ 00–07, 1Y0–1Y3, and 2Y0–2Y3 are active low decoder outputs.

■ E1, E2, and E3 are decoder enable inputs. E1 and E2 are active low, and E3 is active high.

■ A2–A0, A1–A0, and B1–B0 are address or select inputs. They are all active high.

Example 4.5

Use a full decoding scheme to design an address decoder for a computer that has the following address space assignments:

> SRAM1: $2000–$3FFF
> ROM1: $4000–$5FFF
> EEPROM: $6000–$7FFF
> SRAM2: $A000–$BFFF
> ROM2: $C000–$DFFF

Solution: Since each block of this memory module is 8KB, it will be very straightforward to use the full address decoding scheme to implement the address decoder. A 3-to-8 decoder such as the 74LS138 is perfect for this use (LS stands for low power Schottky technology). The three highest address bits of each module (listed below) are used to select the memory modules:

> SRAM1: 001
> ROM1: 010
> EEPROM: 011
> SRAM2: 101
> ROM2: 110

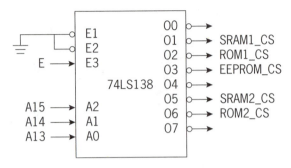

Figure 4.2   Address decoder design for Example 4.5

The address decoder for this computer system is shown in Figure 4.2.

The reader should also note that the E clock signal is used to control the E3 input. The lower eight address signals are invalid most of the time when the E clock is low and are valid when the E clock is high. Using the E clock as an enable input to the decoder makes sure that no decoder output is asserted when the address signals are not valid. The user can also tie the E3 signal to high. There are potential problems in this approach, however, because even if the microcontroller is not driving the address bus, the address inputs to the decoder may have valid values that will assert one of the decoder outputs and hence select one of the memory or peripheral chips. If this approach is taken, then memory and peripheral chips must decode the E clock signal directly to make sure they don't respond when the microcontroller is not performing an external bus cycle.

Example 4.6

You have been assigned to design a 68HC11-based product. This product will have 2KB of external EEPROM and 2KB of external SRAM. Use the partial decoding scheme to design a decoder for this product.

Solution: As the designer of this system, you have a lot of freedom to make the design decision. The inexpensive, off-the-shelf dual 2-to-4 74LS139 decoder can be used. The 2KB external EEPROM and SRAM chips have 11 address inputs. The lowest 11 address bits of the 68HC11 are connected directly to the corresponding pins of these two memory components. The highest 5 address signals are available for controlling the address decoder. One approach is to use the highest two address signals, A15 and A14, as the address inputs to the decoder. The resulting decoding circuitry is shown in Figure 4.3.

The inverter makes sure that the decoder does not enable any component when the E clock signal is low. The address space allocation is as follows:

EEPROM: $4000–$7FFF

SRAM: $8000–$BFFF

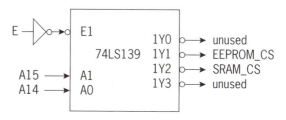

Figure 4.3 Address decoder for Example 4.6

Each component is allocated 16KB of address space. Since address signals A13–A11 are not used in address decoding, each location in the EEPROM or SRAM will respond to eight different addresses. This is an example of *partial decoding*. For example, the following addresses all select the same location in the EEPROM:

$4000, $4800, $5000, $5800, $6000, $6800, $7000, $7800

Partial decoding has the advantage of smaller and simpler decoder circuitry. If partial decoding is not used, this problem would require a 5-to-32 decoder, which is a large package that is not commercially available. However, partial decoding has a significant disadvantage: it prevents full use of the microprocessor's available memory space and produces difficulties if the memory system is expanded at a later time.

4.7 Bus Cycles

In the normal expanded mode, the 68HC11 can access external memories or I/O devices. It performs a read bus cycle to retrieve data from an external memory location or I/O device, and it performs a write bus cycle to output data. The timing requirements of both the 68HC11 bus cycles and the external devices must be met in order to make the data transfer successful.

4.7.1 Conventions of Timing Diagrams

A digital signal has two states, high and low, as shown in Figure 4.4. In Figure 4.4, the transitions from 1 to 0 and from 0 to 1 are instantaneous. In reality, however, a signal transition takes time. The time needed for a signal to go

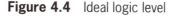

Figure 4.4 Ideal logic level

from 10% of the power supply voltage (V_{DD}) to 90% of the power supply voltage is called the *rise time,* and the time needed for a signal to drop from 90% of the power supply voltage to 10% of the power supply voltage is called the *fall time* (see Figure 4.5).

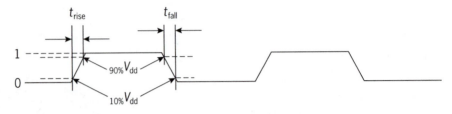

Figure 4.5   Real signal

A single signal is often represented as a set of line segments (for example, see Figure 4.6). The horizontal axis is time and the vertical axis is the magnitude (in volts) of the signal. Multiple signals of the same nature, such as address and data signals, are often grouped together and represented as two parallel lines with crossovers, as illustrated in Figure 4.7. The crossover represents the point when one or multiple signals change values.

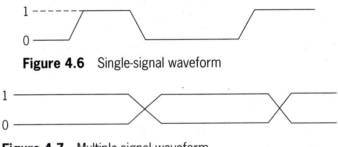

Figure 4.6   Single-signal waveform

Figure 4.7   Multiple-signal waveform

Sometimes a signal value is unknown because the signal is changing. Single and multiple unknown signals are represented by hatched areas in the timing diagram, as shown in Figure 4.8.

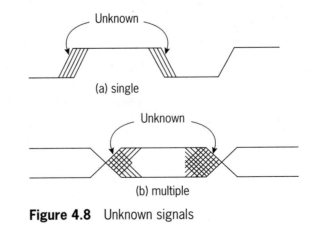

Figure 4.8   Unknown signals

Sometimes one or multiple signal lines are not driven (because their drivers are in a high impedance state) and hence cannot be received. An undriven signal is said to be *floating*. Single or multiple floating signals are represented by a value between high and low, as shown in Figure 4.9.

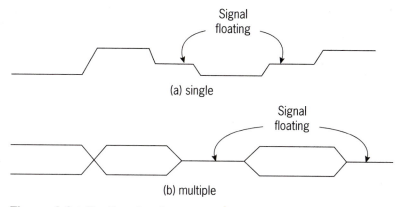

(a) single

(b) multiple

Figure 4.9 Floating signals

Sometimes there are causal relationships between two or more signals. The causal relationship is often represented by an arrow, as shown in Figure 4.10.

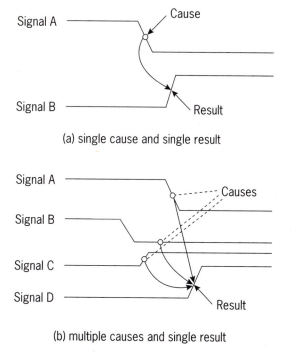

(a) single cause and single result

(b) multiple causes and single result

(*continues*)

Figure 4.10 Causal relationships between signals

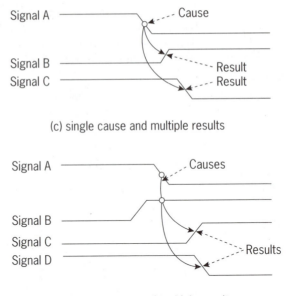

(c) single cause and multiple results

(d) multiple causes and multiple results

Figure 4.10   (*continued*)

4.7.2 The 68HC11 Bus Cycle Timing Diagram

The 68HC11 external bus cycle is controlled by the E clock signal and can be performed only in the expanded mode. There are 16 address and 8 data signals. Each read or write bus cycle takes one E clock cycle. A read bus cycle timing diagram is shown in Figure 4.11, and the corresponding timing parameters are shown in Table 4.5. The lower 8 address signals and the data signals are time-multiplexed. During the first half of the E clock cycle (E clock is low), the lower 8 address signals appear on port C pins. During the second half of the E clock cycle, data is driven on the same port C pins by external devices.

Most memory chips require address signals to be stable during the complete read/write bus cycle. However, the 68HC11 drives the lower 8 address signals only during the first half of the E clock cycle. The external memory system must latch the lower 8 address signals so that they stay stable during the whole bus cycle. The 68HC11 provides the address strobe (AS) signal for external devices to latch the address signals. The AS pulse is activated during the first half of the E clock cycle. Address signals should be latched by the falling (not rising) edge of the AS signal because they are not valid on the rising edge, as shown in the timing diagram. The low address signals become valid t_{ASL} ns before the falling edge of the AS signal and remain valid for t_{AHL} ns after the falling edge of the AS signal.

For convenience, Motorola measures all timing parameters using 20% and 70% of the power supply voltage (V_{DD}) as reference points. Many timing parameters are measured relative to the rising and falling edges of the E clock. The phrase "before the falling edge of the E clock" uses the time when the magnitude of the E clock is $0.7V_{DD}$ as a reference point. The phrase "after the falling edge of the E clock" uses the time when the magnitude of the E clock is

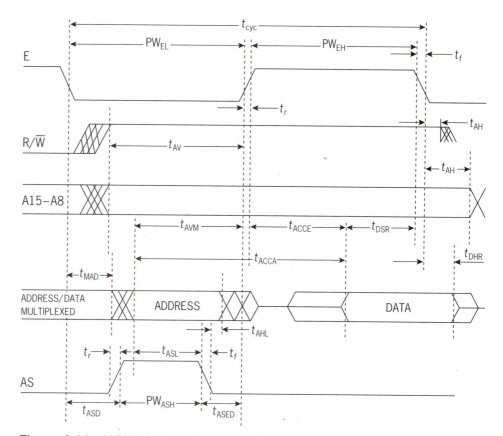

Figure 4.11 MC68HC11 read bus cycle timing diagram

$0.2V_{DD}$ as a reference point. Similarly, "before the rising edge of the E clock" refers to the time when the magnitude of the E clock is $0.2V_{DD}$, and "after the rising edge of the E clock" refers to the time when the magnitude of the E clock is $0.7V_{DD}$. For example, the E clock rise time t_r is the amount of time it takes the E clock to rise from $0.2V_{DD}$ to $0.7V_{DD}$. The E clock fall time t_f is the amount of time that it takes the E clock to drop from $0.7V_{DD}$ to $0.2V_{DD}$. The timing parameter t_{AV}, the *nonmultiplexed address to E rise* time, is the delay from the time when the nonmultiplexed address signals A15–A8 become valid until the E clock rises to $0.2V_{DD}$. In order to latch data correctly, the 68HC11 requires the external devices to drive valid data t_{DSR} ns before the E clock drops to $0.7V_{DD}$ and keep the data valid for t_{DHR} ns after the E clock has dropped to $0.2V_{DD}$. The timing parameter t_{DSR} is called the *read data setup time* and the timing parameter t_{DHR} is called the *read data hold time.* Other timing parameters are defined and measured in a similar way.

The timing diagram of a write bus cycle is shown in Figure 4.12. This timing diagram is very similar to the read bus cycle timing diagram except that the R/$\overline{\text{W}}$ signal goes low and data is driven by the 68HC11. The write data is available t_{DDW} ns after the rising edge of the E clock and will stay valid for at least t_{DHW} ns after the falling edge of the E clock.

Parameter	Symbol	1.0 MHz Min	1.0 MHz Max	2.0 MHz Min	2.0 MHz Max	3.0 MHz Min	3.0 MHz Max
Cycle time	t_{cyc}	1000	—	500	—	333	—
Pulse width, E low	PW_{EL}	477	—	227	—	146	—
Pulse width, E high	PW_{EH}	472	—	222	—	141	—
E and AS rise and fall time	t_r	—	20	—	20	—	20
	t_f	—	20	—	20	—	15
Address hold time	t_{AH}	95.5	—	33	—	26	—
Non-muxed address valid time to E rise	t_{AV}	281.5	—	94	—	54	—
Read data setup time	t_{DSR}	30	—	30	—	30	—
Read data hold time	t_{DHR}	0	145.5	0	83	0	51
Write data delay time	t_{DDW}	—	190.5	—	128	—	71
Write data hold time	t_{DHW}	95.5	—	33	—	26	—
Muxed address valid time to E rise	t_{AVM}	271.5	—	84	—	54	—
Muxed address valid to AS fall	t_{ASL}	151	—	26	—	13	—
Muxed address hold time	t_{AHL}	95.5	—	33	—	31	—
Delay time, E to AS rise	t_{ASD}	115.5	—	53	—	31	—
Pulse width, AS high	PW_{ASH}	221	—	96	—	63	—
Delay time, AS to E rise	t_{ASED}	115.5	—	53	—	31	—
MPU address access time	t_{ACCA}	744.5	—	307	—	196	—
MPU access time	t_{ACCE}	—	442	—	192	—	111
Muxed address delay	t_{MAD}	145.5	—	83	—	51	—

All times are in nanoseconds.
MPU stands for microprocessor unit.

Table 4.5 ■ MC68HC11A series read/write timing parameters

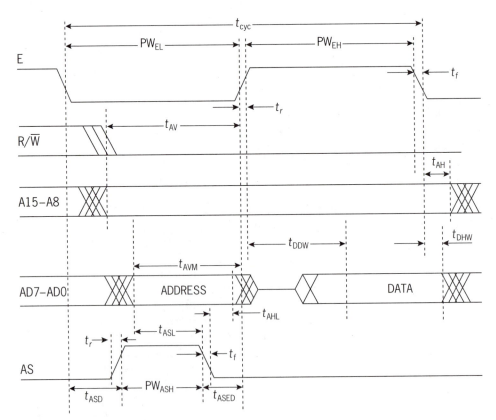

Figure 4.12 68HC11 write bus cycle timing diagram

4.7.3 Adding External Memory to the 68HC11

When adding external memory to the 68HC11, the user has at least five choices of memory technology: DRAM, SRAM, ROM, EPROM, and EEPROM. Since the 68HC11 has a relatively small address space, most external memory needs can be satisfied by one or two memory chips. DRAMs have the lowest per bit cost. However, the following two factors make it the most expensive memory technology for a digital system that needs only a very small amount of external memory:

1. *Periodic refresh requirements.* DRAM chips use capacitors to store information, but without periodic refresh, the information stored in the capacitors will leak away. A timer is needed to generate a periodic refresh request. However, the MPU may request access to the DRAM when the DRAM is requesting the refresh operation—then the memory system must delay the refresh operation because the 68HC11 bus cycle takes one clock cycle and cannot be delayed.

2. *Address multiplexing.* DRAM chips use address multiplexing techniques to reduce the number of address pins. This requires the memory system to include an address multiplexer, further increasing the cost of the memory system.

The control circuit designs for interfacing SRAM, ROM, EPROM, and EEPROM to a microprocessor or microcontroller are similar. In the following discussions we will use an 8KB SRAM HM6264A to illustrate the control logic design of an SRAM memory system.

The HM6264A SRAM

The pin layout of the Hitachi 8KB HM6264A is shown in Figure 4.13. The HM6264A has 13 address pins to address each of the 8192 locations on the chip, and it uses × 8 configuration, i.e., each location has 8 bits. There are two chip enable signals: $\overline{CS1}$ is active low, and CS2 is active high. There are also

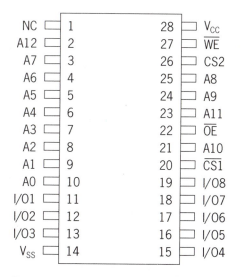

Figure 4.13 Hitachi HM6264A pin assignment

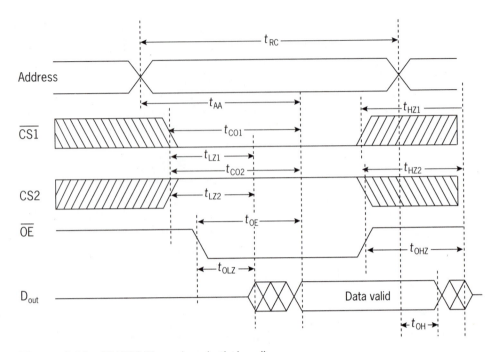

Figure 4.14   HM6264A read cycle timing diagram

active low write enable ($\overline{\text{WE}}$) and active low output enable ($\overline{\text{OE}}$) signals that control the data in and out of the chip. The output enable signal $\overline{\text{OE}}$ can be tied to low permanently.

The read cycle timing diagram of the HM6264A is shown in Figure 4.14. The HM6264A comes in three versions, which can be classified according to access time: one version (HM6264A-10) has a 100-ns access time, the second (HM6264A-12) has a 120-ns access time, and the third (HM6264A-15) has a 150-ns access time. The read cycle timing parameters for the HM6264A-10 and HM6264A-12 are shown in Table 4.6.

Read cycle time is the shortest separation allowed between two consecutive memory read accesses. A read access takes at least as long as the *read cycle time* (t_{RC}). For SRAMs, the read cycle time is as long as the *read access time*. There are four read access times in the HM6264A:

> *Address access time* (t_{AA}): the access time from the moment that the address becomes valid until valid data becomes available at the data pins (I/O_8–I/O_1), if all other control signals ($\overline{\text{CS1}}$, CS2, and $\overline{\text{OE}}$) are asserted
>
> *CS1 access time* (t_{CO1}): the access time from the moment that $\overline{\text{CS1}}$ becomes valid until valid data becomes available at the data pins, if all other control signals are asserted and a valid address is applied at the address pins
>
> *CS2 access time* (t_{CO2}): the delay time from the moment that CS2 becomes valid until valid data appears at the data pins, if all other control signals are asserted and a valid address is applied at the address pins

Parameter	Symbol	HM6264A-10		HM6264A-12	
		Min	Max	Min	Max
Read cycle time	t_{RC}	100	—	120	—
Address access time	t_{AA}	—	100	—	120
CS1 to output valid	t_{CO1}	—	100	—	120
CS2 to output valid	t_{CO2}	—	100	—	120
Output enable to output valid	t_{OE}	—	50	—	60
Output hold from address change	t_{OH}	10	—	10	—
Chip selection to output in low-Z	t_{LZ1}	10	—	10	—
(CS1 and CS2)	t_{LZ2}	10	—	10	—
Output enable to output in low Z	t_{OLZ}	5	—	5	—
Chip selection to output in high Z	t_{HZ1}	0	35	0	40
(CS1 and CS2)	t_{HZ2}	0	35	0	40
Output disable to output high-Z	t_{OHZ}	0	35	0	40

All times are in nanoseconds.

Table 4.6 ■ HM6264A read cycle timing parameters

OE access time (t_{OE}): the access time from the moment that $\overline{OE}$ becomes valid until valid data appears at the data pins, if all other control signals are asserted and a valid address is applied at the address pins.

The address data hold time (t_{OH}) is the length of time that the data from the I/O_8–I/O_1 pins stays valid after the address changes. There are two *chip deselection to output in high Z delay times* (t_{HZ1} and t_{HZ2}, corresponding to $\overline{CS1}$ and CS2, respectively). Another timing parameter of interest to the memory designer is *output disable to output in high Z* (t_{OHZ}).

There are two write cycle timing diagrams for the HM6264A: in one the $\overline{OE}$ signal is permanently tied to low, while in the other the $\overline{OE}$ signal is pulled to high during the write process. Since the next section contains an example in which the output enable signal is grounded permanently, only the write cycle timing diagram with $\overline{OE}$ grounded will be discussed here. This diagram is shown in Figure 4.15, and the corresponding timing parameters are given in Table 4.7. The HM6264A requires the write data to be valid at least t_{DW} ns before the write signal goes to high and to remain valid at least t_{DH} ns after the rising edge of the write signal. The write pulse width (t_{WP}) must be long enough for data to be written into the memory chip correctly.

Interfacing the HM6264A-12 to the 68HC11A1

We will use a 2 MHz (in terms of E clock frequency) 68HC11A1 to illustrate the design process involved in interfacing an HM6264A-12 to a 68HC11A1. In the 68HC11A1, addresses $1100–$B5FF and $B800–$FFBF are not occupied and can be assigned to the SRAM. The HM6264A-12 has a 120-ns access time, and address spaces $4000–$5FFF are assigned to the HM6264A in this example. The 68HC11 must be configured to operate in the expanded mode. Since the lower address signals (A7–A0) and data signals (D7–D0) are time-

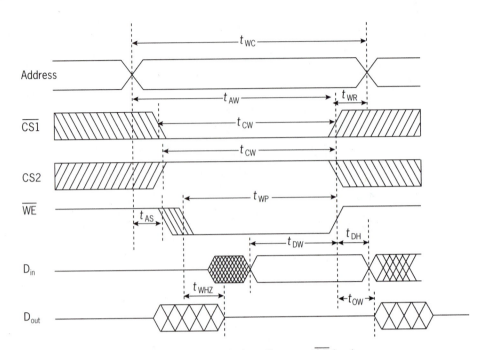

Figure 4.15   HM6264A write cycle timing diagram ($\overline{OE}$ low)

multiplexed, a latch is needed to latch the lower address signals so that all address signals can remain stable during the whole access cycle. The 8-bit latch 74F373 is selected for this purpose. An address decoder is needed to enable one and only one external device (memory or interface chip) to respond to the access request from the 68HC11—we will select the 74F138 3-to-8 decoder. Since the address signals are not valid for most of the time when the E clock signal is low, the E clock is used to qualify the write access. The circuit con-

Parameter	Symbol	HM6264A-10		HM6264A-12	
		Min	*Max*	*Min*	*Max*
Write cycle time	t_{WC}	100	—	120	—
Chip selection to end of write	t_{CW}	80	—	85	—
Address setup time	t_{AS}	0	—	0	—
Address valid to end of write	t_{AW}	80	—	85	—
Write pulse width	t_{WP}	60	—	70	—
Data to write time overlap	t_{DW}	40	—	50	—
Data hold from write time	t_{DH}	0	—	0	—
Write to output in high-Z	t_{WHZ}	0	35	0	40
Output active from end of write	t_{OW}	5	—	5	—
Write recovery time	t_{WR}	0	—	0	—

All times are in nanoseconds.

Table 4.7 ■ HM6264A write cycle timing parameters

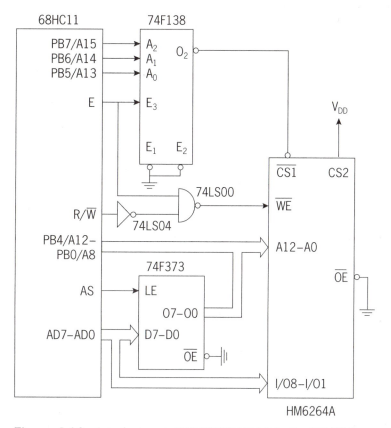

Figure 4.16 Interfacing an 8KB HM6264A-12 to the 68HC11

nection is shown in Figure 4.16. The output enable signal $\overline{OE}$ to the 74F373 is tied to the ground. The latch enable (LE) is connected to the AS signal from the 68HC11. A7–A0 are latched by the falling edge of the AS signal. The 74F373 has a latch delay time of 11.5 ns. The valid address will appear at pins O_7–O_0 11.5 ns after the falling edge of the LE (or AS) signal.

The upper three address signals A15–A13 are decoded by the 74F138 3-to-8 decoder, which has a propagation delay of 8 ns at room temperature. The lower address signals (A12–A0) are connected to the corresponding address inputs of the HM6264A chip.

This computer system may need to access other external devices, which can be selected using a decoder. However, it is very common for a user program to access a nonexistent memory location. The access of a nonexistent memory location is called a *bus error*. When a bus error occurs, a program cannot execute correctly. The design of the 68HC11 does not allow it to detect a bus error because each bus cycle takes exactly one E clock cycle, even if the bus cycle is accessing a nonexistent memory location. If we OR those decoder outputs that map to nonexistent memory locations and connect the result to either the $\overline{IRQ}$ or $\overline{XIRQ}$ input of the 68HC11, the MCU will be interrupted when a nonexistent location is accessed. The user can then identify the problem from the external interrupt signal. Interrupts will be discussed in Chapter 5.

The R/$\overline{\text{W}}$ signal is inverted before being NANDed with the E clock signal. The output of the NAND gate becomes the $\overline{\text{WE}}$ input signal to the HM6264A. NANDing the E and $\overline{\text{W}}$ together allows write access to the SRAM only when the address is valid. The output enable input $\overline{\text{OE}}$ is grounded. The HM6264A will output data only when the $\overline{\text{WE}}$ signal is high. The maximum low-to-high and high-to-low propagation delays of both the 74LS00 and the 74LS04 are 15 ns. In addition, the $\overline{\text{O}}_2$ output may become valid only when the E clock is high because the E3 input to the 74F138 is connected to the E clock signal from the 68HC11.

Another way to connect the enable inputs of the 74F138 and the HM6264A is as follows:

1. Tie the $\overline{\text{E1}}$ and $\overline{\text{E2}}$ inputs to the 74F138 to low and pull E3 to high using a pull-up resistor.
2. Connect the $\overline{\text{O}}_2$ output of the 74F138 to the $\overline{\text{CS1}}$ input of the HM6264A, and connect the E clock of the 68HC11 to the CS2 input of the HM6264A.

This circuit connection can achieve the same goal, and the timing verification is similar to that of the previous method.

Read Access Timing Analysis   To facilitate the read timing analysis, the read cycle timing diagram of the SRAM is overlapped with the 68HC11 read bus cycle timing diagram, as shown in Figure 4.17. During the read access to the HM6264A, control signals CS2 and $\overline{\text{OE}}$ are asserted permanently. The higher address inputs (A12–A8) to the HM6264A become valid 94 ns before the rising edge of the E clock signal. The lower eight address signals are latched by the falling edge of the AS signal, which is valid 53 ns before the rising edge of E. Since the 74F373 latch has a latch delay time of 11.5 ns, A7–A0 will become valid to the HM6264A 41.5 ns before the rising edge of E. Since the address decoder is qualified by the E clock signal, the chip select signal $\overline{\text{CS1}}$ is asserted (goes low) 8 ns after the rising edge of the E clock signal. Because it is the last asserted control signal in a read access, the $\overline{\text{CS1}}$ determines the data setup time to the 68HC11. The data from the HM6264A will become valid 128 ns (8 ns + 120 ns) after the rising edge of the E clock signal or 94 ns (PW_{EH} − 128 ns = 222 ns − 128 ns) before the falling edge of the E clock signal. The data setup time requirement is met because the 68HC11 requires only 30 ns.

The 68HC11 drives the high address signals (A15–A8) for 33 ns (t_{AH}) after the falling edge of the E clock signal. The lower address latch 74F373 does not latch a new value until 174 ns (227 − 53) after the falling edge of the E clock. Therefore the higher address signals determine the actual value of t_{OH}. The $\overline{\text{CS1}}$ is the only control signal that becomes invalid, and it goes high 8 ns after the falling edge of the E clock signal. The data hold time is the smaller of the following two values:

1. The output hold from address change time (t_{OH}). The HM6264A holds data for 10 ns after the address input becomes invalid, which occurs 33 ns after the E clock goes high. Therefore, the data hold for 43 ns (10 ns + 33 ns) after the E clock goes low.
2. Chip selection ($\overline{\text{CS1}}$ in this example) to output in high impedance time (t_{HZ1}): 8–48 ns.

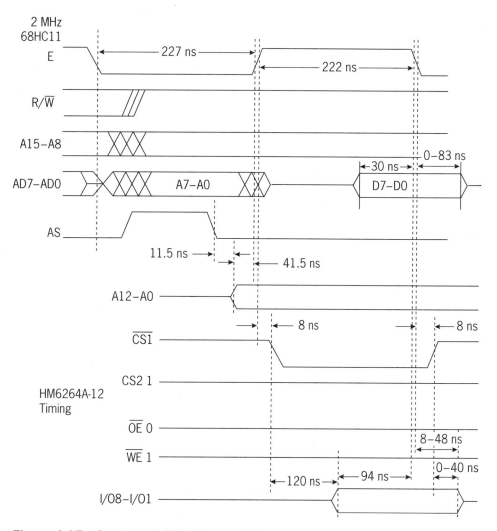

Figure 4.17 Overlapped 68HC11 and HM6264A-12 read timing diagrams

The HM6264A therefore provides only an 8–43 ns data hold time during a read access. The 68HC11 requires a data hold time from 0 to 83 ns (this range is due to IC fabrication process variations). The data hold time requirement is thus violated. However, the 68HC11 does not drive the multiplexed address/data bus until 138 ns after the falling edge of the E clock cycle, and the capacitance of the data bus will hold the data from memory for a short time and thus satisfy the data hold time requirement. The following analysis should give the reader an idea of the duration over which the data bus capacitance can hold the data at a valid level.

On a printed circuit board, each pin of the multiplexed address/data bus and the ground plane form a capacitor. After the memory chip stops driving the data bus, the charge across the capacitor leaks away via

1. input current into the 68HC11 data pin (on the order of 10 μA)
2. input leakage into the memory chip (on the order of 2 μA)

3. other leakage paths on the printed circuit board, depending on the system configuration

The time that it takes the capacitor voltage to degrade from high to low can be estimated as follows. Let

ΔV = the voltage change required for the data bus signal to change its value from 1 to 0. For example, for a 5V power supply, ΔV is equal to 2.5V.

Δt = the time that it takes the voltage across the capacitor to drop by ΔV

I = the total leakage current

C = the capacitance of the data bus on the printed circuit board. C varies with the printed circuit board but is in the range of 20 pF per foot. Usually the length of the data bus of a 68HC11 circuit is shorter than 1 foot. However, we will use this value in our calculation.

Then the elapsed time before the data bus signal degrades to an invalid level can be estimated by the following equation:

$$\Delta t \approx C \Delta V \div I$$

By substituting values into the equation above (and ignoring the third source of leakage), we obtain

$$\Delta t \approx C \Delta V \div I \approx 20 \text{ pF} \times 2.5 \text{V} \div 12 \text{ } \mu\text{A} \approx 4 \text{ } \mu\text{s}$$

Although the equation above is oversimplified, it does give us some idea about the order of the time over which the charge across the data bus capacitor will hold after the memory chips stop driving the data bus. Suppose the other leaking paths have ten times the current leakage (which is very likely) and the data bus length is 6 inches—then Δt becomes 200 ns.

Write Access Timing Analysis   As we did for the read cycle timing analysis, we overlapped the write access timing diagram with the 68HC11 write bus cycle timing diagram in Figure 4.18. The $R/\overline{W}$ signal becomes stable to the NAND gate input 79 (94 − 15) ns before the E clock rising edge. Because of the NAND gate propagation delay, the write enable signal $\overline{WE}$ is valid 15 ns after the rising edge of the E clock and becomes invalid 15 ns after the falling edge of the E clock. The chip enable signal $\overline{CS1}$ is valid 8 ns after the rising edge of the E clock signal. The write pulse is a delayed version of the E pulse and has a width equal to 222 ns. The SRAM requires the write pulse width to be at least 70 ns. The 68HC11 drives the output data 128 ns after the rising edge of the E clock and hence provides 109 ns ($\text{PW}_{\text{EH}} - t_{\text{DDW}}$ + delay of 74LS00) of data setup time, satisfying the 50-ns requirement. The 68HC11 holds the data valid for 33 ns after the falling edge of the E clock. The HM6264A-12 requires 0 ns of data hold time and hence is satisfied. A summary of the write access timing analysis is shown in Table 4.8.

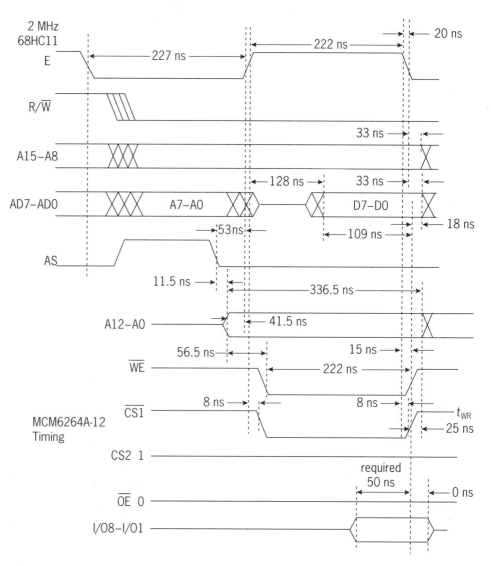

Figure 4.18 Overlapped 68HC11 and HM6264A-12 write cycle timing diagrams

Parameter	Required time (ns)	Actual time (ns)
Write cycle time	≥ 120	336.5
Write pulse width	≥ 70	222
Address setup time	≥ 0	76.5
Address valid to end of write	≥ 85	318.5
Data valid to end of write	≥ 50	109
Data hold time	≥ 0	18
Write recovery time	≥ 0	24

Table 4.8 ■ Summary of write cycle timing analysis

Example 4.7

Calculate the data setup and hold times provided by the 68HC11 to the HM6264A SRAM during the write access.

Solution: The following analysis is based on the 2 MHz E clock signal. The 68HC11 drives the data pins 128 ns (t_{DDW}) after the rising edge of the E clock signal. The high pulse width of the E clock (from 70% of peak level to 70% of peak level) is 222 ns (PW_{EH}). The $\overline{WE}$ signal will become invalid 15 ns after the falling edge of the E clock because the propagation delay of the 74LS00 NAND gate is 15 ns. The write data setup time for SRAM is relative to the rising edge (20% of V_{dd}) of the $\overline{WE}$ signal and has a value given by the following equation:

$$t_{DW} = PW_{EH} - t_{DDW} + \text{delay of 74LS00}$$
$$= 222 - 128 + 15$$
$$= 109 \text{ ns}$$

The data hold time is relative to the rising edge of the $\overline{WE}$ signal (70% of V_{dd}). The $\overline{WE}$ signal becomes high 15 ns after the falling edge of the E clock, and the data holds for 33 ns after the falling edge of the E clock. Therefore, the data hold time is the difference of these two values: 33 ns − 15 ns = 18 ns.

Example 4.8

Compute the remaining timing parameters for the write access for the memory system in Figure 4.16.

Solution: The values of the remaining timing parameters are calculated as follows (readers should refer to the timing diagram in Figure 4.18):

■ Write cycle time (t_{WC}). Write cycle time is the period during which the address inputs to the HM6264A are valid. Lower address inputs (A7–A0) to the HM6264A become valid later than the upper address inputs (A12–A8). Lower address signals become valid 41.5 ns before the rising edge of the E clock. The upper address signals remain valid for 33 ns after the falling edge of the E clock signal. The lower address signals will remain valid for at least 174 ns ($PW_{EL} - t_{ASED} = 227$ ns − 53 ns) after the E clock's falling edge until the AS signal latches a new value into the address latch. The rise time and the fall time of the E clock are 20 ns. Therefore, the actual maximal write cycle time is calculated as follows and is shown in Figure 4.19:

$$t_{WC} = 41.5 \text{ ns} + t_r + PW_{EH} + t_f + t_{AH}$$
$$= 41.5 \text{ ns} + 20 \text{ ns} + 222 \text{ ns} + 20 \text{ ns} + 33 \text{ ns}$$
$$= 336.5 \text{ ns}$$

■ Write pulse width (t_{WP}). Since the write pulse is generated by NAND-ing the complement of the R/$\overline{W}$ signal and the E clock signals, the duration of the write enable signal (asserted low) will be determined

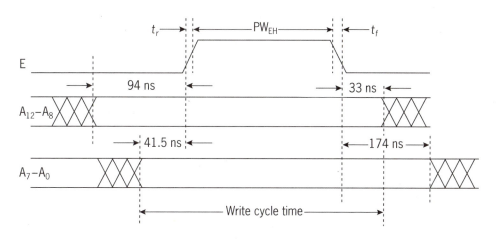

Figure 4.19 Calculation of write cycle time

by the shorter of these two signals. Since the E clock pulse (high period) is shorter than the R/$\overline{\text{W}}$ signal, the write pulse will be equal to the E clock high period, which is 222 ns.

■ Address setup time (t_{AS}). The address setup time is calculated from the moment when all address inputs to the HM6264A become valid until the moment the write enable signal goes low. As we have analyzed before, the address inputs become valid 41.5 ns before the rising edge of the E clock. The write enable signal becomes valid 15 ns after the rising edge of the E clock because the NAND gate has a delay of 15 ns. Therefore, the maximal address setup time can be calculated from the following expression:

$$t_{AS} = 41.5 \text{ ns} + t_r + 15 \text{ ns}$$
$$= 41.5 \text{ ns} + 20 \text{ ns} + 15 \text{ ns}$$
$$= 76.5 \text{ ns}$$

This calculation is illustrated in Figure 4.20.

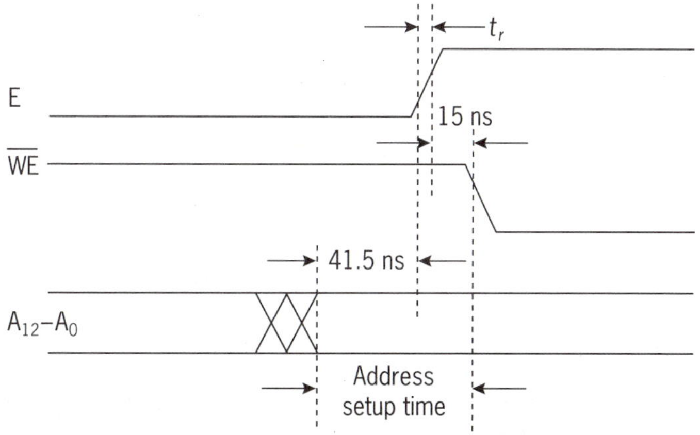

Figure 4.20 Calculation of the address setup time

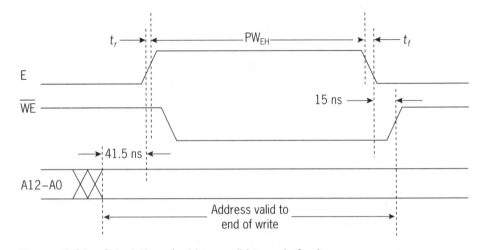

Figure 4.21   Calculation of address valid to end of write

■ Address valid to end of write (t_{AW}). This parameter is calculated from the moment when all address inputs are valid until the write enable signal becomes invalid ($\geq 0.7\ V_{DD}$) to the HM6264A. It is the sum of the following terms:

  1. latest address signals (A7–A0) valid to rising edge of the E clock delay (41.5 ns)

  2. rise time of the E clock (a maximum of 20 ns)

  3. E clock pulse high period (222 ns)

  4. fall time of E clock (20 ns)

  5. propagation delay of NAND gate (15 ns)

Therefore, the actual value of t_{AW} is 318.5 ns. The calculation process is shown in Figure 4.21.

■ Write recovery time (t_{WR}). Write recovery time is the time delay from the moment that the earliest chip select signal ($\overline{CS1}$ in this example) becomes invalid until the moment that the address signals become invalid. Since the $\overline{CS1}$ signal starts to go high when the E clock begins to fall, the parameter t_{WR} can be calculated from the following expression:

$$t_{WR} = t_{AH} - \text{propagation delay of the decoder}$$
$$= 33\ \text{ns} - 8\ \text{ns}$$
$$= 25\ \text{ns}$$

This calculation is shown in Figure 4.22.

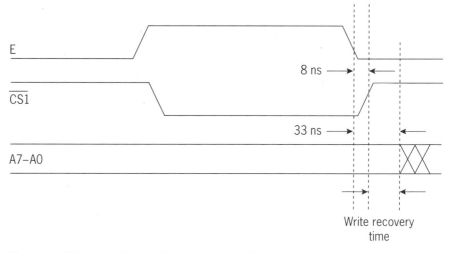

Figure 4.22   Calculation of write recovery time

4.8 Summary

This chapter has thoroughly explored the 68HC11 operation modes, memory component characteristics, and memory interfacing issues. The most important issue in memory interfacing is to match the timing requirements of the microprocessor and those of the memory components. Examples and numerous diagrams have helped to explain the design and analysis process.

4.9 Glossary

Address access time   The amount of time it takes for a memory component to send out valid data to the external data pins after address signals have been applied (assuming that all other control signals have been asserted).

Addressing   The application of a unique combination of high and low logic levels to select a corresponding unique memory location.

Address multiplexing   A technique that allows the same address pin to carry different address signals at different times; used mainly by DRAM technology. Address multiplexing can dramatically reduce the number of address pins required by DRAM chips and reduce the size of the memory chip package.

Bootstrap mode   The operation mode in which the 68HC11 will start to execute the bootloader program residing at $BF40–$BFFF after it gets out of the reset state. The bootloader program will use the serial communication interface to read a 256-byte program into the on-chip RAM at locations $0000–$00FF. After the last byte is received, control is passed to that program at location $0000.

Bus cycle timing diagram   A diagram that describes the transitions of all the involved signals during a read or write operation.

Column address strobe ($\overline{\text{CAS}}$)   The signal used by DRAM chips to indicate that column address logic levels are applied to the address input pins.

Data hold time   The length of time over which the data must remain stable after the edge of the control signal that latches the data arrives.

Data setup time   The amount of time over which the data must remain stable before the edge of the control signal that latches the data arrives.

Dynamic memories   Memory devices that require periodic refreshing of the stored information, even when power is on.

Expanded mode   The operation mode in which the 68HC11 can access external memory components by sending out address signals. A 64KB memory space is available in this mode.

Fall time   The amount of time a digital signal takes to go from logic high to logic low.

Floating signal   An undriven signal.

Memory capacity   The total amount of information that a memory device can store; also called memory density.

Memory organization   A description of the number of bits that can be written into or read from a memory chip during a read or write operation.

Nonvolatile memory   Memory that retains stored information even when power to the memory is removed.

Refresh   An operation performed on dynamic memories in order to retain the stored information during normal operation.

Refresh period   The time interval within which each location of a DRAM chip must be refreshed at least once in order to retain its stored information.

Reset state   The state in which the voltage level of the $\overline{\text{RESET}}$ pin of the 68HC11 is low. In this state, a default value is established for most on-chip registers, including the program counter. The operation mode is established when the 68HC11 exits the reset state.

Rise time   The amount of time a digital signal takes to go from logic low to logic high.

Row address strobe ($\overline{\text{RAS}}$)   The signal used by DRAM chips to indicate that row address logic levels are applied to the address input pins.

Reset handling routine   The routine that will be executed when the microcontroller or microprocessor gets out of the reset state.

Single-chip mode   The operation mode in which the 68HC11 functions without external address and data buses.

Special test mode   The 68HC11 operation mode used primarily during Motorola's internal production testing.

Static memories   Memory devices that do not require periodic refreshing in order to retain the stored information as long as power is applied.

Volatile memory   Semiconductor memory that loses its stored information when power is removed.

4.10 Exercises

E4.1. Can the user change the 68HC11 operation mode after the $\overline{\text{RESET}}$ signal goes high?

E4.2. Write a byte into the INIT register to remap the SRAM and registers blocks to $4000–$40FF and $C000–$C03F, respectively.

E4.3. How many memory chips are required to build a 1MB, 32-bit wide memory system using the following SRAM chips?

    a. 256K × 1 SRAM chips

    b. 256K × 4 SRAM chips

    c. 64K × 8 SRAM chips

    d. 128K × 16 SRAM chips

E4.4. Refer to Example 4.6. Find the addresses that will select the same location as address $5080.

E4.5. Design an address decoder for a 68HC11-based product containing the following memory modules:

    ROM1: 4KB

    SRAM1: 4KB

    ROM2: 4KB

The reset vector is stored in ROM1. Assign the address space so that addresses $FFFE and $FFFF are covered by ROM1.

E4.6. Design the control circuitry to interface the Am27C64 EPROM to the 2 MHz 68HC11A1. The Am27C64 is a 8192 × 8-Bit CMOS EPROM. The pin assignment of this chip is shown in Figure 4E.1, and the normal read access timing diagram and timing parameters are given in Figure 4E.2 and Table 4E.1. Choose the Am27C64-120, which has a 120-ns access time. Perform read access timing analysis to verify that all timing requirements are satisfied.

E4.7. Suppose that the address inputs A2–A0 of the 74138 decoder are connected in order to the address outputs A15–A13 of the 68HC11 and that the E3 input is connected to A12. E1 and E2 are tied to ground permanently. Determine the address ranges controlled by O_0–O_7.

E4.8. Referring to Figure 4.11, identify the intervals during which signals A15–A8 and A7/D7–A0/D0 are unknown and in high impedance state.

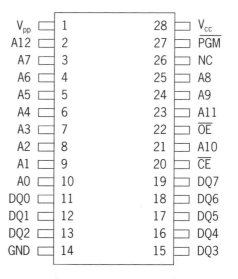

Figure 4E.1 Am27C64 EPROM pin assignment

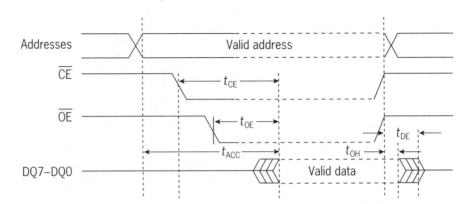

Figure 4E.2 Am27C64 read timing diagram

Parameter	Symbol	Am27C64									
		-55		-70		-90		-120		-150	
		Min	Max	Min	Max	Min	Max	Min	Max	Min	Max
Address to output delay	t_{ACC}	—	55	—	70	—	90	—	120	—	150
Chip enable to output delay	t_{CE}	—	55	—	70	—	90	—	120	—	150
Output enable to output delay	t_{OE}	—	35	—	40	—	40	—	50	—	65
Output enable high to output float	t_{DF}	—	25	—	25	—	25	—	30	—	30
Output hold from address, CE, or OE, whichever occurred first	t_{OH}	0	—	0	—	0	—	0	—	0	—

All times are in nanoseconds.

Table 4E.1 ■ EPROM Am27C64 timing parameters

5

INTERRUPTS

AND

RESETS

5.1 Objectives

After completing this chapter, you should be able to

- explain the difference between interrupts and resets
- describe the handling procedures for interrupts and resets
- raise one of the 68HC11 maskable interrupts to highest priority
- enable and disable maskable interrupts
- use the low-power mode to lower the power consumption
- use COP watchdog timer reset to prevent software failure
- set up the EVB interrupt vector jump table

5.2 Basics of Interrupts

Interrupts and resets are among the most useful mechanisms that a computer system provides. With interrupts and resets, I/O operations are performed more efficiently, errors are handled more smoothly, and CPU utilization is improved. This chapter will begin with a general discussion of interrupts and resets and then focus on the specific features of the 68HC11 interrupts and resets.

5.2.1 What is an Interrupt?

An *interrupt* is an unusual event that requires the CPU to stop normal program execution and perform some service related to the unusual event. An interrupt can be generated internally or externally. An external interrupt is generated when external circuitry asserts an interrupt signal to the CPU. An internal interrupt can be generated by the hardware circuitry in the CPU or caused by software errors. In some microcontrollers, timers, I/O interface functions, and the CPU are incorporated on the same chip, and these subsystems can generate interrupts to the CPU. Abnormal situations that occur during program execution, such as illegal opcodes, overflow, division by zero, and underflow, are called *software interrupts*. The terms *traps* and *exceptions* are both used to refer to software interrupts.

5.2.2 Why Are Interrupts Used?

Interrupts are useful in many applications, such as

- Coordinating I/O activities and preventing the CPU from being tied up during the data transfer process. As we will explain in more detail in Chapter 6, the interrupt mechanism enables the CPU to perform other functions during an I/O activity (when the I/O device is busy). CPU time can thus be utilized more efficiently because of the interrupt mechanism.

- Providing a graceful way to exit from an application when a software error occurs. The service routine for a software interrupt may also output useful information about the error so that it can be corrected.

- Reminding the CPU to perform routine tasks. For example, keeping the time of day is a common function of a computer. Without interrupts, the CPU would need to use program loops to create the delay or check the current time from a dedicated time-of-day chip. However, a timer circuit or dedicated timer chip can generate interrupts to the CPU so that it can update the current time and perform routine tasks such as updating the system resource status. In modern computer-operating systems, multiple user programs are resident in the main memory and the CPU time is divided into slots of about 10 to 20 ms. The operating system assigns a program to be executed for one time slot. At the end of a time slot or when a program is waiting for the completion of I/O, the operating system takes over and assigns another program to be executed. This technique is called *multitasking.* Because input/output operations are quite slow, CPU utilization is improved dramatically by multitasking since the CPU can execute another program when one program is waiting for the completion of input/output. Multitasking is made possible by the timer interrupt. Multitasking computer systems incorporate timers that periodically interrupt the CPU. On a timer interrupt, the operating system updates the system resource utilization status and switches the CPU from one program to another.

5.2.3 Interrupt Maskability

Some interrupts can be ignored by the CPU while others cannot. Interrupts that can be ignored by the CPU are called *maskable interrupts.* Interrupts that the CPU cannot ignore are called *nonmaskable interrupts.* A program can request the CPU to service or ignore a maskable interrupt by setting or clearing an *enable bit.* When an interrupt is *enabled,* it will be serviced by the CPU. When an interrupt is *disabled,* it will be ignored by the CPU. An interrupt request is said to be *pending* when it is active but not yet serviced by the CPU. A pending interrupt may or may not be serviced by the CPU, depending on whether or not it is enabled.

5.2.4 Interrupt Priority

If more than one interrupt is pending at the same time, the CPU needs to decide which interrupt should receive service first. The solution to this problem is to prioritize all interrupt sources. An interrupt with higher priority always receives service before interrupts at lower priorities. In most microprocessors, interrupt priorities are not programmable.

5.2.5 Interrupt Service

The CPU provides service to an interrupt by executing a program called the *interrupt service routine.* An interrupt is an abnormal event for the CPU. After providing service to an interrupt, the CPU must resume normal program execution. How does the CPU stop the execution of a program and resume it later? It does this by saving the program counter and the CPU status information before executing the interrupt service routine and then restoring the saved program counter and CPU status before exiting the interrupt service routine. The complete interrupt service cycle is:

1. When an interrupt occurs, save the program counter value in the stack.

2. Save the CPU status (including the CPU status register and some other registers) in the stack.

3. Identify the cause of interrupt.

4. Resolve the starting address of the corresponding interrupt service routine.

5. Execute the interrupt service routine.

6. Restore the CPU status and the program counter from the stack.

7. Restart the interrupted program.

Another issue is related to the time when the CPU begins to service an interrupt. For all hardware maskable interrupts, the microprocessor starts to provide service when it completes execution of the current instruction. For some nonmaskable interrupts, the CPU may start the service without completing the current instruction. Many software interrupts are caused by an error in instruction execution that prevents the instruction from being completed. The service to this type of interrupt is simply to output an error message and abort the program.

For some other types of interrupts, the CPU resolves the problem in the service routine and then reexecutes the instruction that caused the interrupt. One example of this type of interrupt is the *page fault* interrupt in a *virtual memory operating system*. In a virtual memory operating system, all programs are divided into pages of equal size. The size of a page can range from 512B to 32KB. When a program is started, the operating system loads only a few pages (possibly only one page) from secondary storage (generally hard disk) into main memory and then starts to execute the program. Before long, the CPU may need to execute an instruction that hasn't been loaded into the main memory yet. This causes an interrupt called a page fault interrupt. The CPU then executes the service routine for the page fault interrupt, which loads the page that caused the page fault into main memory. The program that caused the page fault is then restarted. The virtual memory operating system gets its name from the fact that it allows execution of a program without loading the whole program into main memory. With a virtual memory operating system, a program that is much larger than the physical main memory can be executed.

5.2.6 Interrupt Vector

The term *interrupt vector* refers to the starting address of the interrupt service routine. In general, interrupt vectors are stored in a table called an *interrupt vector table.* The interrupt vector table is fixed for some microprocessors and may be relocated for other microprocessors (for example, the Am29000 microprocessor family).

The CPU needs to determine the interrupt vector before it can provide service. One or both of the following methods can be used by a microprocessor to determine the interrupt vector:

1. Fetch the vector from a fixed memory location. For most microprocessors, the interrupt vector of each internal interrupt (or trap or exception) is usually stored at a predefined location in the interrupt vector table, where the microprocessor can get it directly.

2. Execute an interrupt acknowledge cycle to fetch a vector number in order to locate the interrupt vector. Some microprocessors obtain the interrupt vector for external interrupts using this method. During the interrupt acknowledge cycle, the microprocessor performs a read bus cycle and the external I/O device that requested the interrupt places a number on the data bus to identify itself. This number is called the *interrupt vector number.* The address of the memory location that stores the interrupt vector is usually a multiple (2 or 4

are most common) of the vector number. The CPU needs to perform a read cycle in order to obtain it.

Members of the Motorola 68000 family of microprocessors use both methods to determine the interrupt vector, whereas the 68HC11 uses solely the first method.

A microprocessor that uses the second method to determine the interrupt vector usually supports a lot of external interrupts. When multiple interrupts occur, the interrupt with the highest priority gets CPU service, so the various interrupt sources need to be prioritized. In general, interrupt controller circuitry is used to prioritize external interrupt sources. The interrupt controller enables the I/O device (in the form of an interface chip) with the highest priority among all pending interrupts to place its vector number on the data bus during the interrupt acknowledge cycle. The interrupt vector number is then multiplied by a factor (such as 2, 4, or 8) to derive the location where the interrupt vector is stored. The role of an interrupt controller is illustrated in Figure 5.1.

In Figure 5.1, each I/O device has two pins connected to the interrupt controller. One pin is an input (INT_i) to the controller, and its function is to indicate that there is a pending interrupt request from device i. The other pin ($IACK_i$) is the interrupt acknowledge signal to device i, and it allows device i to place the interrupt vector number on the data bus during the vector number fetch cycle. There are also interrupt request and acknowledge pins between the interrupt controller and the CPU. Whenever there is a pending interrupt or interrupts, the interrupt controller asserts the interrupt signal(s) (INT) to the CPU. The CPU then asserts the interrupt acknowledge signal (IACK) to the interrupt controller during the vector number fetch cycle. In response, the interrupt controller asserts the interrupt acknowledge signal $IACK_i$ with the highest priority among all pending interrupts. Device i then places its interrupt vector number on the data bus to be fetched by the processor.

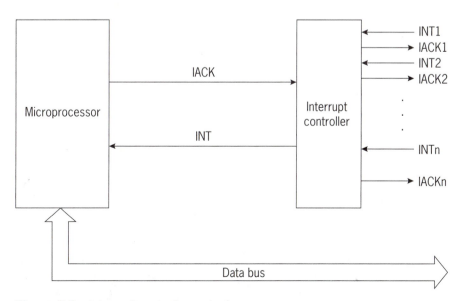

Figure 5.1 Interrupt controller and microprocessor

5.2.7 Interrupt Programming

There are three steps in interrupt programming.

Step 1

Initialize the interrupt vector table (that is, place the starting address of each interrupt service routine in the table). This can be done by using the assembler directive ORG (or its equivalent):

```
OR\G $xxxx        ;xxxx is the vector table base address
FDB  service_1    ;reserve two bytes
FDB  service_2
       .
       .
       .
FDB  service_n
```

where *service_1* is the starting address of the service routine for interrupt source *i*. The assembler syntax and the number of bytes needed to store an interrupt vector on your particular microprocessor may be different. You should refer to the user's manual for your specific microprocessor for the proper syntax.

Step 2

Write the service routine. An interrupt service routine should be as short as possible. For some interrupts, the service routine may only output a message to indicate that something unusual has occurred. A service routine is similar to a subroutine—the only difference is the last instruction. An interrupt service routine uses the *return from interrupt* (or *return from exception*) instruction instead of the *return from subroutine* instruction to return to the interrupted program. The following is an example of an interrupt service routine:

```
SERVICE1:   LDX #MSG
            JSR PUTST    ;PUTST outputs a string pointed to by index
   *                     ;register X
            RTI          ;return from interrupt
MSG:        FCC 'There is an error;'
```

The service routine may or may not return to the interrupted program, depending on the cause of the interrupt. It makes no sense to return to the interrupted program if the interrupt is caused by a software error such as division by zero or overflow, because the program is unlikely to generate correct results under those circumstances. In such cases the service routine would return to the monitor program or the operating system instead. Returning to a program other than the interrupted program can be achieved by changing the saved program counter (in the stack) to the desired value. Execution of the *return from interrupt* instruction will then return CPU control to the new address. The RTI instruction will pull from the stack whatever registers were pushed onto the stack by the interrupt. Most microprocessors (including the 68HC11) set a flag (the interrupt status flag) in a status register when an interrupt occurs. The user must remember to clear the interrupt status flag in the interrupt service routine—otherwise, the same interrupt will occur immediately after execution of the RTI instruction. You will be reminded about this point in later chapters.

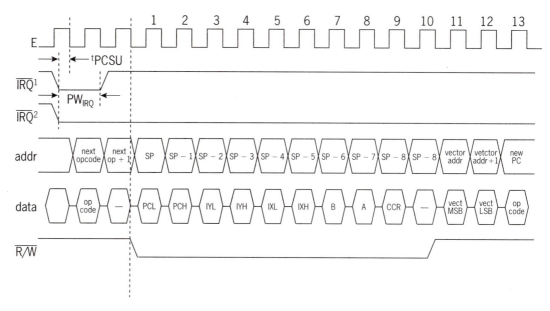

1. Edge sensitive IRQ pin
2. Level sensitive IRQ pin

Figure 5.2   Interrupt timing diagram

Step 3

Enable the interrupts to be serviced. An interrupt can be enabled by clearing the CPU-level interrupt mask and setting the local interrupt enable bit in the I/O control register. It is a common mistake to forget to enable interrupts.

5.2.8 The Overhead of Interrupts

Although the interrupt mechanism provides many advantages, it also involves some overhead. The overhead of the 68HC11 interrupt includes:

1. Saving the CPU registers and fetching the interrupt vector. As shown in Figure 5.2, this takes 12 E clock cycles.
2. The execution time of the RTI instruction. This instruction restores all the CPU registers stored in the system stack and takes 12 E clock cycles to execute.

The total overhead is thus 24 E clock cycles, which amounts to 12 μs for a 2 MHz E clock. You should be aware of the overhead involved in interrupt processing when deciding whether to use the interrupt mechanism.

5.3 Resets

The initial values of some CPU registers, flip-flops, and the control registers in I/O interface chips must be established before the computer can operate properly. Computers provide a reset mechanism to establish initial conditions.

There are at least two types of resets in each microprocessor: the *power-on reset* and the *manual reset*. A power-on reset allows the microprocessor to establish the initial values of registers and flip-flops and to initialize all I/O interface chips when power to the microprocessor is turned on. A manual reset without power-down allows the computer to get out of most error conditions (if hardware hasn't failed) and reestablish the initial conditions. The computer will *reboot* itself after a reset.

The reset service routine has a fixed starting address and is stored in the read-only memory of all microprocessors. At the end of the service routine, control should be returned to either the monitor program or the operating system. An easy way to do that is to push the starting address of the monitor or the operating system into the stack and then execute a return from interrupt instruction.

Like nonmaskable interrupts, resets are unmaskable. However, resets are different from nonmaskable interrupts in that no registers are saved by resets because resets establish the values of registers.

5.4 68HC11 Interrupts

The 68HC11 supports sixteen hardware interrupts, two software interrupts, and three resets, as listed in Table 5.1. Those interrupts can be divided into two basic categories, maskable and nonmaskable. In the 68HC11 A series (A0, A1, and A8), the first fifteen interrupts (from SCI serial system to $\overline{IRQ}$ pin interrupt) in Table 5.1 can be masked by setting the I bit of the condition code register. In addition, all of the on-chip interrupt sources are individually maskable by local control bits. $\overline{IRQ}$ and $\overline{XIRQ}$ are the only external interrupt sources.

The user has the option of selecting edge triggering or level triggering for the $\overline{IRQ}$ input by setting the IRQE bit of the OPTION register. Falling-edge triggering is selected if this bit is set to 1; otherwise, low-level triggering is selected. This option can be set only during the first 64 E clock cycles after the rising edge of the reset input and hence must be done in the reset handling routine. By default, level triggering is selected.

The software interrupt (SWI instruction) is a nonmaskable instruction rather than a maskable interrupt source. The illegal opcode interrupt is also a nonmaskable interrupt. The interrupt source $\overline{XIRQ}$ is considered a nonmaskable interrupt because once enabled, it cannot be masked by software; however, it is masked during a reset.

5.4.1 The 68HC11 Interrupt-Handling Procedure

When an interrupt occurs, the CPU registers are saved in the stack in the order shown in Figure 5.3. After these registers have been saved, the I bit in the CCR register is set to 1 to disable further interrupts. The starting address of the corresponding service routine is fetched from a predefined table, and program execution is resumed from that address. All interrupts and resets are prioritized. Table 5.1 lists the priority of each interrupt and the location where its starting address and reset service routine are stored.

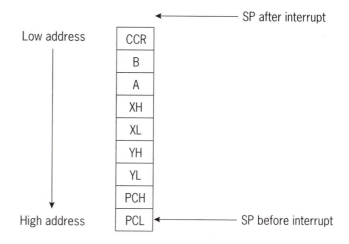

Figure 5.3 68HC11 interrupt stacking order

Vector address	Interrupt source	Priority
FFC0, C1	reserved	
.	.	
.	.	
.	.	
FFD4, D5	reserved	
FFD6, D7	SCI serial system	lowest
FFD8, D9	SPI serial transfer complete	
FFDA, DB	pulse accumulator input edge	
FFDC, DD	pulse accumulator overflow	
FFDE, DF	timer overflow	
FFE0, E1	timer output compare 5	
FFE2, E3	timer output compare 4	
FFE4, E5	timer output compare 3	
FFE6, E7	timer output compare 2	
FFE8, E9	timer output compare 1	
FFEA, EB	timer input capture 3	
FFEC, ED	timer input capture 2	
FFEE, EF	timer input capture 1	
FFF0, F1	real time interrupt	
FFF2, F3	$\overline{\text{IRQ}}$ pin interrupt	
FFF4, F5	$\overline{\text{XIRQ}}$ pin interrupt	
FFF6, F7	SWI	
FFF8, F9	illegal opcode trap	
FFFA, FB	COP failure (reset)	
FFFC, FD	COP clock monitor fail	
FFFE, FF	$\overline{\text{RESET}}$	highest

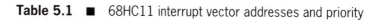

Table 5.1 ■ 68HC11 interrupt vector addresses and priority

Example 5.1

The 68HC11 is executing the TAP instruction in line 6 of the following program segment when an $\overline{\text{IRQ}}$ interrupt occurs. List the stack contents immediately before the interrupt service routine is entered.

Line	Location	Opcode	Instruction mnemonics
			ORG $E000
1	E000	C620	LDAB #$20
2	E002	8E00FF	LDS $#00FF
3	E005	CE1000	LDX #$1000
4	E008	4F	CLRA
5	E009	1830	TSY
6	E00B	06	TAP
7	E00C	AB00	ADDA 0,X

Solution: The interrupt will be serviced after the TAP instruction is completed. The contents of all registers when the TAP instruction is completed are shown in Figure 5.4.

The contents of B are $20 when the first instruction is completed.

The contents of SP are $00FF when the second instruction is completed.

The contents of X are $1000 when the third instruction is completed.

The contents of A are $00 when the fourth instruction is completed.

The contents of Y are $0100 when the fifth instruction is completed.

The contents of CCR are $00 when the sixth instruction is completed.

CCR	0 0
B	2 0
A	0 0
X	1 0 0 0
Y	0 1 0 0
PC	E 0 0 C
SP	0 0 F F

Figure 5.4 Register contents after the TAP instruction is completed

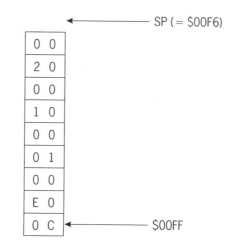

Figure 5.5 Stack contents when TAP is completed

The contents of the stack will therefore be as shown in Figure 5.5 immediately before the service routine is entered (executed).

5.4.2 Priority Structure of the 68HC11 Maskable Interrupts

The priority of each maskable interrupt source is fixed. However, the user can program the HPRIO register to promote one of the maskable interrupts to highest priority among all maskable interrupts. The contents of the HPRIO register are as follows:

7	6	5	4	3	2	1	0	
RBOOT	SMOD	MDA	IRV	PSEL3	PSEL2	PSEL1	PSEL0	at $103C

value after
reset   (refer to Table 5.4)   0   1   0   1

The last four bits of the HPRIO register select the interrupt source, as shown in Table 5.2, while the values of the upper four bits are established during reset and serve the following purposes:

- ■ RBOOT—Read bootstrap ROM

  This bit is writable only when the SMOD bit equals 1. The bootstrap ROM will be enabled and located from $BF40 to $BFFF when the RBOOT bit is 1, but it will be disabled and not present in the memory map when this bit is 0. This 192-byte mask-programmed ROM contains the firmware required to load a user's program through the serial communication interface into the internal SRAM and jump to the loaded program. In all modes other than the special bootstrap mode, this ROM is disabled and does not occupy any space in the 64K-byte memory map.

- ■ SMOD—Special mode

  This bit may be written to 0 but not back to 1. Its value is established on the rising edge of the RESET signal. The 68HC11 opera-

PSEL3	PSEL2	PSEL1	PSEL0	Interrupt source promoted
0	0	0	0	timer overflow
0	0	0	1	pulse accumulator overflow
0	0	1	0	pulse accumulator input edge
0	0	1	1	SPI serial transfer complete
0	1	0	0	SCI serial system
0	1	0	1	reserved (defaults to $\overline{\text{IRQ}}$)
0	1	1	0	IRQ (external pin)
0	1	1	1	real time interrupt
1	0	0	0	timer input capture 1
1	0	0	1	timer input capture 2
1	0	1	0	timer input capture 3
1	0	1	1	timer output compare 1
1	1	0	0	timer output compare 2
1	1	0	1	timer output compare 3
1	1	1	0	timer output compare 4
1	1	1	1	timer output compare 5

Table 5.2 ■ PSEL3–PSEL0 bits for selecting highest priority interrupts

tion mode is determined by the combination of this bit and the MDA bit, as shown in Table 5.3.

■ MDA—Mode A select

This bit is writable only when SMOD equals 1. The value of the MDA bit is established on the rising edge of the RESET signal. The 68HC11 operation mode is determined by the combination of this bit and the SMOD bit (see Table 5.3).

■ IRV—Internal read visibility

The IRV bit is writable only when the SMOD bit equals 1. This bit is forced to 0 if the SMOD bit equals 0. When the IRV bit is set to 1, data can be driven onto an external bus during internal reads (for example, a read from an on-chip I/O control register or status register). Otherwise, data from internal reads will not be visible during internal reads. The IRV control bit is used during factor testing and sometimes during emulation to allow internal read accesses to be visible on the external data bus.

Operation mode	SMOD	MDA
Normal single-chip	0	0
Normal expanded	0	1
Special bootstrap	1	0
Special test	1	1

Table 5.3 ■ Mode select of 68HC11

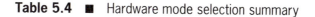

Inputs			Control bits in HPRIO (latched at reset)			
MODB	*MODA*	**Mode description**	*RBOOT*	*SMOD*	*MDA*	*IRV*
1	0	normal single-chip	0	0	0	0
1	1	normal expanded	0	0	1	0
0	0	special bootstrap	1	1	0	1
0	1	special test	0	1	1	1

Table 5.4 ■ Hardware mode selection summary

The establishment of these four bits is shown in Table 5.4. In assigning highest priority interrupts, note that the HPRIO register may be written only when the I bit in the CCR register is a 1. Also note that the interrupt vector address is not affected by reassigning an interrupt source to the highest priority position. The upper four bits may be written only while the 68HC11 is operating in one of the special modes (when SMOD is 1).

5.4.3 Nonmaskable Interrupts

There are three nonmaskable interrupt sources in the 68HC11: an illegal opcode fetch, an SWI instruction, and the $\overline{XIRQ}$ pin interrupt.

ILLEGAL OPCODE TRAPS

Some combinations of the opcode byte are not defined, and if one of those combinations is executed, an illegal opcode trap occurs. For this reason, the illegal opcode vector should never be left uninitialized. It is also a good idea to reinitialize the stack pointer after an illegal opcode interrupt so that repeated execution of illegal opcodes does not cause stack overruns.

THE SOFTWARE INTERRUPT INSTRUCTION (SWI)

The SWI instruction is executed in the same manner as any other instruction and takes precedence over all interrupts maskable by the CCR I bit. Registers are saved in the stack as in any other maskable interrupt. The SWI instruction is not masked by the global interrupt mask bits (the I and X bits) in the CCR register.

SWI instructions are commonly used in the debug monitor to implement *breakpoints* and to transfer control from a user program to the debug monitor. A breakpoint in a user program is the memory location where the programmer wants program execution to be stopped and information about instruction execution (in the form of register contents) to be displayed. To implement breakpoints, the debug monitor sets up a breakpoint table. Each entry of the table holds the address of the breakpoint and the opcode byte at the breakpoint. The monitor also replaces the byte at the breakpoint with the opcode of the SWI instruction. When the instruction at the breakpoint is executed, it thus causes an SWI interrupt. The service routine of the SWI interrupt will look up the breakpoint table and take different actions depending on whether the saved PC value is in the breakpoint table:

Case 1

The saved PC value is not in the breakpoint table. In this case, the service routine will simply replace the saved PC value (in the stack) with the address of the monitor program and return from the interrupt.

Case 2

The saved PC value is in the breakpoint table. In this case, the service routine will

1. replace the SWI opcode with the opcode in the breakpoint table
2. replace the saved PC value (in the stack) with the address of the monitor program
3. display the contents of the CPU registers
4. return from the interrupt

An EVB user can use SWI as the last instruction in a program. When the SWI instruction is executed, CPU control is returned to the Buffalo monitor, and the user can then enter commands to examine the contents of memory locations or registers or do something else.

THE $\overline{\text{XIRQ}}$ PIN INTERRUPT

The $\overline{\text{XIRQ}}$ interrupt can be masked by the X bit in the CCR register. Upon reset, the X bit is 1 and the $\overline{\text{XIRQ}}$ interrupt is thus disabled. The X bit can be cleared within 64 clock cycles after reset, but it cannot be cleared and set after that and hence is nonmaskable. When an $\overline{\text{XIRQ}}$ interrupt occurs, all registers are saved in the stack and the X bit in the CCR register is set to 1 to prevent further $\overline{\text{XIRQ}}$ interrupts.

Several external interrupts can be tied to the $\overline{\text{IRQ}}$ or $\overline{\text{XIRQ}}$ pin. In this application, the user must choose a level-sensitive triggering interrupt for the $\overline{\text{IRQ}}$. The option of an edge-triggering $\overline{\text{IRQ}}$ interrupt can be set only within the first 64 E clock cycles after reset. If the user does need edge-triggering interrupt capability, four other interrupt sources can be utilized: input captures IC1, IC2, and IC3 and pulse accumulator pin PAI. They will be discussed in detail in Chapter 7.

Because of these characteristics, the $\overline{\text{XIRQ}}$ interrupt can be used to detect emergency situations.

5.5 Low-Power Modes

A good embedded microcontroller should consume as little power as possible. When the microcontroller is performing operations, power consumption is unavoidable. However, the microcontroller in an embedded system may not always be performing a useful operation, so it would be nice if the power consumption could be drastically reduced when the microcontroller is not doing something useful. This issue is especially important for battery-powered products.

Microcontroller manufacturers are already investigating methods of reducing power consumption when the microcontroller is idle, and the so-called power-down mode has been incorporated in several microcontrollers. The

Complementary Metal Oxide Semiconductor (CMOS) technology is the most popular technology for implementing microcontrollers, and its power consumption is proportional to the operating frequency of the circuit. By slowing down the circuit operation and/or even turning off part of the circuit, the power consumption of these microcontrollers can be dramatically reduced.

The 68HC11 has two programmable low-power consumption modes: WAIT and STOP.

5.5.1 The WAIT Instruction

The WAIT instruction puts the microcontroller in a low power-consumption mode while keeping the oscillator running. Upon execution of a WAIT instruction, the machine state is saved in the stack and the CPU stops instruction execution. The wait state can be exited only through an unmasked interrupt or reset. The amount of power saved depends on the application and also on the circuitry connected to the microcontroller pins.

The WAIT instruction can be used in situations where the CPU has nothing to do but wait for the arrival of an interrupt. Since all CPU registers are already saved in the stack, the interrupt response time can be much faster.

5.5.2 The STOP Instruction

The STOP instruction places the microcontroller in its lowest power-consumption mode, provided the S bit in the CCR register is 0. If the S bit is 1, the STOP mode is disabled and STOP instructions have no effect. In the STOP mode, all clocks including the internal oscillator are stopped, causing all internal processing to be halted. Exit from the STOP mode can be accomplished by $\overline{RESET}$, an $\overline{XIRQ}$ interrupt, or an unmasked $\overline{IRQ}$.

When the $\overline{XIRQ}$ interrupt is used, the 68HC11 exits from the STOP mode regardless of the state of the X bit in the CCR; however, the actual recovery sequence differs depending on the state of the X bit. If the X bit is 1, then processing will continue with the instruction immediately following the STOP instruction and no $\overline{XIRQ}$ interrupt service routine will be executed. If the X bit is 0, the 68HC11 will perform the sequence that services the $\overline{XIRQ}$ interrupt. A RESET will always result in an exit from the STOP mode, and the restart of microcontroller operations is then determined by the reset vector.

5.6 The 68HC11 Resets

The 68HC11 has four possible sources of resets: an active low external reset pin ($\overline{RESET}$), a power-on reset, a computer operating properly (COP) watchdog timer reset, and a clock monitor failure reset.

5.6.1 The External Reset Pin

The $\overline{RESET}$ pin is used to reset the microcontroller and allow an orderly system start-up. When a reset condition is detected, this pin is driven low by an internal device for four E clock cycles and then released; two E clock cycles

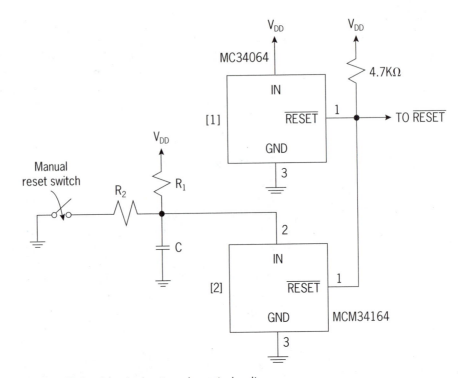

Figure 5.6   A typical external reset circuit

later, it is sampled. If the pin is still low, it means that an external reset has occurred. If the pin is high, it indicates that the reset was initiated internally by either the computer operating properly (COP) watchdog timer or the clock monitor. An externally generated reset should stay active for at least eight E clock cycles.

It is very important to control reset during power transitions. If the reset line is not held low while V_{DD} is below its minimum operating level, the on-chip EEPROM contents could be corrupted; both the EEPROM memories and the EEPROM-based CONFIG register can be affected. A low-voltage inhibit (LVI) circuit that holds reset low whenever V_{DD} is below its minimum operating level is required to protect against EEPROM corruption. The reset circuit shown in Figure 5.6 is recommended by Motorola. The MC34064 is a low-voltage sensing circuit that holds $\overline{RESET}$ low whenever V_{DD} is below operating level. The MC34164 is also a low-voltage sensing device. The MC34164 and the RC circuit on its input provide an external manual and power-on reset delay. LVI circuit [1] is required for virtually every 68HC11 system. The external power-on reset and manual reset switch are optional.

5.6.2 The Power-On Reset

The power-on reset occurs when a positive transition is detected on V_{DD}. The power-on reset is used strictly for power turn-on conditions and should not be used to detect drops in power-supply voltage. The power-on circuitry provides a 4064-cycle time delay from the time of the first oscillator operation.

In a system where E = 2 MHz, the power-on reset lasts about two milliseconds. If the external $\overline{\text{RESET}}$ pin is low at the end of the power-on delay time, the 68HC11 remains in the reset condition until the $\overline{\text{RESET}}$ pin goes high.

5.6.3 The CPU after Reset

After reset, the CPU fetches the reset vector from locations $FFFE and $FFFF ($BFFE and $BFFF if in the special bootstrap or special test operating mode) during the first three cycles and begins instruction execution. The stack pointer and other CPU registers are indeterminate immediately after reset; however, the X and I interrupt mask bits in the CCR are set to mask any interrupt requests, and the S bit is set to 1 to disable the STOP mode.

All I/O control registers are initialized by reset. The initial value of each control register will be discussed in the appropriate chapters.

5.6.4 Establishing the Mode of Operation

During a reset, the basic mode of operation is established, determining whether the 68HC11 will operate as a self-contained single-chip system or as an expanded system that includes external memory resources. There are also special variations of these two basic modes of operation: the bootstrap mode is a variation of the normal single-chip mode, and the special test mode is a variation of the normal expanded mode. The levels of the two mode select pins during reset determine which of these four modes of operation will be selected. The levels of these two mode select pins also set the upper four bits in the HPRIO register, as shown in Table 5.4.

5.6.5 The Computer Operating Properly (COP) Watchdog Timer Reset

The COP watchdog timer system is intended to detect software processing errors. When the COP system is in use, software is responsible for keeping a free-running watchdog timer from timing out. If the watchdog timer times out, it is an indication that software is no longer being executed in the intended sequence and thus a system reset is initiated.

The COP system is enabled by clearing the NOCOP bit in the CONFIG register, and it is disabled by setting the NOCOP bit. The CONFIG register is implemented in EEPROM and is not as easy to change as a normal control register. Even after the NOCOP bit is changed, the 68HC11 must be reset before the new status becomes effective. In the special test and bootstrap operating modes, the COP system is initially inhibited by the disable resets (DISR) control bit in the TEST1 register. The DISR bit can be written to zero to enable COP resets while the 68HC11 is in the special test or bootstrap operating mode.

The COP time-out period is set by the COP timer control bits (CR1 and CR0) in the OPTION register. After reset, these bits are both zero; this combination selects the fastest time-out rate. The 68HC11 internal E clock is first divided by 2^{15} before it enters the COP watchdog system, and then the CR1 and CR0 bits determine a further scaling factor for the watchdog timer (see Table 5.5).

CR1	CR0	E ÷ 2¹⁵ divided by	Crystal frequency	
			8 MHz	4 MHz
			Nominal time-out	
0	0	1	16.384 ms	32.768 ms
0	1	4	65.536 ms	131.07 ms
1	0	16	262.14 ms	524.29 ms
1	1	64	1.049 s	2.1 s
			2 MHz	1 MHz
			Bus Frequency (E Clock)	

Table 5.5 ■ Watchdog rates vs. crystal frequency

The COP timer must be reset by a software sequence prior to time-out to avoid a COP reset. The software COP reset is a two-step sequence. The first step is to write a $55 to the COPRST register (at $103A); this step arms the COP timer-clearing mechanism. The second step is to write a $AA into the same register; this step clears the COP timer. Any number of instructions can be executed between these two steps as long as both steps are performed in the correct sequence before the timer times out.

5.6.6 The Clock Monitor Reset

The clock monitor circuit is based on an internal resistor-capacitor (RC) time delay. If no clock edges are detected within this RC time delay, the clock monitor can optionally generate a system reset. The clock monitor function is enabled/disabled by the CME bit (bit 3) in the OPTION register.

The RC time-out may vary from lot to lot and from part to part because of processing variations. An E clock frequency below 10 KHz will definitely be detected as a clock monitor error. An E clock frequency of 200 KHz or more will prevent clock monitor errors. Any system operating below a 200 KHz E clock frequency should not use the clock monitor function.

The clock monitor is often used as a backup for the COP watchdog system. Since the COP relies on a clock, it is unable to function if the clock stops. In such a situation, the clock monitor system could detect clock failures not detected by the COP system.

Another use for the clock monitor reset is to protect against unintentional execution of the STOP instruction. Some applications view the STOP instruction as a serious problem because it causes the microcontroller clock to stop, thus disabling all software execution and on-chip peripheral functions. The stop disable bit (S) in the CCR register is the first line of defense against unwanted STOP instructions. When the S bit is 1, the STOP instruction acts as a no operation instruction, which does not interfere with clock operation. The clock monitor can provide an additional level of protection by generating a system reset if the 68HC11 clocks are accidentally stopped.

If the 68HC11 clocks slow down or stop when the clock monitor is enabled, a system reset is generated. The bidirectional $\overline{\text{RESET}}$ pin is driven low to reset the external system and the microcontroller. If the 68HC11 is in the special test or bootstrap mode, resets from the COP and clock monitor systems are initially disabled by a 1 in the DISR bit in the TEST1 register. The COP and clock monitor resets can be reenabled while the 68HC11 is still in the special operation modes by clearing the DISR bit to 0. In normal operation modes, the DISR bit is forced to 0 and cannot be set to 1.

5.7 68HC11 Interrupt Programming

As in any other microprocessor, there are three steps in 68HC11 interrupt programming:

Step 1
Write the interrupt handlers.
Step 2
Set up the interrupt vector table.
Step 3
Set the global and local interrupt enable bits.

An interrupt handling routine should have the following form:

```
HNDLER    .
            .
            .
          RTI
```

where HNDLER is the label of the first instruction in the interrupt service routine.

The interrupt vector table can be set up using assembler directives. For example, to add the interrupt vector of $\overline{IRQ}$ into the table, use the following directives:

```
ORG $FFF2
FDB HNDLER
```

where HNDLER is the label of the first instruction of the $\overline{IRQ}$ service routine.

The user may not want to use all of the possible interrupts in his/her application. However, it is a good idea to initialize all interrupt vectors. A common service routine should be written to handle all unwanted interrupts—then if an unwanted interrupt is accidentally triggered, the common service routine can provide an exit. A simple form of a common service routine just outputs a message and quits.

The 68HC11 will service the maskable interrupts when the I bit of the CCR register is 0. The I bit can be cleared by the CLI instruction. Sometimes the user may want to mask all of the maskable interrupts—the instruction SEI can be used to set the I bit of the CCR register to 1 and disable all maskable interrupts.

The EVM user can follow the previous three steps to do interrupt programming. However, the design of the EVB and EVBU precludes the user from using the second step because the memory space from $E000 to $FFFF is in the EPROM, which cannot be written into by the user program. The EVB and EVBU reserve 60 bytes ($00C4–$00FF) of the on-chip SRAM as an *interrupt vector jump table*. Each entry of the table consists of three bytes. The first

Interrupt vector	Field
serial communication interface (SCI)	$00C4-$00C6
serial peripheral interface (SPI)	$00C7-$00C9
pulse accumulator input edge	$00CA-$00CC
pulse accumulator overflow	$00CD-$00CF
timer overflow	$00D0-$00D2
timer output compare 5	$00D3-$00D5
timer output compare 4	$00D6-$00D8
timer output compare 3	$00D9-$00DB
timer output compare 2	$00DC-$00DE
timer output compare 1	$00DF-$00E1
timer input capture 3	$00E2-$00E4
timer input capture 2	$00E5-$00E7
timer input capture 1	$00E8-$00EA
real time interrupt	$00EB-$00ED
IRQ	$00EE-$00F0
XIRQ	$00F1-$00F3
software interrupt (SWI)	$00F4-$00F6
illegal opcode	$00F7-$00F9
computer operating properly (COP)	$00FA-$00FC
clock monitor	$00FD-$00FF

Table 5.6 ■ 68HC11 EVB and EVBU interrupt vector jump table

byte should be set to the opcode ($7E) of the JMP instruction, and the second and the third bytes should be set to the starting address of the corresponding service routine. Each entry (two bytes) of the default vector table of the 68HC11 should contain the address of the first byte of the corresponding entry in the interrupt vector jump table. The EVB interrupt vector jump table is shown in Table 5.6.

Example 5.2

Initialize the interrupt vector jump table entry for the $\overline{\text{IRQ}}$ interrupt in the EVB, given that the label of the first instruction of the $\overline{\text{IRQ}}$ service routine is IRQ_HND.

Solution: Add the following assembler directives to your program:

```
        ORG  $00EE      ;address of the first byte of IRQ interrupt vector
  *                     ;jump table entry
        JMP  IRQ_HND
```

Example 5.3

Write a main program and an interrupt service routine. The main program will initialize a variable *count* to 100, enable the $\overline{\text{IRQ}}$ interrupt, and stay in a loop to check the value of *count*. When the value of *count* is 0, the program should jump to the breakpoint located at $F000. The interrupt service routine will decrement the count and return. The $\overline{IRQ}$ pin is connected to some external device that generates interrupts from time to time. Assume the voltage level of the $\overline{IRQ}$ pin returns to high before the $\overline{IRQ}$ service routine is exited.

Solution: The first step is to set up the interrupt vector entry for the $\overline{IRQ}$ interrupt using the following assembler directives:

```
ORG  $FFF2
FDB  IRQHND    ;IRQHND is the starting address of the IRQ service routine
```

The main program and the $\overline{IRQ}$ service routine are as follows:

```
         ORG  $00
count    RMB  1
         ORG  $C000
         SEI               ;disable all maskable interrupts at the beginning
         LDS  #$00FF        ;initialize the stack pointer
         LDAA #100
         STAA count         ;initialize the value of count to 100
         CLI                ;enable the IRQ interrupt
loop     LDAA count         ;check the value of count and
         BNE  loop          ;wait for the IRQ interrupt
         SEI                ;disable interrupts
         JMP  $F000         ;jump to the breakpoint
IRQHND   DEC  count         ;decrement the value of count
         RTI                ;return from interrupt
         END
```

Most interrupt sources in the 68HC11 are related to I/O subsystems. Examples of I/O subsystem interrupts will be discussed in appropriate chapters.

5.8 Summary

This chapter has explored the 68HC11 interrupt mechanisms. Interrupts allow the CPU to be shared by a variety of tasks. Each task gets CPU time only in response to its request for it. The 68HC11 also allows the programmer to disable all maskable interrupts when necessary, for example, when the CPU is executing a critical section of a code.

Interrupt latency is the delay from the moment of the interrupt request until the interrupt is serviced. If a maskable interrupt request cannot tolerate a long interrupt latency, the 68HC11 allows the user to raise its priority.

Examples of interrupt applications will be discussed in later chapters when I/O subsystems are introduced.

5.9 Glossary

COP (computer operating properly) watchdog timer   A system for detecting software processing errors. If software is written correctly, then it should complete all operations within some time limit. Software problems can be detected by setting a watchdog timer so that the software resets the timer before it times out.

Exception   A software interrupt, such as an illegal opcode, an overflow, division by zero, or an underflow.

Illegal opcode   A binary bit pattern of the opcode byte for which an operation is not defined.

Interrupt   An unusual event that requires the CPU to stop normal program execution and perform some service to the unusual event.

Interrupt priority   The order in which the CPU will service interrupts when all of them occur at the same time.

Interrupt service   The service provided to a pending interrupt by CPU execution of a program called a service routine.

Interrupt vector   The starting address of an interrupt service routine.

Interrupt vector table   A table that stores all interrupt vectors.

Low-power mode   An operation mode in which less power is consumed. In CMOS technology, the low-power mode is implemented by either slowing down the clock frequency or turning off some circuit modules within a chip.

Multitasking   A computing technique in which CPU time is divided into slots that are usually 10 to 20 ms in length. When multiple programs are resident in the main memory waiting for execution, the operating system assigns a program to be executed to one time slot. At the end of a time slot or when a program is waiting for completion of I/O, the operating system takes over and assigns another program to be executed.

Maskable interrupts   Interrupts that can be ignored by the CPU. This type of interrupt can be disabled by setting a mask bit or by clearing an enable bit.

Nonmaskable interrupts   Interrupts that the CPU cannot ignore.

Reset   A signal or operation that sets the flip-flops and registers of a chip or microprocessor to some predefined values or states so that the circuit or microprocessor can start from a known state.

Trap   A software interrupt; an exception.

5.10 Exercises

E5.1. What is the name given to a routine that is executed in response to an interrupt?

E5.2. What are the advantages of using interrupts to handle data input and output?

E5.3. What are the requirements of interrupt processing?

E5.4. How do you enable other interrupts when the 68HC11 is executing an interrupt service routine?

E5.5. Why would there be a need to promote one of the maskable interrupts to highest priority?

E5.6. Write the assembler directives to initialize the $\overline{\text{IRQ}}$ interrupt vector located at $E200.

E5.7. What is the last instruction in most interrupt service routines? What does this instruction do?

E5.8. Suppose the 68HC11 is executing the following instruction segment and the $\overline{IRQ}$ interrupt occurs when the TSY instruction is being executed. What will be the contents of the top ten bytes in the stack?

```
ORG   $E000
LDS   #$FF
CLRA
LDX   #$1000
BSET  10,X $48
LDAB  #$40
INCA
TAP
PSHB
TSY
ADDA  #10
```

E5.9. Suppose the E clock frequency is 4 MHz. Compute the COP watchdog timer time-out period for all the possible combinations of the CR1 and CR0 bits in the OPTION register.

E5.10. Suppose the starting address of the service routine for the timer overflow interrupt is at $D000. Write the assembler directives to initialize its vector table entry on the EVB computer.

E5.11. Write an instruction sequence to clear the X and I bits in the CCR. Write an instruction sequence to set the S, X, and I bits in the CCR.

E5.12. Why does the 68HC11 need to be reset when the power supply is too low?

E5.13. Write the instruction sequence to prevent the COP timer from timing out and resetting the microcontroller.

5.11 Lab Exercises and Assignments

L5.1. *Simple interrupt* Connect the $\overline{IRQ}$ pin of an EVB or EVM to a debounced switch that can generate a negative-going pulse. Write a main program and an $\overline{IRQ}$ service routine. The main program initializes the variable *irq_cnt* to 10, stays in a loop, and keeps checking the value of *irq_cnt*. When *irq_cnt* is decremented to 0, the main program jumps to the breakpoint. The $\overline{IRQ}$ service routine simply decrements *irq_cnt* by 1 and returns.

The lab procedure is as follows:

Step 1
Connect the $\overline{IRQ}$ pin of the EVB or EVM to a debounced switch that can generate a clean negative pulse.

Step 2
Enter the main program and $\overline{IRQ}$ service routine, assemble them, and then download them to the single-board computer. Remember to enable the $\overline{IRQ}$ interrupt in your program.

Step 3
Set a breakpoint and then execute the program.

Step 4
Pulse the switch ten times.

If everything is operating properly, you should get the monitor prompt back.

6

PARALLEL

I/O

PORTS

6.1 Objectives

After completing this chapter, you should be able to:

- define I/O addressing methods
- explain the data transfer synchronization methods between the CPU and I/O interface chip
- explain the data transfer synchronization methods between the I/O interface chip and the I/O device
- explain input and output handshake protocols
- input data from simple switches
- output data to LED and LCD displays
- explain the operation of the Centronics printer interface
- explain keyboard scanning and debouncing
- do I/O programming
- add PIA-type devices to the 68HC11
- do PIA programming

6.2 Basic I/O Concepts

Peripheral devices (also called I/O devices) are pieces of equipment that exchange data with a computer. Examples of peripheral devices include switches, light-emitting diodes, cathode-ray tube (CRT) screens, printers, modems, keyboards, and disk drives. The speeds and characteristics of these peripheral devices differ significantly from those of the CPU, so they are not connected directly to the microprocessor. Instead, interface chips are used to resolve the differences between the microprocessor and these peripheral devices.

The major function of an interface chip is to synchronize data transfer between the CPU and an I/O device. An interface chip consists of control registers, status registers, data registers, latches, data direction registers, and control circuitry. In input operations, the input device places the data in the data register, which holds the data until it is read by the microprocessor. In output operations, the processor writes the data into the data register on the interface chip, and the data register holds the data until it is fetched by the output device.

An interface chip has data pins that are connected to the microprocessor data bus and I/O port pins that are connected to the I/O device. Since a computer may have multiple I/O devices, the microprocessor data bus can be connected to the data buses of multiple interface chips. To avoid data bus contention, the address decoder circuitry allows only one interface chip at a time to react to the microprocessor data transfer. Each interface chip is designed to have a *chip enable* signal input or inputs. When the chip enable signal is asserted, the interface chip is allowed to react to the data transfer request from the CPU. Otherwise, the data pins of the interface chip are isolated from the microprocessor data bus. The data pin drivers of the interface chip are normally designed to be *tristate* devices, so that they remain in a high-impedance state except when they are enabled. When the output pins of a peripheral device are in a high-impedance state, the device is electrically isolated from the data bus.

The simplified block diagram in Figure 6.1 shows a microcomputer system with one RAM module, one ROM module, one input device (not shown) connected to an interface chip, and one output device (not shown) connected to an interface chip. To transfer data to or from one of these modules, the microprocessor sends out an address and the address decoder asserts one and only one of its outputs to enable one module to respond.

Data transfer between the I/O device and the interface chip can proceed bit-by-bit (serial) or in multiple bits (parallel). Data are transferred serially in low-speed devices such as modems and low-speed printers. Parallel data transfer is mainly used by high-speed I/O devices. We will only discuss parallel I/O in this chapter. Serial I/O will be discussed in Chapters 8 and 9. Other issues related to I/O operations will be discussed in the following sections.

6.3 I/O Addressing

An interface chip normally has several registers. Each of these registers must be assigned a separate address in order to be accessed. It is possible to

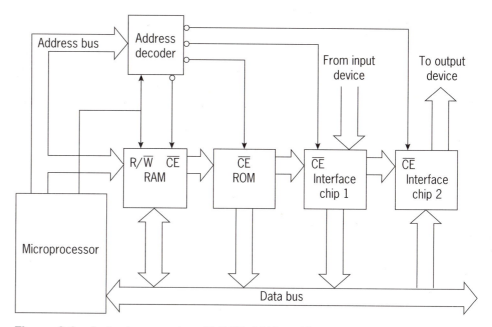

Figure 6.1 A simple computer with RAM, ROM, and input and output ports

have two or more registers share the same address, but in this case some other control signal, such as the R/$\overline{W}$ signal, must then be used to select one of the registers. This approach is very common in the interface chips designed for 8-bit microprocessors.

Different microprocessors deal with I/O devices differently. Some microprocessors have dedicated instructions for performing input and output operations—this approach is called *isolated I/O*. In this method, input instructions, such as IN 1, read from the input devices. This input instruction inputs a word from input device 1 into the accumulator. Similarly, the output instruction OUT 4 outputs a word from the accumulator to the fourth output device. In the isolated I/O method, I/O devices have their own address space, which is separate from the main memory address space. The advantages of this method include:

- The isolated I/O method is not as susceptible to software errors as memory-mapped I/O because different instructions are used to access memory and I/O devices.
- I/O devices do not occupy the limited memory space, which is quite small in an 8-bit microprocessor.
- The I/O address decoder can be smaller because the I/O address space is much smaller.

Its disadvantages are:

- Inflexible I/O instructions—most microprocessors that use the isolated I/O method have only a few I/O instructions.
- Inflexible I/O addressing modes—the memory addressing modes are not available for addressing I/O devices.

Other microprocessors do not have separate instructions for input and output. Instead, these microprocessors use the same instructions for reading from memory and reading from input devices, as well as for writing data into memory and writing data to output devices. I/O devices and main memory share the same address space. This approach is called *memory-mapped I/O*. In this method, an I/O address such as $1000 is considered a byte in memory space. A load instruction like LDAA $1000 is then an input instruction, and a store instruction like STAA $1000 is an output instruction. There is no need for a separate input or output instruction. The major advantage of memory-mapped I/O is that all instructions and addressing modes are available to I/O operations, so I/O programming is very flexible. The disadvantages of memory-mapped I/O are:

- Memory-mapped I/O is more susceptible to software errors. A programmer may accidentally write into the main memory during an I/O operation or into an I/O device during a main memory access, thus corrupting the computer system.
- A smaller memory space is available for main memory.

Traditionally, Intel microprocessors use the isolated I/O method, whereas Motorola microprocessors use the memory-mapped I/O method. The 68HC11 microcontroller has no separate I/O instructions and uses memory-mapped I/O exclusively.

6.4 I/O Transfer Synchronization

The role of an interface chip is illustrated in Figure 6.2. The electronics in the I/O device converts electrical signals into mechanical actions or vice versa, but the microprocessor interacts with the interface chip instead of dealing with the I/O device directly.

When inputting data, the microprocessor reads the data from the interface chip, so there must be some mechanism to make sure that the data is valid when the microprocessor reads it. When outputting data, the microprocessor writes data into the interface chip. Again, there must be some mechanism to make sure that the output device is ready to accept the data when the micro-

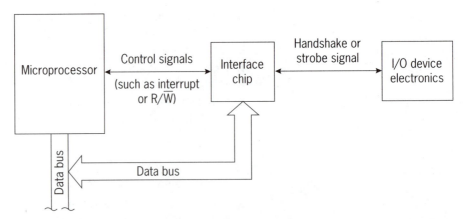

Figure 6.2 The role of an interface chip

processor outputs it. These two aspects of the I/O synchronization process—synchronization between the interface chip and the microprocessor and synchronization between the interface chip and the I/O device electronics—will be discussed in the following subsections.

6.4.1 Synchronizing the Microprocessor and the Interface Chip

To input valid data from an input device, the microprocessor must make sure that the interface chip has correctly latched the data from the input device. There are two ways to do this:

1. *The polling method.* The interface chip uses a status bit to indicate whether it has valid data (normally stored in a data register) for the microprocessor. The microprocessor knows that the interface chip has valid data when the status bit is set to 1. The microprocessor keeps checking (reading) the status register until the status bit is set to 1, and then it reads the data from the data register. When this method is used, the CPU is tied up and cannot do anything else when it is reading the data. However, this method is very simple to implement and is often used if the microprocessor has nothing else to do when waiting for completion of the input.

2. *The interrupt-driven method.* In this method, the interface chip asserts an interrupt signal to the microprocessor when it has valid data in the data register. The microprocessor then executes the service routine associated with the interrupt to read the data.

To output data successfully, the microprocessor must make sure that the output device is not busy. There are also two ways to do this:

1. *The polling method.* The interface chip has a data register that holds data that is to be output to an output device. New data should be sent to the interface chip only when the data register of the interface chip is empty. In this method, the interface chip uses a status bit to indicate whether the output data register is empty. The microprocessor keeps checking the status bit until it indicates that the output data register is empty and then writes the data into it.

2. *The interrupt-driven method.* The interface chip asserts an interrupt signal to the microprocessor when the output data register is empty and can accept more data. The microprocessor then executes the service routine associated with the interrupt and outputs the data.

6.4.2 Synchronizing the Interface Chip and the I/O Devices

The interface chip is responsible for making sure that data is properly transferred to and from I/O devices. The following methods have been used to synchronize data transfer between the interface chip and I/O devices:

■ *The brute-force method.* Nothing special is done in this method. For input, the interface chip returns the voltage levels on the input port pins to the microprocessor. For output, the interface chip makes the data written by the microprocessor directly available on output port pins. This method is useful in situations in which the timing of the data is unimportant. It can be used to test the voltage level of a signal, set the voltage of an output pin to high or low, or drive LEDs. All I/O ports of the 68HC11 can perform brute force I/O.

■ *The strobe method.* This method uses strobe signals to indicate that data are stable on input or output port pins. During input, the input device asserts a strobe signal when the data are stable on the input port pins. The interface chip latches the data into the data register using the strobe signal. For output, the interface chip first places the output data on the output port pins. When the data become stable, the interface chip asserts a strobe signal to inform the output device to latch the data on the output port pins. This method can be used if the interface chip and I/O device can keep up with each other. Port B of the 68HC11 supports the strobed output mode, while port C supports the strobed input mode.

■ *The handshake method.* The previous two methods cannot guarantee correct data transfer between an interface chip and an I/O device when the timing of data is critical. For example, it takes a much longer time to print a character than it does to send a character to the printer electronics, so data shouldn't be sent to the printer if it is still printing. The solution is to use a handshake protocol between the interface chip and the printer electronics. There are two handshake methods: the *interlocked handshake* and the *pulse-mode handshake.* Whichever handshake protocol is used, two handshake signals are needed—one (let's call it H1) is asserted by the interface chip and the other (let's call it H2) is asserted by the I/O device. The handshake signal transactions for input and output are described in the following subsections. Note that the handshake operations of some interface chips may differ slightly from what we describe here. Port C of the 68HC11 supports the handshake I/O method.

INPUT HANDSHAKING PROTOCOL

The signal transaction of the input handshake protocol is illustrated in Figure 6.3.

Step 1
The interface chip asserts (or pulses) H1 to indicate its intention to input a byte.

Step 2
The input device puts valid data on the data port and also asserts (or pulses) the handshake signal H2.

Step 3
The interface chip latches the data and de-asserts the H1 signal. After some delay, the input device also de-asserts the H2 signal.

The whole process will be repeated if the interface chip wants more data.

OUTPUT HANDSHAKE PROTOCOL

The signal transaction for output handshaking is shown in Figure 6.4. It also takes place in three steps:

Step 1
The interface chip places data on the data port and asserts (or pulses) H1 to indicate that it has data to be output.

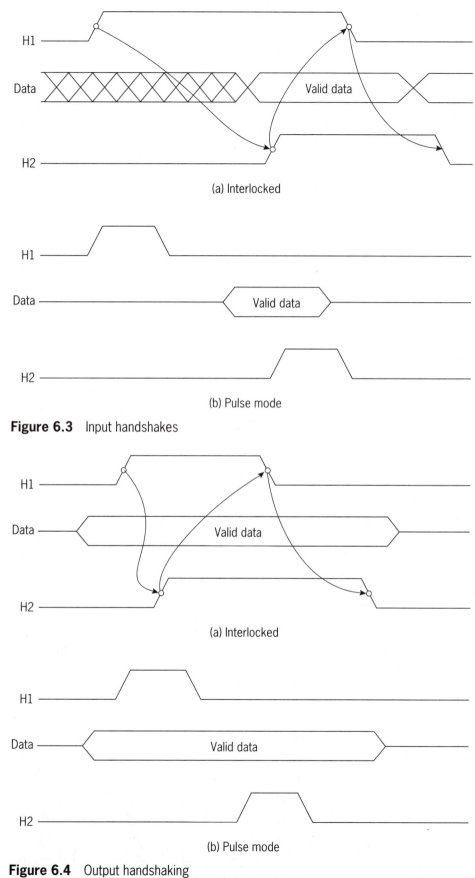

(a) Interlocked

(b) Pulse mode

Figure 6.3 Input handshakes

(a) Interlocked

(b) Pulse mode

Figure 6.4 Output handshaking

Step 2

The output device latches the data and asserts (or pulses) H2 to acknowledge the receipt of data.

Step 3

The interface chip de-asserts H1 following the assertion of H2. The output device then de-asserts H2.

The whole process will be repeated if the microprocessor has more data to output.

6.5 68HC11 Parallel I/O Overview

The 68HC11A8 has 40 I/O pins arranged in five I/O ports. All of these pins serve multiple functions depending on the operation mode and data in the control registers. Ports C and D are used as general-purpose I/O pins under the control of their associated data direction registers. Ports A, B, and E, with the exception of port A pin 7, are fixed-direction inputs or outputs and therefore do not have data direction registers. Port B, port C, the STRA pin, and the STRB pin are used for strobed and handshake parallel I/O, as well as for general-purpose I/O.

Other series of 68HC11 microcontrollers have different numbers of I/O ports. Interested readers should refer to the appropriate technical data book.

6.5.1 Bidirectional I/O (Ports C and D)

Each pin of ports C and D has an associated bit in a specific port data register and another in a data direction register. The data direction register bits are used to specify the primary direction of data for each I/O pin. When an output pin is read, the value at the input to the pin driver is returned. When a pin is configured as an input, that pin becomes a high-impedance input. If a write is executed to an input pin, its value does not affect the I/O pin but is stored in an internal latch. When the pin becomes an output, this value appears at the I/O pin. Data direction register bits are cleared by reset to configure I/O pins as inputs.

The AS and $R/\overline{W}$ pins are used for bus control in the expanded mode and as parallel I/O strobes (STRA and STRB) in the single-chip mode.

Other series of the 68HC11 have more bidirectional I/O ports than the 68HC11A8. A partial list is shown in Table 6.1. A data direction register is associated with each general-purpose bidirectional I/O port.

Series	General I/O port names
A and E	8-bit port: C; 6-bit port: D
D	8-bit ports: B, C, and D
F	8-bit ports: A, C, and G; 6-bit port: D
G	8-bit ports: A, C, G, H, and J; 6-bit port: D
K	8-bit ports: A, B, C, F, G, and H; 6-bit port: D
L	8-bit ports: C and G; 6-bit port: D
N	8-bit ports: A, B, C, F, and H; 6-bit port: D

Table 6.1 ■ 68HC11 bidirectional I/O ports (partial list)

Example 6.1

Write an instruction sequence to set port D as an output port, and output the value $CD to it.

Solution: Since port D is bidirectional, we need to configure it as an output port.

```
REGBAS    equ   $1000      ; base address of I/O registers
PORTD     equ   $08        ; offset of PORTD from REGBAS
DDRD      equ   $09        ; offset of DDRD from REGBAS

          LDX   #REGBAS
          LDAA  #$3F       ; port D has only 6 pins
          STAA  DDRD,X     ; set port D as an output port
          LDAA  #$CD
          STAA  PORTD,X    ; output the data
          END
```

6.5.2 Fixed Direction I/O (Ports A, B, and E)

In the 68HC11A8, the pins of ports A, B, and E (except for port A bit 7) have fixed data directions. When port A is being used for general-purpose I/O, bits 0, 1, and 2 are configured as input only, and writes to these pins have no effect. Bits 3, 4, 5, and 6 of port A are configured as output only, and reads of these pins return the levels sensed at the input to the pin drivers. Port A bit 7 can be configured as a general-purpose input or output using the DDRA7 bit (bit 7) in the pulse accumulator control register (PACTL). Port B is configured as output only, and reads of these pins will return the levels sensed at the input of the pin drivers. Port E contains the eight A/D channel inputs, but these pins can also be used as general-purpose digital inputs. Writes to port E have no effect. The directions of the port A pins vary with the chip version, as shown in Table 6.2.

It is fairly straightforward to do input and output with these fixed-direction I/O ports. For example, the following instruction sequence outputs the value of $FF to port B:

```
REGBAS    equ   $1000      ;base address of I/O registers
PORTB     equ   $04        ;offset of PORTB from REGBAS
          LDX   #REGBAS
          LDAA  #$FF
          STAA  PORTB,X
```

The following instruction sequence reads a byte from port E:

```
PORTE     equ   $0A
          LDAA  PORTE,X
```

6.5.3 Port Registers

The 68HC11 writes to the port data register when outputting data and reads from the port data register when inputting data from an input device. Each I/O port has an associated data register. Port C also has a latched port

Chip version	Pin directions
A series (A0, A1, A9)	1 bidirectional (PA7) 7 fixed directional, PA0–PA2 are input and PA3–PA6 are output
D series (D0, D3, 711D3)	2 bidirectional (PA7 and PA3) 4 fixed directional, PA0–PA2 are input and PA5 is output
E series (E0, E1, E9, E20, 711E9, 711E20, E8, 811E2)	2 bidirectional (PA7 and PA3), 6 fixed directional, PA0–PA2 are input and PA4–PA6 are output
F, G, K, L, M and N series (see Table 4.1)	8 bidirectional (pin directions are to be set by programming the DDRA register)

1. The directions of PA7 and PA3 in the D and E series are set by programming the DDRA7 and DDRA3 bits of the PACTL register.
2. The direction of PA7 in A series is set by programming the DDRA7 bit of the PACTL register.

Table 6.2 ■ 68HC11 chip versions and port A pin directions (a partial list)

data register (PORTCL). The PORTCL register is part of the handshake I/O subsystem and will be explained in Section 6.7. The addresses of these data registers are as follows:

PORTA ($1000)
PORTB ($1004)
PORTC ($1003)
PORTCL ($1005)
PORTD ($1008)
PORTE ($100A)

6.6 68HC11 Strobed I/O

Port B of the 68HC11 supports simple strobed output. Port C supports strobed input as well as handshake input and output. The strobe mode and handshake mode I/O are controlled by the *parallel I/O control register* (PIOC), whose contents are as follows:

	7	6	5	4	3	2	1	0	
	STAF	STAI	CWOM	HNDS	OIN	PLS	EGA	INVB	PIOC at $1002
value after reset	0	0	0	0	0	U	1	1	

STAF: Strobe A flag. This bit is set when a selected edge occurs on the STRA signal.

STAI: Strobe A interrupt enable. When the STAF and STAI bits are both equal to 1, a hardware interrupt request will be made to the CPU.

CWOM: Port C wired-or mode
    0: All port C outputs are normal CMOS outputs.
    1: All port C outputs act as open-drain outputs.

HNDS: Handshake/simple strobe mode select
    0: simple strobe mode
    1: handshake mode

OIN: Output/input handshaking
    0: input handshake
    1: output handshake

PLS: Pulse/interlocked handshake operation
    0: interlocked handshake selected
    1: pulse handshake selected

    When the interlocked handshake operation is selected, STRB, once activated, stays active until the selected edge of STRA is detected. When the pulsed handshake operation is selected, STRB is pulsed for 2 E cycles.

EGA: Active edge for STRA
  0:      falling edge
  1:      rising edge

INVB: Invert strobe B
  0:      STRB active level is logic 0.
  1:      STRB active level is logic 1.

The parallel I/O strobe mode can be selected by setting the handshake bit (HNDS) in the PIOC register to 0. Port C becomes a strobed input port with the STRA pin as the edge-detecting latch command input. Port B becomes a strobed output port with the STRB pin as an output strobe. The active edges of STRA and STRB are defined by the PIOC register.

6.6.1 Strobed Input Port C

In this mode, there are two addresses where port C can be read, the PORTC data register and the latched port C register (PORTCL). Reading the PORTC register returns the current values on the port C pins. Reading the PORTCL register returns the contents of the latch PORTCL. The STRA pin is used as an edge-detecting input, and the edge-select for the strobe A (EGA) bit in the PIOC register defines either the falling or the rising edge as the significant edge. Whenever the selected edge is detected at the STRA pin, the current logic levels at port C are latched into the PORTCL register and the strobe A flag (STAF) in the PIOC register is set. If the strobe A interrupt enable (STAI) bit is also set, an internal interrupt to the CPU is generated. The STRA interrupt shares an interrupt vector with the $\overline{IRQ}$ pin interrupt. The strobe A flag is automatically cleared by reading the PIOC register and then reading the

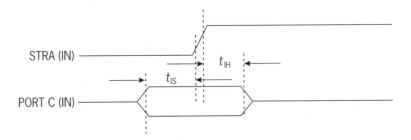

t_{IS}: input setup time (60 ns at 2 MHz)
t_{IH}: input hold time (100 ns at 2 MHz)

Figure 6.5   Port C strobe input timing (Redrawn with permission of Motorola)

PORTCL register. The signal transaction for the port C strobed input (where the active edge of STRA is the rising edge) is shown in Figure 6.5. The parameter t_{IS} is the data setup time required by the port C strobe input, and the parameter t_{IH} is the data hold time required by the port C strobe input.

6.6.2 Strobed Output Port B

In this mode, the STRB pin is a strobed output that is pulsed for two E clock periods each time there is a write to port B. The INVB bit in the PIOC register controls the polarity of the pulse on the STRB pin. The active level is low when the INVB bit is 0. The signal transaction for the port B strobed output is shown in Figure 6.6. The STRB signal is programmed to be active high in this example. The port B data becomes valid t_{PWD} ns after the rising edge of the E clock. The STRB signal is then asserted (high in Figure 6.6) t_{DEB} ns after the falling edge of the E clock. The STRB signal will remain high for two E clock cycles. An external output device will use the STRB signal to latch the port B data.

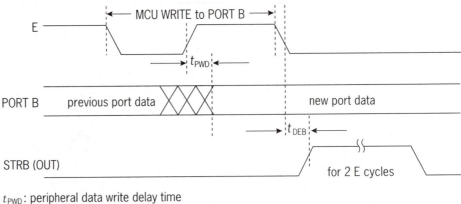

t_{PWD}: peripheral data write delay time
t_{DEB}: E fall to STRB valid delay time

Figure 6.6   Port B strobe output timing (Redrawn with permission of Motorola)

6.7 Port C Handshake I/O

The parallel I/O handshake modes involve port C and the STRA and STRB pins. There are two basic modes (input and output) and two variations on the handshake signals (pulsed and interlocked). In all handshake modes, STRA is an edge-detecting input and STRB is a handshake output pin (which indicates that I/O device is ready for the next I/O transaction).

6.7.1 Port C Input Handshake Protocol

In the input handshake protocol, port C is a latching input port, STRA is an edge-sensitive latch command from the external system that is driving port C, and STRB is a ready output pin controlled by logic in the 68HC11.

When it is ready to accept new data, the 68HC11 asserts (or pulses) the STRB pin, then the external device places data on the port C pins and asserts (or pulses) the STRA pin. The active edge on the STRA pin latches the port C data into the PORTCL register, sets the STAF flag, and de-asserts the STRB pin. De-assertion of the STRB pin automatically inhibits the external device from strobing new data into port C. Reading the PORTCL register asserts the STRB pin, indicating that new data can now be applied to port C. Note that reading the PORTC data register does not assert the STRB pin. The signal transaction for port C handshake signals is shown in Figure 6.7. The STRB signal is active high in this figure.

The STRB pin can be configured (by setting the PLS control bit) to be a pulse output (pulse mode) or a static output (interlocked mode). The port C data direction register bits should be cleared for each pin that is to be used as

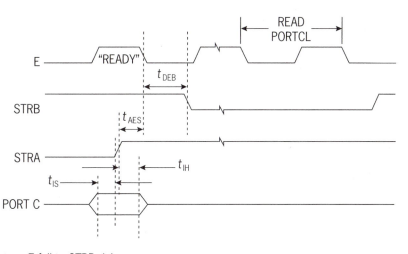

t_{DEB}: E fall to STRB delay
t_{AES}: STRA asserted to E fall
t_{IS}: port C data setup time
t_{IH}: port C data hold time

Figure 6.7 Port C interlocked input handshaking (Redrawn with permission of Motorola)

a latched input pin. However, some port C pins can be used as latched inputs with the input handshake protocol while other port C pins are simultaneously used as static inputs and still others as static outputs. The input handshake protocol has no effect on the use of port C pins as static inputs or static outputs. Reads of the PORTC data register always return the static logic level at the port C pins (for pins configured as inputs).

6.7.2 Port C Output Handshake Protocol

In the output handshake protocol, port C is an output port, STRB is a "ready" output, and STRA is an edge-sensitive acknowledge input signal that is used to indicate to the 68HC11 that the output data has been accepted by the external device. In a variation of this output handshake protocol, STRA can also be used as an output-enable input and as an edge-sensitive acknowledge input. The 68HC11 places data on the port C output pins (by writing into the PORTCL register) and then indicates that stable data is available by asserting (or pulsing) the STRB pin. The external device then processes the available data and asserts (or pulses) the STRA pin to indicate that new data can be placed on the port C output pins. The active edge on the STRA pin causes the STRB pin to be de-asserted and the STAF status flag to be set. In response to the setting of the STAF bit, the program transfers new data out of port C as required. Writing data to the PORTCL register causes the data to appear on the port C pins and asserts the STRB pin. The signal transaction for the port C handshake output is shown in Figure 6.8.

In Figure 6.8, both handshake signals are active high. When the 68HC11 writes a byte into the PORTCL register, the data becomes valid t_{PWD} ns after the rising edge of the E clock. The STRB handshake output is asserted t_{DEB} ns after the rising edge of the E clock signal, and the STRB signal is used by the output device to latch data. After latching the port C data, the output device asserts the STRA signal to inform the 68HC11 that the data has been latched.

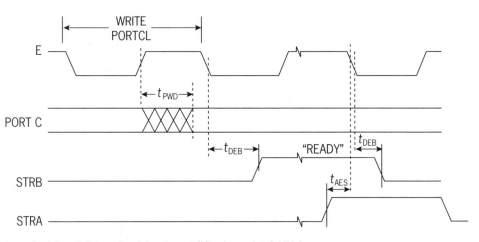

t_{PWD}: Peripheral data write delay time, 150 ns max (at 2 MHz)
t_{DEB}: E fall to STRB delay, 225 ns (at 2 MHz)
t_{AES}: STRA asserted to E fall setup time, 0 ns (at 2 MHz)

Figure 6.8 Port C interlocked output handshake timing (Redrawn with permission of Motorola)

The STRB output is then dropped to tell the output device that the data on the port C pins is not new. The output device also drops the STRA handshake input to complete the handshake cycle. When the STRA input pin is inactive, all port C pins obey the data direction specified by the data direction register so that pins that are configured as inputs are high impedance. When the STRA pin is activated, all port C pins are forced to outputs regardless of the data in the data direction register. Port C pins intended as static outputs or normal handshake outputs should have their corresponding data direction register bits set, and pins intended as tristate handshake outputs should have their corresponding data direction register bits clear. A user must write data to PORTC to change the nonhandshake pins in port C and to PORTCL to change the data on a full-handshake port C pin.

6.8 Simple Input Devices

An input device is any device that sends a binary number to the microprocessor. Examples of input devices include switches, analog-to-digital converters, keyboards, etcetera. A set of eight DIP switches is probably the simplest input device that we can find. Such switches can be connected directly to port E or port C of the 68HC11, as shown in Figure 6.9. When a switch is

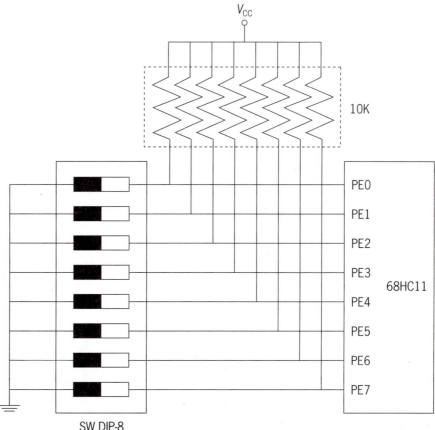

Figure 6.9   Connecting a set of eight DIP switches to port E of the 68HC11

closed, the associated port E pin input is 0. Otherwise, the associated port E pin has an input of 1. Each port E pin is pulled up to high via a 10KΩ resistor when the associated switch is open.

Example 6.2

Write a sequence of instructions to read the value from an eight-DIP switch connected to port E of the 68HC11.

Solution:
```
REGBAS   EQU  $1000       ; base address of I/O register block
PORTE    EQU  $0A         ; offset of PORTE from REGBAS

         LDX  #REGBAS     ; place the base address of I/O register block in X
         LDAA PORTE,X     ; read port E value
         .
         .
         .
```

6.9 Interfacing Parallel Ports to the Keyboard

A keyboard is arranged as an array of switches, which can be mechanical, membrane, capacitive, or Hall-effect in construction. In mechanical switches, two metal contacts are brought together to complete an electrical circuit. In membrane switches, a plastic or rubber membrane presses one conductor onto another; this type of switch can be made very thin. Capacitive switches internally comprise two plates of a parallel plate capacitor; pressing the key cap effectively increases the capacitance between the two plates. Special circuitry is needed to detect this change in capacitance. In a Hall-effect key switch, the motion of a crystal perpendicular to the magnetic flux lines of a permanent magnet is detected as a voltage appearing between the two faces of the crystal—it is this voltage that registers a switch closure.

Because of their construction, mechanical switches have a problem called *contact bounce*. Instead of producing a single, clean pulse output, closing a mechanical switch generates a series of pulses because the switch contacts do not come to rest immediately. This phenomenon is illustrated in Figure 6.10,

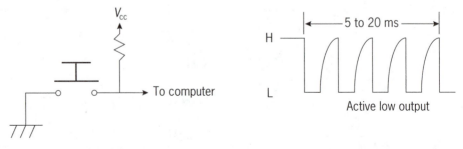

Figure 6.10   Contact bounce

where it can be seen that a single physical push of the button results in multiple electrical signals being generated and sent to the computer. (In Figure 6.10, when the key is not pressed, the voltage output to the computer rises to 5 V.) The response time of the switch is several orders of magnitude slower than that of the computer, so the computer could read a single switch closure many times during the time the switch is operated, interpreting each low signal as a new input when in fact only one input is being sent.

Because of the contact bounce in the keyboard, a debouncing process is needed. A keyboard input program can be divided into three stages, which will be discussed in the following subsections:

1. Keyboard scanning to find out which key was pressed

2. Key debouncing to make sure a key was indeed pressed

3. Table lookup to find the ASCII code of the key that was pressed

6.9.1 Keyboard Scanning Techniques

A keyboard can be arranged as a linear array of switches if it has only a few keys. A keyboard with more than a few keys is often arranged as a matrix of switches that uses two decoding and selecting devices to determine which key was pressed by coincident recognition of the row and column of the key.

As shown in Figure 6.11, a CMOS analog multiplexor MC14051 and a 3-to-8 decoder can be used to scan the keyboard to determine which key is pressed. In this discussion, we will assume that either no key or only one key is pressed. A scanning program is invoked to search for the pressed key, and then the debounce routine is executed to verify that the key is closed. For a 64-key keyboard, six bits are needed to select the rows and columns (three bits for each). The MC14051 has three select inputs (A, B, and C), eight data inputs (X0–X7), one inhibit input, and one common output (X). The common output is active

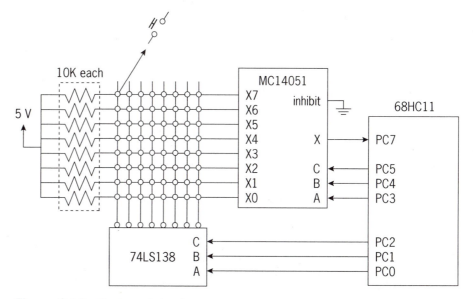

Figure 6.11 Keyboard structure

high. When a data input is selected by the select signals, its voltage will be passed to the common output.

The MC14051 is used to select the row, and a 74LS138 3-to-8 decoder is used to select the column. The data inputs (X7–X0) of the MC14051 are pulled up to high. When a key is pressed, the row and column where the keyboard is pressed are shorted together by the switch. Since the 3-to-8 decoder has low output, the common output of the MC14051 will be driven to low and can be detected by the microcontroller.

Port C is used to handle the keyboard scanning in the following program. Since port C is a bidirectional port, it must be configured properly. In our example, we will use the most significant bit (msb) of port C to detect whether a key has been pressed. Therefore, pin 7 of port C should be configured for input and the lower six bits for output. Bit 6 is not used. The following two instructions will configure port C as desired.

```
DDRC    EQU  $1007
        LDAA #$3F
        STAA DDRC
```

The reader should know that if two adjacent keys in the same matrix row are pressed at the same time, the output pins of the 74LS138 will be shorted together and may possibly damage the decoder. Practical keyboard circuits include diodes to prevent such damage to the interface circuitry.

The ideas behind the keyboard scanning program are illustrated in the flowchart in Figure 6.12. Since the lowest six bits of the PORTC register are used to specify the row number and column number, these two indices can be incremented by a single INC instruction. If the keyboard is not pressed and the lowest six bits of the PORTC register are all 1s, the row number and column number must be reset to 0. The scanning program is as follows:

```
REGBAS  EQU  $1000
KEYBD   EQU  $03              ; use port C for keyboard

        LDX   #REGBAS
RESETC  CLR   KEYBD,X         ; start from row 0 and column 0
SCAN    BRCLR KEYBD,X $80 DEBNCE    ; detect a pressed key
        BRSET KEYBD,X $3F RESETC    ; need to reset the row and column count
        INC   KEYBD,X         ; check the next row or column
        BRA   SCAN
        END
```

Keyboard decoding can be done strictly in software without using any hardware decoder or multiplexer. One such example is given in problem E6.16. A membrane-type keyboard is often used in products that require the user to make the selection. However, programs become much longer when a hardware decoder is not used.

6.9.2 Keyboard Debouncing

Contact bounce is due to the dynamics of a closing contact. The signal falls and rises a few times within a period of about 5 milliseconds as the contacts bounce. Since a human being cannot press and release a switch in less

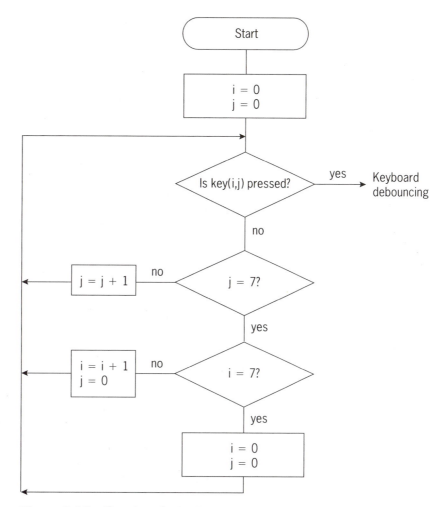

Figure 6.12 Flowchart for keyboard scanning for the circuit in Figure 6.11

than 20 ms, a debouncer will recognize that the switch is closed after the voltage is low for about 10 ms and will recognize that the switch is open after the voltage is high for about 10 ms.

Both hardware and software solutions to the key bounce problem are available. Keyboard bouncing can be reduced by using a good switch. A mercury switch is much faster, and optical and Hall-effect switches are free of bounce. Hardware solutions to contact bounce include an analog circuit that uses a resistor and a capacitor to smooth the voltage and two digital solutions that use set-reset flip-flops or CMOS buffers and double-throw switches.

In practice, hardware and software debouncing techniques are both used (but not at the same time). Dedicated hardware scanner chips are also used frequently—typical ones are the National Semiconductor 74C922 and 74C923. These two chips perform both keyboard scanning and debouncing.

HARDWARE DEBOUNCING TECHNIQUES

■ *Set-reset flip-flops.* Before being pressed, the key is touching either the set or the reset input. When pressed, the key moves to the other position. When the key bounces, it will not return to its original position nor will it move to the opposite position. Both the set and the reset inputs are grounded, and hence the output Q will not change. Since the flip-flop has memory, the key will be recognized as closed. The principle behind this solution is that the flip-flop changes value only once when the key plate moves from one side to the other. This solution is illustrated in Figure 6.13a.

■ *Noninverting CMOS gates with high-input impedance,* as shown in Figure 6.13b. When the wiper moves, the resistor tends to maintain the output voltage at its old value. On the first bounce, the output changes, and it remains at that level as the wiper leaves the plate, because of the resistor. Successive bounces do not change the output. Thus, the output is debounced.

■ *Integrating debouncers.* The RC constant of the integrator (smoothing filter) determines the rate at which the capacitor charges up towards the supply voltage once the ground connection via the switch has been removed. As long as the capacitor voltage does not exceed the logic zero threshold value, the V_{out} signal will continue to be recognized as a logic zero. This solution is shown in Figure 6.13c.

SOFTWARE DEBOUNCING TECHNIQUES

An easy software solution to the key bounce problem is the *wait-and-see technique.* When the input drops, indicating that the switch might be closed, the program waits 10 ms and looks at the input again. If it is still low, the program decides that the key has indeed been pressed. If it is high, the program decides that the input signal was noise or that the input is bouncing—if it is bouncing, it will certainly be clear later if the key was actually pressed. In either case, the program returns to wait for the input to drop. The following sample program uses this technique to do the debouncing:

```
REGBAS  EQU  $1000      ; base address of I/O register block
KEYBD   EQU  $03        ; offset of register PORTC from REGBAS
TEN_MS  EQU  2000       ; 10 ms in number of E clock cycles

DEBNCE  LDY  #REGBAS
        LDX  #TEN_MS
WAIT    NOP             ; wait for 10 ms
        NOP             ;     "
        DEX             ;     "
        BNE  WAIT       ;     "
        LDAA KEYBD,Y    ; recheck the pressed key
        BMI  SCAN       ; rescan the keyboard if the key is not pressed
        JMP  GETCODE    ; the keyboard is indeed pressed
        END
```

6.9.3 ASCII Code Table Lookup

After the key has been debounced, the keyboard should reference the ASCII table and send the corresponding ASCII code to the CPU. The code segment for looking up the ASCII code in a table is:

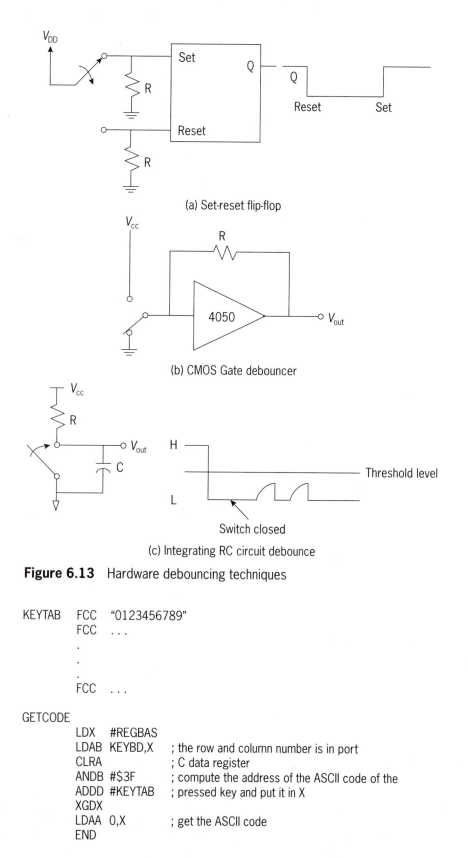

(a) Set-reset flip-flop

(b) CMOS Gate debouncer

(c) Integrating RC circuit debounce

Figure 6.13 Hardware debouncing techniques

```
KEYTAB   FCC   "0123456789"
         FCC   . . .
          .
          .
          .
         FCC   . . .

GETCODE
         LDX    #REGBAS
         LDAB   KEYBD,X      ; the row and column number is in port
         CLRA                ; C data register
         ANDB   #$3F         ; compute the address of the ASCII code of the
         ADDD   #KEYTAB      ; pressed key and put it in X
         XGDX
         LDAA   0,X          ; get the ASCII code
         END
```

6.10 Simple Output Devices

A microprocessor is commonly connected to many more types of output devices than input devices. These output devices include light-emitting diode (LED) indicators, segmented displays constructed from LEDs, liquid crystal displays (LCDs), motors, relays, DACs (digital-to-analog converters), and even vacuum-tube devices such as fluorescent displays.

6.10.1 A Single LED

An LED will illuminate if it is forward-biased and has enough current flowing through it. Depending on the manufacturer, the current required to light an LED brightly may range from a few milliamperes to more than 10 mA. The voltage drop across the LED when it is forward-biased can range from about 1.6 V to above 2 V.

LED indicators are simple to interface to the microprocessor if the interface circuit has at least 10 mA of current available to drive the LED. In the following discussion, we shall assume that the LED requires about 10 mA to light and that the forward voltage drop is about 1.7 V. Figure 6.14 illustrates a single LED indicator connected so that a logic 1 causes it to illuminate. The 74HC04 inverter has a low output voltage of about 0.1 V and a high output voltage of about 4.9 V. The current flow through the LED is about 10 mA when the input to the inverter is 1. When the LED is lit, it glows red, green, or yellow. Notice in this interface that an inverter provides the required 10 mA of current through a series current-limiting resistor. The inverter selected here can sink up to 20 mA of current.

6.10.2 The Seven-Segment Display

A seven-segment display consists of seven LED segments (a, b, c, d, e, f, and g). There are two types of seven-segment displays. In a *common-cathode seven-segment display,* the cathodes of all seven LEDs are tied together and a segment will be lit whenever a high voltage is applied at the corresponding segment input. In a *common-anode seven-segment display,* the anodes of all seven LEDs are tied in common and a segment will be lit whenever a low volt-

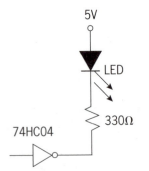

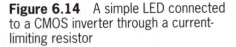

Figure 6.14   A simple LED connected to a CMOS inverter through a current-limiting resistor

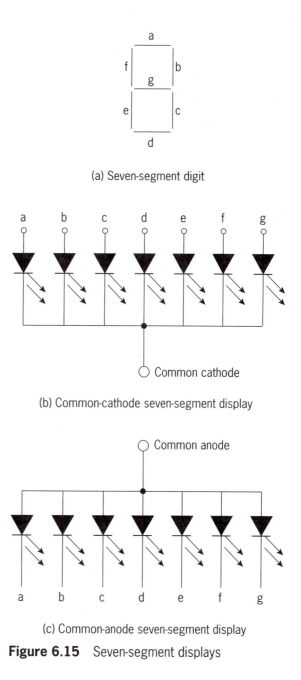

(a) Seven-segment digit

(b) Common-cathode seven-segment display

(c) Common-anode seven-segment display

Figure 6.15 Seven-segment displays

age is applied at the corresponding segment input. Diagrams of a common-cathode seven-segment display and a common-anode seven-segment display are shown in Figure 6.15. The current required to light a segment is similar to that required to light an LED.

6.10.3 LCD Displays

The liquid crystal display (LCD), which is found on many digital wristwatches and in most calculators, is among the most familiar output devices.

The main advantage of the LCD display is that it has very high contrast, unlike the LED display, so it can be seen extremely well in very bright light. The main problem with the LCD display is that it requires a light source in dim or dark areas because it produces no light of its own. Light sources are often included at the edge of the glass to flood the LCD display with light in dimly lit environments. However, the light consumes more power than the entire display.

Figure 6.16 illustrates the basic construction of an LCD display. The LCD type of display that is most common today allows light to pass through it whenever it is activated. Earlier LCD displays absorbed light. Activation of a segment requires a low-frequency bipolar excitation voltage of 30–1000 Hz. The polarity of this voltage must change, or else the LCD will not be able to change very quickly.

The LCD functions in the following manner. When a voltage is placed across the segment, it sets up an electrostatic field that aligns the crystals in the liquid. This alignment allows light to pass through the segment. If no voltage is applied across a segment, the crystals appear to be opaque because they are randomly aligned. Random alignment is assured by the ac excitation voltages applied to each segment. In a digital watch, the segments appear to darken when they are activated because light passes through the segment to a black cardboard backing that absorbs all light. The area surrounding the activated segment appears lighter in color because much of the light is reflected by the randomly aligned crystals. In a *backlit* computer display, the segment appears to grow brighter because of a light placed behind the display; the light is allowed to pass through the segment when it is activated.

A few manufacturers have developed color LCD displays for use with computers and small color televisions. These displays use three segments (dots) for each picture element (pixel), and the three dots are filtered so that they pass red, blue, and green light. By varying the number of dots and the amount of time each dot is active, just about any color and intensity can be displayed. White light consists of 59% green, 30% red, and 11% blue light. Secondary

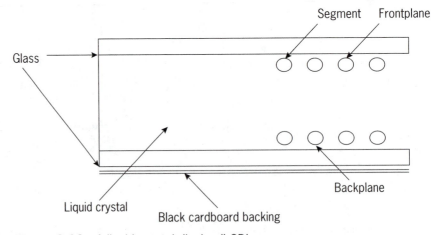

Figure 6.16 A liquid crystal display (LCD)

colors are magenta (red and blue), cyan (blue and green), and yellow (red and green). Any other colors, not just secondary colors, can be obtained by mixing red, green, and blue light.

6.10.4 Driving LED Displays with 6811 Parallel Ports

The 68HC11 can drive LED devices easily. We'll look at several examples that illustrate the interfacing method.

Example 6.3 ■ LED light pattern control with 68HC11.

Use the 68HC11 port B pins PB3, PB2, PB1, and PB0 to drive blue, green, red, and yellow LEDs. Light the blue LED for 2 seconds, then the green LED for 4 seconds, then the red LED for 8 seconds, and finally the yellow LED for 16 seconds. Repeat this operation forever.

Solution: The circuit connection is shown in Figure 6.17. As we have explained, sufficient current must be provided to illuminate the LEDs. The 68HC11 itself does not have the current capability to light the LEDs, so the 74HC04s are used to provide the driving capability required by the LED. The 74HC04 inverter has the capability to sink 20 mA of current.

Port B of the 68HC11 must output the appropriate values to turn the LEDs on and off. A two-level loop is used to create the desired delay. The inner loop creates a delay of $\frac{1}{10}$ second and must be repeated the appropriate number of times. To create 2-, 4-, 8-, and 16-second delays, the inner loop must be repeated 20, 40, 80, and 160 times. The values $08, $04, $02, and $01 should be written into port B in order to turn on the blue, green, red, and yellow lights, respectively. The light pattern and the repetition count are placed in a table. The program reads the loop count and the light pattern until the end of the table is reached and then repeats. The program is shown below:

```
REGBAS    equ   $1000      ; base address of I/O register block
PORTB     equ   $04        ; offset of PORTB from REGBAS
```

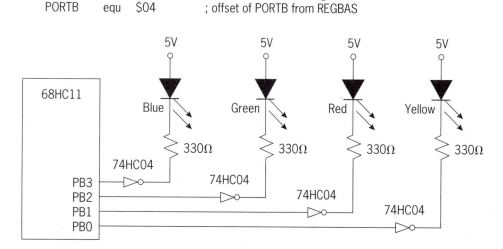

Figure 6.17 LEDs circuit connection

```
            ORG   $00
lt_tab      FCB   20,$08      ; LED light pattern table
            FCB   40,$04      ;     "
            FCB   80,$02      ;     "
            FCB   160,$01     ;     "

            ORG   $C000
LOOP        LDY   #lt_tab
NEXT        LDAB  0,Y          ; get the repetition count
            LDAA  1,Y          ; select the port B pin to be pulled up to 1
            INY                ; move to next entry
            INY                ;     "
            LDX   #REGBAS
            STAA  PORTB,X      ; output the LED light pattern
PT_LP       LDX   #20000
tenth_s     NOP                ; 1/10 second delay loop
            NOP                ;     "
            DEX                ;     "
            BNE   tenth_s      ;     "
            DECB
            BNE   PT_LP        ; repeat the TENTH_S loop
            CPY   #lt_tab+8    ; is this the end of the table?
            BNE   NEXT         ; if not, display the next light pattern
            BRA   LOOP         ; reinitialize
            END
```

Seven-segment displays are mainly used to display binary-coded-decimal (BCD) digits. Any output port of the 68HC11 can be used to drive seven-segment displays. Either port B or port C (port D does not have enough pins—it has only 6) can be used to output the segment pattern. One port will be adequate if there is only one display to be driven. The total supply current of the 68HC11 (shown in Table 6.3) is not enough to drive the seven-segment display directly, but this problem can be solved by adding a buffer chip such as 74ALS244. For example, a common-cathode seven-segment display can be driven by port B of the 68HC11, as shown in Figure 6.18.

The output high voltage (V_{OH}) of the 74ALS244 is about 3 V. Since the voltage drop of one LED segment is about 1.7 V, a 100 Ω current-limiting resistor

Operation mode	Operating frequency	Total supply current
single-chip	DC ≈ 2 MHz	15 mA
	3 MHz	27 mA
expanded	DC ≈ 2 MHz	27 mA
	3 MHz	35 mA
Total supply current is the total current supplied by the power supply.		

Table 6.3 ■ Total supply current of the 68HC11

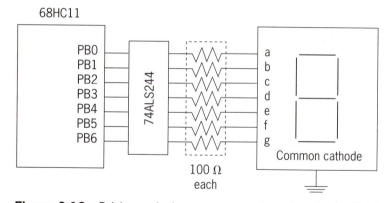

Figure 6.18 Driving a single seven-segment display with 68HC11

could limit the current to about 13 mA, which would be adequate to light an LED segment. The light patterns corresponding to the ten BCD digits are shown in Table 6.4. The numbers in Table 6.4 require that segments g, . . . , a be connected from the most significant bit to the least significant bit of the output port.

There is often a need to display multiple BCD digits at one time, and this can be accomplished using the time-multiplexing technique, which will be discussed shortly. An example of a circuit that can display six BCD digits is shown in Figure 6.19. In Figure 6.19, the common cathode of a seven-segment display is connected to the collector of an NPN 2N2222 transistor. When a port D pin voltage level is high, the connected NPN transistor will be turned on and be driven into the saturation region. The common cathode of the display will then be pulled down to low (about 0.2 V), allowing the display to be lit. By turning the six NPN transistors on and off in turn many times in one second, multiple digits can be displayed. A 2N2222 transistor can sink from 100 mA to 300 mA of current. The maximum current that flows into the common cathode is about 70 mA (7×10 mA = 70 mA) and hence can be sunk by a 2N2222 transistor.

BCD digit	Segments							Corresponding hex number
	g	f	e	d	c	b	a	
0	0	1	1	1	1	1	1	$3F
1	0	0	0	0	1	1	0	$06
2	1	0	1	1	0	1	1	$5B
3	1	0	0	1	1	1	1	$4F
4	1	1	0	0	1	1	0	$66
5	1	1	0	1	1	0	1	$6D
6	1	1	1	1	1	0	1	$7D
7	0	0	0	0	1	1	1	$07
8	1	1	1	1	1	1	1	$7F
9	1	1	0	1	1	1	1	$6F

Table 6.4 ■ BCD to seven-segment decoder

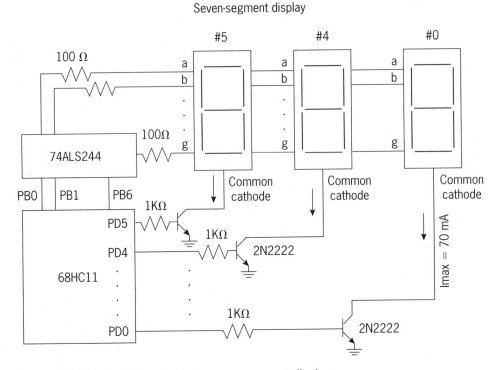

Figure 6.19   Port B controls six seven-segment displays

Example 6.4

Write a sequence of instructions to display a 6 on the #3 seven-segment display.

Solution: To display a 6 on the #3 display, the numbers $7D and $08 should be written into the port B and port D data registers, respectively. The number 7D is the segment pattern of digit 6. The number 08 sets the port D pin PD3 to 1 and other port D pins to 0, allowing only the #3 display to illuminate. The instruction sequence is as follows:

```
REGBAS  equ   $1000      ; I/O register block base address
PORTB   equ   $04        ; offset of PORTB from REGBAS
PORTD   equ   $08        ; offset of PORTD from REGBAS
DDRD    equ   $09        ; offset of DDRD from REGBAS
SIX     equ   $7D        ; segment pattern for 6
THIRD   equ   $08        ; value to set pin PD3
OUTPUT  equ   $3F        ; value to be written into DDRD
  *                      ; to configure port D for output

        ORG   $C000      ; starting address of the program
        LDX   #REGBAS
        LDAA  #OUTPUT
        STAA  DDRD,X     ; configure port D pins for output

        LDAA  #SIX
        STAA  PORTB,X    ; output the segment pattern of the digit 6
```

```
LDAA  #THIRD
STAA  PORTD,X      ; allow the display #3 to light
END
```

The circuit in Figure 6.19 can display six digits simultaneously by using the time-multiplexing technique, in which each seven-segment display is lighted in turn for a short period of time and then turned off. When one display is lighted, all the other displays are turned off. Within one second, each seven-segment display is lighted and then turned off many times. Because of the persistence of vision, the six displays will appear to be lighted simultaneously and continuously. The idea behind this technique is illustrated in the flowchart in Figure 6.20; in this example, each display is lighted for 1 ms at a time.

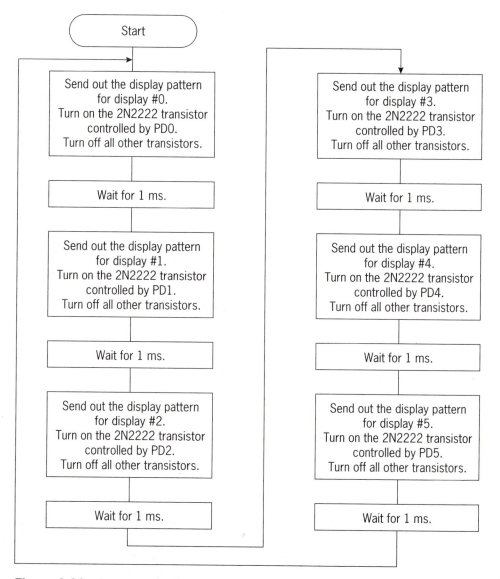

Figure 6.20 Program logic flow for time-multiplexing display technique

Example 6.5

Display 123456 on the six seven-segment displays shown in Figure 6.19.

Solution: Digits 1, 2, . . . , and 6 are displayed on the seven-segment displays #5, #4, . . . , #0. To light these displays, the numbers $20, $10, $08, $04, $02, and $01 should be written into the port D data register of the 68HC11. Build a table that consists of pairs of port B and port D values as follows:

Seven-segment LED	Displayed BCD digit	Port B	Port D
#5	1	$06	$20
#4	2	$5B	$10
#3	3	$4F	$08
#2	4	$66	$04
#1	5	$6D	$02
#0	6	$7D	$01

This table can be created by using the following assembler directives:

```
          ORG  $00
display   FCB  $06,$20   ; table of the display patterns
          FCB  $5B,$10   ;    "
          FCB  $4F,$08   ;    "
          FCB  $66,$04   ;    "
          FCB  $6D,$02   ;    "
          FCB  $7D,$01   ;    "
```

Instead of directly translating the flowchart in Figure 6.20 into the program, we first modify the algorithm slightly, as illustrated in Figure 6.21. If the address of the first byte of the display table is at *display,* then the address of the byte that immediately follows the last byte of the table is *display+12,* since there are six entries in the table.

The following program will turn on each seven-segment display for 1 ms at a time. Thus each seven-segment display will be turned on and off 167 times per second. All six digits will appear to be lighted simultaneously using this technique.

```
REGBAS   equ  $1000      ; I/O register block base address
PORTB    equ  $04        ; offset of PORTB register from REGBAS
PORTD    equ  $08        ; offset of PORTD register from REGBAS
DDRD     equ  $09        ; offset of DDRD register from REGBAS
OUTPUT   equ  $3F        ; value to be written into DDRD
*                        ; to configure port D for output

         ORG  $00
display  FCB  $06,$20    ; table of the display patterns
         FCB  $5B,$10    ;            "
         FCB  $4F,$08
         FCB  $66,$04
```

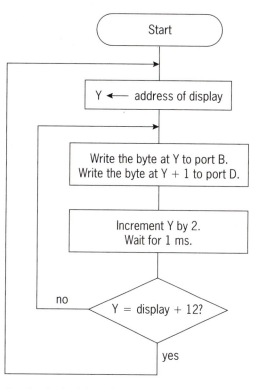

display is the label of the display table

Figure 6.21   Modified time-multiplexed seven-segment display

```
        FCB    $6D,$02
        FCB    $7D,$01

        ORG    $C000              ; starting address of the program
        LDX    #REGBAS
        LDAA   #OUTPUT
        STAA   DDRD,X             ; set port D pins for output

BEGIN   LDY    #display           ; get the base address of the table
NEXT    LDAA   0,Y                ; get the BCD number to be displayed
        LDX    #REGBAS
        STAA   PORTB,X            ; send out the display pattern
        LDAA   1,Y                ; select the seven-segment display to be lighted
        STAA   PORTD,X            ;              "
        INY                       ; move to next entry of the table
        INY                       ;              "
        LDX    #200
```

* The following four instructions will be executed 200 times and hence create a delay
* of 1 ms with a 2MHz 68HC11

```
AGAIN   NOP                      ; 2 E clock cycles
        NOP                      ; 2 E clock cycles
        DEX                      ; 3 E clock cycles
        BNE    AGAIN             ; 3 E clock cycles
```

```
CPY   #display+12    ; reach the last digit?
BEQ   BEGIN          ; yes, start all over again
BRA   NEXT           ; no, get next digit
END
```

6.10.5 Interfacing to the D/A Converter
Using the 68HC11 Output Ports

Digital-to-analog conversion is required when a digital code must be converted to an analog signal. The analog signal can be used to control the output level of another system, for example, the flow level of a fluid system or the volume level of a stereo system. A general D/A converter consists of a network of precision resistors, input switches, and level shifters that activate the switches that convert a digital code to an analog voltage or current. A D/A converter may also contain input or output buffers, amplifiers, and internal references.

D/A converters commonly have a *fixed* or *variable reference* voltage level. The reference voltage level can be generated internally or externally. It determines the switching threshold of the precision switches that form the controlled impedance network that controls the value of the output signal.

Fixed reference D/A converters have current or voltage reference output values that are proportional to the digital input. *Multiplying D/A converters* produce an output signal that is proportional to the product of a varying reference level and a digital code.

The AD557 by Analog Devices is an 8-bit D/A converter that produces an output voltage proportional to the digital input code. The AD557 can operate off a +5 V power supply, the reference voltage is generated internally, and input latches are available for microprocessor interfacing. The pin configuration of the AD557 is shown in Figure 6.22.

The AD557 has a unipolar 0 V to +2.55 V output range when operated as

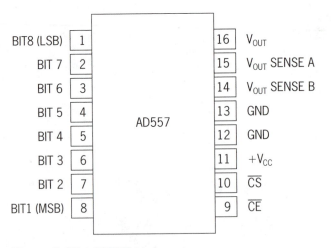

Figure 6.22   AD557 pin layout

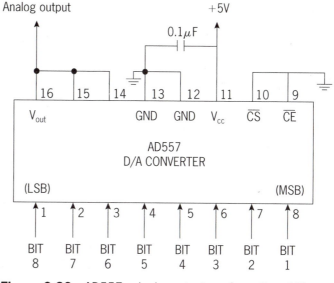

Figure 6.23 AD557 unipolar output configuration, 0 V to 2.55 V operation

shown in Figure 6.23. The output voltage can be calculated using the following equation:

$$V_{out} = (\text{decimal equivalent of input code}) \times 0.01 \text{ V}$$

The AD557 has input latches to simplify interfacing synchronous circuits and microprocessor systems to the AD557 for D/A conversion. The input latches are controlled by the chip select ($\overline{CS}$) and the chip enable ($\overline{CE}$) inputs. If the latches are not required, the chip select and chip enable inputs are grounded. In this mode, the output voltage settles within 1 μs after the digital value is applied at the input pins.

Table 6.5 shows the latched input operation of the AD557. Data is latched into the input latches on the positive edge of either $\overline{CS}$ or $\overline{CE}$. The data is held

Input data	$\overline{CE}$	$\overline{CS}$	D/A converter data	Latch condition
0	0	0	0	"transparent"
1	0	0	1	"transparent"
0	↑	0	0	latching
1	↑	1	1	latching
0	0	↑	0	latching
1	0	↑	1	latching
x	1	x	previous data	latching
x	x	1	previous data	latching

Table 6.5 ■ AD557 input latch operation

in the input latch until both $\overline{CS}$ and $\overline{CE}$ return to a logic low level. When both $\overline{CS}$ and $\overline{CE}$ are at the logic low level, the data is transferred from the input latch to the D/A converter for conversion to an analog voltage. By combining the use of a parallel output port and a D/A converter, a wide variety of waveforms can easily be generated. The following example generates a sawtooth waveform using port B of the 68HC11 and the AD557 D/A converter.

Example 6.6 ■ Sawtooth waveform generation.

Use the 68HC11 port B and an AD557 D/A converter to generate a sawtooth waveform. The circuit connection is shown in Figure 6.24. At the beginning, a 0 is written into the PORTB register, causing the AD557 to output a minimum value close to 0 V. After a short delay, the PORTB register is incremented by 1 and the AD557 output is incremented by one step. This process continues until the hexadecimal value in the PORTB register reaches its maximum value of $FF and the AD557 output reaches its maximum analog output voltage. Incrementing the PORTB register one more time causes the hexadecimal number in the PORTB register to "roll over" to $00, so the AD557 output falls from its maximum value to its minimum value in one step. The process will repeat until the 68HC11 is reset or the power is turned off.

The following program generates the sawtooth waveform:

```
REGBAS   EQU $1000        ; base address of I/O registers
PORTB    EQU $04          ; offset of PORTB from REGBAS

         ORG $C000
         LDX  #REGBAS     ; point X to the register block base
         CLR  PORTB,X     ; send a 0 to AD557
AGAIN    INC  PORTB,X     ; increment the digital code by 1
         BRA  AGAIN       ; repeat
         END
```

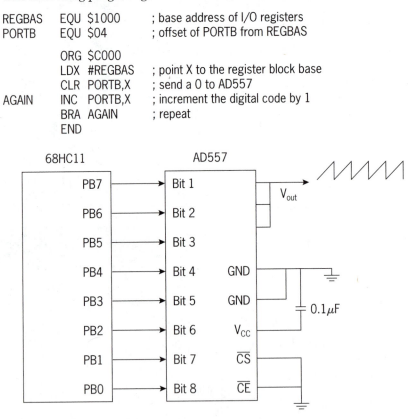

Figure 6.24 Using an AD557 to generate a sawtooth waveform

Example 6.7

Using the same circuit as in Example 6.6, write a program to generate the waveform shown in Figure 6.25.

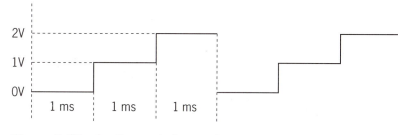

Figure 6.25 Another periodic waveform

Solution: The solution to this problem is fairly straightforward and is outlined by the flowchart shown in Figure 6.26.

The program is:

```
REGBAS    EQU  $1000
PORTB     EQU  $04
          ORG  $C000
```

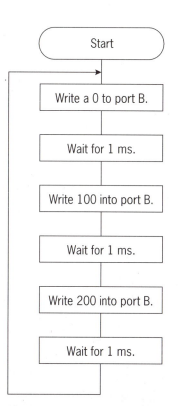

Figure 6.26 Algorithm for waveform generation

```
            LDY   #REGBAS
start       CLR   PORTB,Y    ; output 0V
```

* The following 5 instructions create a delay of 1 ms

```
            LDX   #200
loopms1     NOP
            NOP
            DEX
            BNE   loopms1

            LDAB  #100       ; output 1V
            STAB  PORTB,Y    ;      "
```

* The following 5 instructions create a delay of 1 ms

```
            LDX   #200
loopms2     NOP
            NOP
            DEX
            BNE   loopms2

            LDAB  #200       ; output 2V
            STAB  PORTB,Y    ;      "
```

* The following 5 instructions create a delay of 1 ms

```
            LDX   #200
loopms3     NOP
            NOP
            DEX
            BNE   loopms3

            BRA   start
            END
```

6.11 Centronics Printer Interface

The standard Centronics printer interface allows the transfer of byte-wide, parallel data under the control of two handshake lines, $\overline{\text{DATA STROBE}}$ and $\overline{\text{ACKNLG}}$ (acknowledge). Short data setup (50 ns) and hold (100 ns) times with respect to the active edge of the $\overline{\text{DATA STROBE}}$ input are required. The Centronics interface provides the following signals:

- D7–D1: Data pins.
- BUSY: Printer busy signal (output from printer).
- PE: Printer error. When PE is high, the printer is in check (disabled due to an internal condition such as being out of paper or a malfunction). This signal is an output from the printer.
- SLCT: Printer on line (output from the printer).

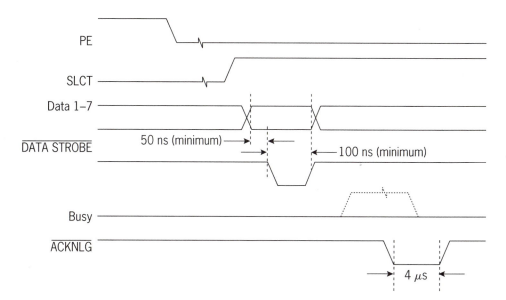

Figure 6.27 Centronics interface timing requirements

■ $\overline{\text{DATA STROBE}}$: This signal is an input to the printer and is active low. Data is latched into the printer by this signal (input to the printer).

■ $\overline{\text{ACKNLG}}$: Acknowledge. $\overline{\text{ACKNLG}}$ is active low and is an output from the printer.

PE, SLCT, and BUSY are status signals and are not all needed in any particular printer design. The Centronics interface timing is shown in Figure 6.27.

Example 6.8 ■ Interface a Centronics printer to port C

The designer is expected to

1. synchronize data transfer between port C and the printer using the pulse-mode output handshake
2. use the polling method to determine whether the printer is ready for another character
3. write routines to initialize the printer, output a character, and output a string terminated by a null character

The BUSY signal is not needed in this example. The hardware connections are shown in Figure 6.28.

Solution: First, we need to verify that the data setup and hold time requirements with respect to the active edge of $\overline{\text{DATA STROBE}}$ are satisfied. A printer using the Centronics interface requires a data setup time of 50 ns and a data hold time no more than 100 ns. The data setup time (see Figure 6.8) is equal to

period of E $\div$ 2 $-$ t_{PWD} + t_{DEB} = 250 ns $-$ 150 ns + 225 ns = 325 ns > 50 ns (required)

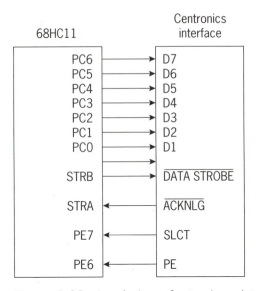

Figure 6.28   Interfacing a Centronics printer to port C

The data hold time (see Figure 6.8) with respect to the active edge of $\overline{\text{DATA}}$ $\overline{\text{STROBE}}$ is also satisfied because port C data does not change during the period when STRB is active, which is two E cycles (1 μs at 2 MHz) in the pulse-mode output handshake.

Port C should be configured as follows:

1. Set port C for output, i.e., write the value $FF into the DDRC register.

2. Program the PIOC register so that the following parameters are selected:

   ■ Output pulse mode handshaking

   ■ Strobe A interrupt disabled

   ■ STRA falling edge as the active edge because $\overline{\text{ACKNLG}}$ is active low

   ■ STRB active low so that the falling edge of STRB can be used to latch data into the printer interface

   ■ Normal port C pins (not wired-ORed)

   With this setting, the value $1C should be written into the PIOC register.

The routines to initialize port C, output a character, and output a string using the polling method follow. A string is terminated by the null character (ASCII code is 0). The following code (or routine) initializes port C.

```
REGBAS   EQU  $1000   ; base address of I/O registers
DDRC     EQU  $07     ; offset of DDRC from REGBAS
PIOC     EQU  $02     ; offset of PIOC from REGBAS
PORTCL   EQU  $05     ; offset of PORTCL from REGBAS
PORTE    EQU  $0A     ; offset of PORTE from REGBAS
OUTPUT   EQU  $FF     ; value to set port C to output
```

The printer initialization routine PRT_INIT requires no incoming parameter; its role is to make sure that the printer is on-line and in working condition before it initializes the port C output function. Printer errors can be caused by paper jams, the printer being out of paper, etcetera. The printer error condition is indicated by a logic high on the PE6 pin. The PE7 pin indicates whether the printer is on-line. Usually a printer is on-line when its power is on and there is a parallel connector between the computer and the printer.

```
prt_init    PSHA
            PSHX
            LDX    #REGBAS
            BRSET  PORTE,X $40 *     ; don't go if printer error
            BRCLR  PORTE,X $80 *     ; don't go if printer is not on-line
            LDAA   #OUTPUT           ; configure port C for output
            STAA   DDRC,X            ;      "
            LDAA   #$1C              ; initialize PIOC
            STAA   PIOC,X            ;      "
            PULX
            PULA
            RTS
```

The routine PRT_CHAR, which outputs the character in the accumulator A, has two incoming parameters:

1. the character to be output, which is passed in accumulator A
2. the I/O register block base address, which is passed in index register X

PRT_CHAR uses the polling method to output a character. It checks bit 7 of the PIOC register, and if bit 7 is a 1, the character in accumulator A is written into the PORTCL register.

```
prt_char
            BRCLR  PIOC,X $80 *      ; poll the STAF flag until it is set to 1
            STAA   PORTCL,X          ; output the character
            RTS
```

The routine PRT_STR, which outputs a string, has two incoming parameters:

1. the I/O register block base address, which is passed in index register X
2. the starting address of the string to be output, which is passed in index register Y

This routine gets a character from the string and then calls the routine PRT_CHAR to output it; it repeats this operation until the character is a NULL character, which indicates the end of the string. After outputting a character, the string pointer Y is incremented by one.

```
prt_str     PSHA
NEXT        LDAA   0,Y               ; get the next character
            BEQ    QUIT              ; is it a null character?
            JSR    prt_char          ; output it if it is not a null character
            INY                      ; increment the string pointer
            BRA    NEXT
QUIT        PULA
            RTS
```

Example 6.9

Rewrite the three subroutines in Example 6.8 using the interrupt-driven output method.

Solution: In the interrupt-driven output method, the routine PRT_STR is called with the same parameters as in Example 6.8. This routine enables the interrupt, gets the next character to be output, and waits for the printer interrupt. When there is an interrupt, the interrupt service routine PRT_CHAR is executed and a character is printed. This process is repeated until the whole string is printed and the routine PRT_STR disables the interrupt and returns to the caller. The printer initialization routine is identical to the routine in the polling version. The routine PRT_STR has two incoming parameters:

1. the I/O register block base address, which is passed in index register X
2. the starting address of the string to be output, which is passed in index register Y

The string pointer (index register Y) will be incremented by the PRT_CHAR routine during the printing process. The interrupt-driven PRT_STR routine is:

```
prt_str    BSET PIOC,X $40    ; enable the STRA interrupt
           CLI                ; enable global interrupts
again      LDAA 0,Y           ; wait for interrupt
           BNE  again         ; at the end of the string?
           BCLR PIOC,X $40    ; disable STRA interrupt when done
           SEI                ; disable the global interrupt
           RTS
```

When the interrupt-driven output method is used, the routine PRT_CHAR is implemented as an interrupt-handling routine and the contents of Y are incremented during the output process. Since all CPU registers are pushed into the system stack and then restored when the interrupt is exited, the interrupt service routine must also update the copy of the Y register in the stack. The new value of Y will be lost if this is not done. The old copy of the index register Y is located five bytes below the top of the stack. The interrupt-driven PRT_CHAR routine is:

```
prt_char   STAA PORTCL,X    ; output the character
           INY              ; increment the string pointer
           TSX              ; point X to the top byte of the stack
           STY  5,X         ; update the Y value in the stack
           RTI
```

To use the interrupt-driven output method, the interrupt vector table should be initialized. The starting address of the STRA interrupt must be stored at address $FFF2.

```
           ORG  $FFF2       ; INT and STRA interrupt vector
           FDB  prt_char    ; printer interrupt handler
```

6.12 Interfacing a 6821 to the 68HC11

In this section, we will discuss the details of interfacing a 6821 to the 68HC11. This section will illustrate the principles of I/O addressing and decoding, I/O interface chip programming, and timing analysis. We will also discuss the advantages of an 8-bit microcontroller over an 8-bit microprocessor in embedded applications.

6.12.1 Overview of the 6821

The Motorola 6821 peripheral interface adaptor (PIA) was designed to work with Motorola 8-bit microprocessors. The 6821 has two 8-bit parallel ports. Each data line on the two ports can be individually configured for either data input or data output. A simplified diagram of the 6821 is shown in Figure 6.29. The 6821 has six addressable registers and 40 pins. The microprocessor (or microcontroller) addresses each of the 6821 registers as if it were a memory location. The pins of the 6821 are divided into a microprocessor side and a peripheral side. The pins on the microprocessor side are connected to the various buses of the microprocessor, and the pins on the peripheral side are connected to one or more peripheral devices. Each of the two 8-bit I/O ports on the peripheral side, port A and port B, has a peripheral data register, a data direction register, and a control register.

A logical question to ask at this point is why we need to discuss the PIA, since the 68HC11 already has parallel ports to interface to parallel devices. The answer is that the 68HC11 has a very limited number of pins and it has only one port (port C) that can perform handshake I/O. In applications that require more than one handshake parallel port, the PIA is a good choice.

6.12.2 The 6821 Signal Pins

The PIA's 8 data pins (D0–D7) are connected to the microprocessor's data bus and are used to transfer data between the 6821 and the microprocessor. These data pins are bidirectional, so the microprocessor can write data to and read data from any one of the six addressable registers by means of the data lines.

The 6821 has three chip select signals. One ($\overline{CS2}$) is active low, and the others (CS0 and CS1) are active high. The 6821 will respond to an external access request only if all three signals are active at the same time (i.e., CS1 = CS0 = 1 and $\overline{CS2}$ = 0). Two register select (RS0 and RS1) signals are used to select the PIA's six internal registers. These two signals are usually connected to the lowest two address signals (A1 and A0) from the microprocessor. The data register and the data direction register of each port share the same address. The register selection scheme is shown in Table 6.6.

The $\overline{RESET}$ pin of the 6821, along with the $\overline{RESET}$ pin of the microprocessor, is usually connected to the computer system's reset button. When the $\overline{RESET}$ pin is grounded, all six 6821 internal registers are cleared.

The read/write (R/$\overline{W}$) pin is connected to the read/write line of the microprocessor. The microprocessor uses the read/write line to notify the 6821 that it wants to read from or write into one of the six registers.

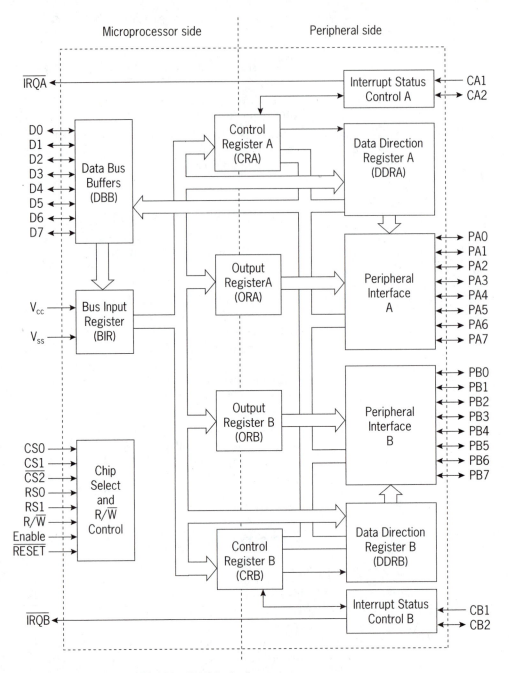

Figure 6.29 Motorola 6821 PIA block diagram (Redrawn with permission of Motorola)

RS1	RS0	Register addressed
0	0	Data register A or data direction register A
0	1	Control register A
1	0	Data register B or data direction register B
1	1	Control register B

Table 6.6 ■ PIA register selection

The interrupt request A ($\overline{\text{IRQA}}$) and interrupt request B ($\overline{\text{IRQB}}$) lines are active low and are connected to the interrupt request ($\overline{\text{IRQ}}$) pin of the microprocessor. The power supply pins are connected to the computer's +5 V power supply and ground.

The 6821 has two I/O ports. Each port has eight data lines. The data lines for port A are labeled PA0–PA7, and the data lines for port B are labeled PB0–PB7. Each port also has two control signals. The port A control lines are CA1 and CA2, and the port B control lines are CB1 and CB2.

The E(nable) pin is the only timing signal supplied to the 6821. Timing of all other signals is referenced to the leading or trailing edge of the E pulse. This pin should be connected to the E clock output pin of the 68HC11.

6.12.3 The 6821 Registers

Each 6821 port contains three addressable registers: a peripheral data register, a data direction register, and a control register. In the following discussion, we will use the acronym of each register to refer to its address. For example, the instruction LDAA PDRA will be used to copy the contents of the port A peripheral data register into accumulator A. Notice that no peripheral data register is shown in the block diagram in Figure 6.29. The combination of the output register (ORA or ORB) and the data bus buffer is referred to as the *peripheral data register*. Both the output register and the data bus buffer are accessed with the same address. The peripheral data register of port A is referred to as PDRA, and that of port B is referred to as PDRB. Each peripheral data register can be configured to act as either an output latch or an input buffer.

Each data direction register (DDRA or DDRB) determines the direction in which data can move through its 6821 port, and each data line can be individually configured for either input or output. Storing a binary 0 in a DDR bit configures its associated data line as an input line, and storing a binary 1 in a DDR bit configures its associated data line as an output line.

Example 6.10

Configure port A as an input port and port B as an output port.

Solution: To configure port A as an input port, each bit of the DDRA register must be set to 0, that is, the value $00 should be written into DDRA. To configure port B as an output port, each bit of the DDRB register must be set to 1, that is, the value $FF should be written into DDRB.

The peripheral data register and the data direction register of each port share the same address. Bit 2 of the port control register determines which of these two registers can be accessed at that address. If bit 2 of a control register is 0, the data direction register is selected; if bit 2 is 1, the peripheral data register is selected.

The bit functions of the two 6821 control registers are shown in Figure 6.30. The eight bits of each control register are divided into two functions, control

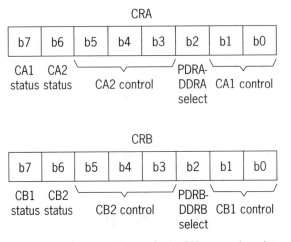

Figure 6.30 Functions of two PIA control registers

and status. Bits 0 through 5 are control bits that are used to configure the 6821 operation. Bits 6 and 7 are status bits that are set by input signals on the control lines (CA1, CA2, CB1, and CB2).

Control lines CA1 and CB1 are inputs. Control lines CA2 and CB2 can be configured as either inputs or outputs. The meanings of control bits 3 to 5 differ depending on whether CA2 (or CB2) is configured as an input or an output. Bit 6 will be set only if CA2 (or CB2) is configured as an input and an active edge is detected (rising or falling, depending on the setting of bit 4). When port A (or port B) is configured to operate in handshake mode, pin CA2 (or CB2) must be configured as an output.

A status bit (bit 6 or 7) will be set when the active edge of the corresponding control signal is detected. These status flags must be cleared after the microprocessor has taken care of the situation. They can be cleared by reading the appropriate port peripheral data register (using a load instruction).

The meanings of the individual bits are summarized in Figure 6.31.

6.12.4 6821 Operation

If there are no timing issues, as is true for some I/O devices, the microprocessor can read directly from or write directly into 6821 peripheral data registers without checking any status bits. When the microprocessor (or microcontroller) needs to make sure that an input device has placed data in the 6821 peripheral data register during an input or that an output device is ready for more data during an output, either the polling or the interrupt method can be used.

- *Polling method.* The user disables the interrupts from the 6821, and the user program checks either bit 7 or bit 6 to determine whether the I/O device is ready.

- *Interrupt method.* Either control input for each port can cause the 6821 to assert the port interrupt signal. The corresponding interrupt enable bit(s) must be set to 1 if this method is to be used.

b7	b6	b5	b4	b3	b2	b1	b0
0 = no active edge detected 1 = active edge detected	0 = no active edge detected 1 = active edge detected	0 = input	0 = falling edge triggered 1 = rising edge triggered	0 = IRQ disabled 1 = IRQ enabled	0 = DDR 1 = PDR	0 = falling edge triggered 1 = rising edge triggered	0 = IRQ disabled 1 = IRQ enabled
		1 = output	0 = handshaking	0 = complete handshaking 1 = partial handshaking			
			1 = Steady state	0 = Low 1 = High			
CA1/CB1 status	CA2/CB2 status	CA2/CB2 control			PDR/DDR select	CA1/CB1 control	

Figure 6.31 Functions of control register CRA/CRB bits

Either the brute force or the handshake method can be used to synchronize data transfer between the 6821 and I/O devices. Port A of the 6821 is designed to perform input handshaking, and port B is designed to perform output handshaking. Output handshaking cannot be used in port A, and input handshaking cannot be used in port B.

Port A of the 6821 can use either complete handshaking (similar to interlock handshaking) or partial handshaking (identical to pulse mode handshaking) to input data from an input device. The control signal CA1 is an input, and the CA2 signal must be configured for output. The connections between the 6821 and the input device and the input handshake signal transactions are shown in Figure 6.32. When input handshaking is used, the input device places data on the port A data lines and asserts the CA1 signal. The active edge sets bit 7 of the CRA register to 1. The CA2 signal is held normally low in complete handshaking and high in partial handshaking.

In *complete input handshaking,* the 6821 outputs a high on the CA2 pin when the active edge of the CA1 signal is received, thus acknowledging the receipt of the CA1 control signal from the input device. The $\overline{\text{IRQA}}$ signal will be pulled to low if the CA1 interrupt is enabled—this informs the microprocessor that a valid byte has been received. If the CA1 interrupt is not enabled, then the microprocessor needs to poll bit 7 of the CRA register to discover this condition. In either case, the microprocessor reads the PDRA register to get the data, forcing the 6821 to return the CA2 signal to the low state to notify the input device that the input data has been read and the input device no longer needs to hold the data on the data lines (PA7–PA0).

In *partial input handshaking,* the 6821 pulls $\overline{\text{IRQA}}$ (if the CA1 interrupt is enabled) to low to interrupt the microprocessor when the active edge of CA1 is

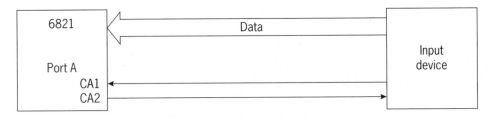

(a) Connection between the 6821 and input device for input handshaking

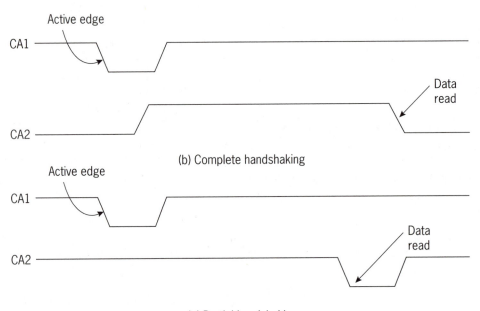

(c) Partial handshaking

Figure 6.32 Complete and partial input handshaking (Redrawn with permission of Motorola)

detected. The 6821 does not change the voltage level of CA2 until the microprocessor reads the PDRA register. When the microprocessor reads the data, the 6821 activates a low pulse on the CA2 pin to acknowledge that the data has been read.

Port B of the 6821 can use either complete handshaking (similar to interlock handshaking) or partial handshaking (identical to pulse mode handshaking) to output data to an output device. The CB1 signal is an input, and the CB2 signal must be configured for output. The connections between the 6821 and the output device and the output handshake signal transactions are shown in Figure 6.33.

In *complete output handshaking,* the output device activates a pulse on the CB1 pin when it is ready to accept a byte of data. The 6821 outputs a high on the CB2 pin when the active edge of the CB1 signal is received, acknowledging receipt of the control signal CB1 from the output device. The $\overline{\text{IRQB}}$ signal will be pulled to low if the CB1 interrupt is enabled—this informs the microprocessor that the output device is ready for another byte. If the CB1 in-

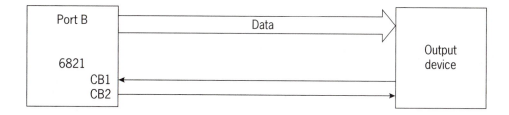

(a) Connections between the 6821 and the output device for output handshaking

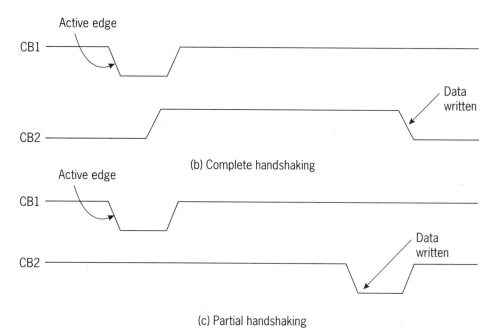

Figure 6.33   Complete and partial output handshaking (Redrawn with permission of Motorola)

terrupt is not enabled, the microprocessor needs to poll bit 7 of the CRB register to detect this condition. In either case, the microprocessor will write a byte into the PDRB register to output the data, forcing the 6821 to return the CB2 signal to the low state to notify the output device that output data is available on the port B data lines. The microprocessor needs to read the PDRB register in order to clear the interrupt flag (bit 6) of the CRB register.

In *partial output handshaking,* the output device sends a pulse on the CB1 pin to notify the microprocessor that it is ready to accept data. When the active edge of CB1 is detected, the 6821 pulls $\overline{IRQB}$ (if the CB1 interrupt is enabled) to low to interrupt the microprocessor. The 6821 does not change the voltage level of CB2 until the microprocessor writes into the PDRB register. When the microprocessor writes a byte into the PDRB register, the 6821 activates a low pulse on the CB2 pin to inform the output device that data are available on the port B data lines. As in complete handshaking, the microprocessor needs to read the PDRB register in order to clear the interrupt flag (bit 6) of the CRB register.

6.12.5 Interfacing the 6821 to the 68HC11

The 6821 can be interfaced to the 68HC11 when the 68HC11 is configured to operate in the expanded mode. The data pins on the microprocessor side can be connected directly to the 68HC11 port C pins. Two register select pins (RS1 and RS0) can be connected to the lowest two address pins (A1 and A0, respectively).

An address space must be assigned to the 6821 registers so that they can be accessed by the 68HC11. Address space assignment will affect the address decoding scheme. As we discussed in Chapter 4, the partial address decoding scheme can simplify the decoder design and will be used here. Before we make the address assignment, we must also consider whether we need to add other devices to the 68HC11.

CASE 1: THE 6821 IS THE ONLY EXTERNAL CHIP CONNECTED TO THE 68HC11.

In this case, we can reduce hardware costs by connecting the highest three address signals (A15–A13) directly to the three chip select pins. Since chip select signal $\overline{CS2}$ is active low, a low voltage is required to activate it. As shown in Table 6.7, there are six possible connection methods, and all of them are acceptable. Since the PIA is the only external chip to the 68HC11, we can make the design simple by using A12 and A11 as register select inputs (RS1 and RS0). The resulting circuit is shown in Figure 6.34.

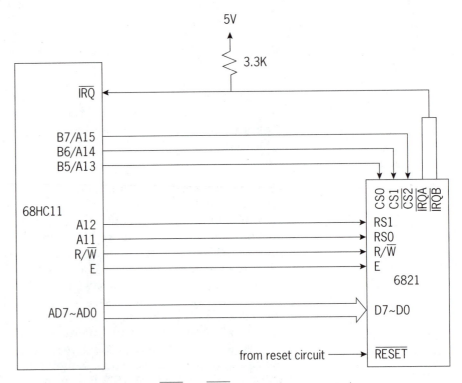

If an interrupt is not required, $\overline{IRQA}$ and $\overline{IRQB}$ can be left unconnected.

Figure 6.34   6821 microprocessor side connections I

#	A15	A14	A13	Address space assigned to 6821
1	$\overline{CS2}$	CS1	CS0	$6000–$7FFF
2	CS1	$\overline{CS2}$	CS0	$A000–$BFFF
3	CS0	CS1	$\overline{CS2}$	$C000–$DFFF
4	$\overline{CS2}$	CS0	CS1	$6000–$7FFF
5	CS0	$\overline{CS2}$	CS1	$A000–$BFFF
6	CS1	CS0	$\overline{CS2}$	$C000–$DFFF

Table 6.7 ■ Possible connections of A15–A13 to $\overline{CS2}$, CS1, and CS0

Example 6.11

Why is the address space $6000–$7FFF assigned to the 6821 in the first connection method in Table 6.7?

Solution: In the first connection method, the value of $A_{15}A_{14}A_{13}$ must be 011 so that the 6821 can be selected. The lower 13 address signals can be any value. Therefore, any address from %0110000000000000 ($6000) to %0111111111111111 ($7FFF) selects the 6821.

Two port interrupt request ($\overline{IRQA}$ and $\overline{IRQB}$) outputs are wired-ORed to the $\overline{IRQ}$ input of the 68HC11. The 6821 should share a reset input with the 68HC11 and obtain its E(nable) input from the 68HC11.

CASE 2: THE 68HC11 HAS SEVERAL EXTERNAL PERIPHERAL DEVICES AND MEMORY COMPONENTS.

In this case an address decoder is needed to select one and only one device to respond to the 68HC11's external access request. Proper address space assignment must be made. The microprocessor side circuit connections will be similar to those for case 1 except for the chip select inputs.

Example 6.12

Make an appropriate connection between the 6821 and the 68HC11 to meet the following operation requirements:

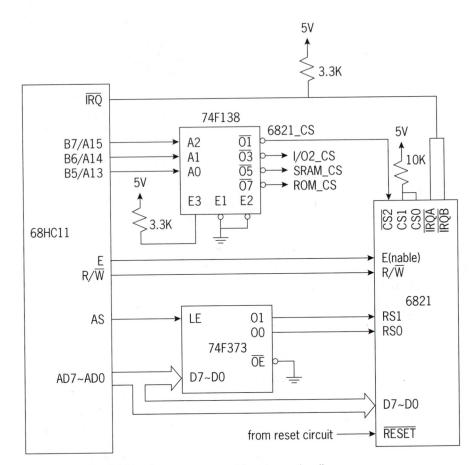

Figure 6.35   6821 microprocessor side connection II

1. The system has four external devices with the following address space assignments:

   6821: $2000–$3FFF

   I/O2: $6000–$7FFF

   SRAM: $A000–$BFFF

   ROM: $E000–$FFFF (contains the interrupt vector table)

2. Interrupt-driven I/O is to be used in the 6821.

Solution: Since each external device is assigned a block of 8KB of address space, a 3-to-8 decoder is used to decode the address signals and generate the chip select signals. The circuit connection is shown in Figure 6.35.

■

In the address decoder, only outputs that are used are shown. The decoder output $\overline{O1}$ is used as the $\overline{CS2}$ chip select input to the 6821. The output $\overline{O1}$ will be pulled down to low when A15–A13 are 001 and hence selects the 6821. Both CS1 and CS0 are pulled up to high through a 10KΩ resistor. The microproces-

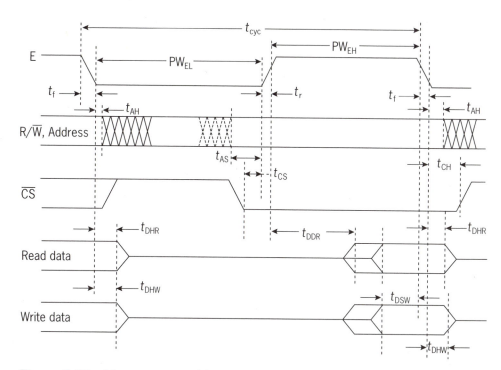

Figure 6.36 Microprocessor side bus timing diagram for a 2-MHz 6821

Characteristic	Symbol	Minimum (ns)	Maximum (ns)
Cycle time	t_{cyc}	500	10,000
Pulse width, E low	PW_{EL}	210	—
Pulse width, E high	PW_{EH}	220	—
Clock rise and fall time	t_r, t_f	—	20
Address hold time	t_{AH}	10	—
Address setup time before E	t_{AS}	40	—
Chip select time before E	t_{CS}	40	—
Chip select hold time	t_{CH}	10	—
Read data hold time	t_{DHR}	20	50
Write data hold time	t_{DHW}	10	—
Output data delay time	t_{DDR}	—	150
Input data setup time	t_{DSW}	60	—

Table 6.8 ■ Timing characteristics of a 2-MHz 6821

sor side bus timing diagram and the timing characteristics of the 2 MHz 6821 are given in Figure 6.36 and Table 6.8.

Example 6.13

Perform a timing analysis to verify that the timing requirements for both the 6821 and the 68HC11 are satisfied in the circuit connections shown in Figure 6.35. The timing requirements for the 68HC11 are given in Table 4.5 of Chapter 4.

Solution: The address-input-to-output-valid propagation delay of the 74F138 address decoder is 8 ns at room temperature. Since all three enable signals are permanently enabled, their propagation delay time can be ignored.

With regard to the 74F373 latch, the propagation delay of concern to us is the delay from the falling edge of LE to output-valid. This delay is 11.5 ns at room temperature.

The timing requirements for the 6821 and the 68HC11 are checked as follows:

E clock:
The E(nable) input to the 6821 is from the 68HC11. It is easy to see that the rise/fall time and pulse width requirements are satisfied by comparing Tables 6.8 and 4.5. The results of the comparison are given in Table 6.9.

Address (RS1 and RS0) and R/$\overline{W}$ Inputs:
The multiplexed low address signals can be latched and will propagate to the output of the 74F373 41.5 ns before the rising edge of the E(nable) input. This value is derived as follows: the delay time from AS valid to E Rise is 53 ns, and the propagation delay of 74F373 from LE valid to output-stable is 11.5 ns (53 ns − 11.5 ns = 41.5 ns). The timing requirement is satisfied since 41.5 ns is greater than timing requirement t_{AS} of the 6821. The 2 MHz 6821 requires an address hold time (t_{AH}) of 10 ns after the falling edge of the E clock. The address latch won't latch a new value until 53 ns before the rising edge of the next E clock or 174 ns = (227 ns − 53 ns) after the falling edge of the E clock if a new bus cycle is started in the next E clock cycle. Therefore, the address hold time requirement is satisfied.

The R/$\overline{W}$ output from the 68HC11 is valid 94 ns before the rising edge of the E clock, and the 6821 requires this signal to be valid 40 ns before the rising edge of the E input. The R/$\overline{W}$ signal stays valid for

Parameter	Required time (ns)	Actual time (ns)
E rise time	≤20	20
E fall time	≤20	20
Pulse width high	≥220	222
Pulse width low	≥210	227
E cycle time	≥500	500

Table 6.9 ■ Checking of E(nable) input timing

Parameter	Required (in ns)	Actual value (in ns)
RSx signals setup time (t_{AS})	≥ 40	41.5
RSx signals hold time (t_{AH})	≥ 10	174
R/$\overline{W}$ setup time (t_{AS})	≥ 40	94
R/$\overline{W}$ hold time (t_{AH})	≥ 10	33
$x = 0, 1$		

Table 6.10 ■ Comparison of R/$\overline{W}$ and address timing requirements

33 ns after the falling edge of the E clock. Thus this requirement is also satisfied.

These comparisons are summarized in Table 6.10.

Chip Select ($\overline{CS2}$):

High address signals (A15–A13) from the 68HC11 are stable 94 ns before the rising edge of the E clock. The address decoder output will become stable 94 ns − 8 ns = 86 ns before the rising edge of the E clock. The 6821 requires the chip select ($\overline{CS2}$) to be valid (low) 40 ns before the rising edge of the E clock, so the t_{CS} timing requirement is therefore satisfied. The 68HC11 holds these address signals for at least 33 ns after the falling edge of the E clock, and hence the chip select output will not be changed until 41 ns after the falling edge of the E clock. The 6821 requires a chip select hold time of 10 ns for $\overline{CS2}$. Hence, the t_{CH} timing requirement is also satisfied.

Read Data Setup Time and Hold Time (required by 68HC11):

The 6821 will present the data 150 ns after the rising edge of the E clock or 72 ns before the falling edge (222 ns − 150 ns = 72 ns) of the E clock; this satisfies the read data setup time requirement for the 68HC11 (30 ns is required). The 6821 stops driving the microprocessor side data bus 20–50 ns after the falling edge of the E clock. The 68HC11 requires a data hold time of up to 83 ns, so this timing requirement appears to be violated. However, for the same reasons discussed in Chapter 4, the capacitance on the data bus would hold the data for an extra amount of time and the data hold time requirement can be met. These comparisons are shown in Table 6.11.

Parameter	Required (ns)	Actual value (ns)
t_{DSR}, read data setup time	≥ 30	72
t_{DHR}, read data hold time	0–83	20–50 and held by bus

Table 6.11 ■ Comparison of read data timing requirements

Write Data Setup Time and Hold Time (required by 6821):
On a write bus cycle, the 68HC11 presents the data 128 ns after the rising edge of the E clock or 94 ns $(222 - 128 = 94)$ before the falling edge of the E clock. The 6821 requires a write data setup time (t_{DSW}) of 60 ns. This requirement is easily satisfied. The 68HC11 holds the write data for at least 33 ns, which also exceeds the requirement (t_{DHW}) of the 6821 (10 ns is required).

After this analysis, we conclude that the circuit connection shown in Figure 6.35 satisfies all the timing requirements of the 68HC11 and the 6821.

Example 6.14

Write instructions to output the value of $69 to port A of the circuit shown in Figure 6.35.

Solution: The 6821 base address in Figure 6.35 is $2000. Note that the port data direction and the data registers have the same address, and bit 2 of the port control register decides which register is addressed.

This problem can be solved in four steps:

Step 1
Set bit 2 of PIACRA to 0 so that the PIADDRA register is selected.
Step 2
Store the value $FF into PIADDRA to configure port A for output.
Step 3
Set bit 2 of PIACRA to 1 so that the PIAPDRA register is selected.
Step 4
Write the value of $69 into the PIAPDRA register.

The instruction sequence is as follows:

```
PIA      equ   $2000         ; base address of 6821
PIADDRA  equ   $00           ; offset of data direction register A from base of 6821
PIAPDRA  equ   $00           ; offset of port A data register from base of 6821
PIACRA   equ   $01           ; offset of control register A from base of 6821
OUTPUT   equ   $FF

         LDX   #PIA          ; place the base address of the 6821 into X
         BCLR  PIACRA,X $04   ; select the port A data direction register
         LDAA  #OUTPUT       ; configure port A for output
         STAA  PIADDRA,X      ;     "
         BSET  PIACRA,X $04   ; select the port A data register
         LDAA  #$69          ; output the value of $69 to port A
         STAA  PIAPDRA,X      ;     "
         END
```

Example 6.15

Write an instruction sequence to initialize port B of the 6821 in Figure 6.35 to operate using the handshake protocol with the following parameters:

1. Use complete handshaking.
2. Enable the CB1 interrupt.

3. Disable the CB2 interrupt.

4. Select the falling edge of the CB1 signal as the active edge.

Solution: Please refer to Figure 6.31. First we need to decide on the values to be written into the PIA control register CRB (only bits 5–0 need to be programmed):

For complete handshaking, pin CB2 must be configured as an output. Set bit 5 to 1.

Set bit 4 to 0 to enable port B output handshaking.

Set bit 3 to 0 to select complete handshaking.

Set bit 2 to 0 so that the data direction register is selected.

Set bit 1 to 0 to select the falling edge of CB1 as the active edge.

Set bit 0 to 1 to enable CB1 interrupt.

Since bits 7 and 6 are status bits, we can just write a 0 to them. The value of %00100001 should be written into the port B control register to set the specified operation parameters. The following instruction sequence will initialize port B as a complete handshake output port:

```
PIA       equ   $2000      ; base address of 6821
PIADDRB   equ   $02        ; offset of data direction register B from base of 6821
PIACRB    equ   $03        ; offset of control register B from base of 6821
OUTPUT    equ   $FF        ; value to set a port for output
INIT_B    equ   $21        ; value to initialize the PIA CRB register

          LDX   #PIA            ; place the base address of the 6821 into X
          BCLR  PIACRB,X $04    ; select the port B data direction register
          LDAA  #OUTPUT         ; set port B as an output port
          STAA  PIADDRB,X       ;          "
          BSET  PIACRB,X $04    ; from now on, select the port B data register
          LDAA  #INIT_B         ; initialize port B control
          STAA  PIACRB,X        ;          "
          END
```

The 6821 PIA implements all the I/O capabilities available in the 68HC11. It can interface with any I/O device that can be interfaced to the 68HC11. Since the interface method and the I/O programming of the 6821 are very similar to those of the 68HC11 I/O ports, they will be left as exercises at the end of the chapter.

■

6.13 Microprocessor or Microcontroller?

By now, you should be convinced that interfacing a 6821 to a microprocessor to drive parallel devices is not very difficult. You might now ask whether we should use the combination of an 8-bit microprocessor and an interface chip or an 8-bit microcontroller alone to drive I/O devices. Let's compare these two approaches:

- Product size: A 6821 in the DIP package is as big as an 8-bit microprocessor such as a Motorola 6800 in the DIP package. An 8-bit microcontroller like the 68HC11 is no bigger than an 8-bit microprocessor alone. The interconnection of signals also occupies a significant area on the printed circuit board. Therefore, a product using an 8-bit microcontroller would be much more compact than the combination of an 8-bit microprocessor and interface chips.

- Power consumption: A 6821 consumes about 550 mW of power. An 8-bit microprocessor such as the Motorola 6802 consumes about 1 W. The 68HC11A8 consumes no more than 150 mW at room temperature. Therefore, using the 68HC11 would be much more energy-frugal.

- Product reliability: A product using the combination of an 8-bit microprocessor and interface chips would have many more wires in the design and therefore would be more likely to fail.

- Design time: A microcontroller-based product is easier to design because the designer does not need to worry about the timing and loading issues between the microprocessor and the interface chips.

It is not difficult to conclude from these comparisons that using 8-bit microcontrollers is a better approach than using 8-bit microprocessors in an embedded system.

6.14 The 68HC24 Port Replacement Unit

When developing an embedded product using the 68HC11, the designer could design the prototype in the expanded mode and use an external EPROM to store the application program. After verifying that the prototype works correctly, the designer would move the application program from the external EPROM to the internal ROM (or EPROM) and operate the 68HC11 in single-chip mode.

In the expanded mode, ports B and C are not available for parallel I/O. To make the conversion from prototype to finished end-product smoother, we need something that emulates ports B and C when their registers and handshake functions are accessed. The MC68HC24 PRU is a gate array that emulates the single-chip mode functions of ports B and C, which are lost to the expanded bus function when the MCU is operated in the expanded mode, which permits program development in an external EPROM or RAM. A system consisting of an MC68HC11 in expanded mode, an MC68HC24, an HC373 octal latch, and an external EPROM performs like the 68HC11A8 operating in the single-chip mode, thus allowing an application program to be developed and tested before a masked ROM pattern is ordered.

The logic in the 68HC11 was specifically designed to permit emulation of single-chip functions with the 68HC24. First, the addresses associated with

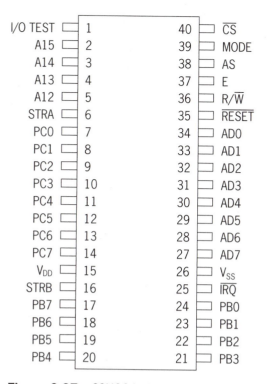

Figure 6.37 68HC24 pin configuration (Redrawn with permission of Motorola)

ports B and C and their handshake I/O functions are treated as external addresses when the 68HC11 is operating in the expanded mode. Furthermore, the interrupts associated with the handshake I/O system are vectored to the same address as $\overline{IRQ}$ interrupts. Thus, the interrupt output of the 68HC24 can be connected to the $\overline{IRQ}$ interrupt input of the 68HC11, and handshake interrupts are treated as internal handshake functions. The 68HC11 allows registers and/or internal RAM to be remapped to any 4K boundary. The 68HC24 copies this logic so that the registers in the 68HC24 automatically track the internal remapping logic. Software written on an expanded system that includes a 68HC24 will operate exactly as it would in the internal ROM of an 68HC11A8 in single-chip mode.

The pin diagram of the 68HC24 is shown in Figure 6.37. The circuit connections for the 68HC11 and the 68HC24 in the EVB are shown in Figure 6.38. The 68HC24 is designed to be a memory-mapped device, and the circuit in Figure 6.38 uses the partial decoding method. Any data written into the PORTB register will appear on the PB7–PB0 pins of the MC68HC24. When port C is configured for input, a read from the PORTC or PORTCL register will receive the data from pins PC7–PC0 of the MC68HC24. When configured for output, any data written into the PORTC or PORTCL register will appear on pins PC7–PC0.

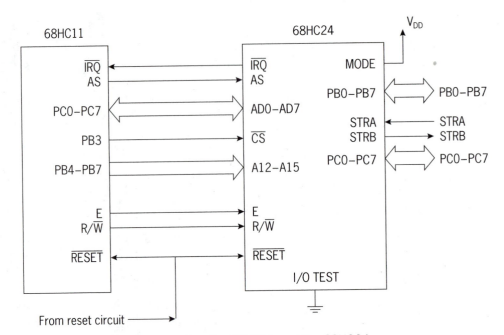

Figure 6.38   Circuit connecting the 68HC11 and the 68HC24

6.15 Summary

This chapter has discussed in detail many of the issues and techniques related to the I/O subsystem. When performing an I/O operation, the microprocessor manipulates the registers or latches in the interface chip. Address space assignment must be made for the I/O devices, and since there are very few registers in an I/O interface chip, the partial decoding scheme is often used for I/O devices to simplify the decoder design.

Both polling and interrupt methods can be used to synchronize data transfer between the processor and the interface chip. Brute force, strobe, and handshaking methods synchronize data transfer between an interface chip and I/O devices.

I/O programming deals mainly with the setting or selection of data transfer direction, active signal levels, synchronization methods, interrupts or alternative systems, etcetera. This information is all stored in control registers inside the interface chip.

The I/O designer also needs to consider the timing requirements of the microprocessor and the interface chip (just as he or she must do when adding external memory to the microcomputer). Data transfer can be successful only if all timing requirements are met.

This chapter has discussed switches, LED and LCD displays, D/A converters, and keyboards to demonstrate I/O methods for the 68HC11 I/O ports and the 6821.

6.16 Glossary

Centronics printer interface   A widely adopted printer interface protocol that uses pulse mode handshake protocol to achieve synchronization of data transfer.

Chip enable signal   A signal that can be driven to allow and disallow data to be read from or written into the I/O interface chip or memory components.

Contact bounce   A problem that occurs in mechanical-type key switches when the switch contacts do not come to rest immediately after the key is pressed.

Debounce   The process of minimizing the effect of key bounce so that the processor can read the correct data.

Handshake   A method of synchronizing data transfer between the interface chip and the I/O device. Two handshake signals are required—one handshake signal is asserted by the interface chip, and the other is asserted by the I/O device. The interface chip initiates the handshake cycle by asserting one of the handshake signals to indicate that it wants more data during an input operation or that it has data on the data pins during an output operation. The input device places data on the data pins and asserts the other handshake signal to indicate the availability of data. The output device asserts the second handshake signal to indicate that it has accepted the data. Both handshake signals are then de-asserted one after the other according to the predefined order.

Interrupt-driven I/O method   A method of synchronizing data transfer between the microprocessor and the interface chip. The interface chip interrupts the microprocessor when it is ready. The microprocessor reads data from or writes data into the interface chip in the interrupt service routine.

Isolated I/O   An I/O method in which microprocessors have separate instructions for performing I/O operations and have separate memory space for I/O devices.

Keyboard scanning   A process for detecting which key is pressed in a keyboard device. The key switches of a keyboard are organized into rows and columns. Keyboard scanning proceeds row by row and column by column.

LCD (liquid crystal display)   A display in which liquid crystals are organized into segments or pixels (for color). Light can pass through a segment or a pixel when it is activated and aligned.

LED (light emitting diode)   A diode that emits light when there is enough current flowing through it.

Memory-mapped I/O   An I/O method in which the same instruction set is used to perform I/O operations and memory references. I/O registers and memory components occupy the same memory space.

Peripheral devices   Pieces of equipment that exchange data with a computer.

Polling   A method of synchronizing data transfer between a microprocessor and an interface chip. The microprocessor keeps reading the interface chip status register until it is sure that the interface chip has valid data or is ready to receive more data and then it proceeds with the I/O operation.

Strobe method   A method of synchronizing data transfer between the interface chip and the I/O device. During input, the input device asserts a strobe signal to latch data into the interface chip data register. During output, the interface chip asserts a strobe signal to latch data into the output device.

6.17 Exercises

E6.1. What is isolated I/O? What is memory-mapped I/O?

E6.2. Describe interlocked input handshaking.

E6.3. Write an instruction sequence to configure port C to operate with the following parameters:

> enable STRA interrupt
>
> no port C wired-or
>
> input handshaking
>
> pulse mode handshake protocol
>
> STRA falling edge as active edge
>
> STRB active low

E6.4. Write a sequence of instructions to read in the current port C pin logic levels.

E6.5. *Traffic Light Controller.* Use the 68HC11EVM or 68HC11EVB and green, yellow, and red LEDs to simulate a traffic light controller. The traffic light patterns and durations for the east-west and north-south traffic are given in Table E6.1. Write a program to control the light patterns, and connect the circuit to demonstrate the changes in the lights.

E6.6. Compute the average maximum current that flows into the collector of the 2N2222 transistor in Figure 6.19. Note that each seven-segment display is illuminated for only $\frac{1}{6}$ second every second.

E6.7. Calculate the period of the sawtooth waveform generated by the circuit in Example 6.6. How would you double the period of the sawtooth waveform?

E6.8. Operate the AD557 D/A converter in the transparent mode (the chip enable and chip select are tied to low). Set V_{ref} to 5 V. What will the output voltage be when the digital input is 0, 16, 32, 64, 128, and 192?

E6.9. Write a program to control the circuit shown in Figure 6.24 to generate a *triangular waveform*. Observe the waveform using an oscilloscope. Answer the following questions based on the evaluation board that you are using:

> a. What is the highest frequency of the waveform that can be achieved?

East-west			North-south			Duration
green	*red*	*yellow*	*green*	*red*	*yellow*	*(seconds)*
1	0	0	0	1	0	20
0	0	1	0	1	0	4
0	1	0	1	0	0	15
0	1	0	0	0	1	3

Table E6.1 ■ Traffic light pattern

    b. How can you double and quadruple the period of the generated triangle waveform?

E6.10. How many different addresses refer to the same register in Figure 6.34?

E6.11. Write an instruction sequence to read a byte from port A of the 6821. Assume that the base address of the 6821 is $A000 and that A1 and A0, the lowest two address signals of the 68HC11, are connected to RS1 and RS0 of the 6821 input lines, respectively.

E6.12. Referring to Figure 6.35, connect the O1 output of the 74F373 to the RS0 input of the 6821 and the O0 output of the 74F373 to the RS1 input of the 6821. What addresses will address PDRA? What addresses will address CRA? What addresses will select CRB?

E6.13. The typical input leakage current of the CS1 and CS0 inputs of the 6821 is 1 μA. What will be the voltage drop on the 10-KΩ pull-up resistor in Figure 6.35?

E6.14. Use a 6821 to drive six seven-segment displays using a circuit similar to the one in Figure 6.19. Use port A to control the segment inputs and port B to control the common cathodes of these displays. Write a program to display the decimal digits 123456 forever. Assume the base address of the 6821 is $2000.

E6.15. You are to design a computer system that uses the 68HC11 as the main CPU and a 6821 to drive a Centronics printer. Draw the circuit connections, and write the routines to perform printer initialization, output one character, and output one string. Assume the address space $2000–$3FFF is assigned to the 6821. Provide both the polling and the interrupt-driven versions of the routines.

E6.16. Write programs to perform keyboard scanning, debouncing, and ASCII code table lookup for the circuit connections shown in Figure E6.1. Port C pins PC7–PC4 select the rows, and pins PC3–PC0 select the columns. The

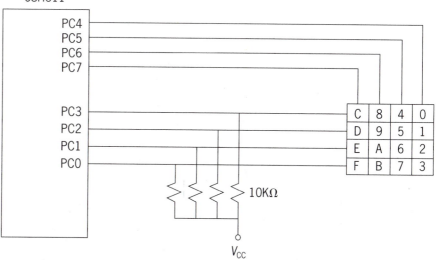

Figure E6.1   Sixteen-key keypad connected to 68HC11

PC7	PC6	PC5	PC4	Selected keys
1	1	1	0	0, 1, 2, and 3
1	1	0	1	4, 5, 6, and 7
1	0	1	1	8, 9, A, and B
0	1	1	1	C, D, E, and F

Table E6.2 ■ Sixteen-key keypad row selections

row selection of the 16-key keypad is shown in Table E6.2. When a key is pressed, the corresponding row and column are shorted together. For example, the PC4 output should be set to low in order to detect whether key 0, 1, 2, or 3 is pressed. If key 3 is pressed, the PC0 input would read a low value. The 16-key membrane-type keypad can be purchased from many electronics re-sellers, such as Jameco or Jade Electronics.

E6.17. Use port B to drive the D/A converter AD557 and write a program to generate the waveform shown in Figure 6.E2. The circuit connections are shown in Figure 6.22.

6.18 Lab Experiments and Assignments

L6.1. *Interrupt-driven input experiment.* The purpose of this lab is to practice interrupt-driven input. You will apply three interrupt pulses to the $\overline{\text{IRQ}}$ pin. Each interrupt pulse will cause one byte to be read from an 8-DIP switch. Connect an 8-DIP switch to the EVB or EVM port E as shown in Figure 6.9. The procedure for the experiment is as follows:

Step 1
Write a main program that performs the following operations:

a. initializes an $\overline{\text{IRQ}}$ interrupt count *ircnt* to 3

b. initializes the buffer pointer *buf_ptr* to $00

c. stays in a wait loop and keeps checking the interrupt count *ircnt*; the main program will disable the $\overline{\text{IRQ}}$ interrupt and jump to the EVB or EVM monitor when *ircnt* is 0

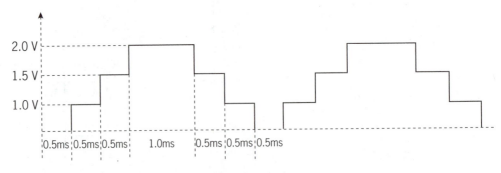

Figure E6.2   Periodic waveform to be generated

Step 2

Write an $\overline{\text{IRQ}}$ service routine that performs the following operations:

- a. reads the 8-DIP switch and saves the value at the memory location pointed to by *buf_ptr*
- b. increments the buffer pointer *buf_ptr* by 1
- c. decrements the $\overline{\text{IRQ}}$ interrupt count *ircnt* by 1

Step 3

Set up the 8-DIP switch properly so that it represents the value that the user wants to input to the EVB.

Step 4

Apply a debounced pulse to the $\overline{\text{IRQ}}$ pin of the EVB so that the $\overline{\text{IRQ}}$ service routine reads the 8-DIP switch value.

Step 5

Repeat steps 3 and 4 three times and then stop.

Examine the contents of memory locations $00–$02 to see if the values have been read correctly.

L6.2. *Seven-segment display experiment.* Use ports B and D to drive two seven-segment displays. Port B pins drive the segment inputs and port D pins PD1 and PD0 drive the bases of two 2N2222 npn transistors. The collectors of these two 2N2222 transistors are connected to the common cathodes of the seven-segment displays. Write a program to perform the following operations:

Step 1

Display the BCD digits 00 on these two seven-segment displays.

Step 2

Use program delay loops to create delays and increment the seven-segment displays every second.

Step 3

Display the decimal values from 00–59 and then roll back to 00.

Step 4

Repeat step 3 three times and then jump to the EVM or EVB monitor.

L6.3 Time-of-day input and display. Use port E to read time-of-day from an 8-DIP switch and use port B and port D to drive six seven-segment displays. Use the circuits in Figure 6.9 and Figure 6.19 and follow this procedure:

Step 1

Write a main program that performs the following operations:

- a. initializes an $\overline{\text{IRQ}}$ interrupt count *ircnt* to 3.
- b. initializes the buffer pointer *buf_ptr* to $00.
- c. stays in a wait loop and keeps checking the interrupt count *ircnt*; the main program will disable the $\overline{\text{IRQ}}$ interrupt and jump to the EVB or EVM monitor when *ircnt* is 0

Step 2

Write an $\overline{\text{IRQ}}$ service routine that performs the following operations:

- a. reads the 8-DIP switch and saves the value at the memory location pointed to by *buf_ptr*

  b. increments the buffer pointer *buf_ptr* by 1

  c. decrements the $\overline{\text{IRQ}}$ interrupt count *ircnt* by 1

Step 3
Set up the 8-DIP switch properly so that it represents the value that
the user wants to input to the EVB.

Step 4
Apply a debounced pulse to the $\overline{\text{IRQ}}$ pin of the EVB (or EVM) so that
the $\overline{\text{IRQ}}$ service routine reads the 8-DIP switch value.

Step 5
Repeat steps 3 and 4 three times to input the current hours, minutes,
and seconds digits.

Step 6
The main program should use a delay loop to create delays to update
the time-of-day. The time-of-day is displayed in 24-hour format. The
main program does not request the user to reenter the current time-of-
day if the user enters incorrect values. However, it will reset the time
to a valid value when the time is to be updated. For example, if out-of-
range seconds digits (i.e., ≥ 60) are entered, they will be reset to 00 in
the next second; if out-of-range minutes digits (i.e., ≥ 60) are entered,
they will be reset to 00 in the next update of the minutes digits; if out-
of-range hours digits are entered, they will be reset to 00 in the next
update of the hours digits.

7

68HC11

TIMER

FUNCTIONS

After completing this chapter, you should be able to:

- use the input-capture function to measure the duration of a pulse or the period of a square wave
- use the output-compare function to create a delay
- use the output-compare function to generate a pulse or a square waveform
- use multiple output-compare functions to control the same pin
- use the forced output-compare function
- use the input-capture function and the pulse accumulator as edge-sensitive interrupt sources
- use the pulse accumulator function to count the number of events that occur within a time interval
- use the real timer function or the output-compare function to generate periodic interrupts

7.2 Introduction

There are many applications that require a dedicated timer system, including

- delay creation and measurement
- period measurement
- event counting
- time-of-day tracking
- periodic interrupt generation to remind the processor to perform routine tasks
- waveform generation

Without a dedicated timer system, some of these applications would be very difficult to implement. The 68HC11 includes a powerful timer system to support these applications. At the heart of the 68HC11 timer functions is the 16-bit free-running main timer. One of the 68HC11 timer functions is *input capture*, which can latch the main timer value into a register when the rising or falling edge (selected by the user program) of a signal arrives. This capability allows the user to measure the width or period of an unknown signal, and it can also be used as a time reference for triggering other operations. There are three input-capture functions in the 68HC11.

Another 68HC11 timer function is *output compare*, which compares the value of the main timer with that of an output-compare register once every E clock cycle and performs the following operations when they are equal:

1. (optional) triggers an action on a pin (set to high, set to low, or toggle its value)

2. sets a flag in a register

3. (optional) generates an interrupt to the microcontroller

The output-compare function is typically used to create a delay or generate a digital waveform with a specified frequency and duty cycle. The key to using the output-compare function is to make a copy of the main timer, add a delay to this copy, and then save this sum into an *output-compare register*. The delay that is added determines when the main timer value will be equal to that of the output-compare register. The 68HC11 provides five output-compare functions.

The third 68HC11 timer function is the *real-time interrupt (RTI)* function. There are some routine tasks that must be performed periodically and cannot tolerate two successive performances being separated by more than some predetermined time limit; checking the temperature and pressure in some process control plants is an example of such a task. These applications can be implemented as interrupt service routines to be invoked by the timer interrupt. The 68HC11 RTI function can generate periodic interrupts and hence is ideal for these applications. Because of the nature of the RTI interrupt, the required task will be performed within a specified time limit (hence the name *real-time*).

The fourth function is the *computer operating properly* (COP) subsystem. This function can be used to reset the computer system if a software program has bugs in it and fails to take care of the COP system within a time limit. The COP system was discussed in detail in section 5.6.5.

The fifth 68HC11 timer function is the 8-bit *pulse accumulator.* This function is often used to count events occurring within some time limit or to measure the duration of a pulse.

The timer system involves more registers and control bits than any other subsystem in the 68HC11. Each of the three input-capture functions has its own 16-bit time-capture latch (input-capture register), and each of the five output-compare functions has a 16-bit compare register. All timer functions, including the timer overflow and RTI, have their own interrupt controls and separate vectors.

7.3 The Free-Running Main Timer

The main timer (TCNT), shown in Figure 7.1, is clocked by the output of a four-stage prescaler (divided by 1, 4, 8, and 16), which in turn is driven by the E clock signal. The main timer is cleared to 0 during a reset and is a read-only register except in test mode. When the count changes from $FFFF to $0000, the timer overflow flag (TOF) bit in *timer flag register 2* (TFLG2) is set. The timer overflow flag can be cleared by writing a 1 to it. An interrupt can be enabled by setting the timer overflow interrupt enable bit (TOI) of *timer interrupt mask register 2* (TMSK2). The contents of TFLG2 and TMSK2 are shown in Figure 7.2. Bits that are related to the free-running timer are in boldface.

The prescale factor for the main timer is selected by bits 1 and 0 of TMSK2, as shown in Table 7.1. The prescale factor can be changed only once within the first 64 E clock cycles after a reset, and the resulting count rate stays in effect until the next reset.

The timer counter (TCNT) register is meant to be read using a double-byte read instruction such as LDD or LDX. During a 16-bit read, the upper byte is accessed in the first bus cycle, while the lower byte is returned in the next bus cycle. The low-order half of the counter is momentarily frozen when the upper

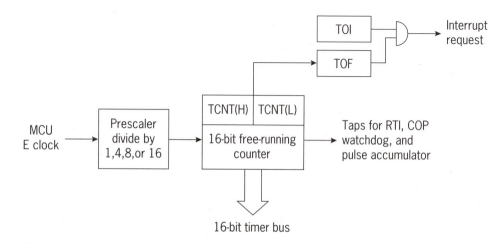

Figure 7.1 68HC11 Main timer system

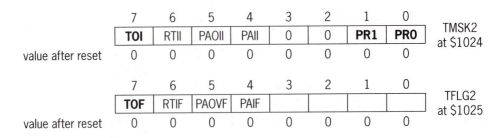

7	6	5	4	3	2	1	0	
TOI	RTII	PAOII	PAII	0	0	**PR1**	**PR0**	TMSK2 at $1024
value after reset 0	0	0	0	0	0	0	0	

7	6	5	4	3	2	1	0	
TOF	RTIF	PAOVF	PAIF					TFLG2 at $1025
value after reset 0	0	0	0	0	0	0	0	

Figure 7.2   TMSK2 and TFLG2 registers

			Overflow period	
PR1	PR0	Prescale factor	2 MHz E clock	1 MHz E clock
0	0	1	32.77 ms	65.54 ms
0	1	4	131.1 ms	262.1 ms
1	0	8	262.1 ms	524.3 ms
1	1	16	524.3 ms	1.049 s

Table 7.1   ■   Main timer clock frequency vs. PR1 and PR0 values

byte is accessed. This procedure assures that the two bytes read from TCNT belong with each other. If the user accesses TCNT with two 8-bit reads, the result might not be correct, because the lower byte of TCNT would be incrementing when the upper byte is accessed.

Example 7.1

What values will be in accumulators A and B after execution of the following three instructions if TCNT contains $5EFE when the upper byte is accessed? Assume the PR1 and PR0 bits of the TMSK1 register are 00.

```
REGBAS   EQU   $1000      ; base address of I/O register block
TCNTH    EQU   $0E        ; offset of TCNTH from REGBAS
TCNTL    EQU   $0F        ; offset of TCNTL from REGBAS

         LDX   #REGBAS
         LDAA  TCNTH,X    ; copy the upper byte of the main timer to A
         LDAB  TCNTL,X    ; copy the lower byte of the main timer to B
```

Solution: Since the PR1 and PR0 bits (in TMSK2) are set to 00, the prescale factor of TCNT is 1. The E clock is used as the clock input to the main timer. The instruction LDAA TCNTH,X accesses the upper byte of TCNT. Therefore, accumulator A will get the value $5E. The instruction LDAB TCNTL,X accesses the lower byte four E clock cycles later. At that time TCNT will be incremented to $5F02. Accumulator B will therefore get the value $02.

If the instruction LDD TCNT,X is used to access TCNT, then accumulators A and B will receive $5E and $FE, respectively. This example shows that the value returned from two 8-bit accesses can be very different from the actual value in TCNT.

The TCNT register can be used to keep track of time as well as to create time delays. However, implementing time delays using TCNT may not be easy. For example, to create a delay of 10 ms at the 2-MHz clock rate, a very naive approach would be to write the following program:

```
REGBAS    EQU   $1000
TCNT      EQU   $0E
```

```
*A 10-ms delay at the 2-MHz clock rate is equivalent to 20,000 clock cycles. By adding
*20000 to the current value of TCNT and waiting until this value is smaller than the
*incrementing TCNT, a 10-ms delay is created.
```

```
(1)          LDD   TCNT,X
(2)          ADDD  #20000
(3) LOOP     CPD   TCNT,X
(4)          BGT   LOOP
             .
             .
             .
```

However, there are several potential problems in this approach:

1. Adding 20000 to the current value of TCNT could cause the value to go from positive to negative, causing instruction 4 to fall through because the BGT instruction performs a sign comparison.

2. The first problem can be solved by using the BHI instruction, which performs an unsigned comparison.

3. However, for some other delay, the resulting sum may be smaller than the current value in TCNT because of the limit of 16-bit addition. In this situation, overflow occurs. For example, adding a delay count of $9000 to the current value of TCNT ($8000) will result in a 16-bit sum of $1000. This situation makes the previous solution unusable because the BHI instruction will also fall through. This problem can be solved by checking the carry flag.

4. The longest delay that can be generated by this approach is limited to about 32.7 ms ($2^{16} - 1$ E clock cycles) at 2 MHz. The user program must keep track of the number of timer overflows in order to generate a longer delay.

The solution of this problem will be left to the reader as an exercise.

7.4 Input-Capture Functions

Some applications need to know the arrival time of events. In a computer, *physical time* is often represented by the count in a counter, while the

Figure 7.3 Events represented by signal edges

occurrence of an event is represented by a signal edge (either the rising or falling edge). The time when an event occurs can be recorded by latching the count when a signal edge arrives, as illustrated in Figure 7.3.

The 68HC11 timer system has three input-capture channels to implement this operation. Each input-capture channel includes a 16-bit input-capture register, input edge-detection logic, and interrupt generation logic, as shown in Figure 7.4. In the 68HC11, physical time is represented by the count in the 16-bit free-running timer counter (TCNT). An input-capture function is used to record the time when an external event occurs. When a selected edge is detected at the input pin, the contents of TCNT are latched into a specified input-capture register. The user can select the signal edge to be captured and may optionally generate an interrupt to the microcontroller when the selected edge is detected. The captured 16-bit timer value is stored in the input-capture register. There are three input-capture registers: TIC1 (located at $1010 and $1011), TIC2 (located at $1012 and $1013), and TIC3 (located at $1014 and $1015). The TIC$x$ ($x = 1$, 2, or 3) registers are not affected by reset and cannot be written in by software. Pins PA2, PA1, PA0 are input only and are used as input-capture channels 1, 2, and 3, respectively, when they are enabled. If an input-capture channel is disabled, the corresponding pin becomes a general-purpose input.

The user can select the signal edge to capture by programming the TCTL2 register. The TCTL2 register is 6-bit, and its contents are shown in Figure 7.5. The signal edge to be captured is specified by two bits. The user can select to capture the rising or falling edge or both edges. When the edge-select bits are 00, the corresponding input-capture channel is disabled.

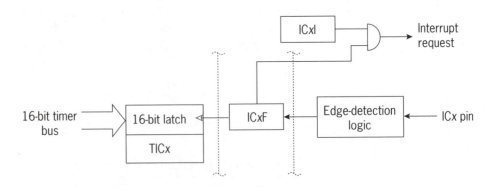

Figure 7.4 Input-capture function block diagram

	5	4	3	2	1	0	
	EDG1B	EDG1A	EDG2B	EDG2A	EDG3B	EDG3A	TCTL2 at $1021
value after reset	0	0	0	0	0	0	

```
EDGxB  EDGxA
  0      0      capture disabled
  0      1      capture on rising edge
  1      0      capture on falling edge
  1      1      capture on any edge
       x = 1, . . . , 3
```

Figure 7.5 Contents of TCTL2 (Reprinted with permission of Motorola)

Example 7.2

Write an instruction sequence to capture the rising edge of the signal connected to PA0 (IC3).

Solution: The PA0 pin is used as the input-capture channel IC3. To capture the rising edge, the edge-select bits (bits 1 and 0 of TCTL2) of channel IC3 must be set to 01. The following instruction sequence will set up TCTL2 to capture the time of a rising edge on PA0.

```
REGBAS    EQU   $1000
TCTL2     EQU   $21

          LDX   #REGBAS
          BCLR  TCTL2,X %00000010    ; clear bit 1 to 0 without affecting other bits
          BSET  TCTL2,X %00000001    ; set bit 0 to 1 without affecting other bits
```

The input-capture status bit in the TFLG1 register (at $1023) will be set when a selected edge is detected at the corresponding input-capture pin. The contents of the TFLG1 register are shown in Figure 7.6. Bits 7 to 3 are output-compare status bits, and bits 2 to 0 are input-capture status bits.

A status flag can be cleared by writing a 1 to it. Two methods are most often used:

1. Load an accumulator with a mask that has a 1 (or 1s) in the bit(s) corresponding to the flag(s) to be cleared; then write this value to TFLG1 or TFLG2.

2. Use a bit clear (BCLR) instruction with a mask having 0s in the positions corresponding to the flags to be cleared and 1s in all other bits. For example, the instruction BCLR TFLG1,X %01111111 will clear the OC1 flag.

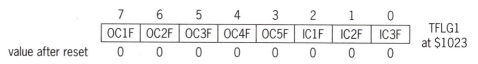

	7	6	5	4	3	2	1	0	
	OC1F	OC2F	OC3F	OC4F	OC5F	IC1F	IC2F	IC3F	TFLG1 at $1023
value after reset	0	0	0	0	0	0	0	0	

Figure 7.6 The contents of the TFLG1 register (Reprinted with permission of Motorola)

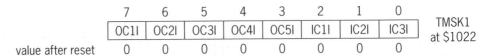

value after reset

Figure 7.7   Contents of the TMSK1 register (Reprinted with permission of Motorola)

It is not appropriate to use the bit set (BSET) instruction to clear the flags in the timer flag registers because this could inadvertently clear one or more of the other flags in the register. The BSET instruction is a read-modify-write instruction that reads the operand, ORs this value with a mask having 1(s) in the bit(s) to be set, and writes the resulting value back to the operand address. When this instruction is used, all flags that are set at the time the flag register is read will be cleared.

The input-capture subsystem can optionally generate an interrupt request when a selected edge is detected. A *mask register* (TMSK1 at $1022) is provided to enable and disable interrupt requests. When a bit is set, the corresponding interrupt request can be generated. The bit assignment of the TMSK1 register is as follows: bits 2 to 0 enable or disable interrupts from input-capture channels 1 to 3, and bits 7 to 3 enable or disable interrupt requests from output-compare channels 1 to 5. After reset, all bits are cleared to 0, which disables all output-compare and input-capture interrupts. The contents of TMSK1 are shown in Figure 7.7. The memory locations that hold the interrupt vectors of channels IC1, IC2, and IC3 are listed in Table 7.2.

Channel	Pin	68HC11 vector location	EVB pseudo vector location
IC1	PA2	$FFEE–$FFEF	$E8–$EA
IC2	PA1	$FFEC–$FFED	$E5–$E7
IC3	PA0	$FFEA–$FFEB	$E2–$E4

Table 7.2 ■ Locations that hold input-capture interrupt vectors

7.5 Applications of the Input-Capture Function

The input-capture function has many applications. Examples include:

■ *Event arrival time recording.* Sometimes an application needs to compare the arrival times of several different events. The input-capture function is very suitable for this. The number of events that can be compared is limited by the number of input-capture channels.

■ *Period measurement.* To measure the period of an unknown signal, the input-capture function should be configured to capture the main timer values corresponding to two consecutive rising or falling edges, as illustrated in Figure 7.8.

■ *Pulse-width measurement.* To measure the width of a pulse, the rising and falling edges are captured, as shown in Figure 7.9. Since

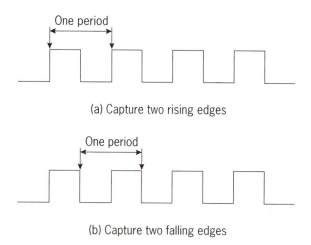

(a) Capture two rising edges

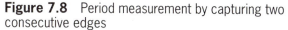

(b) Capture two falling edges

Figure 7.8 Period measurement by capturing two consecutive edges

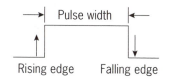

Figure 7.9 Pulse-width measurement using input capture

the free-running timer is 16-bit, it can only measure a signal period (or pulse width) no longer than 2^{16} E clock cycles. If the period of a slower signal needs to be measured, the number of timer overflows that occur between the two edges must be taken into account.

■ *Interrupt generation.* The three input-capture inputs can serve as three edge-sensitive interrupt sources. Once enabled, interrupts will be generated on the selected edge.

■ *Event counting.* An event can be represented by a signal edge. An input-capture channel can be used in combination with an output-compare function to count the number of events that occur during an interval. Whenever an edge arrives, an interrupt is generated. An event counter can be set up and incremented by the input-capture interrupt service routine. This application is illustrated in Figure 7.10.

Events (represented by signal edges)

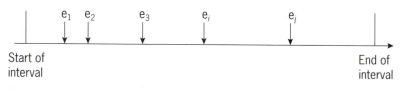

Figure 7.10 Using an input-capture function for event counting

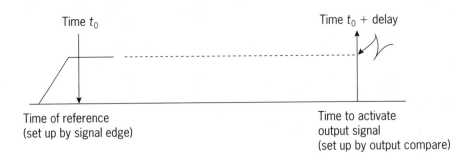

Time t_0

Time t_0 + delay

Time of reference
(set up by signal edge)

Time to activate
output signal
(set up by output compare)

Figure 7.11 A time reference application

■ *Time reference.* In this application, an input-capture function is used in conjunction with an output-compare function. For example, if the user wishes to activate an output signal a certain number of clock cycles after detecting an input event (a rising or falling edge), the input-capture function would be used to record the time at which the edge is detected. A number corresponding to the desired delay would be added to this captured value and stored in an output-compare register. This application is illustrated in Figure 7.11.

Example 7.3

Use the input-capture function IC1 of the 68HC11 to measure the period of an unknown signal. The period is known to be shorter than 32 ms, and the signal is connected to the IC1 pin. Write a program to set up the input-capture function to perform the measurement and store the result in memory.

Solution: The circuit connection is shown in Figure 7.12. Since the period of the unknown signal is less than 32 ms, we don't need to take the main timer overflow into account. To capture the rising edge of IC1, we need to set the value of register TCTL2 to %00010000. Either the polling or the interrupt method can be used to measure the signal frequency. The algorithm of the polling version of the measurement program is shown in Figure 7.13. In the polling method, the user program checks whether the IC1F flag of TFLG1 is set. When the IC1F flag (bit 2 of the TFLG1 register) is set to 1, a signal edge has been captured. The following conditional branch instruction is often used to poll the IC1F:

register to check

specify bit 2 is to be checked

BRCLR TFLG1,X $04 *

branch target is this instruction

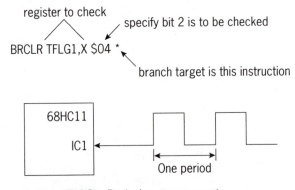

68HC11

IC1

One period

Figure 7.12 Period measurement

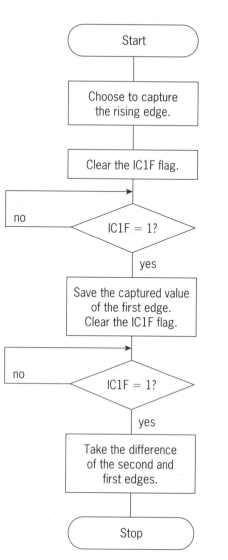

Figure 7.13 Logic flow of the period-measurement program (polling version)

The polling version of the period-measurement program is as follows:

```
REGBAS    EQU    $1000           ; I/O register block base address
TFLG1     EQU    $23             ; offset of TFLG1 from REGBAS
TIC1      EQU    $10             ; offset of TIC1 from REGBAS
TCTL2     EQU    $21             ; offset of TCTL2 from REGBAS
IC1rise   EQU    $10             ; value to select the IC1 rising edge
CLEAR     EQU    $04             ; value to be written into TFLG1 to clear IC1F flag

          ORG    $0000
edge1     RMB    2               ; the captured first edge value
period    RMB    2               ; frequency value

          ORG    $C000           ; starting address of the program
          LDX    #REGBAS
```

```
              LDAA   #CLEAR            ; clear the IC1F flag
              STAA   TFLG1,X          ;      "
```

* The following two instructions configure TCTL2 to capture the rising edge of IC1

```
              LDAA   #IC1rise          ; capture the rising edge of IC1
              STAA   TCTL2,X          ;      "

              BRCLR  TFLG1,X $04 *     ; wait for the arrival of the first rising edge
              LDD    TIC1,X           ; save the captured first edge
              STD    EDGE1            ;      "
              LDAA   #CLEAR           ; clear the IC1F flag
              STAA   TFLG1,X          ;      "
              BRCLR  TFLG1,X $40 *     ; wait for the arrival of the second rising edge
              LDD    TIC1,X           ; subtract the first edge from the second one
              SUBD   edge1            ;      "
              STD    period           ; save the period
              SWI                     ; break to monitor
              END
```

To use the interrupt method, we need to set the IC1I bit of the TMSK1 register to 1 and clear the global interrupt mask, i.e., the I bit of the CCR register. The principles of the interrupt-driven version of the period-measurement program are shown in Figure 7.14. The dotted lines in Figure 7.14 indicate program control transfers caused by interrupts and returns from interrupts.

The interrupt-driven version of the period-measurement program using the input-capture function is:

```
REGBAS    EQU    $1000            ; base address of the I/O register block
TFLG1     EQU    $23              ; offset of TFLG1 from REGBAS
TMSK1     EQU    $22              ; offset of TMSK1 from REGBAS
TIC1      EQU    $10              ; offset of TIC1 from REGBAS
TCTL2     EQU    $21              ; offset of TCTL2 from REGBAS
IC1rise   EQU    $10              ; value to select the rising edge of IC1 to capture
IC1I      EQU    $04              ; mask to select the IC1I bit of the TMSK1 register
IC1FM     EQU    $FB              ; mask to clear IC1F flag using the BCLR instruction

          ORG    $00
edge_cnt  RMB    1                ; edge count
EDGE1     RMB    2                ; captured first edge value
period    RMB    2                ; period in number of E clock cycles

          ORG    $FFE8
          FDB    IC1HND           ; set up IC1 interrupt vector
```

* *
* Use the following assembler directives to replace the previous two directives if
* this program is to be run on the EVB
*
```
*         ORG    $E8
*         JMP    IC1HND
```
* *
```
          ORG    $C000            ; starting address of the main program
          LDS    #$00FF           ; set up stack pointer (set to $3F for EVB)
          LDX    #REGBAS
```

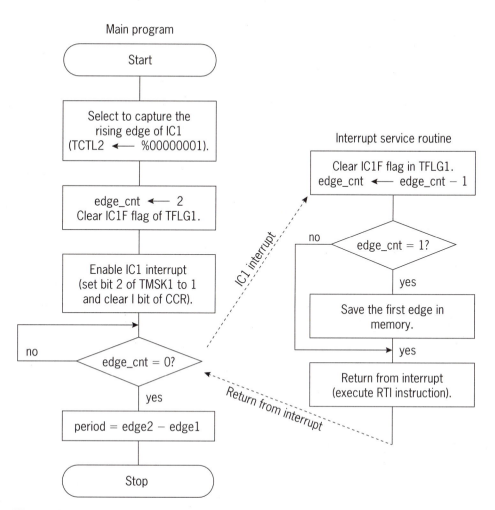

Figure 7.14 Flowchart of the period-measurement program (interrupt-driven version)

```
        LDAA  #IC1rise        ; prepare to capture the rising edge of IC1
        STAA  TCTL2           ;     "
        BCLR  TFLG1,X IC1FM   ; clear the IC1F flag of the TFLG1 register
        LDAA  #2              ; initialize edge count to 2
        STAA  edge_cnt        ;     "
        BSET  TMSK1,X IC1I    ; set the IC1 interrupt enable bit
        CLI                   ; enable interrupt to 68HC11
```

* The following two instructions wait until the variable edge_cnt has been decremented
* to 0

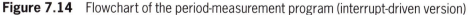

```
WAIT    TST   edge_cnt       ; check if two edges have been captured
        BNE   WAIT
        LDD   TIC1,X         ; get the second edge time
        SUBD  EDGE1          ; take the difference of edge1 and edge 2
        STD   period
         .                   ; do something else
         .
         .
```

* IC1 interrupt service routine is as follows

```
IC1HND      LDX    #REGBAS
            BCLR   TFLG1,X IC1FM     ; clear the IC1F flag
            DEC    edge_cnt
            BEQ    SKIP              ; is this the second edge?
            LDD    TIC1,X
            STD    EDGE1             ; save the arrival time of the first edge
SKIP        RTI
            END
```

Example 7.4

Write a program that uses the input-capture function IC1 to measure the pulse width of an unknown signal connected to the IC1 pin. Leave the pulse width in D. The main timer prescale factor is 1.

Solution: To measure the pulse width, first the rising and then the falling edge must be captured. The program first sets up IC1 to capture the rising edge. After the arrival of the first edge, IC1 is reconfigured to capture the falling edge. The pulse width is the difference of these two edges. The polling approach is easier to implement, and the corresponding program is as follows:

```
REGBAS     EQU    $1000            ; base address of the I/O register block
TFLG1      EQU    $23              ; offset of TFLG1 from REGBAS
TIC1       EQU    $10              ; offset of TIC1 from REGBAS
TCTL2      EQU    $21              ; offset of TCTL2 from REGBAS
IC1rise    EQU    $10              ; value to select the rising edge of IC1
IC1fall    EQU    $20              ; value to select the falling edge of IC1
IC1F       EQU    $04              ; mask to select the IC1F flag

           ORG    $0000
TEMP       RMB    2
           ORG    $C000            ; starting address of the program
           LDX    #REGBAS
```

* The following two instructions configure TCTL2 to capture the rising edge of IC1

```
           LDAA   #IC1rise
           STAA   TCTL2,X
           LDAA   #IC1F            ; clear the IC1F flag
           STAA   TFLG1,X          ;        "
rise       BRCLR  TFLG1,X IC1F rise ; wait for the arrival of the rising edge
           LDD    TIC1,X           ; save the rising edge
           STD    TEMP             ;        "
```

* The following two instructions configure TCTL2 to capture the falling edge of IC1

```
           LDAB   #IC1fall
           STAB   TCTL2,X          ; select to capture the falling edge of IC1
           LDAA   #IC1F            ; clear the IC1F flag
```

```
                    STAA   TFLG1,X          ;    "
          fall      BRCLR  TFLG1,X IC1F fall ; wait for the arrival of the falling edge
                    LDD    TIC1,X           ; get the captured second edge
                    SUBD   TEMP             ; take the difference of the falling and the rising
          *                                 ; edges
                    END
```

The programs in Examples 7.3 and 7.4 work for any pulse period or pulse width less than $(2^{16} - 1)$ E clock cycles. If the period or pulse width is longer, the TCNT register will completely roll over at least once. We must then count the number of times TCNT rolls around by counting the number of times TOF sets. Let

$ovcnt$ = the main timer overflow count
$diff$ = the difference of two edges
$edge1$ = the captured time of the first edge
$edge2$ = the captured time of the second edge

Then the period or pulse width can be calculated using the following equations:

Case 1
edge2 ≥ edge1

$$\text{period (or pulse width)} = ovcnt \times 2^{16} + diff$$

Case 2:
edge2 < edge1

$$\text{period (or pulse width)} = (ovcnt - 1) \times 2^{16} + diff$$

In case 2, the timer overflows at least once even if the period or pulse width is shorter than $(2^{16} - 1)$ E clock cycles. Therefore, we need to subtract 1 from the timer overflow count in order to get the correct result.

Example 7.5

Write a program that uses input-capture channel IC1 to measure a period longer than 2^{16} E clock cycles (32.77 ms for 2 MHz E clock).

Solution: We will use the polling method to capture two consecutive edges and include a timer overflow interrupt-handling routine to keep track of the number of main timer overflows. Two bytes are reserved for this purpose so that we can measure a period or pulse width as long as 35.8 minutes (65536 × 32.77 ÷ 1000 ÷ 60 ≈ 35.8). The period is obtained by appending the difference of the two captured edges to the overflow count. The idea of the program logic is illustrated in Figure 7.15.

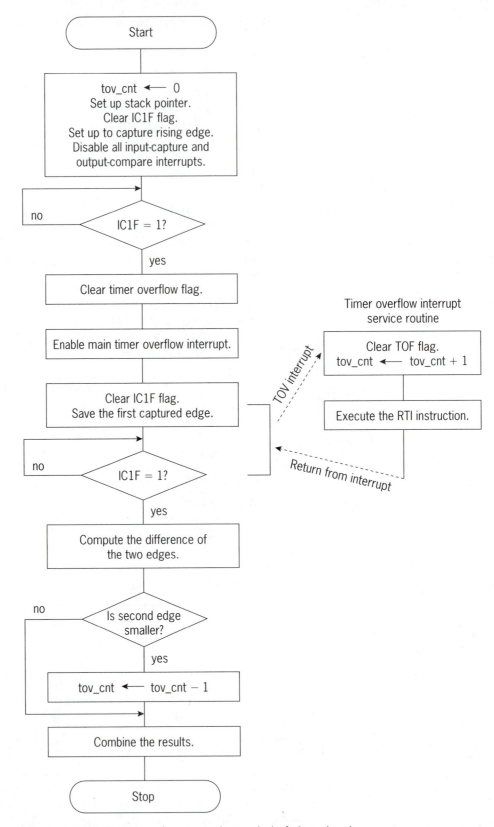

Figure 7.15 Flowchart for measuring period of slow signal

The program is:

```
REGBAS    EQU    $1000              ; base address of I/O register block
TFLG1     EQU    $23                ; offset of TFLG1 from REGBAS
TIC1      EQU    $10                ; offset of TIC1 from REGBAS
TCTL2     EQU    $21                ; offset of TCTL2 from REGBAS
TMSK1     EQU    $22                ; offset of TMSK1 from REGBAS
TMSK2     EQU    $24                ; offset of TMSK2 from REGBAS
TOF       EQU    $80                ; mask to select timer overflow flag
IC1rise   EQU    $10                ; value to select the rising edge of IC1
IC1F      EQU    $04                ; mask to select the IC1F flag
TOI       EQU    $80                ; mask to select the timer overflow interrupt

          ORG    $0000

EDGE1     RMB    2                  ; the captured first edge value
overflow  RMB    2                  ; number of main timer overflows
PERIOD    RMB    2                  ; period value

          ORG    $FFDE              ; timer overflow interrupt vector
          FDB    tov_hnd            ;      "

* * * * * * * * * * * * * * * * * * * * * * * * * * * * * * * * * * * * * * * * * * * * * * * * * * * * *
* Use the following assembler directives to set up pseudo interrupt
* vector so that this program can be run on EVB
*
*         ORG    $00D0
*         JMP    tov_hnd
* * * * * * * * * * * * * * * * * * * * * * * * * * * * * * * * * * * * * * * * * * * * * * * * * * * * *

          ORG    $C000
          LDS    #$00FF             ; set stack pointer (set to $3F for EVB)
          SEI                       ; disable interrupt before initialization
          CLR    overflow           ; initialize main timer overflow count to 0
          CLR    overflow+1         ;      "
          LDX    #REGBAS
          LDAA   #IC1rise           ; select to capture the rising edge
          STAA   TCTL2,X            ;      "
          LDAA   #IC1F              ; clear the IC1F flag in TFLG1 register
          STAA   TFLG1,X            ;      "
          BCLR   TMSK1,X $FF        ; disable all input-capture and output-compare
*                                   ; interrupts
          BRCLR  TFLG1,X IC1F *     ; wait for the arrival of the first rising edge

* The timer overflow flag must be cleared before enabling the timer overflow interrupt

          LDAA   #TOF
          STAA   TFLG2,X            ; clear the TOF flag in TFLG2 register
          BSET   TMSK2,X TOI        ; enable the main timer overflow interrupt
          CLI                       ; enable interrupt to the 68HC11
          LDD    TIC1,X             ; save the captured rising edge
          STD    EDGE1              ;      "
          LDAA   #IC1F              ; clear the IC1F flag
          STAA   TFLG1,X            ;      "
```

```
              BRCLR  TFLG1,X IC1F *   ; wait for the second rising edge
              LDD    TIC1,X           ; take the difference of the two edges
              SUBD   EDGE1            ;       "
              STD    PERIOD           ; save the period
              BCC    next             ; check whether the captured first edge is larger
```

* When the second edge is smaller, we need to subtract 1 from the timer overflow count
* using the following 4 instructions

```
              DEC    overflow+1
              LDAA   overflow
              SBCA   #0
              STAA   overflow
next          .                        ; do something else
              .
              .
```

* the timer overflow interrupt service is in the following

```
tov_hnd  LDX    #REGBAS
         LDAA   #TOF                   ; clear the TOF flag
         STAA   TFLG2,X                ;       "
```

* the following three instructions increment the timer overflow count by 1

```
              LDD    overflow
              ADDD   #1
              STD    overflow
              RTI
              END
```

■

7.6 Output-Compare Functions

The 68HC11A8 has five output-compare functions. Output-compare functions can be used to inform the 68HC11 that an event has occurred and to control the level on output pins PA7–PA3. As shown in Figure 7.16, each output-compare function consists of

1. a 16-bit comparator
2. a 16-bit compare register (TOCx)
3. an output action pin (OCx—can be pulled down to 0, pulled up to 1, or toggled)
4. an interrupt request circuit
5. a forced-compare function (FOCx)
6. control logic

The five output-compare registers occupy memory locations from $1016 through $101F, as shown in Table 7.3.

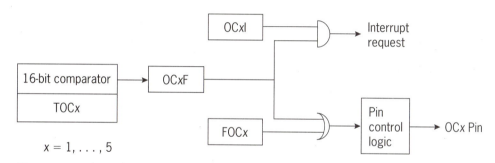

Figure 7.16 Output-compare function block diagram

7.6.1 Operation of the Output-Compare Function

The main application of an output-compare function is performing an action at a specific time in the future (when the 16-bit free-running counter reaches a specific value). The action might be to toggle a signal, turn on a switch, etcetera. To use an output-compare function, the user

1. makes a copy of the current contents of the free-running main timer (TCNT)
2. adds to this copy a value equal to the desired delay
3. stores the sum into an output-compare register.

The user has the option of specifying the action to be activated on the selected output-compare pin by programming the TCTL1 register. The comparator compares the value of TCNT and that of the specified output-compare register in every E clock cycle. If they are equal, the specified action on the output-compare pin is activated and the associated status bit in TFLG1 is set to 1. An interrupt request will be generated if it is enabled. The 16-bit output-compare register for each output-compare function can be read and written at any time. The TOCx registers are forced to $FFFF during a reset.

The following actions can be activated on an output-compare (OC) pin:

- pull up to high
- pull down to low
- toggle

Register	Address
TOC1	$1016–$1017
TOC2	$1018–$1019
TOC3	$101A–$101B
TOC4	$101C–$101D
TOC5	$101E–$101F

Table 7.3 ■ The output-compare registers in the 68HC11A8

	7	6	5	4	3	2	1	0	
	OM2	OL2	OM3	OL3	OM4	OL4	OM5	OL5	TCTL1 at $1020
value after reset	0	0	0	0	0	0	0	0	

OMx	OLx	
0	0	OCx does not affect pin
0	1	Toggle OCx pin on successful compare
1	0	Clear OCx pin on successful compare
1	1	Set OCx pin on successful compare

Figure 7.17 The contents of the TCTL1 register (Reprinted with permission of Motorola)

The action of an output-compare pin can be selected by programming the TCTL1 register (at $1020) as shown in Figure 7.17.

A successful compare will set the corresponding flag bit in the TFLG1 register. An interrupt will be generated if it is enabled. An output-compare interrupt is enabled by setting the corresponding bit in the TMSK1 register. The memory locations that hold interrupt vectors for output-compare interrupts are listed in Table 7.4. Output-compare functions are not disabled at any time. Therefore, the output-compare flags in the TFLG1 register will be set most of the time. It is very important to clear those flags before using output-compare functions. The output-compare 1 (OC1) function is different from other output-compare functions and will be discussed later.

7.6.2 Applications of the Output-Compare Function

An output-compare function can be programmed to perform a variety of functions. Generation of a single pulse, a square-wave, and a specific delay are among the most popular applications.

Output compare	Pin	68HC11 vector location	EVB pseudo vector
OC1	PA7	$FFE8–$FFE9	$00DF–$00E1
OC2	PA6	$FFE6–$FFE7	$00DC–$00DE
OC3	PA5	$FFE4–$FFE5	$00D9–$00DB
OC4	PA4	$FFE2–$FFE3	$00D6–$00D8
OC5	PA3	$FFE0–$FFE1	$00D3–$00D5

Table 7.4 ■ Interrupt vector locations for output-compare functions

Example 7.6

Generate an active high 1 KHz digital waveform with 40% duty cycle from output-compare pin OC2. Use the polling method to check the success of the output compare. The frequency of the E clock is 2 MHz, and the prescale factor to the free-running timer is 1.

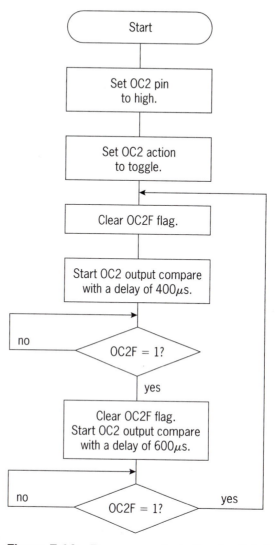

Figure 7.18 The program logic flow for digital waveform generation

Solution: An active high 1 KHz waveform with 40% duty cycle looks like this:

The logic flow of this problem is illustrated in the flowchart in Figure 7.18. At 2 MHz, the 800 and 1200 clock cycles correspond to delays of 400 μs and 600 μs, respectively.

The following program implements this algorithm using the OC2 function:

```
REGBAS    EQU    $1000            ; base address of I/O register block
PORTA     EQU    $00              ; offset of PORTA from REGBAS
```

```
        TOC2    EQU    $18              ; offset of TOC2 from REGBAS
        TCNT    EQU    $0E              ; offset of TCNT from REGBAS
        TCTL1   EQU    $20              ; offset of TCTL1 from REGBAS
        TFLG1   EQU    $23              ; offset of TFLG1 from REGBAS
        LOTIME  EQU    1200             ; value to set low time to 600 μs
        HITIME  EQU    800              ; value to set high time to 400 μs
        TOGGLE  EQU    $40              ; value to select the toggle action
        OC2     EQU    $40              ; mask to select OC2 pin and OC2F flag
        CLEAR   EQU    $40              ; value to clear OC2F flag
        sethigh EQU    $40              ; value to set OC2 pin to high

                ORG    $C000            ; starting address of the program
                LDX    #REGBAS
                BSET   PORTA,X OC2      ; set OC2 pin to high
                LDAA   #CLEAR           ; clear OC2F flag
                STAA   TFLG1,X          ;      "
                LDAA   #TOGGLE          ; select output compare action to toggle
                STAA   TCTL1,X          ;      "

    * the following 3 instructions start the OC2 output compare with a delay of 800 E clock
    * cycles

                LDD    TCNT,X           ; start an OC2 operation with a delay of 400 μs
                ADDD   #HITIME          ;      "
                STD    TOC2,X           ;      "

    high        BRCLR  TFLG1,X OC2 high ; wait until OC2F flag is set to 1
                LDAA   #CLEAR           ; clear the OC2F flag
                STAA   TFLG1,X          ;      "
                LDD    TOC2,X           ; start another OC2 operation which toggle
                ADDD   #LOTIME          ; (pulled to high) OC2 pin 1200 E clock cycles
                                        ; later
    *           STD    TOC2,X           ;      "

    low         BRCLR  TFLG1,X OC2 low  ; wait until OC2F flag is set to 1
                LDAA   #CLEAR           ; clear OC2F flag
                STAA   TFLG1,X          ;      "
                LDD    TOC2,X           ; start the next OC2 compare operation
                ADDD   #HITIME          ; which will toggle OC2 pin 400 μs later
                STD    TOC2,X           ;      "
                BRA    high
                END
```

Example 7.7

Generate a delay of one second using the OC2 function. The E clock frequency is 2 MHz, and the prescale factor to the free-running timer is 1.

Solution: Since the output-compare register is only 16-bit, the longest delay that can be generated in one output-compare operation is only 32.7 ms at 2 MHz E clock frequency. In order to generate a delay of one second, we will need to perform multiple output-compare operations. For example, we can perform

40 such output-compare operations, with each output-compare operation creating a delay of 25 ms. The following program implements this idea with the OC2 function. A memory location is used to keep track of the number of output-compare operations remaining to be performed.

```
        REGBAS   EQU    $1000        ; base address of I/O register block
        TOC2     EQU    $18          ; offset of TOC2 from REGBAS
        TCNT     EQU    $0E          ; offset of TCNT from REGBAS
        TFLG1    EQU    $23          ; offset of TFLG1 from REGBAS
        OC2      EQU    $40          ; mask to select OC2 pin and OC2F flag
        CLEAR    EQU    $40          ; value to clear OC2F flag
        DLY25ms  EQU    50000        ; the number of E clock cycles to generate 25 ms
      *                              ; delay
        onesec   EQU    40           ; the number of output compare operations to be
      *                              ; performed
                 ORG    $0000
        asec     RMB    1            ; the memory location to keep track of the
      *                              ; number of
      *                              ; output-compare operations remaining to be
      *                              ; performed

                 ORG    $C000        ; starting address of the program
                 LDX    #REGBA
                 LDAA   #OC2         ; clear OC2 flag
                 STAA   TFLG1,X      ;      "
                 LDAA   #onesec      ; initialize the output-compare count
                 STAA   asec         ;      "
                 LDD    TCNT,X       ; prepare to start the OC2 operation
        wait     ADDD   #DLY25ms     ; add 25 ms delay
                 STD    TOC2,X       ; start the OC2 compare operation
                 BRCLR  TFLG1,X OC2 *; wait until OC2F is set
                 LDAA   #OC2         ; clear the OC2F flag
                 STAA   TFLG1,X      ;      "
                 DEC    asec         ; decrement the output compare counter
                 BEQ    exit         ; is one second expired?
                 LDD    TOC2,X       ; prepare to start the next OC2 compare
      *                              ; operation
                 BRA    wait
        exit     .                   ; other operations may follow here
      *          .
      *          .
                 END
```

Example 7.8

Suppose an alarm device is already connected properly and the subroutine to turn on the alarm is also available. Write a program to implement the alarm timer—it should call the given alarm subroutine when the alarm time is reached.

Solution: We will use the OC2 function to update the current time. Each OC2 output-compare operation generates a delay of 20 ms. In order to generate a delay of one minute, 3000 such output-compare operations (call each such operation a *tick*, with one tick equal to 20 ms) must be performed. With a 2 MHz E clock frequency, the delay is equal to the duration of 40000 E clock cycles. An interrupt will be requested when the OC2 output-compare succeeds. The OC2 interrupt service routine decrements the ticks by one and updates the current time. The interrupt service routine also compares the current time with the predefined alarm time. When they are equal, a subroutine is called to invoke the alarm device.

```
        REGBAS   EQU    $1000      ; base address of I/O register block
        TOC2     EQU    $18        ; offset of TOC2 from REGBAS
        TCNT     EQU    $0E        ; offset of TCNT from REGBAS
        TFLG1    EQU    $23        ; offset of TFLG1 from REGBAS
        TMSK1    EQU    $22        ; offset of TMSK1 from REGBAS
        CLEAR    EQU    $40        ; value to clear OC2F flag
        DLY20ms  EQU    40000      ; the number of E clock cycles needed to generate a
*                                  ; 25 ms delay
        one_min  EQU    3000       ; number of output compares to be performed to
*                                  ; generate a delay of one minute

                 ORG    $0000
hours            RMB    1
minutes          RMB    1
ticks            RMB    2
alarm            RMB    2
routine          FDB    handler    ; the starting address of the alarm handler

                 ORG    $FFE6      ; interrupt vector location for OC2
                 FDB    oc2_hnd    ; set up OC2 interrupt vector
```

****
* Use the following assembler directives to replace the previous two
* directives if the program is to be run on the EVB
*
```
*                ORG    $00DC
*                JMP    oc2_hnd
```
****

```
                 ORG    $C000      ; starting address of the program
                 LDS    #$00FF     ; initialize the stack pointer (set to $003F for EVB)
                 SEI               ; disable interrupt to 68HC11 at the beginning
                 LDD    #one_min   ; initialize the output-compare count to generate
                 STD    ticks      ; a one-minute delay
                 LDX    #REGBAS
                 LDAA   #CLEAR     ; clear the OC2F flag
                 STAA   TFLG1,X    ;       "
                 STAA   TMSK1,X    ; enable the OC2 interrupt
                 LDD    TCNT,X     ; start an OC2 output-compare operation with a
                 ADDD   #DLY20ms   ; delay equal to 20 ms
                 STD    TOC2,X     ;       "
                 CLI               ; enable interrupt to 68HC11
loop             BRA    loop       ; loop forever and wait for OC2 interrupts
```

* the interrupt service routine of OC2 follows

```
oc2_hnd   LDX    #REGBAS
          LDAA   #CLEAR     ; clear OC2F flag
```

```
                    STAA  TFLG1,X      ;      "
                    LDD   TOC2,X
                    ADDD  #DLY20ms     ; start the next 20 ms delay
                    STD   TOC2,X       ;      "
                    LDY   ticks        ; decrement the minute ticks
                    DEY                ;      "
                    STY   ticks        ;      "
                    BNE   CASE2        ; has one minute expired?
                    LDD   #one_min     ; reinitialize the one-minute counter
                    STD   TICKS        ;      "

                    LDD   hours        ; load the hours and minutes
                    INCB               ; increment the minute
                    CMPB  #60          ; is it time to increment the hour?
                    BNE   CASE1        ; no need to update hours yet
                    CLRB               ; clear minutes to 0
                    INCA               ; increment the hour
                    CMPA  #24          ; is it time to reset hours to 0
                    BNE   CASE1        ; no need to reset hours to 0 yet
                    CLRA               ; reset hours to 00
        CASE1       STD   hours        ; save the current time in memory
                    CPD   alarm        ; compare to alarm time
                    BNE   CASE2
                    LDX   routine      ; if match, then get the subroutine address
                    JSR   0,X          ; call the alarm subroutine
        CASE2       RTI
                    END
```

7.7 Using OC1 to Control Multiple OC Functions

The output-compare function OC1 is special because it can control up to five output-compare functions at the same time. The register OC1M is used to specify the output-compare functions to be controlled by OC1. The value that any OCx $(x = 1, \ldots, 5)$ pin is to assume when the value of TOC1 equals TCNT is specified by the OC1D register. To control an output-compare pin using OC1, the user sets the corresponding bit of OC1M. When a successful OC1 compare is made, each affected pin assumes the value of the corresponding bit of OC1D. The contents of the OC1M and OC1D registers and the pin controlled by each bit are shown in Figure 7.19.

	7	6	5	4	3	2	1	0	
	M7	M6	M5	M4	M3	0	0	0	OC1M at $100C
Value after reset	0	0	0	0	0	0	0	0	
	PA7	OC2	OC3	OC4	OC5	--------	--------	--------	pin controlled

	7	6	5	4	3	2	1	0	
	D7	D6	D5	D4	D3	0	0	0	OC1D at $100D
Value after reset	0	0	0	0	0	0	0	0	

Figure 7.19 Contents of the OC1M and OC1D registers (Reprinted with permission of Motorola)

Example 7.9

What values should be written into OC1M and OC1D if we want pins OC2 and OC3 to assume the values of 0 and 1 when the OC1 compare succeeds?

Solution: Bits 6 and 5 of OC1M must be set to 1, and bits 6 and 5 of OC1D should be set to 0 and 1, respectively. The following instruction sequence will set up these values:

```
REGBAS   EQU  $1000        ; base address of I/O register block
OC1M     EQU  $0C          ; offset of OC1M from REGBAS
OC1D     EQU  $0D          ; offset of OC1D from REGBAS

         LDX  #REGBAS
         LDAA #$60
         STAA OC1M,X        ; specify pins to be controlled by OC1
         LDAA #$20
         STAA OC1D,X        ; specify the values for OC2 and OC3 pins to assume on
                            ; a successful OC1 compare
```

The OC1 function is especially suitable when several functions are to be controlled by one signal. The PA7 pin is a bidirectional pin that can be controlled by OC1. In order for PA7 to be controlled by OC1, it must be configured for output by setting the DDRA7 bit of the PACTL register to 1. OC1 can also be used in conjunction with one or more other output-compare functions to achieve even more timing flexibility. OC1 can control a timer output even when one of the other output-compare functions is already controlling the same pin, thus allowing the user to schedule two successive edges of each output signal at once. Example 7.11 illustrates this application.

Example 7.10   Using the OC1 function to control multiple OC pins.

An application requires control of five valves. The first, third, and fifth valves should be opened for five seconds and then closed for five seconds. When these three valves are closed, the second and fourth valves must be opened, and vice versa. This process is repeated forever. The requirements of this application can be satisfied by using the OC1 function. Pins OC1–OC5 are used to control valves 1 to 5, respectively. When the OCx (x = 1, . . . , 5) pin is high, the corresponding valve will be opened. Write a program to perform the specified operations.

Solution: For this application, all output-compare pins must be controlled by OC1. The PA7 pin must be set for output. The value %11111000 should be written into OC1M. In order to open valves 1, 3, and 5, the value %10101000 should be written into OC1D. To open valves 2 and 4, the value %01010000 should be written into OC1D. To create a delay of five seconds, 200 output-compare operations will be performed, with each output compare creating a delay of 25 ms. The following program implements these ideas.

```
REGBAS      EQU     $1000           ; base address of I/O register block
PACTL       EQU     $26             ; offset of PACTL from REGBAS
OC1D        EQU     $0D             ; offset of OC1D from REGBAS
OC1M        EQU     $0C             ; offset of OC1M from REGBAS
TOC1        EQU     $16             ; offset of TOC1 from REGBAS
TCNT        EQU     $0E             ; offset of TCNT from REGBAS
TFLG1       EQU     $23             ; offset of TFLG1 from REGBAS
oc1m_in     EQU     $F8             ; value to initialize the OC1M register
oc1d_in1    EQU     $A8             ; value to initialize the OC1D register to open
*                                   ; valves 1,3, and 5
oc1d_in2    EQU     $50             ; value to initialize the OC1D register to open
*                                   ; valves 2 and 4
five_sec    EQU     200             ; number of OC1 compares to be performed to
*                                   ; create a 5-second delay
CLEAR       EQU     $40             ; value to clear OC2F flag
DLY25ms     EQU     50000           ; the number of E clock cycles to generate 25 ms
*                                   ; delay
PA7         EQU     $80             ; mask to select PA7 pin
OC1         EQU     $80             ; mask to specify OC1
OC1F        EQU     $7F             ; value to select the OC1F flag

            ORG     $0000
oc1_cnt     RMB     1               ; keep track of the number of OC1 operations
*                                   ; remaining to be performed

            ORG     $C000           ; starting address of the program
            LDX     #REGBAS
            BSET    PACTL,X PA7     ; set the PA7 pin for output
            LDAA    #oc1m_in        ; allow OC1 to control 5 pins
            STAA    OC1M,X          ;         "
            BCLR    TFLG1,X OC1F    ; clear OC1F flag
            LDAA    #five_sec       ; initialize the OC1 count
            STAA    oc1_cnt         ;         "
            LDAA    #oc1d_in1       ; set pins OC1, OC3, and OC5 to high after 25 ms
            STAA    OC1D,X          ;         "
            LDD     TCNT,X          ; start an OC1 output compare
repeat1     ADDD    #DLY25ms        ; to create a delay of 25 ms
            STD     TOC1,X          ;         "
            BRCLR   TFLG1,X OC1 *   ; wait until the OC1F flag is set to 1
            BCLR    TFLG1,X OC1F    ; clear the OC1F flag
            DEC     oc1_cnt         ; decrement the output-compare count
            BEQ     change          ; at the end of 5 seconds change the valve setting
            LDD     TOC1,X          ; prepare to start the next OC1 compare
            BRA     repeat1
```

* The following two instructions change the valve setting

```
change      LDAA    #oc1d_in2       ; set to open OC2, OC4
            STAA    OC1D,X          ;         "

            LDAA    #five_sec       ; re-initialize the output-compare count
            STAA    oc1_cnt         ;         "
repeat2     LDD     TOC1,X          ; start the next OC1 operation
            ADDD    #DLY25ms        ;         "
```

```
                    STD    TOC1,X          ;      "
                    BRCLR  TFLG1,X OC1  *  ; wait until 25 ms expires
                    BCLR   TFLG1,X OC1F    ; clear OC1F flag
                    DEC    oc1_cnt         ; decrement the number of OC1 compares to be
            *                              ; performed
                    BEQ    switch          ; switch the valve setting
                    BRA    repeat2
    switch          LDAA   #five_sec       ; reinitialize the output-compare count
                    STAA   oc1_cnt         ;      "
                    LDAA   #oc1d_in1       ; change the valve setting
                    STAA   OC1D,X          ;      "
                    LDD    TOC1,X          ; prepare to start another OC1 operation
                    BRA    repeat1         ;      "
                    END
```

■

Two output-compare functions can control the same pin simultaneously. Thus OC1 can be used in conjunction with one or more other output compares to achieve even more timing flexibility. For example, we can generate a digital waveform with a given duty cycle by using OC1 and any other output-compare function.

Example 7.11

Use OC1 and OC2 together to generate a 5 KHz digital waveform with 40% duty cycle (active high).

Solution: The idea in using OC1 and OC2 together to generate the specified waveform is to use the OC1 function to pull the OC2 pin to high every 200 μs and use the OC2 function to pull the OC2 pin to low 80 μs later. The waveform is shown in Figure 7.20. Output-compare interrupts for OC1 and OC2 will be enabled. Each interrupt service routine clears the interrupt flag and then starts another output-compare operation with a 200-μs delay.

```
REGBAS    EQU    $1000        ; base address of I/O register block
TMSK1     EQU    $22          ; offset of TMSK1 from REGBAS
PORTA     EQU    $00          ; offset of PORTA from REGBAS
OC1D      EQU    $0D          ; offset of OC1D from REGBAS
OC1M      EQU    $0C          ; offset of OC1M from REGBAS
TOC1      EQU    $16          ; offset of TOC1 from REGBAS
```

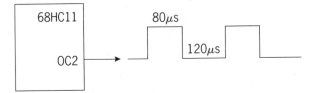

Figure 7.20   Using OC1 and OC2 together to generate a 40% duty cycle waveform

```
TOC2        EQU    $18              ; offset of TOC2 from REGBAS
TCTL1       EQU    $20              ; offset of TCTL1 from REGBAS
TCNT        EQU    $0E              ; offset of TCNT from REGBAS
TFLG1       EQU    $23              ; offset of TFLG1 from REGBAS
tctl1_in    EQU    $80              ; value to set the OC2 action to pull the OC2 pin to low
oc1m_in     EQU    $40              ; value to allow OC1 to control OC2 pin
oc1d_in     EQU    $40              ; value to be written into OC1D to pull OC2 pin to high
fiveKHz     EQU    400              ; timer count for 5 KHz
diff        EQU    160              ; the count difference between two OC functions

            ORG    $FFE6
            FDB    OC2_HND          ; set up OC2 interrupt vector
            FDB    OC1_HND          ; set up OC1 interrupt vector
```

```
* * * * * * * * * * * * * * * * * * * * * * * * * * * * * * * * * * * * * * * * * * * * * * * * * * * * * * * * * * * * *
* Replace the previous three assembler directives with the following directives if
* the program is to be run on the EVB
*
*           ORG    $00DC
*           JMP    OC2_HND          ; pseudo interrupt vector for OC2
*           JMP    OC1_HND          ; pseudo interrupt vector for OC1
* * * * * * * * * * * * * * * * * * * * * * * * * * * * * * * * * * * * * * * * * * * * * * * * * * * * * * * * * * * * *
```

```
            ORG    $C000            ; starting address of the program
            LDS    #$00FF           ; set up stack pointer (set to $003F for EVB)
            LDX    #REGBAS
            BCLR   TFLG1,X $3F      ; clear OC1F and OC2F flags
            LDAA   #tctl1_in        ; define OC2 action to pull OC2 pin to low
            STAA   TCTL1,X          ;       "
            LDAA   #oc1d_in         ; define OC1 action to pull the OC2 pin to high
            STAA   OC1D,X           ;       "
            LDAA   #oc1m_in         ; allow OC1 to control OC2 pin
            STAA   OC1M,X           ;       "
            BSET   TMSK1,X $C0      ; enable OC1 and OC2 interrupt
            BCLR   PORTA,X $40      ; pull OC2 pin to low
            LDD    TCNT,X
            ADDD   #fiveKHz         ; add 200 μs delay
            STD    TOC1,X           ; start OC1 compare operation
            ADDD   #diff
            STD    TOC2,X           ; start OC2 compare operation
            CLI                     ; enable interrupt to 68HC11
loop        BRA    loop             ; infinite loop to wait for OC1 and OC2 interrupts
```

* The following five instructions are the OC1 service routine:

```
oc1_hnd     BCLR   TFLG1,X $7F      ; clear OC1F flag
            LDD    TOC1,X
            ADDD   #fiveKHz         ; start the next OC1 compare operation
            STD    TOC1,X           ;       "
            RTI
```

* The service routine for the OC2 interrupt is:

```
oc2_hnd     BCLR   TFLG1,X $BF      ; clear the OC2 flag
            LDD    TOC2,X
```

```
        ADDD  #fiveKHz          ; start the next OC2 compare operation
        STD   TOC2,X            ;      "
        RTI
        END
```

7.8 Forced Output Compare

There may be applications in which the user requires an output compare to occur immediately instead of waiting for a match between TCNT and the proper TOC register. This situation arises in spark-timing control in some automotive engine control applications. To use the forced output-compare mechanism, the user would write to the CFORC register with 1s in the bit positions corresponding to the output-compare channels to be forced. At the next timer count after the write to CFORC, the forced channels will trigger their programmed pin actions to occur. The forced actions are synchronized to the timer counter clock, which is slower than the E clock signal if a prescale factor has been specified. The forced output-compare signal causes pin actions but does not affect the OCxF bit or generate interrupts. Normally, the force mechanism would not be used in conjunction with the automatic pin action that toggles the corresponding output-compare pin. The contents of CFORC (at $100B) are shown in Figure 7.21. CFORC always reads as all 0s.

Figure 7.21   Contents of the CFORC register (Reprinted with permission of Motorola)

Example 7.12

Suppose that the contents of the TCTL1 register are %10011000. What would occur on pins PA6–PA3 on the next clock cycle if the value %01111000 is written into the CFORC register?

Solution: The contents of the TCTL1 register configure the output-compare actions shown in Table 7.5.

Bit positions	Value	Action to be triggered
7 6	10	clear PA6
5 4	01	toggle PA5
3 2	10	clear PA4
1 0	00	no effect

Table 7.5 ■ Pin actions on OC2–OC5 pins

Since the CFORC register specifies that output-compare functions 2–5 are to be forced immediately, the actions specified in the third column of Table 7.5 will occur immediately.

■

7.9 Real-Time Interrupt (RTI)

When enabled, the real-time interrupt (RTI) function of the 68HC11 can be used to generate periodic interrupts, and these interrupts can be used to perform program switching in a multitasking operating system (see Chapter 5).

In a multitasking operating system, a *task* is the execution image of a program. Therefore, two tasks can run the same program. The operating system places all the tasks waiting for execution in a queue data structure. Recall that the queue data structure has a head and a tail. Elements can be appended only to the tail and removed only from the head; the queue is therefore a first-in-first-out data structure. When a timer interrupt occurs, the running task is suspended and the operating system takes over the CPU control. The operating system appends the suspended task to the tail of the queue and then removes the task from the head of the queue for execution.

The period of the real-time interrupt is programmable. Its setting is controlled by the lowest two bits of the PACTL register (located at $1026), as shown in Table 7.6. If the divide factor is 1, the real-time interrupt will be generated every 4.1 ms with a 2 MHz E clock. Since the RTI generates interrupts periodically, it is suitable for any application that requires periodic processing, such as updating the time-of-day or generating a delay. However, the user should know that the RTI is not as good as the output-compare function in terms of the resolution of the delays it generates.

RTR1 bit 1	RTR0 bit 0	$(E/2^{13})$ divided by
0	0	1
0	1	2
1	0	4
1	1	8

Table 7.6 ■ RTI clock source prescale factor

■

Example 7.13

Use RTI to generate a delay of 10 seconds.

Solution: To generate a delay of 10 seconds using the RTI function, we need to deal with three registers:

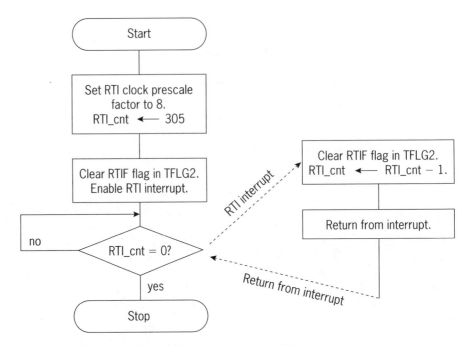

Figure 7.22 Using RTI to create a 10-second delay

1. We need to enable the real-time interrupt, so we need to set the RTI bit (bit 6) of TMSK2 to 1.

2. Since the interrupt interval of the RTI is much shorter than 10 seconds, many RTI interrupts are required to generate a delay of 10 seconds. The RTIF flag (bit 6) of the TFLG2 register will need to be cleared many times.

3. We choose a divide factor of 8, so a delay of about 32.77 ms is generated by each RTI interrupt. The total number of RTI interrupts required to generate a delay of 10 seconds is $1000 \times 10 \div 32.77 = 305.15 \approx 305$. The value to be written into the lowest two bits of the PACTL register is 11_2.

The key ideas of the solution to this problem are illustrated in the flowchart in Figure 7.22.

```
REGBAS    EQU   $1000        ; base address of I/O register block
TMSK2     EQU   $24          ; offset of TMSK2 from REGBAS
TFLG2     EQU   $25          ; offset of TFLG2 from REGBAS
PACTL     EQU   $26          ; offset of PACTL from REGBAS
TENSEC    EQU   305          ; total RTI interrupts in 10 seconds
RTIF      EQU   $40          ; mask to select the RTIF flag

          ORG   $0000
rti_cnt   RMB   2            ; reserve two bytes for RTI interrupt count
          ORG   $FFF0        ; set up RTI interrupt vector
          FDB   rti_hnd      ;      "
```

```
* * * * * * * * * * * * * * * * * * * * * * * * * * * * * * * * * * * * * * * * * * * * * * * * * * * *
* The following directives must be used to replace the previous two directives if
* this program is to be executed on the EVB
*
*             ORG   $00EB
*             JMP   rti_hnd
* * * * * * * * * * * * * * * * * * * * * * * * * * * * * * * * * * * * * * * * * * * * * * * * * * * *

              ORG   $C000              ; starting address of the program
              LDS   #$00FF             ; set up stack pointer (set to $003F for the EVB)
              LDX   #REGBAS
              LDD   #TENSEC            ; initialize the RTI interrupt count
              STD   rti_cnt            ;      "
              BSET  PACTL,X $03        ; set the RTI clock source prescale factor to 8
              LDAA  #RTIF
              STAA  TFLG2,X            ; clear the RTI flag
              STAA  TMSK2,X            ; enable the RTI interrupt
              CLI                      ; enable the interrupt to the 68HC11

* The following two instructions wait until the rti_cnt is decremented to 0

loop          LDD   rti_cnt
              BNE   loop
              .                        ; do something else
              .
              .

* The following instruction sequence is the RTI interrupt service routine

rti_hnd       LDX   #REGBAS
              LDAA  #RTIF
              STAA  TFLG2,X
              LDX   rti_cnt
              DEX
              STX   rti_cnt
              RTI
              END
```

7.10 The Pulse Accumulator

The 68HC11 has an 8-bit pulse accumulator (PACNT) that can be configured to operate in either of two basic modes:

1. *Event-counting mode.* The 8-bit counter increments at each active edge of the PA7 (PAI) pin. Many microcontroller applications involve counting things. These things are called *events*, but in real applications they might be anything: pieces on an assembly line, cycles of an incoming signal, or units of time. To be counted by the pulse accumulator, these things must be translated into rising or falling edges on the PAI pin. A trivial example of event counting

might be to count pieces on an assembly line: a light emitter/detector pair could be placed across a conveyor so that as each piece passes the sensor, the light beam is interrupted and a logic-level signal that can be connected to the PAI pin is produced.

2. *Gated accumulation mode.* The 8-bit counter is clocked by a free-running E-divided-by-64 clock, subject to the PAI signal being active. One common use of this mode is to measure the duration of single pulses. The counter is set to 0 before the pulse starts, and the resulting pulse time is read directly when the pulse is finished.

The PAI pin must be configured for input and the PAI must be enabled before the PAI system can be used. Interrupts are often used in pulse accumulator applications. There are two independent interrupt sources for the pulse accumulator:

1. detection of a selected edge on the PAI pin
2. rollover (overflow) of the 8-bit counter, i.e., from $FF to $00

7.10.1 Pulse-Accumulator–Related Registers

The four registers described below are related to the pulse accumulator. The contents of these registers are shown in Figure 7.23, where the bits related to the pulse accumulator are in boldface. These bits are also discussed below:

1. PACNT: an 8-bit pulse accumulation counter.
2. TMSK2: timer mask register 2. Bits 5 and 4 are related to the operation of the pulse accumulator.

   PAOVI: pulse accumulator overflow interrupt enable. When this bit is set to 1, the pulse accumulator overflow interrupt is enabled.

   PAII: pulse accumulator input interrupt enable. When this bit is set to 1, the pulse accumulator input edge interrupt is enabled.
3. TFLG2: timer flag register 2. Bits 5 and 4 are related to the operation of the pulse accumulator.

7	6	5	4	3	2	1	0	
TOI	RTII	**PAOVI**	**PAII**	0	0	PR1	PR0	TMSK2 at $1024

Value after reset: 0 0 0 0 0 0 0 0

TOF	RTIF	**PAOVF**	**PAIF**	0	0	0	0	TFLG2 at $1025

Value after reset: 0 0 0 0 0 0 0 0

DDRA7	**PAEN**	**PAMOD**	**PEDGE**	0	0	RTR1	RTR0	PACTL at $1026

Value after reset: 0 0 0 0 0 0 0 0

bit 7	bit 6	bit 5	bit 4	bit 3	bit 2	bit 1	bit 0	PACNT at $1027

Value after reset: – – – – – – – –

Figure 7.23 Registers related to the pulse accumulator function (Reprinted with permission of Motorola)

PAOVF: pulse accumulator overflow flag. This bit is set to 1 whenever PACNT rolls over from $FF to $00.

PAIF: pulse accumulator input edge detect flag. This bit is set to 1 each time a selected edge is detected at the PA7/PAI pin.

4. PACTL: pulse accumulator control register. Bits 7 to 4 are related to the operation of the pulse accumulator.

DDRA7: data direction control for port A pin 7. This bit must be set to 0 when the pulse accumulator is enabled.

PAEN: pulse accumulator enable. When this bit is set to 1, the pulse accumulator is enabled.

PAMOD: pulse accumulator mode select. This bit sets the operation mode of the pulse accumulator:

0: external event-counting mode

1: gated time accumulation mode

PEDGE: pulse accumulator edge select. This bit has different meanings depending on the state of the PAMOD bit:

PAMOD	PEDGE	Action on clock
0	0	PAI falling edge increments the counter.
0	1	PAI rising edge increments the counter.
1	0	A 0 on PAI inhibits counting.
1	1	A 1 on PAI inhibits counting.

7.10.2 Pulse Accumulator Applications

The 68HC11 pulse accumulator has several interesting applications, such as interrupting after N events, pulse duration measurement, and frequency measurement.

Example 7.14   Interrupt after N events

In this application, external events are converted into signals and connected to the PAI pin. N must be smaller than 255. Write a program to interrupt the 68HC11 after N events.

Solution: By writing the two's complement of N into PACNT, the counter will overflow after N events and generate an interrupt. The program is as follows:

```
REGBAS   EQU   $1000        ; base address of I/O register block
TMSK2    EQU   $24          ; offset of TMSK2 from REGBAS
TFLG2    EQU   $25          ; offset of TFLG2 from REGBAS
PACTL    EQU   $26          ; offset of PACTL from REGBAS
PACNT    EQU   $27          ; offset of PACNT from REGBAS
PA_INI   EQU   $50          ; value to be written into PACTL to enable PA and
*                           ; select event-counting mode,
*                           ; falling edge as active edge
N        EQU   ...          ; event count
```

```
            ORG   $C000              ; starting address of the program
            LDX   #REGBAS
            BCLR  TFLG2,X $20        ; clear the PAOVF flag to make sure it is not set
            LDAA  #N
            NEGA
            STAA  PACNT,X            ; place the two's complement of N in PACNT
            LDAA  #PA_INI
            STAA  PACTL,X            ; initialize the pulse accumulator
            BSET  TMSK2,X $20        ; enable the PACNT overflow interrupt
            CLI                      ; enable the interrupt to the 68HC11
            END
```

The pulse accumulator can also be used to measure the frequency of an unknown signal. The idea is as follows:

1. Set up the pulse accumulator to operate in event-counting mode.
2. Connect the unknown signal to the PAI pin.
3. Select the active edge (rising or falling).
4. Use one of the output-compare functions to create a delay of one second.
5. Use a memory location to keep track of the number of active edges arriving during one second.
6. Enable the pulse accumulator interrupt. The interrupt-handling routine increments the active edge count by 1.
7. Disable the pulse accumulator interrupt at the end of one second.

Example 7.15

Write a program to measure the frequency of an unknown signal connected to the PAI pin.

Solution: We will use the OC2 function to perform 40 output-compare operations, with each operation creating a delay of 25 ms. This will create a time base of one second. The PAI interrupt is enabled, and a service routine is written that will increment the frequency count by 1 whenever it is invoked. The memory location used as the frequency counter must be initialized to 0. Whenever an active edge arrives at the PAI pin, an interrupt that causes the frequency counter to be incremented by 1 is requested. The number of active edges arriving in one second is the frequency of the unknown signal.

```
REGBAS   EQU   $1000              ; base address of I/O register block
TCNT     EQU   $0E                ; offset of TCNT from REGBAS
TOC2     EQU   $18                ; offset of TOC2 from REGBAS
TFLG1    EQU   $23                ; offset of TFLG1 from REGBAS
TMSK2    EQU   $24                ; offset of TMSK2 from REGBAS
TFLG2    EQU   $25                ; offset of TFLG2 from REGBAS
PACTL    EQU   $26                ; offset of PACTL from REGBAS
```

```
PACNT      EQU   $27               ; offset of PACNT from REGBAS
CLEAR      EQU   $40               ; mask to clear the OC2F flag
OC2        EQU   $40               ; mask to select the OC2F flag
OC2DLY     EQU   50000             ; output-compare delay count to generate a
*                                  ; 25-ms delay
PA_IN      EQU   $50               ; value to be written into PACTL to enable PA
*                                  ; function, select event-counting mode,
*                                  ; rising edge as active edge,
*                                  ; and set PA7 pin for output
onesec     EQU   40                ; number of output compares to be performed
STOP       EQU   $10               ; value to disable PAI interrupt

           ORG   $00
oc2_cnt    RMB   1
FREQCY     RMB   2                 ; memory location to keep track of the number of
*                                  ; active edges arriving in one second
           ORG   $FFDA
           FDB   pa_hdler          ; set up PAI interrupt vector

***********************************************************************
* Use the following directives to replace the above two directives if this program is
* to be executed on the EVB
*
*          ORG   $00CA
*          JMP   PA_HDLER          ; set up PAI pseudo interrupt vector on EVB
***********************************************************************

           ORG   $C000             ; starting address of the program
           LDS   #$00FF            ; set up the stack pointer (set to $3F for EVB)
           LDX   #REGBAS
           LDAA  #onesec           ; set up the number of OC2 operations
           STAA  oc2_cnt           ; to be performed
           LDD   #0
           STD   FREQCY            ; initialize the frequency counter to 0
           LDAA  #PA_IN
           STAA  PACTL,X           ; initialize the PA function
           BCLR  TFLG2,X $EF       ; clear the PAIF flag
           BSET  TMSK2,X $10       ; enable the PAI interrupt
           CLI                     ; enable the interrupt to the 68HC11
           LDD   TCNT,X
sec_loop   ADDD  #OC2DLY           ; start the OC2 operation to create a 25-ms delay
           STD   TOC2,X            ;        "
           LDAA  #CLEAR
           STAA  TFLG2,X           ; clear the OC2F flag
wait25ms   BRCLR TFLG1,X OC2 *     ; wait until the OC2F flag is set
           LDD   TOC2,X
           DEC   oc2_cnt
           BNE   sec_loop          ; repeat if one second has not yet passed
           LDAA  #STOP
           STAA  TMSK2,X           ; disable the PAI interrupt
*                                  ; other operations may follow here
           .
           .
           .
```

* The PAI interrupt service routine is in the following

```
pa_hdler    LDX     #REGBAS
            BCLR    TFLG2,X $EF      ; clear the PAIF flag
            LDD     FREQCY
            ADDD    #1               ; increase the frequency count
            STD     FREQCY           ;      ”
            RTI
            END
```

The pulse accumulator can be set up to measure the duration of an unknown signal using the gated pulse accumulation mode. Remember that the pulse accumulator can count only if the PAI input is asserted, i.e., only when the pulse is present. By counting the number of clock cycles that occur when the pulse is present, the pulse duration can be measured. The clock input to the pulse accumulator is E ÷ 64. The procedure for measuring the pulse width of an unknown signal (positive pulse) is as follows:

Step 1
Set up the pulse accumulator to operate in the gated pulse accumulation mode, and initialize the pulse accumulator counter (PACNT) to 0.
Step 2
Select the falling edge as the active edge. In this setting, the pulse accumulator counter will increment when the signal connected to the PAI pin is high and generate an interrupt to the 68HC11 on the falling edge.
Step 3
Enable the PAI active edge interrupt and wait for the arrival of the active edge of PAI.
Step 4
Stop the pulse accumulator counter when the interrupt arrives.

Since the pulse accumulator counter is 8-bit, it can only measure the width of a pulse with a duration no longer than $2^8 \times 64 = 16386$ E clock cycles. If you need to measure the duration of a very slow pulse, you can keep track of the number of pulse accumulator counter overflows. Let n be the number of PA overflows and m be the contents of the PACNT when the PAI falling edge arrives. Then the duration of the unknown signal can be calculated using the following expression:

$$\text{pulse width} = (2^8 \times n + m) \times 64 \times T_E$$

where T_E is the period of the E clock.

Example 7.16

Write a program to measure the duration of an unknown signal connected to the PAI pin using the idea just described.

Solution:

```
REGBAS   EQU   $1000          ; base address of I/O register block
TMSK2    EQU   $24            ; offset of TMSK2 from REGBAS
TFLG2    EQU   $25            ; offset of TFLG2 from REGBAS
PACTL    EQU   $26            ; offset of PACTL from REGBAS
PACNT    EQU   $27            ; offset of PACNT from REGBAS
STOP     EQU   $00            ; value to stop pulse accumulator
PA_IN    EQU   %01100000      ; value to be written into PACTL to enable PA, select
*                             ; gated accumulation mode, and select active high
         ORG   $00
ov_cnt   RMB   2              ; reserve two bytes to keep track of the PA overflow
*                             ; count
pa_cnt   RMB   1              ; reserve one byte to hold the count of PACNT
edge     RMB   1              ; PAI edge interrupt count


         ORG   $FFDA
         FDB   pa_hnd         ; set up PAI edge interrupt vector
         FDB   pa_over        ; set up PA overflow interrupt vector


* * * * * * * * * * * * * * * * * * * * * * * * * * * * * * * * * * * * * * * * * * * * * * * * * *
* Use the following directives to replace the previous three directives if the program
* is to be run on the EVB
*
*        ORG   $CA
*        JMP   pa_hnd         ; set up EVB PAI pseudo vector
*        JMP   pa_over        ; set up EVB PAOV pseudo vector
* * * * * * * * * * * * * * * * * * * * * * * * * * * * * * * * * * * * * * * * * * * * * * * * * *


         ORG   $C000          ; starting address of the program
         LDS   #$00FF         ; set up the stack pointer (set to $3F for EVB)
         LDX   #REGBAS
         LDD   #0
         STD   ov_cnt         ; initialize the PA overflow count to 0
         LDAA  #1
         STAA  edge           ; initialize the PAI edge interrupt count to 1
         BCLR  TFLG2,X $CF    ; clear the PAOVF and PAIF flags to 0
         LDAA  #PA_IN
         STAA  PACTL,X        ; initialize the pulse accumulator
         CLR   PACNT,X        ; initialize PACNT to 0
         BSET  TMSK2,X $30    ; enable the PAI and PAOV interrupts
         CLI                  ; enable the interrupts to the 68HC11
wait     TST   edge           ; wait for the falling edge of PAI
         BNE   wait           ;          "
         BCLR  PACTL,X $40    ; disable the pulse accumulator when the falling edge
*                             ; arrives
         LDAA  PACNT,X
         STAA  pa_cnt         ; append the contents of PACNT to the overflow
*                             ; count
*        .                    ; do something else
*        .
*        .
```

* The PAI and PAOV interrupt service routines are in the following

```
pa_hnd      BCLR  TFLG2,X $EF    ; clear the PAIF flag
            DEC   edge           ; reset the edge flag to 0
            RTI
pa_over     BCLR  TFLG2,X $DF    ; clear the PAOVF flag
            LDD   ov_cnt
            ADDD  #1             ; increment the PACNT overflow count
            STD   ov_cnt         ;           "
            RTI
            END
```

7.11 Summary

Port A pins PA2–PA0 can be used as general-purpose input pins or as input-capture channels IC1–IC3. Port A pins PA6–PA3 can be used as general-purpose output pins or as output-compare pins OC2–OC5. The PA7 pin is bidirectional; when set for input, it can be used as a general-purpose input pin or as the pulse accumulator input.

Physical time for the 68HC11 is the value in the 16-bit free-running timer. When the selected edge is detected on an input-capture pin, the value of the free-running timer is latched into a 16-bit input-capture register. This feature is often used to measure pulse width and signal frequency. Input-capture channels can also be used as additional interrupt sources or as time references.

Each output-compare function has a 16-bit register, a 16-bit comparator, and an output-compare action pin. Output-compare functions are often used to generate a delay, to create a time base for measurements, or to generate digital waveforms. The steps in using an output-compare function include making a copy of the current contents of TCNT, adding a value equal to the desired delay to the copy of TCNT, and storing the sum back to an output-compare register. An output-compare pin can activate any of the following actions: pull up to high, pull down to low, and toggle.

An output-compare operation can be forced to take effect immediately without setting an output-compare flag. Output compare 1 can control up to five OC pins simultaneously. Each of the output-compare pins OC2–OC5 can be controlled by up to two output-compare functions simultaneously.

The RTI function can be used to generate periodic interrupts so that the microcontroller can be reminded to perform routine tasks.

The 8-bit pulse accumulator can be enabled to operate in either the event-counting or the gated pulse accumulation mode.

The memory locations that hold the timer interrupt vectors are listed in Table 7.7.

Interrupt source	68HC11 vector location	EVB pseudo vector location
pulse accumulator input edge	$FFDA–$FFDB	$CA–$CC
pulse accumulator overflow	$FFDC–$FFDD	$CD–$CF
timer overflow	$FFDE–$FFDF	$D0–$D2
timer output compare 5	$FFE0–$FFE1	$D3–$D5
timer output compare 4	$FFE2–$FFE3	$D6–$D8
timer output compare 3	$FFE4–$FFE5	$D9–$DB
timer output compare 2	$FFE6–$FFE7	$DC–$DE
timer output compare 1	$FFE8–$FFE9	$DF–$E1
timer input capture 3	$FFEA–$FFEB	$E2–$E4
timer input capture 2	$FFEC–$FFED	$E5–$E7
timer input capture 1	$FFEE–$FFEF	$E8–$EA
real time interrupt	$FFF0–$FFF1	$EB–$ED

Table 7.7 ■ 68HC11 Timer interrupt vector locations

7.12 Glossary

Input capture   The 68HC11 function that captures the value of the 16-bit free-running timer into a latch when the falling or rising edge of the signal connected to the input-capture pin arrives.

Output compare   A 68HC11 timer function that allows the user to make a copy of the value of the 16-bit free-running timer, add a delay to the copy, and then store the sum in a register. The output-compare function compares the sum with the free-running timer in each of the following E clock cycles. When these two values are equal, the circuit can trigger a signal change on an output-compare pin and can also generate an interrupt request to the 68HC11.

Physical time   In the 68HC11 timer system, the time represented by the count in the 16-bit free-running timer counter.

Pulse accumulator   A 68HC11 timer function that uses an 8-bit counter to count the number of events that occur or measure the duration of a single pulse.

7.13 Exercises

A 2-MHz E clock is used in all of the following problems.

E7.1. What value will be loaded into accumulator D if the value of TCNT is $1000 when the instruction LDD $100E is being fetched?

E7.2. What value should be written into TCTL1 to toggle the voltage on the PA5 pin on successful output compares?

E7.3. Rewrite the program in Example 7.6 using the interrupt-driven method to generate the waveform.

E7.4. Write a program to use the OC2 function to generate a 5-KHz square waveform with 50% duty cycle.

E7.5. Write a program to use the OC2 function to generate a 2-KHz square waveform with 30% duty cycle.

E7.6. Use the OC1 and OC2 functions together to generate a 1-KHz digital waveform with 20% duty cycle.

E7.7. Write a program to generate ten pulses from the OC3 pin. Each pulse has 1 ms low time and 0.5 ms high time.

E7.8. Write a program to turn a light on for 5 seconds and then turn it off for 5 seconds, repeating this cycle continuously. The light is controlled by pin PB0. The light is turned on when PB0 is high and off when it is low. Use the RTI interrupts to implement this operation.

E7.9. Write a program to interrupt 10 ms after a positive edge is detected on the PA2 pin.

E7.10. How often will the pulse accumulator overflow if PA7 is constantly high in the gated counting mode? Assume the pulse accumulator counter has been programmed to count when the PA7 pin is high.

E7.11. Write a program that uses the OC2 function to create a delay of 20 ms after the rising edge of a signal has been detected on the IC1 pin.

E7.12. What will happen after execution of the following instructions?

```
TCTL1   EQU   $20
CFORC   EQU   $08
TFLG1   EQU   $23
TOC2    EQU   $12
TOC3    EQU   $14
TOC4    EQU   $16
TOC5    EQU   $18
TCNT    EQU   $0E


        LDX   #$1000
        LDD   TCNT,X
        ADDD  #40000
        STD   TOC2,X
        STD   TOC3,X
        STD   TOC4,X
        STD   TOC5,X
        LDAA  #%01011011
        STAA  TCTL1,X
        LDAA  #$FF
        STAA  TFLG1,X
        LDAA  #%01111000
        STAA  CFORC,X
```

E7.13. Write a program to output a 10-ms positive pulse on the PA3 pin every time a rising edge is detected on PA1.

7.14 Lab Experiments and Assignments

L7.1. *Frequency Measurement.* Use one of the input-capture functions and the pulse accumulator function to measure the frequency of an unknown signal. The procedure is as follows:

1. Connect the signal to one of the input-capture pins and to the pulse accumulator input pin.

2. Also connect the signal to an oscilloscope or a frequency counter.

3. Set up a one-second time base using one of the output-compare functions for frequency measurement.

4. Allocate two bytes of on-chip SRAM to accumulate the frequency counts for each method. The frequency counter must be reset to 0 before the program is executed.

5. Use the interrupt-driven method to detect the rising (or falling) edge of the unknown signal.

6. Increment the frequency count when there is an interrupt.

7. At the end of the one-second period, examine the frequency count resulting from each method.

8. Repeat the same measurement in the following frequency ranges:
   a. 10 to 100 Hz
   b. 100 to 1000 Hz
   c. 1 KHz to 10 KHz
   d. 10 KHz to 60 KHz

   Make two measurements in each range.

9. Compare the results of the two methods. Make a conclusion regarding which method is more suitable in each range. Explain why the frequency measurement is not accurate for some frequency ranges.

L7.2. *Waveform Generation.* Use an output-compare function pin to generate an active high digital waveform. The duty cycle and the frequency must be programmable. The procedure is as follows:

1. Connect one of the output-compare function pins to the oscilloscope input.

2. Enter the frequency (as a multiple of one hundred) in the on-chip SRAM at location $00. For example, the number 20 stands for the frequency $20 \times 100 = 2$ KHz. The frequency should be no higher than 20 KHz, for which the number 200 would be entered.

3. Enter the duty cycle in the on-chip SRAM at $01. For example, the number 30 stands for duty cycle 30%.

4. Write a program to
   ■ read the frequency and duty cycle from the SRAM
   ■ use the OC2 pin to generate the specified waveform
   ■ set a time base of 1 minute using the OC3 function
   ■ jump back to the monitor at the end of the minute

5. Download the program into the EVM or EVB and execute the program. The waveform should appear on the oscilloscope screen.

6. Repeat the same experiment with several different frequencies.

L7.3. *Time-of-day Display.* Use an output-compare function, an input-capture function, two parallel ports, and seven-segment displays to implement the time-of-day clock. The procedure is as follows:

Step 1

Write a main program that invokes the Buffalo library routines to put out a prompt that asks the user to enter the current time in the format *hh mm ss*. The main program then invokes the Buffalo input routine to read the time entered by the user. The main program will perform input checking until a valid time is entered.

Step 2

Use one of the output-compare functions to create delays, for example, a 25-ms delay. Each successful output compare will generate an interrupt request to the microcontroller. The output-compare interrupt service routine will check to see if one second has passed. If so, it will update the current time.

Step 3

Use port B to drive the segment pattern inputs for the seven-segment displays. Six seven-segment displays are required to display the time-of-day. Use port D to drive the display select signals as shown in Figure 6.19.

8

68HC11

SERIAL

COMMUNICATION

INTERFACE

8.1 Objectives

After completing this chapter, you should be able to:

- define different types of data links
- explain the four aspects of the RS232 interface standard
- explain the data transmission process for the RS232 interface standard
- explain the types of data transmission errors that occur in serial data communication
- establish a null modem connection
- explain the operation of the SCI subsystem
- wire the SCI pins to an RS232 connector
- program the SCI subsystem to perform terminal I/O
- explain the operation of the Motorola 6850 ACIA chip
- interface the 6850 ACIA to the 68HC11 and an RS232 connector
- program the 6850 ACIA to do data transfer

8.2 Introduction

Asynchronous serial data transfer is widely used for data communication between two computers or between a computer and a terminal (or printer), either with or without a modem. A dedicated or a public phone line can be used as a medium for data transmission.

A basic data communication link is shown in Figure 8.1. A terminal at one end of the link communicates with a computer at the other end. The communication link consists of data terminal equipment (DTE) and an associated modem (data communication equipment (DCE)) at each end. The term *modem* is derived from the process of accepting digital bits, changing them into a form suitable for analog transmission (*mod*ulation), receiving the signal at the other station, and transforming it back to its original digital representation (*demod*ulation).

There are two types of communication links, point-to-point and multidrop, as illustrated in Figure 8.2. In the former, the two end stations communicate as peers, while in the latter, one device is designated as the master (primary) and the others as slaves (secondaries). Each multi-drop station has its own unique *address*, with the primary station controlling all data transfers over the communication link.

A communication link can consist of either two or four wires. A two-wire link provides a signal line and a ground, and a four-wire link provides two such pairs. A communication link can be used in three different ways:

1. *Simplex link.* The line is dedicated to either transmission or reception, but not both.

2. *Half-duplex link.* The communication link can be used for either transmission or reception, but in only one direction at a time.

3. *Full-duplex link.* Transmission and reception can proceed simultaneously. This link requires four wires.

Most data communication takes place over the telephone line or over lines that are engineered to telephone-line specifications. The various standards organizations have therefore published many recommendations defining how these connections and communications should be made. The International Telegraph and Telephone Consultative Committee (CCITT) V series defines these connections. The International Standards Organization (ISO) also publishes many standards, some of which describe the mechanical connectors

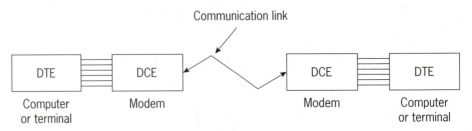

Figure 8.1   A data communication system

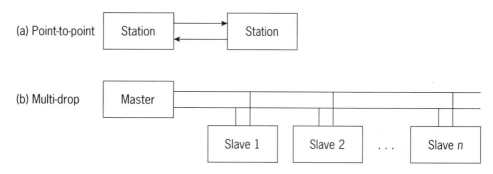

Figure 8.2 Point-to-point and multi-drop communication links

used by computers, terminals, modems, and other devices. Readers who are interested in these standards should refer to books on data communication and computer networking such as:

Uyless D. Black, *Data Communications and Distributed Networks*, 2d ed. Englewood Cliffs, N.J.: Prentice-Hall, 1987.

Uyless Black, *Data Networks, Concepts, Theory, and Practice*. Englewood Cliffs, N.J.: Prentice-Hall, 1989.

William Stallings, *Handbook of Computer Communications Standards*, Vol. I. Carmel, Ind.: Howard W. Sams & Company, 1987.

Data communication engineers and technicians use the term *mark* to indicate a binary 1 and *space* to indicate a binary 0. The terms *high* and *low* are confusing when applied to serial data communication because the logic levels used on a serial communication circuit may not be those used in standard TTL or CMOS digital circuits. The RS232 standard, for example, represents a binary 0, or space, with a positive voltage and a binary 1, or mark, with a negative voltage.

8.3 The RS232 Standard

The RS232 standard published by the Electronic Industry Association (EIA) is one of the most widely used physical level interfaces in the world and is most prevalent in North America. It specifies 25 interchange circuits for DTE/DCE use. In 1960 the EIA established the RS232 interface standard for interfacing between a computer and a modem. It was revised into the RS232-C version in 1969. In January 1987, the RS232-C standard was renamed EIA-232-D. The acronym RS stands for *recommended standard*. There are four aspects of the RS232 standard:

1. The electrical specifications specify the voltage level, rise time, and fall time of each signal.

2. The functional specifications specify the function of each signal.

3. The mechanical specifications specify the number of pins and the shape and dimensions of the connector.

4. The procedural specifications specify the sequence of events for transmitting data, based on the functional specifications of the interface.

8.3.1 The RS232-D Electrical Specifications

The RS232-D electrical specifications use the CCITT V.28 standard and include the following points:

- The interface is rated at a signal rate of <20 kbps.
- The signal can transfer correctly within 15 meters. Greater distances and higher data rates are possible with better design.
- The maximum driver output voltage (with circuit open) is −25 V to +25 V.
- The minimum driver output voltage (loaded output) is −25 V to −5 V or +5 V to +25 V.
- The minimum driver output resistance when power is off is 300 Ω.
- The maximum driver output current (short circuit) is 500 mA.
- The maximum driver output slew rate is 30 V/μs.
- The receiver input resistance is 3–7 KΩ.
- The receiver input voltage range is −25 V to +25 V.
- The receiver output is high when input is open circuit.
- A voltage more negative than −3 V at the receiver's input is interpreted as a logic 1.
- A voltage more positive than +3 V at the receiver's input is interpreted as a logic 0.

8.3.2 The RS232-D Functional Specifications

The EIA-232-D functional specifications are derived from the CCITT V.24 standard. Table 8.1 summarizes the functional specifications of each circuit. There is one data circuit in each direction, so full-duplex operation is possible. One ground lead is for protective isolation; the other serves as the return circuit for both data leads. Hence transmission is unbalanced, with only one active wire. The timing signals provide clock pulses for synchronous transmission. When the DCE is sending data over circuit BB, it also sends 1-0 and 0-1 transitions on circuit DD, with transitions timed to the middle of each BB signal element. When the DTE is sending data, either the DTE or DCE can provide timing pulses, depending on the circumstances. The control signals are explained in the procedural specifications.

8.3.3 The RS232-D Mechanical Specifications

The RS232-D uses a 25-pin D-type connector (see Figure 8.3). This 25-pin connector is available from many vendors. Its exact dimensions can be found in William Stallings, *Handbook of Computer Communications Standards*, Vol. I. Carmel, Inc.: Howard W. Sams & Company, 1987, or in the standard document published by EIA.

Name	Pin[3]	EIA name	Direction to[6]	Function
transmitted data	2	BA	DCE	Data generated by DTE.
received data	3	BB	DTE	Data received by DTE.
request to send	4	CA	DCE	DTE wishes to transmit.
clear to send	5	CB	DTE	DCE is ready to transmit (response to request to send).
data set ready	6	CC	DTE	DCE is ready to operate.
data terminal ready	20	CD	DCE	DTE is ready to operate.
ring indicator	22	CE	DTE	Indicates that DCE is receiving a ringing signal on the communication channel.
carrier detect	8	CF	DTE	Indicates that DCE is receiving a carrier signal.
signal quality detector	21[1]	CG	DTE	Asserted when there is reason to believe that there is an error in the received data.
data signal rate selector	23[5]	CH	DCE	Asserted to select the higher of two possible data rates.
data signal rate selector	23[5]	CI	DTE	Asserted to select the higher of two possible data rates.
transmitter signal element timing	24	DA	DCE	Clocking signal, transitions to ON and OFF occur at the center of each signal element.
transmitter signal element timing	15	DB	DTE	Clocking signal, as above; both leads relate to signals on BA.
receiver signal element timing	17	DD	DTE	Clocking signal, as above, for circuit BB.
protective ground	1	AA	NA[2]	Attached to machine frame and possibly external grounds.
signal ground	7	AB	NA[2]	Establishes common ground reference for all circuits.
secondary transmitted data	14	SBA	DCE	Equivalent to circuit BA but used to transmit data via the secondary channel.
secondary received data	16	SBB	DTE	Equivalent to circuit BB but used to receive data on the secondary channel.
secondary request to send	19	SCA	DCE	Request to transmit data on secondary channel.
secondary clear to send	13	SCB	DTE	DCE is ready to send data via the secondary channel.
secondary received line signal detector	12	SCF	DTE	Indicates proper reception of the secondary channel line signal.
local loopback	18	LL[3]	NA[2]	Controls LL test condition in local DCE. The ON condition causes the local DCE to transmit to the receiving signal converter at the same DCE.
remote loopback	21[1]	RL[3]	NA[2]	Controls RL test condition in remote DCE. The ON condition causes the local DCE to signal the test condition to the remote DCE.
test mode	25	TM[3]	NA[2]	Indicates whether the local DCE is in a test condition.

1. The CG and RL circuits share the same pin.
2. NA stands for not applicable.
3. Pins 9, 10, and 11 are not used.
4. RS232-C does not have RL, LL, and TM circuits.
5. Circuits CH and CI share the same pin.
6. *Direction to* means that the signal in question is an input to the equipment listed below.

Table 8.1 ■ RS232-D circuit definitions

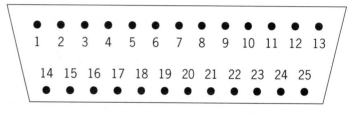

Figure 8.3 RS232-D connector

8.3.4 The RS232-D Procedural Specifications

The sequence of events that occurs during data transmission using the RS232-D is easiest to understand by studying examples. We will use two examples to explain the procedure.

In the first example, two DTEs are connected with a point-to-point link using an asynchronous dedicated-line modem. The modem requires only the following circuits to operate:

- signal ground (AB)
- transmitted data (BA)
- received data (BB)
- request to send (CA)
- clear to send (CB)
- data set ready (CC)
- carrier detect (CF)

Before the DTE can transmit any data, the data-set-ready circuit must be asserted to indicate that the modem is ready to operate. This lead should be asserted before the DTE attempts the request-to-send. The DSR pin can simply be connected to the power supply of the DCE to indicate that it is switched on and ready to go. When the DTE is ready to send data, it asserts request-to-send. The modem responds, when ready, with clear-to-send, indicating that data can be transmitted over circuit BA. If the arrangement is half-duplex, then request-to-send also inhibits the receive mode. The DTE sends data to the local modem bit-serially. The local modem modulates the data into the carrier signal and transmits the resulting signal over the dedicated communication lines. Before sending out modulated data, the local modem sends out a carrier signal to the remote modem so that the remote modem is ready to receive the data. The remote modem detects the carrier and asserts the carrier-detect signal. Assertion of the carrier-detect signal tells the remote DTE that the local modem is transmitting. The remote modem receives the modulated signal from the local modem, demodulates it to recover the data, and sends it to the remote DTE over the received-data pin. The circuit connections are illustrated in Figure 8.4.

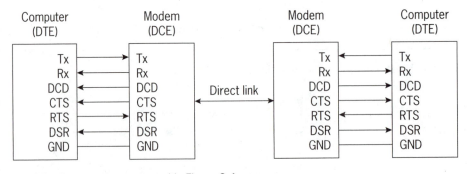

The following acronyms are used in Figure 8.4:

Tx: transmit data   CTS: clear to send
Rx: receive data   RTS: request to send
DCD: data carrier detect   DSR: data set ready

Figure 8.4   Point-to-point asynchronous connection

The second example involves two DTEs exchanging data through a public telephone line. Now one of the DTEs (the initiator) must dial the phone (automatically or manually) to establish the connection. Two additional leads are required for this application:

- data terminal ready (CD)
- ring indicator (CE)

Data transmission in this configuration can be divided into three steps:

Step 1

Establishing the connection. The following events occur during this step:

1. The transmit DTE asserts the data-terminal-ready (DTR) signal to indicate to the local modem that it is ready to make a call.
2. The local modem opens the phone circuit and dials the destination number. The number can be stored in the modem or transmitted to the modem by the computer via the transmit-data pin.
3. The remote modem detects a ring on the phone line and asserts ring-indicator (RI) to indicate to the remote DTE that a call has arrived.
4. The remote DTE asserts the DTR signal to accept the call.
5. The remote modem answers the call by sending a carrier signal to the local modem via the phone line. It also asserts the DSR signal to inform the remote DTE that it is ready for data transmission.
6. The local modem asserts both the DSR and DCD signals to indicate that the connection is established and it is ready for data communication.
7. For full-duplex data communication, the local modem also sends a carrier signal to the remote modem. The remote modem then asserts the DCD signal.

Step 2

Data transmission. The following events occur during this step:

1. The local DTE asserts the RTS signal when it is ready to send data.
2. The local modem responds by asserting the CTS signal.
3. The local DTE sends data bit-serially to the local modem over the transmit-data pin. The local modem then modulates its carrier signal to transmit the data to the remote modem.
4. The remote modem receives the modulated signal from the local modem, demodulates it to recover the data, and sends it to the remote DTE over the received-data pin.

Step 3

Disconnection. Disconnection requires only two steps:

1. When the local DTE has finished the data transmission, it drops the RTS signal.
2. The local modem then de-asserts the CTS signal and drops the carrier (equivalent to hanging up the phone).

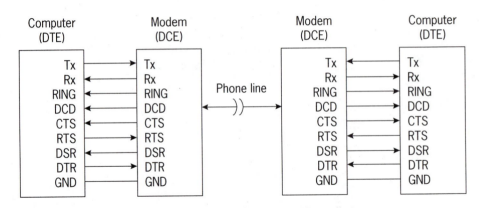

Figure 8.5   Asynchronous connection over public phone line

The circuit connection for this example is shown in Figure 8.5.

A timing signal is not needed in an asynchronous transmission, but the following rules must be observed to guarantee that data is received correctly:

1. Data must be transferred character by character.

2. A character must consist of one start bit, 7 or 8 data bits, an optional parity bit, and 1, or $1\frac{1}{2}$, or 2 stop bits.

3. The start bit must be low.

4. The least significant bit (LSB) must be transferred first, and the most significant bit (MSB) must be transferred last.

5. The stop bit must be high.

6. A clock with a frequency equal to 16 (or 64) times the data bit rate must be used to detect the arrival of the start bit and determine the value of each data bit. A clock with 16 times the bit rate can tolerate a difference of about 5% in the clocks at the transmitter and receiver. The methods for detecting the start bit and determining the values of the data bits vary with different communication chips. This issue will be addressed in later sections.

With this protocol, the format of a character is shown in Figure 8.6. The start bit and stop bit identify the beginning and end of a character. They also permit a receiver to resynchronize a local clock to each new character.

	Start bit	0	1	2	3	4	5	6	7	Stop bit 1	Stop bit 2

Figure 8.6   The format of a character

Example 8.1

The letter A is to be transmitted in the format with eight data bits, no parity, and one stop bit. Sketch the output.

Solution: Letters are transmitted in ASCII code. The ASCII code for the letter A is $41, or %01000001, which will be followed by a stop bit. The output is shown in Figure 8.7.

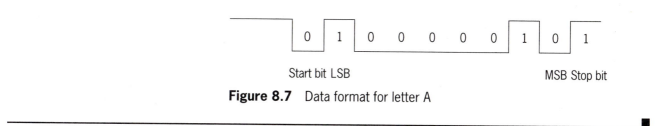

Figure 8.7 Data format for letter A

The RS232-D standard also supports synchronous transmission. Transmitter timing and receiver timing signals are needed in a synchronous transmission. However, synchronous RS232 serial communication is beyond the scope of this text and will not be discussed here.

Example 8.2

How long does it take to transmit one character at a speed of 9600 bauds? Each character is transmitted using a format with seven data bits, even parity, and two stop bits.

Solution: Each character consists of 11 bits (1 start bit, 7 data bits, 1 parity bit, and 2 stop bits). Each bit requires 104 μs (= 1 sec ÷ 9600). Thus each character will require 11 × 104 μs = 1.145 ms to transmit.

8.3.5 Data Transmission Errors

The following types of errors may occur during data transfer using asynchronous serial transmission:

- Framing errors: A framing error occurs when a received character is improperly framed by the start and stop bits; it is detected by the absence of the first stop bit. This error indicates a synchronization error, faulty transmission, or a break condition.
- Receiver overrun: One or more characters in the data stream were lost. That is, a character or a number of characters were received but were not read from the buffer before subsequent characters were received.
- Parity errors: A parity error occurs when an odd number of bits change value. It can be detected by a parity error detector circuit.

8.3.6 The Null Modem Connection

Probably one of the most popular applications of the RS232 interface in today's microprocessor laboratories is to connect a terminal (a PC running a terminal program) to a single-board computer. Both the computer and the terminal are DTEs, but the RS232 standard is not for a direct connection between two DTEs. In order to make this scheme work, a null modem is needed—the null modem interconnects leads in such a way as to fool both DTEs into thinking that they are connected to modems. The null modem connection is shown

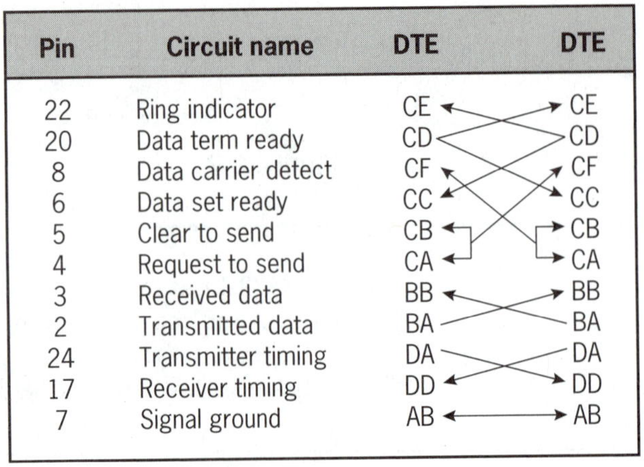

Pin	Circuit name	DTE	DTE
22	Ring indicator	CE	CE
20	Data term ready	CD	CD
8	Data carrier detect	CF	CF
6	Data set ready	CC	CC
5	Clear to send	CB	CB
4	Request to send	CA	CA
3	Received data	BB	BB
2	Transmitted data	BA	BA
24	Transmitter timing	DA	DA
17	Receiver timing	DD	DD
7	Signal ground	AB	AB

Figure 8.8   Null modem connection

in Figure 8.8. The transmitter-timing and receiver-timing signals are not needed in asynchronous data transmission. Since the data transmission is not through a phone line, the ring indicator is not needed either.

8.4 The 68HC11 Serial Communication Interface (SCI)

A full-duplex asynchronous serial communication interface (SCI) is incorporated on the chip of the 68HC11 microcontroller. The SCI transmitter and receiver are functionally independent but use the same data format and bit rate.

The 68HC11 SCI subsystem can be used to communicate with multiple pieces of data communication equipment or just one DTE. The first situation is called a multi-drop environment. In a multi-drop environment, an SCI receiver can be put to sleep when a message is not being sent to it. The sleeping receiver can be configured to wake up using either the idle line or an address mark. When the correct address is detected, the SCI receiver will start to receive the data message and will continue receiving until the end of the message, which will be indicated by the idle line.

8.4.1 Data Format

The SCI subsystem can handle a character that consists of one start bit, eight or nine data bits, and one stop bit. A *break* is defined as the transmission or reception of a low for at least one complete character frame time. An *idle line* is defined as a continuous logic high on the RxD line for at least a complete character time.

8.4.2 The Wake-Up Feature

In a multi-drop environment, the receiver wake-up feature reduces SCI service overhead for the multiple-receiver system. Software in each SCI receiver evaluates the first character(s) of each message. If the message is intended for a different receiver, the receiver is placed in a sleep mode so that the rest of the message will not generate requests for service. Whenever a new message is started, logic in the sleeping receivers causes them to wake up so they can evaluate the initial character(s) of the new message.

A sleeping SCI receiver can be configured to wake up using either of two methods: *idle-line wake-up* or *address mark wake-up:*

1. In *idle-line wake-up,* a sleeping receiver wakes up as soon as the RxD line becomes idle. Systems using this type of wake-up must provide at least one character time of idleness between messages in order to wake up sleeping receivers, but they must not allow any idle time between characters within a message.

2. In *address mark wake-up,* the most significant bit (MSB) of the character is used to indicate whether the character is an address (1) or a data (0) character. Sleeping receivers will wake up whenever an address mark character is received. Systems using this method set the MSB of the first character in each message and leave it clear for all other characters in the message. Idle periods may occur within messages, and no idle time is required between messages for this wake-up method.

8.4.3 Start Bit Detection

The SCI uses a clock signal that is 16 times the bit rate to detect the arrival of a valid start bit or decide the logic level of a data bit. When RxD is high for at least three sampling clock cycles and is then followed by a low voltage, the SCI will look at the third, fifth, and seventh samples after the first low sample (these are called verification samples) to determine whether a valid start bit has arrived. This process is illustrated in Figure 8.9. If the majority of these three samples are low, then a valid start bit is detected. Otherwise, the SCI will repeat the process. If all three samples are 1s, then the noise flag will be set. After detecting the start bit, the SCI will begin to shift in the data bit.

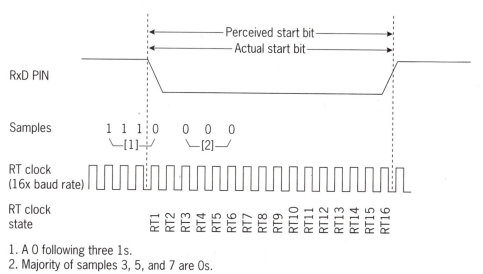

1. A 0 following three 1s.
2. Majority of samples 3, 5, and 7 are 0s.

Figure 8.9   Detection of start bit (ideal case)

8.4.4 Determining the Logic Value of Data Bits

The SCI also uses a clock that is 16 times the bit rate to decide the logic value of any data bit and the stop bit. If the majority of the 8th, 9th, and 10th samples are 1, then the data bit is determined to be 1. Otherwise, the data bit is determined to be 0. If one sample differs from the other two samples, the noise flag will also be set.

8.4.5 Receiving the Stop Bit

The stop bit is a 1. The noise flag will be set during the sampling of the stop bit if those three samples are not all identical.

8.4.6 Receiving Errors

The 68HC11 SCI detects the following receiving errors:

■ *Framing errors (FE):* A framing error occurs when the expected stop bit is not detected during the process of receiving a character. If a framing error occurs without a break (10 zeros for an 8-bit format or 11 zeros for a 9-bit format) being detected, the circuit will continue to operate as if there actually were a stop bit and the start edge will be placed artificially. The FE flag in the SCSR register will also be set.

■ *Receiver overrun (OR):* The overrun bit is set when the next byte is ready to be transferred from the receive shift register to the receive data register (SCDR) but this register is already full (the RDRF bit is set). When an overrun error occurs, the data that caused the overrun is lost and the data that was already in SCDR is not disturbed. The OR bit will be set.

■ *Noise error (NE):* The noise flag bit is set if there is noise on any of the received bits, including the start and stop bits. The NE bit is not set until the RDRF flag is set.

All error flags will be cleared when the SCSR is read and then the SCDR is read.

8.4.7 Receiving a Character

The receiving process in an SCI includes the following steps:

1. Detect the start bit.
2. Shift in eight (or nine) data bits.
3. Shift in the stop bit.
4. When a correct stop bit is detected, the data bits will be loaded into the receive data register from the receive shift register.

8.4.8 Transmitting a Character

The SCI transmitter uses an internally generated bit-rate clock to serially shift data out of the TxD pin. A normal transmission is initiated by enabling the transmitter and then writing data into the transmit data register (TDR). The SCI is double-buffered, so a new character can be written into the transmitter queue whenever the TDRE status flag is set to 1.

The transmitter logic adds a start bit (0) and a stop bit (1) to the data characters presented by the CPU for transmission. The transmitter can be configured to send characters with eight or nine data bits. When the TDR register is able to accept a new data character, the TDRE status flag (bit 7 of the SCSR register) is set and an interrupt can be optionally generated. The transmit complete status flag (bit 6 of the SCSR register) is set and an optional interrupt is produced when the transmitter has finished sending everything in its queue. The SCI transmitter is also capable of sending idle-line characters and break characters, which are useful in a multi-drop SCI network.

8.4.9 SCI Registers

The SCI receiver has a receive shift register and a receive data register (RDR). The SCI transmitter has one transmit data register (TDR) and one transmit shift register. The SCI status register (SCSR) indicates the receive status, the transmit status, and errors that have occurred. Although RDR and TDR are separate registers, they share one address ($102F). Two control registers set up all SCI operation parameters. The data rate of the SCI subsystem is controlled by the BAUD register. The addresses and contents of these registers are described in the following paragraphs.

SERIAL COMMUNICATION DATA REGISTER (SCDR)

RDR and TDR together are referred to as the SCDR register. The SCDR register is 8-bit and is located at $102F.

SERIAL COMMUNICATION STATUS REGISTER (SCSR)

The contents of the SCSR register are as follows:

	7	6	5	4	3	2	1	0	
	TDRE	TC	RDRF	IDLE	OR	NF	FE	0	SCSR at $102E
value after reset	1	1	0	0	0	0	0	0	

TDRE: transmit data register empty. When set to 1, this bit indicates that the transmit data register is empty.

TC: transmit complete. When set to 1, this bit indicates that the transmitter is finished sending and has reached an idle state. This bit is useful in systems where the SCI is driving a modem. When TC is set at the end of a transmission, the modem can be disabled without any delay.

RDRF: receive data register full. When set to 1, this bit indicates that the receive data register is full.

IDLE: idle-line condition. An idle-line condition is detected when this bit is set to 1. When this bit is 0, the RxD line is either active or has never been active since IDLE was cleared.

OR: receiver overrun flag. When set to 1, this bit indicates the receiver overrun condition.

NF: noise flag. When set to 1, this flag indicates that the communication line is noisy.

FE: framing error. When set to 1, this bit indicates that a framing error occurred.

The condition flags related to transmission (TDRE and TC) can be cleared by reading the SCSR register followed by writing a byte into the SCDR register. Condition flags related to reception (RDRF, IDLE, OR, NF, and FE) can be cleared by reading the SCSR register followed by reading the SCDR register.

SERIAL COMMUNICATION CONTROL REGISTER I (SCCR1)

The contents of the SCCR1 register are as follows:

	7	6	5	4	3	2	1	0	
	R8	T8	0	M	WAKE	0	0	0	SCCR1 at $102C
value after reset	u	u	0	0	0	0	0	0	

R8: receive data bit 8. When the SCI system is configured for 9-bit data characters, this bit acts as an extra bit of the RDR. The MSB of the received character is transferred into this bit at the same time that the remaining 8 bits are transferred from the serial receive shifter to the SCDR register.

T8: transmit data bit 8. If the M bit is set, this bit provides a storage location for the 9th bit in the transmit data character. It is not necessary to write to this bit for every character transmitted, only when the value of this bit is to be different from that of the previous character.

M: data format bit.

0 = 1 start bit, 8 data bits, 1 stop bit

1 = 1 start bit, 9 data bits, 1 stop bit

WAKE: wake-up method select.

0 = idle line is selected as wake-up method

1 = address mark is selected as wake-up method

Bits 5, 2, 1, and 0 are not implemented.

SERIAL COMMUNICATION CONTROL REGISTER 2 (SCCR2)

The contents of the SCCR2 register are as follows:

	7	6	5	4	3	2	1	0	
	TIE	TCIE	RIE	ILIE	TE	RE	RWU	SBK	SCCR2 at $102D
value after reset	0	0	0	0	0	0	0	0	

TIE: transmit interrupt enable. When this bit is set to 1, an SCI interrupt is generated if the TDRE flag in the SCSR register is set to 1.

TCIE: transmit complete interrupt enable. When TCIE is set to 1, an SCI interrupt is requested if the TC flag in the SCSR register is set to 1.

RIE: receive interrupt enable. When RIE is set to 1, an SCI interrupt is requested if either the RDRF or the OR flag in the SCSR register is set to 1.

ILIE: Idle-line interrupt enable. When ILIE is set to 1, an SCI interrupt is requested if the IDLE flag in the SCSR register is set to 1.

TE: transmit enable. When this bit is set, the transmit shift register output is applied to the TxD pin. Depending on the state of control bit M in SCCR1, a preamble of 10 (M = 0) or 11 (M = 1) consecutive 1s (representing the idle line) is transmitted when the software sets the TE bit from a cleared state.

RE: receive enable. The SCI receiver is enabled when this bit is set to 1. When the SCI receiver is disabled, the RDRF, IDLE, OR, NF, and FE status flags cannot be set. If these bits are already set, turning off the RE bit does not cause them to be cleared.

RWU: receiver wake-up. When the receiver wake-up bit is set, it puts the receiver to sleep and enables the "wake-up" function.

SBK: send break. If the send-break bit is set and then cleared, the transmitter sends 10 (M = 0) or 11 (M = 1) 0s and then reverts to idle or to sending data. If SBK remains set, the transmitter will continuously send whole blocks of 0s until cleared. At the completion of break code, the transmitter sends at least one high bit to guarantee recognition of a valid start bit.

BAUD RATE REGISTER (BAUD)

The baud rate of the SCI system is derived by dividing the E clock signal by two factors specified in the BAUD register, as illustrated in Figure 8.10. The contents of the BAUD register are:

7	6	5	4	3	2	1	0	
TCLR	0	SCP1	SCP0	RCKB	SCR2	SCR1	SCR0	BAUD at $102B

value after reset   0   0   0   0   0   u   u   u

TCLR: clear baud rate counter. This bit is used to clear the baud rate counter chain during factory testing. TCLR is zero and cannot be set while in normal operating modes.

SCP1–SCP0: SCI baud rate prescale selects. The E clock is divided by the factors shown in Table 8.2, which are determined by the settings of SCP0 and SPC1. This prescaled output provides an input to a divider that is also controlled by SCR2–SCR0.

RCKB: SCI baud rate clock check. This bit is used during factory testing to enable the exclusive-OR of the receiver clock and transmitter

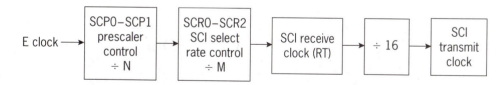

Figure 8.10   SCI rate generator division (Redrawn with permission of Motorola)

SCP1	SCP0	Divide processor clock by
0	0	1
0	1	3
1	0	4
1	1	13

Table 8.2 ■ Baud rate prescale factor

clock to be driven out the TxD pin. The RCKB bit is 0 and cannot be set during normal operation.

SCR2–SCR0: SCI baud rate selects. The output of the prescaler is further divided by a number selected by these three bits, as shown in Table 8.3.

SCR2	SCR1	SCR0	Divide prescale output by
0	0	0	1
0	0	1	2
0	1	0	4
0	1	1	8
1	0	0	16
1	0	1	32
1	1	0	64
1	1	1	128

Table 8.3 ■ SCI baud rate select

Example 8.3

Determine the baud rate prescale and baud rate select factors needed to set the baud rate to 9600 with a 2-MHz E clock.

Solution: The frequency used to determine the bit value is $16 \times 9600 = 153,600$. Multiplying this value by 13, we get $153,600 \times 13 = 1996800 \approx 2$ MHz. Therefore we need a prescale factor of 13 and a baud rate select value of 0. To set these two parameters, the value \$30 should be written into the BAUD register.

8.5 SCI Interfacing

Since the SCI circuit uses 0 V and 5 V to represent logic 0 and 1, respectively, it cannot be connected directly to the RS232 interface. A voltage translation circuit is needed to translate the voltage levels of the SCI signals (RxD and TxD) to and from those of the corresponding RS232 signals. In this section, we will discuss the use of the MAX232 chip from MAXIM to do the voltage translation. A MAX232 has two RS232 drivers and two RS232 receivers. The MAX232 requires only a 5 V power supply to convert a 0 and 5 V input to a +10 V and −10 V output. The pin assignments and the use of each pin of the MAX232 are shown in Figure 8.11. Adding an RS232 driver/receiver chip will

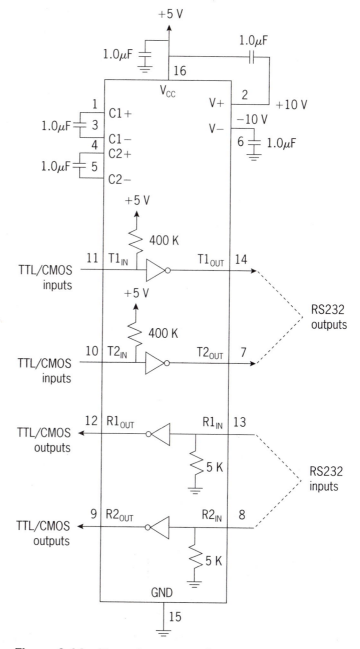

Figure 8.11 Pin assignments and connections of the MAX232

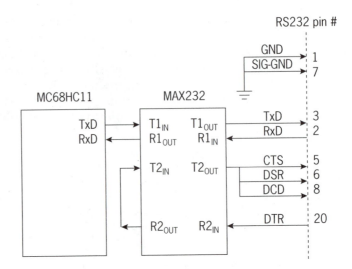

Figure 8.12   Diagram of the SCI's RS232 side circuit connection

then allow the 68HC11 SCI subsystem to communicate to the RS232 interface circuit. An example of such a circuit is shown in Figure 8.12.

Example 8.4

Write a subroutine to initialize the 68HC11 SCI subsystem (E clock frequency is 2 MHz) to operate with the following parameters:

- 9600 baud
- one start bit, eight data bits, and one stop bit
- no interrupt for receive and transmit
- enable receive and transmit
- idle-line wake-up
- do not send break

Solution: According to Example 8.3, the value $30 should be written into the BAUD register to set the specified prescale and baud rate select factors. The data format and wake-up method are set by the SCCR1 register, so the value $00 should be written into the SCCR1 register. For the remaining parameters, the value $0C should be written into the SCCR2 register. The initialization of the SCI system is performed by the routine ONSCI.

```
REGBAS   EQU   $1000        ; I/O register block base address
BAUD     EQU   $2B          ; offset of BAUD from REGBAS
SCCR1    EQU   $2C          ; offset of SCCR1 from REGBAS
SCCR2    EQU   $2D          ; offset of SCCR2 from REGBAS
CR1_INI  EQU   $00          ; 8 data bits, address mark wake-up
CR2_INI  EQU   $0C          ; enable transmitter and receiver but disable all interrupts
baud_ini EQU   $30          ; set baud rate to be 9600
```

```
ONSCI    PSHX                ; save X on the stack
         PSHA                ; save A on the stack
         LDX   #REGBAS
         LDAA  #baud_ini
         STAA  BAUD,X        ; initialize BAUD register
         LDAA  #CR1_INI
         STAA  SCCR1,X       ; initialize SCCR1
         LDAA  #CR2_INI
         STAA  SCCR2,X       ; initialize SCCR2
         PULA                ; restore A from stack
         PULX                ; restore X from stack
         RTS
```

Example 8.5

Write a subroutine to send a break to the communication port controlled by the SCI subsystem. The duration of the transmitted break should be approximately 200,000 E clock cycles or 100 ms at 2.0 MHz.

Solution: A break character is represented by 10 or 11 consecutive 0s and can be sent out by setting bit 0 of the SCCR2 register. As long as bit 0 of the SCCR2 register remains set, the SCI will keep sending out the break character.

```
REGBAS   EQU  $1000         ; I/O register base address
SCCR2    EQU  $2D           ; offset of SCCR2 from REGBAS

sendbrk  PSHX               ; save register
         LDX   #$1000
         BSET  SCCR2,X $01   ; set send-break bit
```

* The following three instructions create a delay of about 100 ms.

```
         LDY   #28571       ; delay for 100 ms
txwait1  DEY                ; 7 clock cycles each loop
         BNE   txwait1

         BCLR  SCCR2,X $01   ; clear send-break bit
         PULX                ; restore X
         RTS
```

Example 8.6

Write a subroutine to output a character from the SCI subsystem using the polling method. The character to be output will be passed in accumulator A.

Solution: A new character should be sent out only if the transmit data register is empty. When the polling method is used, the subroutine will wait until the TDRE bit (bit 7) of the SCSR register is set before sending out the new character. The ASCII code of the character will be sent out, and since the ASCII code has only 7 bits, bit 7 of accumulator A must be cleared to 0 before the code is sent out. The subroutine is as follows:

```
REGBAS    equ    $1000           ; I/O register block base address
SCSR      equ    $23             ; offset of SCSR from REGBAS
SCDR      equ    $2F             ; offset of SCDR from REGBAS
TDRE      equ    $80             ; mask to check transmit data register empty
SCIputch  PSHX                   ; save X
          PSHA                   ; save A
          LDX    #REGBAS
          BRCLR  SCSR,X TDRE *    ; is transmit data register empty?
```

* The following instructions clear bit 7 to make sure the most significant bit of the ASCII
* code is a 0.

```
          ANDA   #$7F            ; mask parity bit
          STAA   SCDR,X          ; send character out from the Tx pin
          PULA                   ; restore A
          PULX                   ; restore X
          RTS
```

Example 8.7

Write a subroutine to input a character from the SCI subsystem using the polling method. The character will be returned in accumulator A.

Solution: When the polling method is used, this subroutine will wait until the RDRF bit (bit 5) of the SCSR register is set and then read the character held in the SCDR register.

```
REGBAS    equ    $1000           ; I/O register block base address
SCSR      equ    $23             ; offset of SCSR from REGBAS
SCDR      equ    $2F             ; offset of SCDR from REGBAS
RDRF      equ    $20             ; mask to check receive data register full

SCIgetch  PSHX
          LDX    #REGBAS
          BRCLR  SCSR,X RDRF *   ; is receive data register full?
          LDAA   SCDR,X          ; read the SCI data register
          PULX
          RTS
```

8.6 Serial Communication Interface Chips

Unlike the 68HC11, a general-purpose microprocessor does not have a serial communication interface to allow communication with another computer using a phone line or modem, so a peripheral chip is required to interface the microprocessor to a DCE. This peripheral chip is called a UART (universal asynchronous receiver transmitter). The function of the UART chip is to perform serial-to-parallel and parallel-to-serial conversions and also to handle the serial communication protocol such as the RS232. The SCI subsystem is the on-chip UART for the 68HC11. The Motorola MC6850, Rockwell R6551, Intel i8251, and Zilog Z8530 are among the most popular UART chips. These chips operate on a 5 V power supply. However, the RS232 signal levels are different from the voltage levels of these chips. Thus, when interfacing a serial interface chip to a DCE, we need some form of voltage level translation or buffering. Two such widely used line drivers (receivers) are the Motorola MC1488 and MC1489. The MC1488 requires a ±12 V power supply to operate, and the MC1489 requires only a 5 V power supply. Most single-board microcomputers require a power supply that can supply −12 V, +5 V, and +12 V outputs simply because the line drivers require a ±12 V power supply to operate. However, several companies have introduced RS232 line drivers (for example, the MAX232 from MAXIM and the MC145403 from Motorola) that can operate with a single 5 V power supply and generate RS232-compatible outputs. These line drivers thus eliminate the need for a ±12 V power supply.

In some applications, the user may need more than one serial communication interface—a requirement that cannot be satisfied by the 68HC11 SCI. One possible solution is to use an external UART chip, as is done in the EVB and EVM. The EVB board has a terminal and a host port. The terminal port is controlled by an MC6850 chip (ACIA), and the host port is controlled by the serial communication interface (SCI) on the 68HC11. In the following section we will discuss the operation of the MC6850 and the terminal port design of the EVB board.

8.7 The MC6850 ACIA

The MC6850 is a UART chip designed to work with the 8-bit MC6800 family microprocessors, and it can also work with the 68000 microprocessor and the 68HC11 microcontroller. The block diagram for the MC6850 is shown in Figure 8.13. The MC6850 has two 8-bit data registers, one each for receiving and transmitting. It also has a programmable control register and a read-only status register.

There are three chip select inputs, two active high (CS0 and CS1) and one active low ($\overline{CS2}$). The register select input, which is usually connected to address line A0 in MC6800 systems, is used in conjunction with the R/$\overline{W}$ input to select one of the four on-chip registers, as illustrated in Table 8.4.

8.7.1 The ACIA Registers

The operation of the ACIA is controlled by four 8-bit registers: the control register, the status register, the transmit data register, and the receive data register. The functions of these registers are discussed in the following subsections.

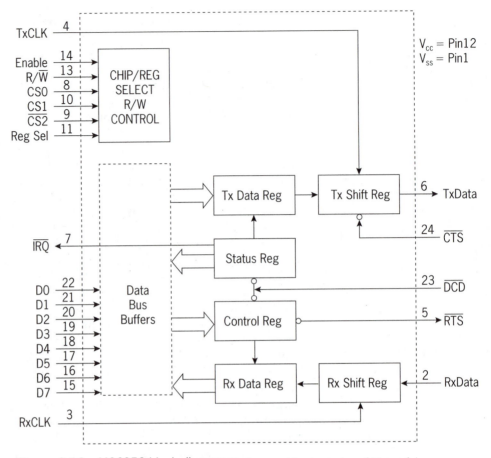

Figure 8.13 MC6850 block diagram (Redrawn with permission of Motorola)

Register Select Input	R/$\overline{W}$	Register
1	0	Tx data register
1	1	Rx data register
0	0	control register
0	1	status register

Table 8.4 ■ MC6850 register selection

THE CONTROL REGISTER

The 8-bit control register is divided into four fields. The meaning of each field is shown in Figure 8.14.

The *counter divide field* selects the divide ratio used in both the transmit and the receive sections of the ACIA. The receive bit rate is set equal to the frequency of RxCLK divided by the factor selected by this field. The transmit bit rate is set equal to the frequency of TxCLK divided by the factor selected by this field.

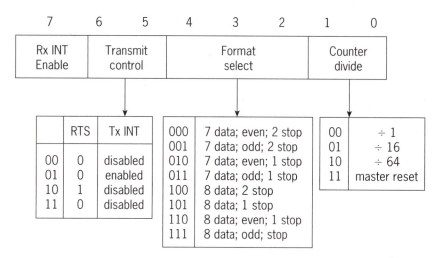

Figure 8.14 MC6850 control register

The *format select field* selects the data character length, the number of stop bits, and the parity. The *transmit control field* provides control over the interrupt resulting from the transmit data register empty condition, the $\overline{RTS}$ output, and the transmission of a break level. When this field is 10, the $\overline{RTS}$ signal will not be asserted. When it is 01, the transmit interrupt will be enabled, and when it is set to 11, a break will be transmitted by the TxData pin. A *break* is a condition in which the transmitter output is held at the active level (low) continuously. A break can be used to force an interrupt at a distant receiver, because the asynchronous serial format precludes the existence of a space level (low) for longer than about 10 bit periods. The *receive interrupt enable bit* will enable all receive interrupts when set to 1. Interrupts will be requested only when the conditions for interrupt are satisfied.

THE STATUS REGISTER

The contents of this register are shown in Figure 8.15.

IRQ: interrupt request flag. The IRQ flag is set to 1 when the $\overline{IRQ}$ pin is active low to indicate that an interrupt has been requested. This bit is cleared by a read operation to the receive data register or a write operation to the transmit data register.

PE: parity error. This flag indicates that the number of 1s in the character does not agree with the preselected odd or even parity.

OVRN: overrun flag. This flag indicates that a character or a number of characters in the data were received but not read from the receive data register before the subsequent characters were received. This situation occurs primarily at higher baud rates. The overrun condition is cleared by reading data from the receive data register or by a master reset.

Figure 8.15 The MC6850 status register

FE: framing error flag. The FE bit is set to 1 when a framing error occurs. A framing error indicates that the received character is improperly framed by the start and stop bits; it is detected by the absence of the first stop bit. This error indicates a synchronization error, faulty transmission, or a break condition normally caused by pressing the Break key on the terminal. The Break key causes the transmission of a continuous stream of null characters for a period of 250–500 milliseconds. A framing error can occur only during receive time.

$\overline{\text{CTS}}$: clear-to-send flag. The $\overline{\text{CTS}}$ bit reflects the current level of the $\overline{\text{CTS}}$ input (active low) from a modem. A high level on the $\overline{\text{CTS}}$ pin will also inhibit the TDRE bit from going to 1. A master reset to the ACIA does not affect the $\overline{\text{CTS}}$ bit.

$\overline{\text{DCD}}$: data carrier detect flag. The $\overline{\text{DCD}}$ bit will go high when the $\overline{\text{DCD}}$ input from the modem has gone high to indicate that a carrier is not present. The $\overline{\text{DCD}}$ bit's being high causes an interrupt request to be generated when the receive-interrupt-enable bit is set. After the $\overline{\text{DCD}}$ input is returned low, the $\overline{\text{DCD}}$ bit remains high until it is cleared by first reading the status register and then the data register or until a master reset occurs. If the $\overline{\text{DCD}}$ input remains high after read status and read data or master reset has occurred, the interrupt is cleared and the $\overline{\text{DCD}}$ status bit will remain high and will follow the $\overline{\text{DCD}}$ input.

TDRE: transmit data register empty. The TDRE bit is set to 1 when the character in the transmit data register has been shifted out from the TxData pin. This bit will be cleared when there is a new character written into the transmit data register.

RDRF: receive data register full. The RDRF bit is set to 1 when there is a character in the receive data register. It is cleared when the CPU reads the receive data register, and it is also cleared by a master reset. A cleared or 0 value for this bit indicates that the contents of the receive data register are not current. The data-carrier-detect bit being high also causes RDRF to indicate empty.

THE TRANSMIT DATA REGISTER

Data is written into the transmit data register during the negative transition of the E(nable) signal when the ACIA has been addressed with the RS pin high and the R/$\overline{\text{W}}$ pin low. Writing data into the register causes the TDRE bit in the status register to be cleared. If the transmitter is idle, the data will be loaded into the transmit shift register (not accessible to the programmer) and then shifted out from the TxData pin. The TDRE bit in the status register will be set at the moment when the data is loaded into the transmit shift register.

THE RECEIVE DATA REGISTER

Data is automatically transferred to the empty receive data register (RDR) from the receive shift register upon receiving a complete character. This event causes the RDRF bit in the status register to be set to 1. Data can be read through the data bus by addressing the ACIA and setting the RS and R/$\overline{\text{W}}$ pins to high. The RDRF bit will be reset to 0 when the receive data register is read. When the receive data register is full, the automatic transfer of data from the receive shift register to the receive data register is inhibited.

8.7.2 ACIA Operations

At the bus interface, the ACIA appears as two addressable memory locations. Internally, there are four addressable registers: two read-only and two write-only registers. The read-only registers are status and receive data registers; the write-only registers are control and transmit data registers.

The master reset (counter control field) in the control register must be set during system initialization to ensure the reset condition and to prepare for programming the ACIA functional configuration when communication is required. During the first master reset, the $\overline{IRQ}$ and $\overline{RTS}$ outputs are held at the 1 level. On all other master resets, $\overline{RTS}$ output can be programmed high or low with the $\overline{IRQ}$ output high.

THE TRANSMIT OPERATION

A typical transmitting sequence consists of reading the ACIA status register either as the result of an interrupt or in a polling sequence. A character may be written into the transmit data register if the TDRE bit in the status register is 1. This character is transferred to the transmit shift register, where it is serialized and transmitted to the TxData pin, preceded by a start bit and followed by one or two stop bits. A parity bit can be optionally added to the character and will occur between the last data bit and the first stop bit. As soon as the character in the transmit data register is transferred to the shift register, the next character can be written into the transmit data register.

THE RECEIVE OPERATION

The receiving sequence starts with the detection of the start bit of a character. The RxCLK clock, which is either 1, 16, or 64 times the bit rate, is used to detect the arrival of a start bit. A divide-by-one clock is used when the received data is externally synchronized with the RxCLK input. For a divide-by-16 or divide-by-64 counter divide factor, the start bit is detected by 8 or 32 consecutive low samples from the RxData pin. A character will be shifted in after the detection of the start bit. As the character is received, parity is checked and the error is available in the status register along with any framing error, overrun error, or receive data register full condition. The receiver is also double-buffered so that a character can be read from the data register as another character is being received in the shift register.

ACIA INTERRUPTS

Only one *transmit interrupt* can be requested by the ACIA. A transmit interrupt will be requested when it is enabled (when bits 6 and 5 of the control register are 01) and when the transmit data register is empty. When the $\overline{RTS}$ pin is high, a transmit interrupt cannot be requested.

A *receive interrupt* will be requested when the receive interrupt is enabled and the receive data register is full or a receiver overrun or data carrier detect goes high (signifying a loss of carrier from a modem). An interrupt caused by the receive data register being full can be cleared by reading the receive data register. An interrupt caused by overrun or loss of carrier can be cleared by reading the status register and then reading the receive data register. Note that bit 7 of the ACIA control register enables all receive interrupts at once.

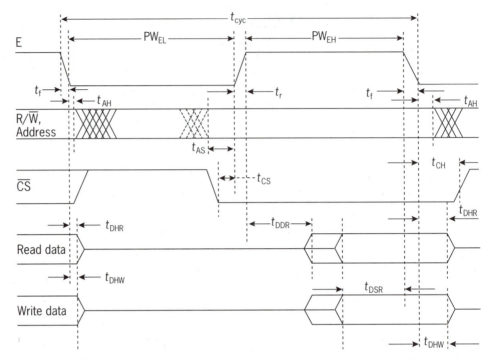

Figure 8.16   Microprocessor side bus timing diagram for the MC6850
(Redrawn with permission of Motorola)

8.7.3 The Bus Timing of the ACIA

The ACIA pins can be divided into two categories: processor side pins and RS232 side pins. Processor side pins allow the microprocessor to access and control the operation of the ACIA. The processor side bus timing characteristics for a 2-MHz MC6850 are shown in Figure 8.16, and the timing parameter values are listed in Table 8.5.

Characteristic	Symbol	Minimum (ns)	Maximum (ns)
cycle time	t_{cyc}	500	10000
pulse width, E low	PW_{EL}	210	9500
pulse width, E high	PW_{EH}	220	9500
clock rise and fall time	t_r, t_f	—	20
address hold time	t_{AH}	10	—
address setup time before E	t_{AS}	40	—
chip select time before E	t_{CS}	40	—
chip select hold time	t_{CH}	10	—
read data hold time	t_{DHR}	20	50
write data hold time	t_{DHW}	10	—
output data delay time	t_{DDR}	—	150
input data setup time	t_{DSW}	60	—

Table 8.5 ■ Timing characteristics of 2-MHz MC6850 (Redrawn with permission of Motorola)

The MC6850 requires the $\overline{CS}$ signal to be valid 40 ns before the rising edge of the E input (at 2 MHz). When read by the processor, the data from the ACIA will be available on the data bus 150 ns after the rising edge of the E input and will remain valid for at least 20 ns (possibly up to 50 ns). When being written to by the processor, the ACIA requires the input data to be valid 60 ns before the falling edge of the E input and remain valid for at least 10 ns after the falling edge of the E input. The address input (register select) to the ACIA must be valid 40 ns before the rising edge of the E input and remain valid 10 ns after the falling edge of the E input.

8.7.4 Interfacing an ACIA to the 68HC11

The MC6850 can be interfaced to the 68HC11 when the 68HC11 is configured in expanded mode. The data pins on the microprocessor side can be connected directly to the 68HC11 port C pins. The register select pin (reg sel) can be connected to the lowest address pin, A0. An address space must be assigned to the MC6850 registers so that they can be accessed by the 68HC11, and this address space assignment will affect the address decoding scheme. As we discussed in Chapter 4, the partial address decoding scheme can simplify the decoder design and will be used here. Before making the address assignment, we must also consider whether we need to add other devices to the 68HC11. There are two sides for interfacing an MC6850: the processor side and the RS232 (DCE) side. The following two examples will demonstrate interfacing the MC6850 to the 68HC11.

Example 8.8

Given the following address space assignments for a 68HC11-based single-board computer, design a circuit to interface the MC6850 to the 68HC11A8 and verify that all the timing requirements are met.

> $0000–$1FFF: part of this space is occupied by on-chip memories and registers, and the rest of this space is not assigned
>
> $2000–$3FFF: 6840 programmable timer chip
>
> $4000–$5FFF: 8 KB SRAM HM6264
>
> $6000–$7FFF: 6821 PIA
>
> $8000–$9FFF: ACIA
>
> $A000–$BFFF: not assigned
>
> $C000–$DFFF: not assigned
>
> $E000–$FFFF: on-chip 8 KB ROM

Solution: An address decoder is required to generate chip enable signals for these external components, and a latch is needed to latch the lowest eight address signals (A7–A0). The AS signal is used to latch these address signals, while address signals A12 and A11 are tied to the chip select signals CS1 and CS0, as shown in Figure 8.17, thus reducing the address space allocated to the ACIA to $9800–$9FFF.

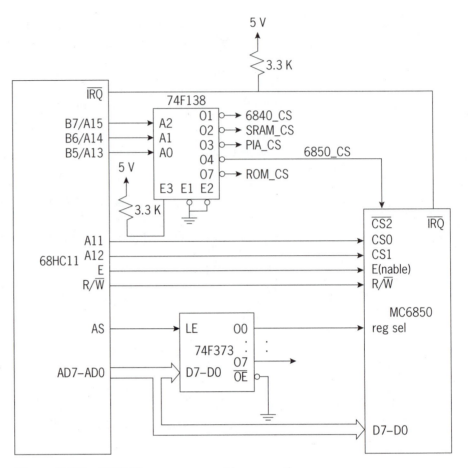

Figure 8.17 6850 Microprocessor side connections

We now need to compare the timing requirements for the MC6850 and the 68HC11:

E clock: The E(nable) input to the 6850 is from the 68HC11. By comparing Tables 8.5 and 4.5, it is easy to see that the rise/fall time and E clock low and high pulse-width requirements are satisfied. These comparisons are summarized in Table 8.6.

Parameter	Required time (ns)	Actual time (ns)
cycle time	500 (min.)	500
rise and fall time	20 (max.)	20
clock pulse width (high)	210 (min.)	227
clock pulse width (low)	220 (min.)	222

Table 8.6 ■ Comparison of E clock timing requirements

Address (reg sel) and R/$\overline{W}$ inputs: The multiplexed low address signals
can be latched and propagated to the output of the 74F373 41.5 ns
before the rising edge of the E(nable) input. This value (41.5 ns) is derived as follows: the delay time from AS-valid to E-rise is 53 ns, and
the propagation delay of the 74F373 from LE-valid to output-stable is
11.5 ns; 53 ns − 11.5 ns = 41.5 ns, and hence the timing requirement
t_{AS} of the MC6850 is satisfied. The 2-MHz 6850 requires a 10-ns address hold time (t_{AH}) after the falling edge of the E clock. The 74F373
will not latch a new value sooner than 53 ns before the next E clock
rising edge, thus providing at least 174 ns (227 ns − 53 ns) of address
hold time. The address hold time requirement is therefore satisfied.
The R/$\overline{W}$ output from the 68HC11 is valid 94 ns before the rising edge
of the E clock, and the MC6850 requires this signal to be valid 40 ns
before the rising edge of the E input, so this requirement is also satisfied. These comparisons are summarized in Table 8.7.

Parameter	Required time (ns)	Actual time (ns)
address setup time before E	40 (min.)	41.5
address hold time	10 (min.)	≥174
R/W setup time	40 (min.)	94
R/W hold time	10 (min.)	30

Table 8.7 ■ Address (reg sel) and R/$\overline{W}$ timing comparisons

Chip select ($\overline{CS2}$, CS1, and CS0): High address signals (A15–A13) from
the 68HC11 are stable 94 ns before the rising edge of the E clock, and
the address decoder output becomes stable 86 ns before the rising edge
of the E clock. The MC6850 requires all three chip select signals to
be valid 40 ns before the rising edge of the E clock. The 68HC11 holds
these address signals for at least 33 ns after the falling edge of the
E clock, and hence chip select output $\overline{CS2}$ will not be changed until
41 ns after the falling edge of the E clock. The chip select signals CS1
and CS0 are connected to A12 and A11, respectively, and their timing
requirements (t_{CS}) are easily satisfied. These timing comparisons are
listed in Table 8.8.

Parameter	Required time (ns)	Actual time (ns)
chip select time before E for $\overline{CS2}$	40 (min.)	86
chip select hold time for $\overline{CS2}$	10 (min.)	41
chip select time before E for CS1	40 (min.)	94
chip select hold time for CS1	10 (min.)	33
chip select time before E for CS0	40 (min.)	94
chip select hold time for CS0	10 (min.)	33

Table 8.8 ■ Comparisons of chip select signals

Read data setup time and hold time (68HC11 requirements): The MC6850 presents the data 150 ns after the rising edge of the E clock (72 ns before the falling edge of the E clock), satisfying the read data setup time requirement of the 68HC11 (30 ns is required). The 6850 stops driving the microprocessor side data bus 20 to 50 ns after the falling edge of the E clock, but the 68HC11 requires a data hold time of 0–83 ns, so this timing requirement is violated. However, as we discussed in Chapter 4, the capacitance on the data bus would hold the data for a longer length of time, and thus the data hold time requirement can be met. The timing comparisons are shown in Table 8.9.

Parameter	Required time (ns)	Actual time (ns)
68HC11 read data setup time	30 (min.)	72
68HC11 read data hold time	0–83 (min.)	20–50, but held by data bus capacitance

Table 8.9 ■ Timing comparisons for 68HC11 read data requirements

Write data setup time and hold time (MC6850 requirements): On a write bus cycle, the 68HC11 presents the data 128 ns after the rising edge of the E clock (or 94 ns before the falling edge of the E clock). The MC6850 requires a write data setup time of 60 ns, so this requirement is easily satisfied. The 68HC11 holds the write data for at least 33 ns, which also exceeds the MC6850 requirement (10 ns). These comparisons are summarized in Table 8.10.

Parameter	Required time (ns)	Actual time (ns)
6850 write data setup time	60	94
6850 write data hold time	10	33

Table 8.10 ■ Timing comparisons for MC6850 write data requirements

From this analysis, we conclude that the circuit connections shown in Figure 8.17 satisfy all the timing requirements of the 68HC11 and the MC6850.

Example 8.9

Implement the null modem connection to the ACIA so that the 68HC11 can talk to a PC via a straight-through RS232 cable and connector. This implementation should allow the user to select from the following baud rates: 300, 600, 1200, 2400, 4800, and 9600.

Solution: To provide the required range of baud rates, appropriate components must be combined with a selection of ACIA transmit and receive clock input

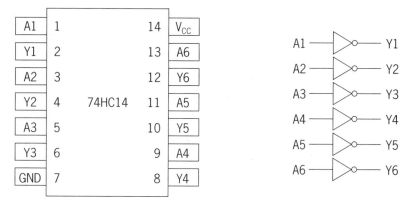

(a) Pin assignments
(b) Logic diagram

Figure 8.18   74HC14 pin assignments and logic diagram

divide factors. In this example, we will use a 2.4576-MHz crystal oscillator to generate the transmit and receive clock signals. Since the output from a crystal oscillator is a sine wave, we need to convert it to a square wave. A Schmitt-Trigger inverter such as a 74HC14 is perfect for this job because it can "square up" slow input rise and fall times. The pin assignments of the 74HC14 are shown in Figure 8.18.

In order to allow the user to select the appropriate baud rate, a ripple counter such as the 74HC4024 can be used. This device consists of seven master-slave flip-flops. The output of each flip-flop feeds the next, and the frequency of each output is half that of the preceding one. The state of the counter advances on the falling edge of the clock input. The pin assignment of the 74HC24 is shown in Figure 8.19. By feeding the output of the 74HC14 Schmitt-Trigger inverter to the Clock input of the 74HC4024, the frequencies of Q1–Q7 will be 1.2288 MHz, 0.6144 MHz, 0.3072 MHz, 0.1536 MHz, 0.0768 MHz, 0.0384 MHz, and 0.0192 MHz, respectively. By dividing these frequencies by 64, baud rates of 9600, 4800, 2400, 1200, 600, and 300 can be derived. A jumper can be used to allow the user to select the baud rate.

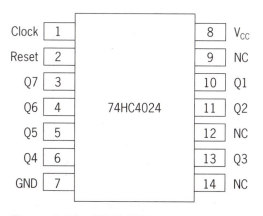

Figure 8.19   74HC4024 pin assignment

A MAX232 is used to translate voltage from the normal CMOS level to the RS232 level or vice versa. The null-modem connection shown in Figure 8.8 should be followed to allow direct connection of the ACIA and a PC using a straight-through RS232 cable. The RS232 side connection of the ACIA is shown in Figure 8.20. Because both the $\overline{DCD}$ and $\overline{CTS}$ inputs to the ACIA are tied to low, the MC6850 will be ready to transmit or receive data immediately after power-on. Since the terminal port is to be connected to a terminal or a PC, the TxD output should be connected to the RxD input to the terminal, and the RxD input should be connected to the TxD output from the terminal. The $\overline{RTS}$ signal is not needed in this connection.

8.7.5 Configuring the ACIA

In Figure 8.20, the TxCLK and RxCLK inputs to the ACIA are tied together and derived by dividing a 2.4576-MHz clock signal by four. The 9600 baud rate is chosen (other baud rates can be chosen by changing the jumper), so the counter divide factor is 64. The configuration of the ACIA is fairly straight-forward.

Example 8.10

Write an instruction sequence to configure the ACIA in Figure 8.20 to operate with the following parameters:

1. disable receive and transmit interrupts
2. counter divide factor is 64
3. data format is eight data bits, one stop bit, and no parity bit

Solution: The user must write a byte into the ACIA control register to set up these parameters accordingly.

1. To disable the receive interrupt, set bit 7 to 0.
2. To disable the transmit interrupt, set bits 6 and 5 to 00 or 10.
3. To select the data format of eight data bits with one stop bit, set the format select field (bits 4, 3, and 2) to 101.
4. To select the counter divide factor of 64, set the counter divide field to 10.
5. The addresses $9800 and $9801 will be used to access the control/status and transmit/receive data registers, respectively.

To set up these parameters, a value of $16 or $56 should be written into the ACIA control register. The following instruction sequence will initialize the ACIA accordingly:

```
acia_ini   EQU  $16
           LDAA #acia_ini
           STAA $9800
```

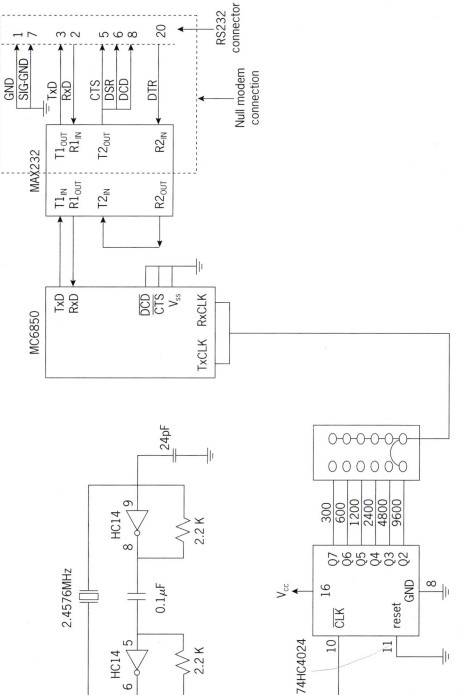

Figure 8.20 Diagram of RS232 side circuit connections

8.7.6 A Terminal I/O Package for ACIA

A set of flexible I/O routines is very useful during the process of software and hardware development, particularly for making the following operations possible:

■ *Checking the results of program execution:* information can be displayed on the PC monitor to allow the user to examine the results of the program execution.

■ *Interaction with the prototype:* the user can provide different data to the application via the PC keyboard to check the correctness of the program logic.

The following list includes some very useful I/O routines:

ONACIA: initializes the ACIA control register.

GETCH: returns a character in accumulator A from the ACIA receive data register.

PUTCH: places the contents of accumulator A in the ACIA transmit data register.

GETSTR: inputs a string that is terminated by a carriage return (CR) character. The input string is to be stored in a buffer pointed to by Y.

PUTSTR: outputs a null-terminated string that is pointed to by Y.

NEWLINE: outputs a CR/LF character to the terminal (LF stands for line-feed).

CHPRSNT: clears the Z bit of the CCR register if a character is found in the ACIA receive data register. Otherwise, it sets the Z bit.

PUTHEX: prints the 8-bit contents of A in 2 hex digits.

ECHOFF: eliminates the character echo on input from the terminal port.

ECHON: causes characters to be echoed on input from the terminal port.

LFON: expands a CR into CR/LF pair, and expands a LF into LF/CR pair.

LFOFF: turns off newline expansion.

The I/O package given in this section can be run on an EVB without modification. The constant definitions for the I/O package are as follows:

```
ACIA      EQU $9800   ; base address of the ACIA
CR        EQU $0D     ; ASCII code of carriage return
LF        EQU $0A     ; ASCII code of line feed
TDRE      EQU $02     ; position of the transmit data register empty bit
RDRF      EQU $01     ; position of the receive data register full bit
control   EQU $0      ; offset of the ACIA control register from the ACIA base
STATUS    EQU $0      ; offset of the ACIA status register from the ACIA base
XMIT      EQU $01     ; offset of the ACIA transmit data register from the ACIA base
RCV       EQU $01     ; offset of the ACIA receive data register from the ACIA base
masterst  EQU $03     ; value to master reset the ACIA
CTL_INI   EQU $16     ; value to configure ACIA to 8 data bits, 1 stop bit, no parity,
   *                  ; no receive and transmit interrupt, counter divide factor 64
```

The subroutine ONACIA first resets the ACIA and then initializes the ACIA control register by writing $16 into it. This specifies the transmission timing (divide by 64), 8 data bits with 1 stop bit and no parity, and all interrupts disabled.

```
ONACIA   PSHX                    ; save registers
         PSHA                    ;    "
         LDX   #ACIA
         LDAA  #masterst         ; reset ACIA
         STAA  control,X         ;    "
         LDAA  #CTL_INI          ; set up ACIA
         STAA  control,X         ;    "
         PULA                    ; restore registers
         PULX                    ;    "
         RTS
```

The logic flow of the subroutine GETCH is shown in the flowchart in Figure 8.21. Its instruction sequence is as follows:

```
GETCH    PSHX                               ; save X
         PSHB                               ; save B
         LDX   #ACIA
retry    BRCLR STATUS,X $70 noerr           ; are there parity, framing, and overrun errors?
         JSR   ONACIA                       ; if there is an error then reinitialize and try
*                                           ; again
         BRA   RETRY                        ;    "
noerr    BRCLR STATUS,X RDRF retry          ; is receive data register full?
         LDAA  RCV,X                        ; read the character
         ANDA  #$7F                         ; clear the parity bit
         LDAB  ECHO                         ; test to see if echo flag is on
         BEQ   quit                         ; branch if the echo flag is not set
         JSR   PUTCH                        ; echo the character to the screen
quit     PULB
         PULX
         RTS                                ; return input character in A
```

The subroutine PUTCH needs to check whether the auto line-feed expansion flag is set. If this flag is set, then both the CR and the LF character will be expanded to the CR/LF pair. The effect is to move the screen cursor to the first column of the next line of the monitor screen. The flowchart of the PUTCH routine is shown in Figure 8.22. The PUTCH subroutine is as follows:

```
PUTCH    PSHB                    ; save B
         BSR   OUTCH             ; output the character
         TST   AUTOLF            ; check to see if auto line-feed flag is set
         BEQ   OUT2              ; prepare to return if not set
         CMPA  #CR               ; check to see if the output character is a
*                                ; carriage return
         BNE   OUT1
         LDAA  #LF               ; also output a line feed after a carriage return
         BSR   OUTCH             ;    "
         BRA   OUT2
OUT1     CMPA  #LF               ; check to see if the output character is a line feed
         BNE   OUT2              ; branch if the character is not a line feed
         LDAA  #CR               ; if yes, also output a carriage return
         BSR   OUTCH             ;    "
```

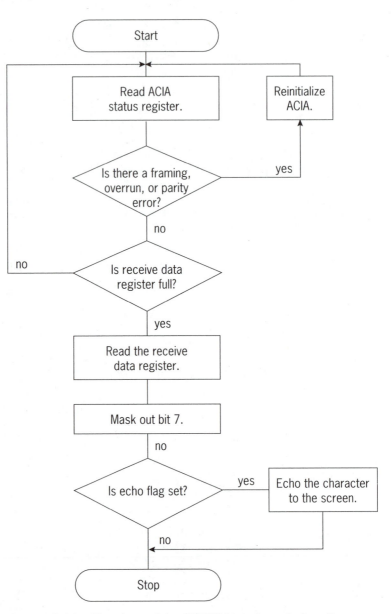

Figure 8.21   Flowchart of the GETCH (get character) routine

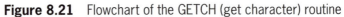

```
OUT2     PULB                          ; restore B
         RTS

OUTCH    LDX     #ACIA
         BRCLR   STATUS,X TDRE *       ; is transmit data register empty ?
* The ASCII code is 7-bit. The following instructions clear bit 7 to make sure a 7-bit
* ASCII code is sent out
         ANDA    #$7F                  ; mask parity bit before output
         STAA    XMIT,X                ; output the character
         RTS
```

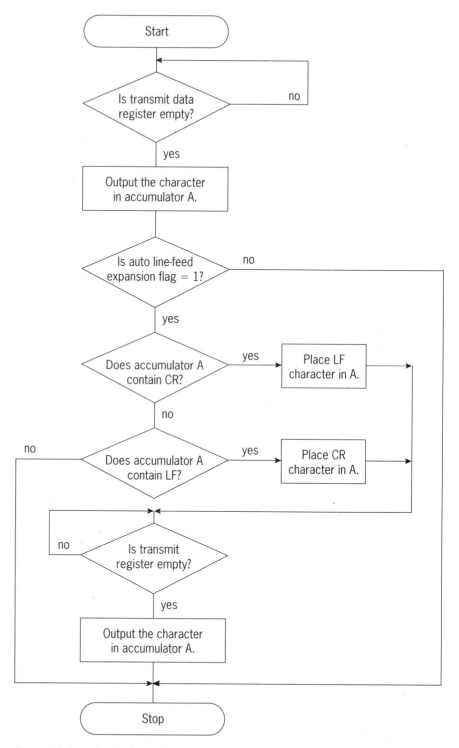

Figure 8.22 Logic flow of the subroutine PUTCH

The subroutine GETSTR inputs a string terminated by the CR character from the keyboard. The index register Y contains the address of the input buffer where the string is to be stored. The subroutine appends a null character to the end of the string.

```
GETSTR    PSHA
GSLOOP    JSR    GETCH
          CMPA #CR          ; is it a carriage return?
          BEQ  GFINIS        ; end of get string
          STAA 0,Y           ; save the character in the buffer
          INY                ; increment the buffer pointer
          BRA  GSLOOP        ; return to get string loop
GFINIS    LDAA #$00          ; terminate the string with a NULL character
          STAA 0,Y           ;      "
          PULA               ; restore A
          RTS
```

The subroutine PUTSTR outputs the string pointed to by the index register Y. This subroutine outputs the string character by character by calling the PUTCH subroutine until the null character is encountered.

```
PUTSTR    PSHA              ; save A
PSLOOP    LDAA 0,Y          ; get a character
          BEQ  PFINIS       ; is this the end of the string ?
          JSR  PUTCH        ; if not, output the character
          INY               ; increment the pointer
          BRA  PSLOOP
PFINIS    PULA              ; restore A
          RTS
```

```
* The following subroutine outputs a CR/LF pair.
```

```
newline   PSHA              ; save A
          LDAA #$01
          STAA AUTOLF       ; turn on auto line-feed expansion
          LDAA #CR          ; output a CR which will be expanded into CR/LF pair
          JSR  PUTCH        ;      "
          PULA              ; restore A
          CLR  AUTOLF       ; clear the auto line-feed flag
          RTS
```

The following subroutine checks to see if a character is present at the input terminal port. The Z bit of the CCR register is updated.

```
CHPRSNT
          PSHA              ; save A
          PSHX              ; save X
          LDX  #ACIA
          LDAA STATUS,X     ; check the status register
          BITA #RDRF        ; test the RDRF bit to update the Z bit of CCR
          PULX              ; restore X
          PULA              ; restore A
          RTS
```

* The following routine outputs the contents of A as two hex digits.

```
PUTHEX    PSHX                ; save registers
          PSHB                ;      "
          PSHA                ;      "
          TAB                 ; make a copy of A in B
          LSRB                ; shift the upper hex digit into the lower half of B
          LSRB                ;      "
          LSRB                ;      "
          LSRB                ;
          LDX   #HEXDIG       ; load the hex digit table address into X
          ABX                 ; X points to the ASCII code of the upper hex digit in A
          LDAA  0,X           ; get the ASCII code
          JSR   PUTCH         ; output the upper hex digit
          PULB                ; put A to B again
          PSHB                ; restore the stack
          ANDB  #$0F          ; mask out the upper 4 bits
          LDX   #HEXDIG       ; load the hex digit table address into X
          ABX                 ; compute the address of the ASCII code of the lower
                              ; hex digit
          LDAA  0,X           ; get the ASCII code of the lower hex digit
          JSR   PUTCH         ; output the lower hex digit
          PULA                ; restore registers
          PULB                ;      "
          PULX                ;      "
          RTS
HEXDIG    FCC   "0123456789ABCDEF"    ; hex digit ASCII code table
```

* The following routine turns off the echo on input:

```
ECHOFF    CLR   ECHO
          RTS
```

* The following routine turns on the echo on input:

```
ECHON     LDAA  #1
          STAA  ECHO
          RTS
```

* The following routine turns on the newline expansion:

```
LFON      LDAA  #1
          STAA  AUTOLF
          RTS
```

* The following routine turns off the newline expansion:

```
LFOFF     CLR   AUTOLF
          RTS

ECHO      RMB   1             ; input echo flag
AUTOLF    RMB   1             ; line-feed expansion flag
          END
```

Example 8.11

Write an instruction sequence to output the following message to the terminal screen: "Hello, world!"

```
        ORG  $C000        ; user RAM starting address
        LDS  #$3F         ; initialize the stack pointer
        LDY  #MSG         ; put the starting address into the Y register
        JSR  PUTSTR       ; call put string routine
          .
          .
          .
MSG     FCC  "Hello, world!"   ; message to be output
        END
```

8.8 Summary

This chapter started by introducing the basic concepts of data communication and then discussed the RS232 standard in detail. A null modem connection allows a direct connection between a computer and a terminal. A voltage level shifter chip (RS232 driver/receiver) is required to interface the SCI to the RS232 connector. The 6850 ACIA is a serial communication chip designed to be used with Motorola 8-bit microprocessors; several examples explain the interfacing of the 6850 to the 68HC11 and the RS232 connector in detail. A complete terminal I/O package is provided and explained in the last section of this chapter.

8.9 Glossary

Break   The transmission or reception of a low for at least one complete character time.

DCE   The acronym for data communication equipment. DCE usually refers to equipment such as a modem, concentrator, or router.

DTE   The acronym for data terminal equipment. DTE usually refers to a computer or terminal.

EIA   The acronym for electronics industry association.

Framing error   A data communication error in which a received character is not properly framed by the start and stop bits.

Full-duplex link   A four-wire communication link that allows both transmission and reception to proceed simultaneously.

Half-duplex link   A communication link that can be used for either transmission or reception, but only in one direction at a time.

Idle   A continuous logic high on the RxD line for one complete character time.

ISO   The acronym for the International Standards Organization.

Mark   A term used to indicate a binary 1.

Modem   A device that can accept digital bits and change them into a form suitable for analog transmission (modulation) and can also receive a modulated signal and transform it back to its original digital representation (demodulation).

Multi-drop   A data communication scheme in which more than two stations share the same data link. One station is designated as the master, and the other stations are designated as slaves. Each station has its own unique address, with the primary station controlling all data transfers over the link.

Null modem   A circuit connection between two DTEs in which the leads are interconnected in such a way as to fool both DTEs into thinking that they are connected to modems. A null modem is only used for short-distance interconnections.

Parity error   An error in which an odd number of bits change value; it can be detected by a parity checking circuit.

Point-to-point   A data communication scheme in which two stations communicate as peers.

Receiver overrun   A data communication error in which a character or a number of characters were received but not read from the buffer before the following characters were received.

RS232   An interface standard recommended for interfacing between a computer and a modem. This standard was established by EIA in 1960 and has since been revised several times.

Simplex link   A line dedicated to either transmission or reception, but not both.

Space   A term used to indicate a binary 0.

UART   The acronym for universal asynchronous receiver and transmitter; an interface chip that allows the microprocessor to perform asynchronous serial data communication.

8.10 Exercises

E8.1. What is a DTE? a DCE?

E8.2. What is multi-drop serial communication?

E8.3. What is a simplex link? a half-duplex link? a full-duplex link?

E8.4. With regard to the voltage of the RS232 driver output, what voltage is considered a mark (1)? What voltage is considered a space (0)?

E8.5. Sketch the TxD pin output of a UART chip for the letter K in RS232 format. The letter K is to be transmitted using a format with eight data bits, even parity, and one stop bit.

E8.6. Compute the time required to transmit 100 characters using the data format of eight data bits, one stop bit, and no parity. The transmit speed is 4800 baud.

E8.7. How does the SCI system determine the logic level of data bits?

E8.8. Write a subroutine to output a null-terminated string using the polling method to the SCI. The string to be output is pointed to by index register Y.

E8.9. Write a subroutine to output a null-terminated string from the SCI using the interrupt-driven method. The string is pointed to by index register Y.

E8.10. Write a subroutine to input a string from the SCI using the polling method. The buffer to hold the string is pointed to by index register Y.

E8.11. Write a subroutine to input a string from the SCI using the interrupt-driven method. This routine will enable the receive interrupt and stay in a loop until the CR character is received. The starting address of the buffer to hold the string is in index register Y.

E8.12. What is a framing error? What is a receiver overrun error?

E8.13. Write a routine that can input two hex digits from the ACIA, store the equivalent binary number in accumulator A, and echo the number to the screen. This program is to be run on the 68HC11EVB using the terminal port.

E8.14. How many addresses are assigned to the ACIA registers in Figure 8.17?

E8.15. Write a byte into the BAUD register to select a baud rate of 4800 at 2 MHz E clock frequency.

8.11 Lab Exercises and Assignments

L8.1. Write a program to simulate the log-in session on any mini- or mainframe computer. Your program should perform the following steps:

1. The program begins by outputting the following message:

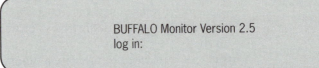

```
BUFFALO Monitor Version 2.5
log in:
```

2. After putting out this message, the program waits for you to type in your user name, which should be terminated by the CR character.

3. After you enter your user name, the program outputs the following prompt:

```
password:
```

4. The program then waits for you to enter the password. After the user name and password are both entered, the program starts to search the username/password table to see if the user name and password match. The table contains 15 pairs of user names and passwords. If the user name and password are correct, the program outputs the following message:

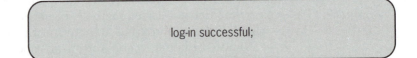

```
log-in successful;
```

If the log-in is not successful, the program should output the following message and repeat the log-in process:

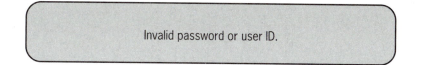

Invalid password or user ID.

L8.2. Write a program to be run on the EVB to compute the mean and median of an array with fifteen 8-bit numbers. Each number consists of two hex digits and will be entered from the keyboard. The program outputs the following prompt:

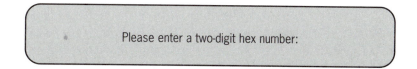

Please enter a two-digit hex number:

The user then enters the hex number. If the number contains an invalid digit, the program will ask the user to reenter a valid number. Each number is terminated by a carriage return.

Each prompt should appear on a separate line. After fifteen 8-bit numbers have been entered, the program computes the mean and median of the array and outputs them to the screen. The program also sorts the given array and displays it on the screen. The following message should appear on the screen at the end of program execution:

The given array is: xx xx xx xx xx xx xx xx xx xx xx xx xx xx xx
The mean of the array is: yy
The median of the array is: zz

9

68HC11

SERIAL

PERIPHERAL

INTERFACE

After completing this chapter, you should be able to:

- describe the operation of the 68HC11 SPI subsystem
- connect an SPI master with one or more SPI slave devices
- simulate the operation of the SPI
- use the SPI system and multiple HC589s to add parallel input ports to the 68HC11
- use the SPI system and multiple HC595s to add parallel output ports to the 68HC11
- use the SPI and M14499(s) to drive multiple seven-segment displays
- use the SPI, the MC145000, and the MC145001 to drive LCDs
- interface the SPI to the MC68HC68T1 time-of-day chip

9.2 Introduction

The number of pins on an 8-bit microcontroller is quite limited. However, it is desirable to implement as many I/O functions as possible on that limited number of pins. Many I/O devices have a low data rate, and using parallel interfaces can certainly satisfy data transfer needs. However, parallel data transfers require too many data pins, and thus fewer functions can be implemented on the same number of signal pins. The solution is to use serial data transfer for low-speed I/O devices. The Motorola 68HC11 provides a serial peripheral interface (SPI) and a serial communication interface (SCI) for low-speed I/O devices. The SPI is a synchronous interface that requires the same clock signal to be used by the SPI subsystem and the external device. (The SCI is an asynchronous interface that allows different clock signals to be used by the SCI system and the external device. The SCI subsystem was discussed in Chapter 8.)

The SPI subsystem allows several microcontrollers with the SPI function or SPI-type peripherals to be interconnected. In the SPI format, two types of devices are involved in data transfer: a master and one or more slaves. The master can initiate a data transfer, but a slave can only respond. A clock signal is required to synchronize the data transfer but is not included in the data stream and must be furnished as a separate signal by the master of the data transfer. The 68HC11 SPI subsystem can be configured as a master or as a slave. Data transfers in the SPI format are initiated by the master SPI device. The SPI subsystem is usually used for I/O port expansion and for interfacing with peripherals, but it can also be used for multiprocessor communication. The SPI is often used to interface with peripheral devices such as TTL shift registers, LED/LCD display drivers, phase-locked loop (PLL) chips, or A/D converter systems that do not need a very high data rate.

9.3 Signal Pins

Four of the 68HC11 port D pins are associated with SPI transfers: $\overline{SS}$/PD5, SCK/PD4, MOSI/PD3, and MISO/PD2. Port D pins are bidirectional and must be configured for either input or output. All SPI output lines must have their corresponding data direction register bits set to 1. If one of these bits is 0, the line is disconnected from the SPI logic and becomes a general-purpose input line. All SPI input lines are forced to act as inputs regardless of what is in their corresponding data direction register bits.

1. MISO/PD2: Master in slave out. The MISO pin is configured as an input in a master device and as an output in a slave device. It is one of the two lines that transfer serial data in one direction with the most significant bit sent first. The MISO line of a slave device is placed in a high-impedance state if the slave device is not selected.

2. MOSI/PD3: Master out slave in. The MOSI pin is configured as an output in a master device and as an input in a slave device. It is the second of the two lines that transfer serial data in one direction with the most significant bit sent first.

3. SCK/PD4: Serial clock. The serial clock is used to synchronize data movement both in and out of a device through its MOSI and MISO

lines. The master and slave devices are capable of exchanging one byte of information during a sequence of eight SCK clock cycles. Since SCK is generated by the master device, this line becomes an input on a slave device.

4. $\overline{SS}$/PD5: Slave select. The *slave select* ($\overline{SS}$) input line is used to select a slave device. For a slave SPI device, it has to be low prior to data transactions and must stay low for the duration of the transaction. The $\overline{SS}$ line on the master device must be tied high. If it goes low, a mode fault error flag (MODF) is set in the serial peripheral status register (SPSR). The $\overline{SS}$ pin can be configured as a general-purpose output by setting bit 5 of the DDRD register to 1, thus disabling the mode fault circuit. The other three SPI lines are dedicated to the SPI whenever the SPI subsystem is enabled.

9.4 SPI-Related Registers

Three registers in the serial peripheral interface provide control, status, and data storage functions: the serial peripheral control register (SPCR), the serial peripheral status register (SPSR), and the serial peripheral data I/O register (SPDR). Because the directions of the SPI pins must be set for proper operation, the DDRD register must also be programmed.

9.4.1 The Serial Peripheral Control Register (SPCR)

The meaning and function of each bit in the SPCR register and the value of each bit after reset are as follows:

	7	6	5	4	3	2	1	0	SPCR
	SPIE	SPE	DWOM	MSTR	CPOL	CPHA	SPR1	SPR0	located at
Value after reset =	0	0	0	0	0	1	u	u	$1028

u = undefined

SPIE: SPI interrupt enable. When this bit is set to 1, the SPI interrupt is enabled.

SPE: SPI enable. When this bit is set to 1, the SPI system is enabled.

DWOM: Port D wired-or mode select. When this bit is set to 1, all port D pins become open-drain. Otherwise, they are normal CMOS output pins.

MTSR: Master mode select. When this bit is set to 1, the SPI master mode is selected. Otherwise, the slave mode is selected.

CPOL: Clock polarity. When the clock polarity bit is 0 and data is not being transferred, a steady-state low value is produced at the SCK pin of the master device. Conversely, if this bit is 1, the SCK pin will idle high. This bit is also used in conjunction with the clock phase control bit to produce the desired clock-data relationship between the master

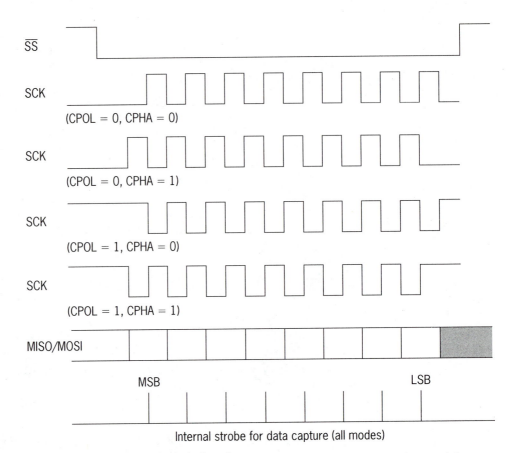

Figure 9.1 SPI data clock timing diagram (Redrawn with permission of Motorola)

and the slave. The relationships between the CPOL bit, the CPHA bit, MOSI, and MISO are shown in Figure 9.1. Data is captured in the center of a bit time. Depending on the combination of the CPOL and CPHA settings, data bits can be captured on the rising (CPOL and CPHA = 00 or 11) or the falling (CPOL and CPHA = 01 or 10) edge of the SCK clock.

CPHA: Clock phase. The clock phase bit, in conjunction with the CPOL bit, controls the clock-data relationship between the master and the slave. The CPHA bit selects one of two fundamentally different clocking protocols. When CPHA = 0, the shift clock is the OR of SCK with $\overline{SS}$. As soon as $\overline{SS}$ goes low, the transaction begins and the first edge on the SCK signal invokes the first data sample. In this clock phase mode, $\overline{SS}$ must go high between successive characters in an SPI message. When CPHA = 1, the $\overline{SS}$ pin may be left low for several SPI characters. This setting requires fewer instructions during the data transfer phase.

SPR1 and SPR0: SPI clock rate select. These two bits control the frequency of the SCK clock output according to the scheme given by the following table:

SPR1	SPR0	E divided by
0	0	2
0	1	4
1	0	16
1	1	32

The highest data rate supported by the SPI is 1 Mbits per second at the 2 MHz E clock rate. The lowest data rate is 62.5 Kbits per second with a 2 MHz E clock frequency.

9.4.2 The Serial Peripheral Status Register (SPSR)

The meaning and function of each bit in the SPSR register and the value of each bit after reset are as follows:

	7	6	5	4	3	2	1	0	SPCR
	SPIF	WCOL	0	MODF	0	0	0	0	located at $1029
Value after reset =	0	0	0	0	0	0	0	0	

SPIF: SPI transfer complete flag. The SPIF flag bit is set upon completion of data transfer between the processor and the external device. Clearing the SPIF bit is accomplished by reading the SPSR register followed by an access of the SPDR register. Unless the SPSR register is read first, attempts to write to the SPDR register are inhibited.

WCOL: Write collision. The write collision bit is set when an attempt is made to write to the SPDR register while data transfer is taking place. Clearing the WCOL bit is accomplished by reading the SPSR register followed by an access to the SPDR register.

MODF: mode fault. The mode fault flag indicates that there may have been a multi-master conflict for system control. It allows a proper exit from system operation to a reset or default system state. The MODF bit is normally clear and is set only when the master device has its $\overline{SS}$ pin pulled low. Setting the MODF bit affects the internal serial peripheral interface system in the following ways:

1. An SPI interrupt is generated if SPIE = 1.
2. The SPE bit is cleared, disabling the SPI.
3. The MSTR bit is cleared, forcing the device into the slave mode.
4. The DDRD bits for the four SPI pins are forced to zeros.

Clearing the MODF bit is accomplished by reading the SPSR, followed by a write to the SPCR. Control bits SPE and MSTR can be restored by user software to their original state after the MODF bit has been cleared. The DDRD register must also be restored after a mode fault.

9.4.3 The Port D Data Direction Register (DDRD)

The meaning and function of each bit in the DDRD register and the value of each bit after reset are as follows:

	7	6	5	4	3	2	1	0	DDRD
	0	0	DDRD5	DDRD4	DDRD3	DDRD2	DDRD1	DDRD0	located at
Controlled pin			$\overline{SS}$	SCK	MOSI	MISO	TxD	RxD	$1009
Value after reset =	0	0	0	0	0	0	0	0	

0 = input
1 = output

DDRD5: Data direction control for port D bit 5 ($\overline{SS}$). When the SPI system is enabled as a slave, the PD5/$\overline{SS}$ pin is the slave select input, regardless of the value of DDRD5. When the SPI system is enabled as a master, the function of the PD5/$\overline{SS}$ pin depends on the value in DDRD5. If DDRD5 is set to 0, the $\overline{SS}$ pin is used as an input to detect mode-fault errors. A low on this pin indicates that some other device in a multiple-master system has become a master and is trying to select this MCU as a slave. When a low occurs on this pin, all SPI pins will be changed to high impedance to prevent damage to the SPI pins. When DDRD5 is set to 1, the $\overline{SS}$ pin is used as a general-purpose output pin and is not affected by the SPI system.

DDRD4: Data direction control for port D bit 4 (SCK). When the SPI system is enabled as a slave, the PD4/SCK pin acts as the SPI serial clock input, regardless of the state of DDRD4. When the SPI system is enabled as a master, the DDRD4 bit must be set to 1 to enable the SCK output.

DDRD3: Data direction control for port D bit 3 (MOSI). When the SPI system is enabled as a slave, the PD3/MOSI pin acts as the slave serial data input, regardless of the state of DDRD3. When the SPI system is enabled as a master, the DDRD3 bit must be set to 1 to enable the master serial data output. If a master device wants to initiate an SPI transfer to receive a byte of data from a slave without transmitting a byte, it might purposely leave the MOSI output disabled. SPI systems that tie MOSI and MISO together to form a single, bidirectional data line also need to selectively disable the MOSI output.

DDRD2: Data direction control for port D bit 2 (MISO). When the SPI system is enabled as a slave, the DDRD2 bit must be set to 1 to enable the slave serial data output. A master SPI device can simultaneously broadcast a message to several slaves as long as no more than one slave tries to drive the MISO pin. SPI systems that tie MOSI and MISO together to form a single, bidirectional data line also need to selectively disable the MISO output.

9.5 SPI Operation

The SPI circuit connections, data transfer timing, and the data transfer process are discussed in the following subsections.

9.5.1 SPI Circuit Connections

In a system that uses the SPI function, one device (normally the 68HC11) is configured as the master and the other devices are configured as slaves. Either a peripheral chip or a 68HC11 can be configured as a slave device. The master SPI device controls data transfer and can control one or more SPI slave devices simultaneously.

In a single-slave configuration, the circuit would be connected as shown in Figure 9.2. The $\overline{SS}$ pin of the master is tied to 5 V and the $\overline{SS}$ pin on the slave SPI device is tied to ground.

In a multiple-slave configuration, all corresponding SPI pins except the $\overline{SS}$ pin can be tied together, as shown in Figure 9.3. The master uses port pins to select a specific slave for data transfer. The $\overline{SS}$ input of the selected slave device should be set to low, but the $\overline{SS}$ input of a slave should be set to high when it is not selected.

Another way to connect multiple SPI slaves to one master is shown in Figure 9.4. Figure 9.4 differs from Figure 9.3 in three ways:

1. The MISO pin of each slave is wired to the MOSI pin of the slave device next to it. The MOSI pins of the master and slave 0 are still wired together.

2. The MISO pin of the master is wired to the MISO pin of the last slave device.

3. The $\overline{SS}$ inputs of all slaves are tied to ground to enable all slaves. No port pin from the master is needed to enable any individual slave device.

Thus the shift registers of the SPI master and slaves become a ring. The data of slave k are shifted to the master SPI, the data of the master are shifted to slave 0, the data of slave 0 are shifted to slave 1, etcetera. In this configuration, a minimal number of signal pins are used to control a larger number of devices. However, the master does not have the freedom to select an arbitrary slave device for data transfer without going through other slave devices.

This type of configuration is often used to extend the capability of the SPI slave. For example, suppose there is an SPI-compatible seven-segment display

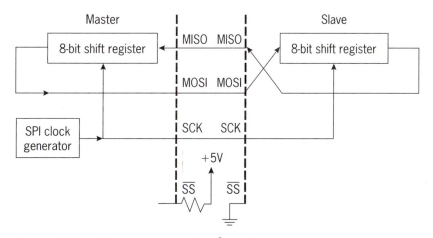

Figure 9.2 SPI master-slave interconnection

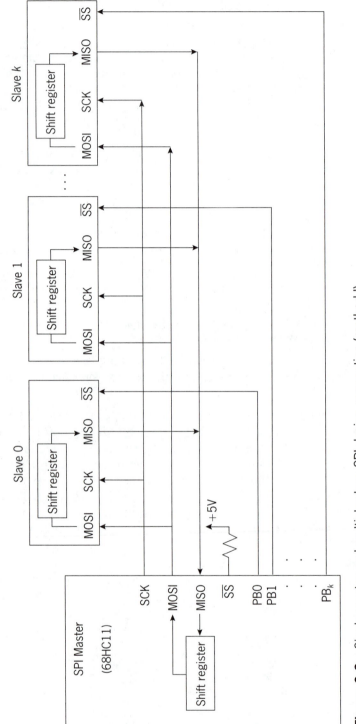

Figure 9.3 Single-master and multiple-slave SPI device connection (method I)

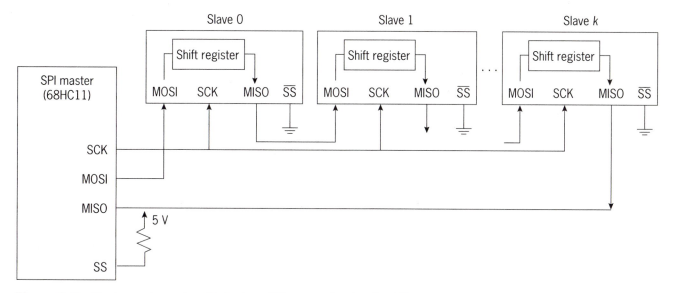

Figure 9.4 Single-master and multiple-slave SPI connection (method II)

driver/decoder that can drive only four digits. By using this configuration, up to $4 \times k$ digits can be displayed when k driver/decoders are cascaded together. An example of this configuration will be discussed in section 9.9.3.

Depending on the capability and role of the slave device, either the MISO or the MOSI pin may not be used in the data transfer. Many SPI-compatible peripheral chips do not have the MISO pin.

9.5.2 SPI Data Transfer

An SPI transfer is initiated by writing data to the shift register in the master SPI device. During an SPI transfer, data is circulated 8 bit positions; thus data is exchanged between the master and the slave. Eight pulses are sent out from the master SCK pin to synchronize the data transfer between the master and the slave SPI devices.

The SPI is double-buffered on read but not on write. If a write is performed during a data transfer, the transfer occurs uninterrupted and the write will be unsuccessful. This condition sets the write collision (WCOL) status bit in the SPSR register. After a data byte is shifted, the SPIF flag of the SPSR register is set to 1.

In the master mode, the SCK pin is an output. The SCK clock idles high (or low) if the CPOL bit in the SPCR register is 1 (or 0). When data is written into the shift register, eight clocks are generated to shift eight bits of data, and then the SCK pin goes idle again.

In the slave mode, the slave start logic receives a logic low at the $\overline{SS}$ pin and a clock input at the SCK pin. Thus the slave is synchronized with the master. Data from the master is received serially at the slave MOSI line and loaded into the 8-bit shift register. After the 8-bit shift register is loaded, its data is transferred in parallel to the read buffer. During a write cycle, data is written into the shift register, then the slave waits for a sequence of clock pulses from the master to shift the data out on the slave's MISO line.

In a single-slave SPI environment, the following instruction sequence will transfer a character from the master to the slave:

```
REGBAS   EQU  $1000     ; base address of I/O register block
SPDR     EQU  $2A       ; offset of SPDR from REGBAS
SPCR     EQU  $28       ; offset of SPCR from REGBAS
SPSR     EQU  $29       ; offset of SPSR from REGBAS
DDRD     EQU  $09       ; offset of DDRD from REGBAS
SPI_DIR  EQU  $38       ; value to set SPI pin directions; this value sets SS,
*                       ; SCK, and MOSI pins for output. Other port D pins
*                       ; are configured as input.
SPI_INI  EQU  $54       ; value to initialize the SPI. This value enables SPI
*                       ; function, disables the SPI interrupt, selects port D pins
*                       ; as normal CMOS output, selects master mode, chooses
*                       ; the falling edge to synchronize data transfer, and sets
*                       ; the data rate to 1 Mbits/sec
         ORG  $00
data     rmb  $10
         .
         .
         .
         LDX  #REGBAS
         LDAA #SPI_DIR
         STAA DDRD,X    ; set directions of SPI pins
         LDAA #SPI_INI
         STAA SPCR,X    ; initialize the SPI system
         LDAA data      ; get a byte to be sent out
         STAA SPDR,X    ; start SPI transfer
wait     LDAB SPSR,X    ; check bit 7 of SPSR to see if the SPI transfer is
         BPL  wait      ; completed
         .
         .
         .
```

To read data from a slave SPI device, the master SPI device also writes a byte into the SPDR register. However, the byte written into the SPDR register is unimportant. The data from the slave is shifted into the SPDR register from the MISO pin. The following instruction sequence reads a byte from the slave SPI device:

```
         LDX  #REGBAS
         LDAA #SPI_DIR
         STAA DDRD,X    ; set directions of SPI pins
         LDAA #SPI_INI
         STAA SPCR,X    ; initialize the SPI system
         STAA SPDR,X    ; start the SPI transfer
here     LDAB SPSR,X    ; wait until a character has been shifted in
         BPL  here      ;     "
         LDAA SPDR,X    ; place the character in A
         .
         .
         .
```

9.5.3 Simulating the SPI

The SPI function is convenient to use when the number of data bits to be transferred is a multiple of eight. When the data to be transferred is not a multiple of eight bits, the SPI might be clumsy to use. In this situation, a software technique can be used to simulate the SPI operation. The peripheral device would use either the falling or the rising clock edge to shift data in/out. The following procedure can be used to shift data into (or out from) a peripheral device on the falling clock edge:

Step 1
Set the clock to high.
Step 2
Apply the data bit on the port pin that is connected to the serial data input pin of the peripheral device.
Step 3
Pull the clock to low.
Step 4
Repeat steps 1 to 3 as many times as needed.

If the rising edge is used to latch data, then the following changes should be made: set the clock to low in step 1, and pull the clock to high in step 3.

9.6 SPI-Compatible Peripheral Chips

The SPI is a protocol developed by Motorola to interface peripheral devices to a microcontroller. As long as a peripheral device supports the SPI interface protocol, it can be used with any microcontroller that implements the SPI function. Several manufacturers produce SPI-compatible peripheral chips. A partial list is given in Table 9.1.

Vendor	Part number	Description
Motorola	MC28HC14	256-byte EEPROM in 8-pin DIP
	MC14021	8-bit input port
	MC74HC165	8-bit input port
	MC74LS165	8-bit input port
	MC74HC589	8-bit parallel-in/serial-out shift register
	MC74HC595	8-bit serial-in/parallel-out shift register
	MC74LS673	16-bit output port
	MC144115	16-segment non-muxed LCD driver
	MC144117	4-digit LCD driver
	MC14549	A/D converter successive approximation register
	MC14559	A/D converter successive approximation register
	MC14489	5-digit 7-segment LED display decoder/driver
	MC14499	4-digit 7-segment LED display decoder/driver
	MC145000	serial input multiplexed LCD driver (master)
	MC145001	serial input multiplexed LCD driver (slave)
	MC145453	LCD driver with serial interface

Table 9.1 ■ SPI-compatible devices   (continues)

Vendor	Part number	Description
	MC144110	6 six-bit D/A converter
	MC144111	4 six-bit D/A converter
	MC144040	8-bit A/D converter
	MC144051	8-bit A/D converter
	MC145050	10-bit A/D converter
	MC145051	10-bit A/D converter
	MC145053	10-bit A/D converter
	MC68HC68T1	real-time clock
	MCCS1850	serial real-time clock
	MC145155	serial input PLL FS
	MC145156	serial input PLL FS
	MC145157	serial input PLL FS
	MC145158	serial input PLL FS
	MC145159	serial input PLL FS
	MC145170	serial input PLL FS
RCA	CDP68HC68A1	10-bit A/D converter
	CDP68HC68R1	128×8-bit SRAM
	CDP68HC68R2	256×8-bit SRAM
	CDP68HC68T1	real-time clock
Signetics	PCx2100	40-segment LCD duplex driver
	PCx2110	60-segment LCD duplex driver
	PCx2111	64-segment LCD duplex driver
	PCx2112	32-segment LCD duplex driver
	PCx3311	DTMF generator with parallel inputs
	PCx3312	DTMF generator with I^{2}C bus inputs
	PCx8570A	256×8 SRAM
	PCx8571	128×8 SRAM
	PCx8573	clock/timer
	PCx8474	8-bit remote I/O expander
	PCx8476	1:4 mux LCD driver
	PCx8477	32/64 segment LCD driver
	PCx8491	8-bit, 4-channel A/D converter and D/A converter
	*SAA1057	PLL tuning circuit: 512 KHz to 120 MHz
	*SAA1060	32-segment LED driver
	*SAA1061	16-segment LED driver
	*SAA1062A	20-segment LED driver
	*SAA1063	fluorescent display driver
	SAA1300	switching circuit
	*SAA3019	clock/timer
	SAA3028	I/R transcoder
	*SAB3013	hex 6-bit D/A converter
	SAB3035	PLL digital tuning circuit with 8 D/A converters
	SAB3036	PLL digital tuning circuit
	SAB3037	PLL digital tuning circuit with 4 D/A converters
	SAA5240	teletext controller chip—625 line system
	TDA3820	digital stereo sound control IC
	TEA6000	MUSTI: FM/RF system
	TDA1534A	14-bit A/D converter—serial output
	TDA1540P,D	14-bit A/D converter—serial input
	NE5036	6-bit A/D converter—serial output
Sprague	UCN4810/5810	power driving output ports
	UAA2022/2023	16-bit power output port

*requires an extra enable line

Table 9.1 ■ (continued)

The Motorola SPI protocol is compatible with National Semiconductor's Microwire protocol. Therefore, any peripheral device that can be interfaced to the SPI can also be interfaced to the Microwire protocol.

9.7 Interfacing the HC589 to the SPI

One application of the HC589 is to expand the number of input ports to the 68HC11. The full name of the HC589 is the 74HC589. This device is available from several manufacturers, each of whom may add a prefix for identification; for example, Motorola adds the prefix MC. In this text, we will use the generic name HC589 to refer to this chip.

The HC589 is an 8-bit serial- or parallel-input/serial-output shift register. Its block diagram and pin assignments are shown in Figure 9.5. The HC589 can accept serial or parallel data input and shift it out serially. The maximum shift clock rate for the HC589 is 6 MHz at room temperature.

9.7.1 HC589 Pins

The functions of the HC589 signal pins are as follows:

A, B, C, D, E, F, G, and H: Parallel data inputs. Data on these inputs are stored in the data latch on the rising edge of the latch clock input.

S_A: Serial data input. Data on this pin are shifted into the shift register on the rising edge of the shift clock input if the serial-shift/parallel-load is high. Data on this pin are ignored when the serial-shift/parallel-load is low.

Serial shift/Parallel load: Shift register mode control. When a high level is applied to this pin, the shift register is allowed to shift data serially.

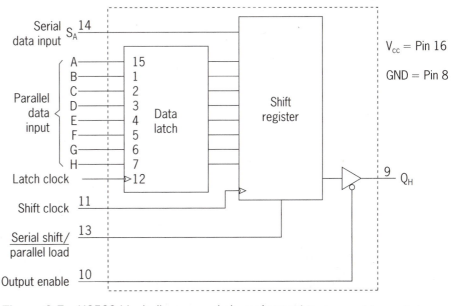

Figure 9.5 HC589 block diagram and pin assignment (Redrawn with permission of Motorola)

When a low level is applied, the shift register accepts parallel data from the data latch.

Shift clock: Serial shift clock. A low-to-high transition on this input shifts data on the serial data input into the shift register; data on stage H is shifted out from Q_H, where it is replaced by the data previously stored in stage G.

Latch clock: Data latch clock. A low-to-high transition on this input loads the parallel data on inputs A–H into the data latch.

Output enable: Active-low output enable. A high level applied to this pin forces the Q_H output into a high-impedance state. A low level enables the output. This control does not affect the state of the input latch or the shift register.

Q_H output: serial data output. This is a three-state output from the last stage of the shift register.

9.7.2 Circuit Connections of the HC589 and the SPI

There are several different ways to interface HC589s to the SPI subsystem. One connection method is shown in Figure 9.6. With this method, multiple bytes can be loaded into HC589s at one time and then shifted into the 68HC11 bit-serially. The procedure for shifting in multiple bytes is as follows:

Step 1
Program the DDRD register to configure the SCK, TxD, and $\overline{SS}$ pins as output and the MISO pin as input.
Step 2
Program the SPCR register to enable the SPI function.

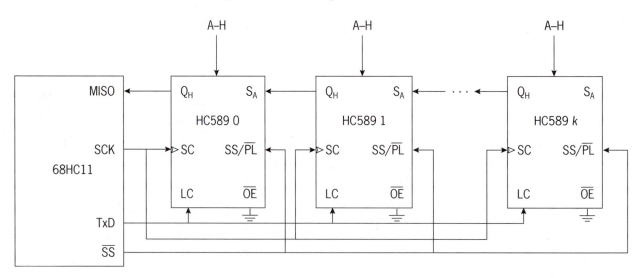

SC: shift clock
LC: latch clock
$\overline{OE}$: output enable
SS/$\overline{PL}$: serial shift/parallel load

Figure 9.6 Serial connection of multiple HC589s to an SPI

Step 3

Set the LC pin to low and then pull it to high; this will load the external data into the data latch in parallel.

Step 4

Set the $\overline{SS}$ pin to low to select the parallel load mode, which will load the contents of the data latch into the shift register.

Step 5

Set the $\overline{SS}$ pin to high to select the serial shift mode.

Step 6

Write a byte into the SPDR to trigger eight SCK clock pulses to shift in eight bits.

Step 7

Repeat step 6 as many times as needed, and save the data in a buffer.

■

Example 9.1

Write a program to input eight bytes from eight external HC589s connected as shown in Figure 9.6 and store the data at locations starting at $0000.

Solution The program should

1. write the value %00111010 into the DDRD register so that the $\overline{SS}$, SCK, MOSI, and TxD pins are configured for output and the MISO and RxD pins are configured for input

2. write the value %01010000 into the SPCR register to:
   a. disable the SPI interrupt
   b. enable the SPI
   c. set port D pins to be normal (not open-drain) CMOS output pins
   d. use the rising edge of the SCK signal to shift data
   e. select a 1-Mbit/sec data rate

The program is as follows:

```
REGBAS    EQU  $1000         ; base address of I/O register block
PORTD     EQU  $08           ; offset of PORTD from REGBAS
DDRD      EQU  $09           ; offset of DDRD from REGBAS
SPCR      EQU  $28           ; offset of SPCR from REGBAS
SPSR      EQU  $29           ; offset of SPSR from REGBAS
SPDR      EQU  $2A           ; offset of SPDR from REGBAS
SPCR_INI  EQU  $50           ; value to initialize the SPCR register
SPI_DIR   EQU  $3A           ; value to initialize the DDRD register

          ORG  $C000         ; starting address of the program
          LDX  #REGBAS
          LDAA #SPI_DIR
          STAA DDRD,X        ; set directions of port D pins
          LDAA #SPCR_INI
          STAA SPCR,X        ; initialize the SPI system
```

* The following two instructions latch data into HC589s in parallel

```
          BCLR PORTD,X $02   ; pull TxD (LC) pin to low
          BSET PORTD,X $02   ; pull TxD (LC) pin to high
```

```
                           BCLR  PORTD,X $20   ; pull SS pin to low to select parallel load mode and
              *                                ; load the contents of the data latch into the shift
              *                                ; register
                           BSET  PORTD,X $20   ; pull SS pin to high to select serial shift mode
                           LDAB  #8            ; loop count for transferring 8 bytes
                           LDY   #$0000        ; initialize buffer pointer Y
           LOOP            STAA  SPDR,X        ; trigger an SPI data transfer
           WAIT            LDAA  SPSR,X        ; wait until SPI transfer is completed
                           BPL   WAIT          ;       "
                           LDAA  SPDR,X        ; get one byte and
                           STAA  0,Y           ; store it in the buffer
                           INY                 ; move the buffer pointer
                           DECB                ; decrement the loop count
                           BNE   LOOP
                             .
                             .
                             .
                           END
```

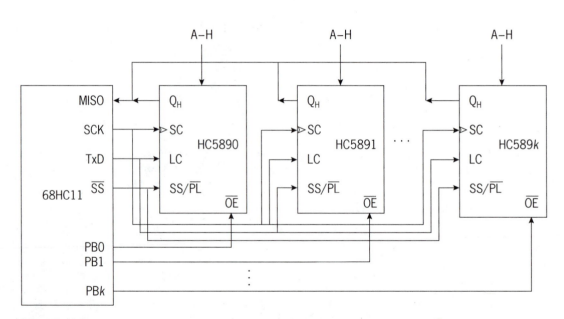

SC: shift clock
LC: latch clock
SS/PL: serial shift/parallel load mode select
OE: output enable

Figure 9.7   Parallel connection of multiple HC589s to an SPI

Another way to connect the HC589s is to tie all the Q_H pins together and then use other port pins to selectively enable one of the HC589s to shift out data to the 68HC11. The circuit connection for this method is shown in Figure 9.7. This circuit configuration allows the user to selectively input data from a specific HC589. The procedure for loading a byte from a specified HC589 into the SPDR register is as follows:

Step 1

Program the DDRD register to set the directions of the MISO, SCK, $\overline{SS}$, and TxD pins.

Step 2

Program the SPCR register to configure the SPI function.

Step 3

Set the TxD pin to low and then pull it to high to load external data into the data latch in parallel.

Step 4

Set the $\overline{SS}$ pin to low to select the parallel load mode, which will load the contents of the data latch into the shift register.

Step 5

Pull the $\overline{SS}$ pin to high to select the serial shift mode.

Step 6

Set the port B pin that controls the specified HC589 to low to enable the shift register to output serial data. The remaining port B pins are set to high.

Step 7

Write a byte into the SPDR register to trigger eight pulses from the SCK pin to shift in the serial data. The external data is now in the SPDR register and ready for use.

Step 8

Repeat steps 6 and 7 as many times as needed.

9.8 Interfacing the HC595 to the SPI

The full name of the HC595 is the 74HC595. Its block diagram and pin assignments are shown in Figure 9.8. The HC595 consists of an 8-bit shift reg-

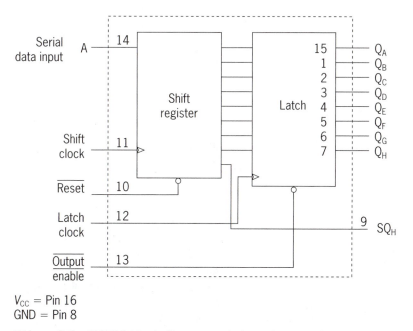

V_{CC} = Pin 16
GND = Pin 8

Figure 9.8 HC595 block diagram and pin assignment (Redrawn with permission of Motorola)

ister and an 8-bit D-type latch with three-state parallel outputs. The shift register accepts serial data and provides a serial output. The shift register also provides parallel data to the 8-bit latch. The shift register and the latch have independent clock inputs. This device also has an asynchronous reset input. The maximum shift clock rate for the HC595 at room temperature is 6 MHz.

9.8.1 HC595 Pins

The functions of the HC595 pins are as follows:

A: Serial data input. The data on this pin is shifted into the 8-bit serial shift register.

Shift clock. A low-to-high transition on this input causes the data at the serial input pin to be shifted into the 8-bit shift register.

$\overline{\text{Reset}}$. Active low. A low on this pin resets the shift register portion of this device only. The 8-bit latch is not affected.

Latch clock. A low-to-high transition on this pin latches the shift register data.

$\overline{\text{Output enable}}$. Active-low output enable. A low on this pin allows the data from the latches to be presented at the outputs. A high on this input forces the outputs $(Q_A–Q_H)$ into high-impedance states. The serial output is not affected by this pin.

$Q_A–Q_H$. noninverted, three-state, latch outputs.

SQ_H. Serial data output. This is the output of the eighth stage of the 8-bit shift register. This output does not have tristate capability.

9.8.2 Circuit Connections of the HC595 and the SPI

One application of the HC595 is to expand the number of parallel output ports of the 68HC11. Many parallel ports can be added to the 68HC11 by using the SPI subsystem and multiple HC595s. Figure 9.9 shows a method of connecting multiple HC595s to the SPI subsystem. The procedure for outputting multiple bytes using this connection is as follows:

Step 1
Program the DDRD register to configure each SPI pin and the TxD pin to appropriate directions. The $\overline{\text{SS}}$, MOSI, SCK, and TxD pins are configured for output, and the MISO and RxD pins for input. Write the value %00111010 into DDRD.

Step 2
Program the SPCR register to enable the SPI function, disable the SPI interrupt, use the rising edge of the SCK signal to shift data in and out, select master mode, select normal port D pins, and set the data rate to 1 Mbits/sec; to do this, write the value %01010000 into SPCR.

Step 3
Write a byte into the SPDR register to trigger eight output pulses from the SCK pin.

Step 4
Repeat step 3 as many times as needed.

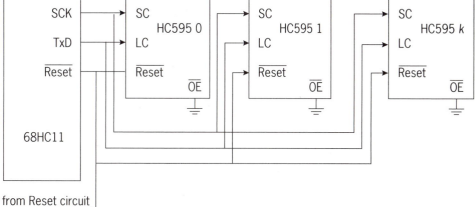

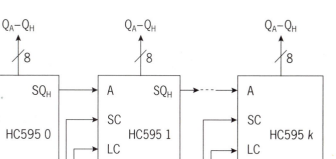

SC: shift clock
LC: latch clock

Figure 9.9   Serial connection of multiple HC595s to the SPI

Step 5
Set the TxD pin to low and then pull it to high to latch the byte in the
shift register of each HC595 into the output latch. After this step, the
Q_A–Q_H pins of each HC595 contain valid data.

Example 9.2

Write a program to output three bytes to the first three HC595s in
Figure 9.8.

Solution   The following program is based on the preceding algorithm:

```
REGBAS    EQU   $1000        ; base address of the I/O register block
PORTD     EQU   $08          ; offset of PORTD from REGBAS
DDRD      EQU   $09          ; offset of DDRD from REGBAS
SPCR      EQU   $28          ; offset of SPCR from REGBAS
SPDR      EQU   $2A          ; offset of SPDR from REGBAS
SPSR      EQU   $29          ; offset of SPSR from REGBAS
SPI_DIR   EQU   $3A          ; value to be written into DDRD to set up SPI pin
*                            ; directions
SPCR_IN   EQU   $50          ; value to initialize SPCR
K         EQU   $3

          ORG   $00
BUFFER    FCB   $11,$22,$33   ; values to be output from SPI
```

```
                ORG  $C000          ; starting address of the program
                LDX  #REGBAS
                LDAA #SPI_DIR
                STAA DDRD,X         ; set the directions of the SPI pins
                LDAA #SPCR_IN
                STAA SPCR,X         ; initialize the SPI system
                LDAB #K             ; initialize the loop count
                LDY  #BUFFER        ; load data buffer base address
ch_loop         LDAA 0,Y            ; get one character
                STAA SPDR,X         ; output the character
wait_ch         LDAA SPSR,X         ; wait until eight bits have been shifted out
                BPL  wait_ch
                INY                 ; move the buffer pointer
                DECB                ; decrement the loop count
                BNE  ch_loop

* The following two instructions create a rising edge from pin TxD to load the data
* into the output latch

                BCLR PORTD,X $02
                BSET PORTD,X $02

                END
```

Another method of expanding the number of parallel output ports is to connect multiple HC595s in parallel to the SPI subsystem, as shown in Figure 9.10.

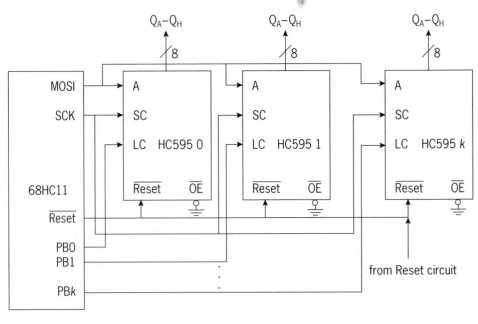

SC: shift clock
LC: latch clock

Figure 9.10 Parallel connection of multiple HC595s to the SPI

With this method, the user can selectively output data to any HC595 by setting the corresponding port B pin after the data has been shifted to that HC595. The procedure for outputting data to I/O devices connected to the HC595s is as follows:

Step 1
Program the DDRD to set up the SPI pin directions.
Step 2
Program the SPCR register to enable the SPI subsystem, select the data rate, select the rising edge of the SCK signal to shift data, select master mode, and disable the interrupt.
Step 3
Write a data byte into the SPDR register to trigger SPI data transfer.
Step 4
Set the PB_i pin to low and then pull it to high to load the byte from the shift register of HC595 i into the output latch.

9.9 Interfacing the SPI to the MC14499

The MC14499 is a seven-segment display decoder/driver with a serial interface port to provide communication with CMOS microprocessors and microcontrollers. This device features NPN output drivers, which allow interfacing to common-cathode seven-segment displays through external series resistors.

9.9.1 MC14499 Pins

The block diagram and the pin assignments of the MC14499 are shown in Figure 9.11. The 16-bit shift register allows data to be shifted in serially from the Data pin. The 4-bit decimal point D flip-flops hold the decimal points of the corresponding four digits. The 16-bit latch holds the values of four 4-bit digits to be displayed. The 4-bit digit to be displayed is selected by Muliplxer, a multiplexer with four 4-bit inputs. The output of the 4-bit multiplexer is decoded by a segment decoder that generates seven-segment outputs. The segment drivers provide the current required to drive seven-segment displays. The Decoder block generates digit-select signals (I, II, III, and IV) to select the seven-segment displays to be lighted. The output of the four-input AND gate is used as the select input to the 2:1 MUX. The input of the four-input AND gate comes from the output of the 4-bit latch. When the output of the 4-bit latch is 1111, the value Q_1 from the decimal point latch is selected. Otherwise, the value of the block designated as MUX is selected.

The MC14499 pins and their functions are:

Data. An input data pin.

Clock. The data on the data pin is sampled and shifted on the falling edge of this signal.

$\overline{\text{Enable}}$. External data can be entered when the $\overline{\text{Enable}}$ pin is low. The data are loaded from the shift register to the latches when $\overline{\text{Enable}}$ goes to high. While the shift register is being loaded, the previous data

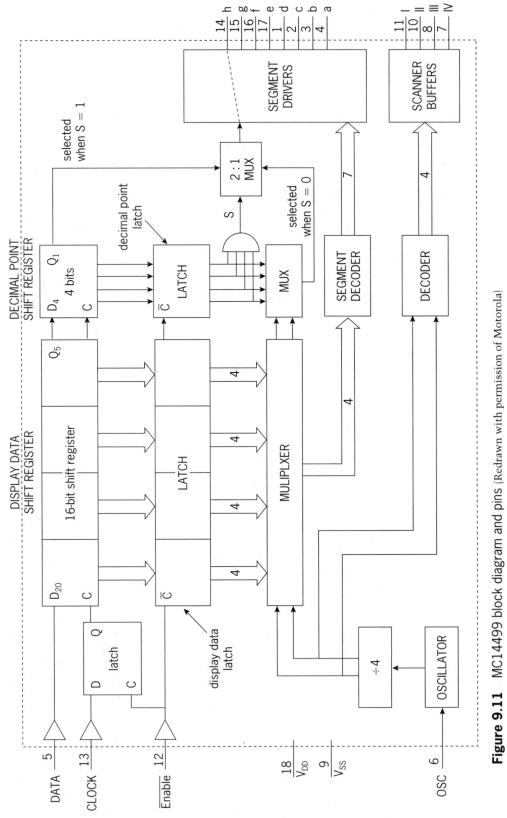

Figure 9.11 MC14499 block diagram and pins (Redrawn with permission of Motorola)

are stored in the latches. If a decimal point is used, the system requires 20 clock pulses to load data; otherwise, only 16 clock pulses are required.

a, b, c, d, e, f, g, and h. Seven-segment outputs for driving the seven-segment and decimal point LEDs.

I, II, III, and IV. Character-select pins. The MC14499 can drive up to four seven-segment displays, but only one seven-segment display is lighted at any time. The display to be lighted is selected by these pins—only one of these four pins goes to high at any time.

OSC. Oscillator input. The MC14499 has an on-chip oscillator that generates clock signals to multiplex four characters to the same seven-segment output and also generates character-select signals. The user should connect a capacitor to this pin to set the oscillator frequency; a capacitor of 0.015 μF is recommended.

9.9.2 MC14499 Operations

The MC14499 accepts a 20-bit input—16 bits for the four-digit display plus 4 bits for the decimal point (these latter four bits are optional). The input sequence is the decimal points followed by the four digits, as shown in Figure 9.12.

In order to enter data, the Enable input, $\overline{\text{Enable}}$, must be active low. The sample and shift are accomplished on the falling clock edge. Data are loaded from the shift register into the latches when $\overline{\text{Enable}}$ goes high. While the shift register is being loaded, the previous data are stored in latches. If the decimal point is used, the system requires 20 clock cycles to load data; otherwise, only 16 are required. The decimal point is shifted out first, followed by four digits. For each digit, the most significant bit is shifted out first and the least significant bit is shifted out last. The highest shift clock rate is 250 KHz.

Using a multiplexing technique allows the MC14499 to drive four seven-segment displays. The on-chip oscillator generates a clock signal that selects 4 bits from the 16-bit latch and also controls the decoder to activate the corresponding digit-select signal. The seven-segment signals are written to the corresponding segment inputs of four seven-segment displays, so one of the four seven-segment displays is lighted at any given time. A time-multiplexing technique turns each seven-segment display on and off many times in a second,

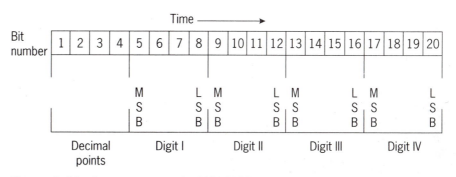

Figure 9.12   Input sequence for MC14499

0000	0	1000	8
0001	1	1001	9
0010	2	1010	A
0011	3	1011	¦
0100	4	1100	¦¦
0101	5	1101	¦_¦
0110	6	1110	-	dash
0111	7	1111		blank

Figure 9.13 Segment code

and hence all four displays appear to be lighted simultaneously. The segment code is shown in Figure 9.13.

9.9.3 Interfacing the MC14499 to the 68HC11

The SPI subsystem of the 68HC11 interfaces to the MC14499 nicely. One possible circuit connection is shown in Figure 9.14. The $\overline{SS}$ signal is used to en-

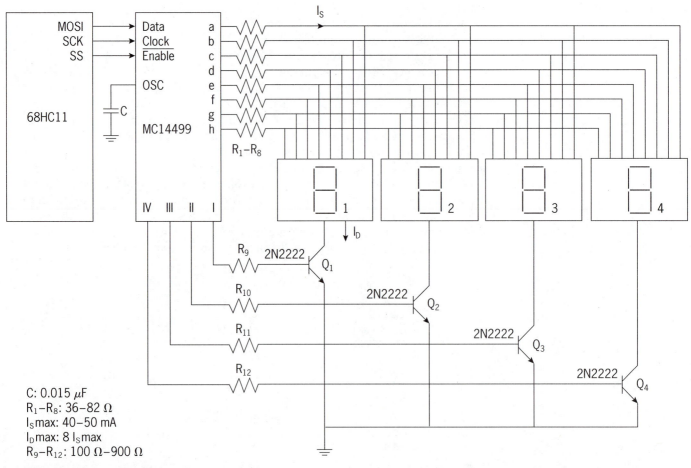

C: 0.015 μF
R_1–R_8: 36–82 Ω
$I_S$max: 40–50 mA
$I_D$max: 8 $I_S$max
R_9–R_{12}: 100 Ω–900 Ω

Figure 9.14 MC14499 to SPI connection

able and disable the data shifting in the MC14499. The $\overline{SS}$ pin must be set to low before the 68HC11 writes the digits into the MC14499 and then set to high so that the data can be loaded into the latch. The $\overline{SS}$ pin must therefore be configured as a general-purpose output port pin. The maximum segment output current is 40–50 mA (the corresponding output voltage is one volt below the power supply, i.e., 4 V for a 5-V power supply) at 25°C. The segment output voltage is 4.25 V for a 10-mA current output at room temperature. The average maximal current that flows into the transistors connected to the common cathodes is $8 \times I_s \max/4 = 80–100$ mA and can be sunk by a 2N2222 transistor.

Example 9.3

Write a program to display the BCD digits 1234 on four seven-segment displays arranged as shown in Figure 9.14. Use the following operation parameters:

1. 125-KHz shift clock rate (set SPR1 and SPR0 to 10).
2. The falling edge should shift in data (set CPOL and CPHA to 01).
3. Master mode.
4. Normal port D pins, i.e., no wired-or capability.
5. Enable the SPI subsystem.
6. Disable the SPI interrupt (use the polling method to check the SPIF flag bit).

Solution As in other SPI applications, the user must set up the SPI pin directions. The $\overline{SS}$, SCK, and MOSI pins must be configured for output, while the other port D pins should be configured for input. Write the value %00111000 into the DDRD register to set up this configuration. Also write the value %01010110 into the SPCR to set up the specified operating parameters. Since no decimal point is to be displayed, only 16 bits will be output. The Enable pin must be pulled down to low so that data can be shifted into the MC14499 and then pulled up to high to load the value into the 16-bit latch for display. The program is:

```
REGBAS   EQU  $1000          ; base address of I/O register block
PORTD    EQU  $08            ; offset of PORTD register from REGBAS
DDRD     EQU  $09            ; offset of DDRD register from REGBAS
SPCR     EQU  $28            ; offset of SPCR register from REGBAS
SPSR     EQU  $29            ; offset of SPSR register from REGBAS
SPDR     EQU  $2A            ; offset of SPDR register from REGBAS
SP_DIR   EQU  %00111000      ; SPI pin directions
SPCR_IN  EQU  %01010110      ; value to be written into SPCR

         ORG  $00
BYTES    FCB  $12,$34

         ORG  $C000          ; starting address of the program
         LDX  #REGBAS
         LDAA #SP_DIR
         STAA DDRD,X         ; set the SPI pin directions
         BCLR PORTD,X $20    ; pull the SS pin to low to allow digits shifted in
```

```
                LDAA  #SPCR_IN
                STAA  SPCR,X        ; initialize the SPI control register
                LDAA  BYTES         ; get digits $12
                STAA  SPDR,X        ; send out the first two digits
        WAIT1   LDAA  SPSR,X        ; wait until 8 bits have been transferred
                BPL   WAIT1         ;      "
                LDAA  BYTES+1       ; get digits $34
                STAA  SPDR,X        ; send out the last two digits
        WAIT2   LDAA  SPSR,X        ; wait until 8 bits have been transferred
                BPL   WAIT2         ;      "
                BSET  PORTD,X $20   ; pull the SS pin to high to load the value into
                END                 ; the latch of MC14499
```

9.9.4 Cascading MC14499s

To display more than four digits, multiple MC14499s must be used. The following procedure should be used to cascade multiple MC14499s (refer to Figure 9.11):

1. Connect the h pin of each MC14499 to the Data pin of the next MC14499.

2. Load 1111 to the decimal point latch of each MC14499 except the last one, which does not connect to another MC14499. This sets up the *cascade mode* of the MC14499s. In the cascade mode, the Q_1 output from the decimal point shift register will be selected by the 2:1 MUX and sent to the h pin.

The circuit for cascading multiple MC14499s is shown in Figure 9.15. In this figure, no decimal point is displayed. However, if the decimal point needs to be displayed, the h pin of the MC14499 that controls the decimal point

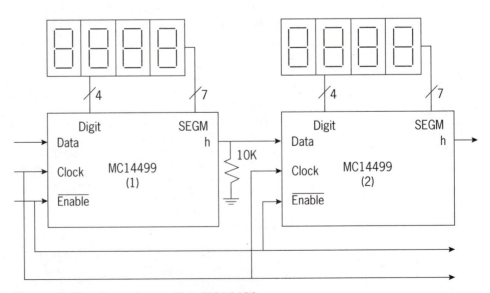

Figure 9.15   Cascading multiple MC14499s

should be connected to the h input of the seven-segment displays instead of being connected to a 10-KΩ resistor. When *n* MC14499s are driven by the SPI function, the update of the seven-segment displays can be divided into two stages:

Stage one: Set up the cascade mode. That is, load 1111 into the decimal point latch of the first *n* − 1 MC14499s using the following procedure:

Step 1
Set the Enable signal to 0.

Step 2
Shift $FFFFF into the first MC14499. This is normally done by sending three bytes of $FF into the MC14499 using the SPI function. Note that four 1s are wasted because the MC14499 requires only 20 bits.

Step 3
Pull the Enable signal to high to load the data in the 20-bit shift register into the display data latch and the decimal point latch. This establishes the cascade mode of the first MC14499, which will allow the signal from Q of the decimal point shift register to be shifted out from the h pin.

Step 4
Repeat steps 1 to 3 (*n* − 2) times. This will set up the cascade mode of the first *n* − 1 MC14499s.

Stage two: Send out new display data. Because each MC14499 requires 20 bits of display data, 4 extra bits may need to be sent when using the SPI transfer process. The procedure is as follows:

Step 1
Set Enable to 0.

Step 2
Send out as many bytes of display data as required using the SPI transfer. Be careful—to leave the cascade mode unaffected, don't pull Enable to high in this process.

Step 3
Set Enable to 1 to load the latch.

The following example illustrates the process of cascading three 14499s.

Example 9.4

Suppose you need to cascade three MC14499s to drive 12 seven-segment displays using circuit connections similar to those in Figure 9.15. The Data, Clock, and Enable pins of the leftmost MC14499 are connected to the MOSI, SCK, and SS pins of the 68HC11. Write a program to display the number 222233.334444.

Solution   The first step is to load the value %1111 into the decimal latches of the two leftmost MC14499s using the following procedure:

Step 1
Pull the Enable pin to low to set up the cascade mode.

Step 2
Use SPI transfer to shift the value $FFFFF into the leftmost MC14499. This can be done by sending three bytes of $FF to the leftmost MC14499 so that the decimal point shift register contains 1111.
Step 3
Pull the Enable pin to high to load %1111 into the 4-bit decimal point latch. This will allow data to be shifted into the middle MC14499.
Step 4
Repeat steps 1–3 to load the value %1111 into the 4-bit decimal point latch of the middle MC14499 and allow data to be shifted into the third MC14499.

In order to display the number 222233.334444, the following binary pattern should be shifted and loaded into the latches of the three MC14499s:

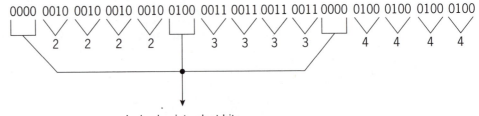

decimal point select bits

However, the number of bits in this bit pattern is not a multiple of 8. To simplify the SPI transfer, four extra 0s are added on the left. The following assembler directives can set up the display data properly:

    data_disp: fcb $00,$22,$22,$43,$33,$30,$44,$44

The program to display the number 222233.334444 is:

```
REGBAS    EQU    $1000           ; base address of I/O register block
PORTD     EQU    $08             ; offset of PORT register from REGBAS
DDRD      EQU    $09             ; offset of DDRD register from REGBAS
SPCR      EQU    $28             ; offset of SPCR register from REGBAS
SPSR      EQU    $29             ; offset of SPSR register from REGBAS
SPDR      EQU    $2A             ; offset of SPDR register from REGBAS
SP_DIR    EQU    %00111010       ; value to set up the directions of the SPI pins
SPCR_IN   EQU    %01010110       ; value to be written into SPCR

          ORG    $00
data_disp FCB    $00,$22,$22,$43,$33,$30,$44,$44    ; data to be displayed

          ORG    $C000           ; start address of the program
          LDX    #REGBAS
```

* The following eight instructions load the binary value %1111 into the decimal point
* latch of the leftmost MC14499s, which enables the data to be shifted into
* the second MC14499

```
          BCLR   PORTD,X $20     ; set the Enable pin to low to enable data shifting
          LDAB   #3              ; use accumulator B as the loop count
          LDAA   #$FF            ; output the first $FF
```

```
again0      STAA    SPDR,X          ;        "
            BRCLR   SPSR,X $80 *    ; wait until the SPIF flag is set
            DECB                    ; decrement the loop count
            BNE     again0          ; need to send next byte of $FF
            BSET    PORTD,X $20     ; load the value in the shift register into the latch
   *                                ; of the leftmost MC14499
```

* The following eight instructions load the binary value %1111 into the decimal point
* latch of the middle MC14499s, which enables the data to be shifted into the
* third MC14499

```
            BCLR    PORTD,X $20     ; set the Enable pin to low to enable data shifting
            LDAB    #3              ; prepare to send three bytes of $FF
            LDAA    #$FF
again1      STAA    SPDR,X          ; send out the byte $FF
            BRCLR   SPSR,X $80 *    ; wait until the SPIF flag is set
            DECB                    ; decrement the loop count
            BNE     again1          ; need to send next byte of $FF?
            BSET    PORTD,X $20     ; load the value in the shift register into the latches
   *                                ; of the first two MC14499s
```

* The following ten instructions send out $0022224333304444 to three MC14499s

```
            BCLR    PORTD,X $20     ; pull the Enable pin to low to allow data shifting
            LDAB    #8              ; set up the byte count to 8
            LDY     #data_disp      ; set Y to point to the data buffer loop
loop        LDAA    0,Y             ; get a byte to be output to the MC14499
            STAA    SPDR,X          ; send out a byte using the SPI function
            BRCLR   SPSR,X $80 *    ; wait until the byte is shifted out
            INY                     ; move the buffer pointer to next byte
            DECB                    ; decrement the loop count by 1
            BNE     loop
            BSET    PORTD,X $20     ; pull the Enable pin to high to load data into the
   *                                ; LATCH
            END
```

9.10 Interfacing LCDs to the SPI

Liquid crystal displays (LCDs) have found broad use in instrument and lap-top computer designs. LCDs can easily handle the display of numeric variables, units (for example, MHz and DM), and annunciators (for example, FREQ and AMPTD).

LCDs are often multiplexed to save connection pins. Several multiplex formats have become popular during the last few years. The Motorola LCD drivers that will be discussed here employ 1/4 multiplexing. In a 1/4-multiplexing LCD, each character is represented by a multiple of four segments. Like LED displays, an LCD digit consists of seven segments and an optional decimal point. Each LCD segment is turned on or off by controlling the backplane and frontplane voltages. Each digit is controlled by two frontplane and four backplane signals. The four backplane signals are multiplexed. The frontplane and

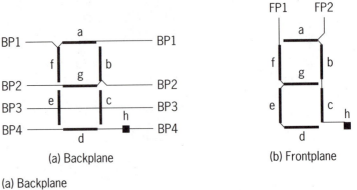

(a) Backplane
(b) Frontplane

Figure 9.16   LCD backplane and frontplane connections

backplane connections to a seven-segment LCD are shown in Figure 9.16. The segment truth table is given in Table 9.2. Note that there is no standard for backplane and frontplane connections on multiplexed LCD displays. Readers should refer to the manufacturer's literature for the actual connections.

LCD drivers are available from many manufacturers. An LCD driver accepts the display pattern that specifies the segments to be turned on (segments corresponding to a particular digit) and generates the appropriate frontplane and backplane signals to drive the LCD.

In the following section we will discuss the Motorola 1/4-multiplexing LCD drivers MC145000 and MC145001.

9.10.1 The MC145000 and MC145001 LCD Drivers

Motorola provides two LCD drivers for multiplexed LCDs. The MC145000 is a master unit, and the MC145001 is a slave unit.

OPERATIONS

The block diagrams of the MC145000 and MC145001 are shown in Figures 9.17 and 9.18. The driver is composed of two independent circuits: the

	FP1	FP2
BP1	f	a
BP2	g	b
BP3	e	c
BP4	d	h

*Because there is no standard for backplane and frontplane connections on multiplexed displays, this truth table can be used only for the connection shown in Figure 9.16.

Table 9.2 ■ LCD segment truth table*

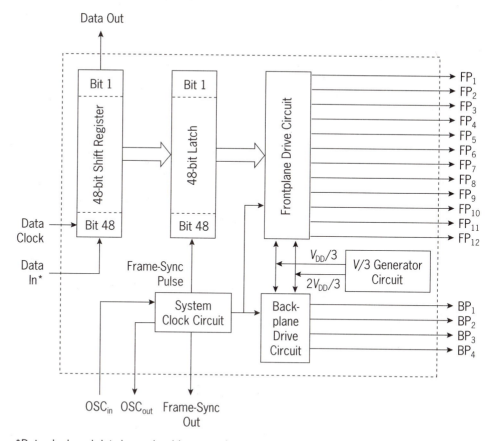

Data Out

Figure 9.17 Block diagram of the MC145000 (master) LCD driver (Redrawn with permission of Motorola)

data input circuit with its associated data clock and the LCD driver circuit with its associated system clock.

In the MC145000, 48 bits of data are serially clocked into the shift register on the falling edge of the external data clock. Data in the shift register are latched into the 48-bit latch at the beginning of each frame period. The frame period, t_{frame}, is the time during which all the LCD segments are set to the desired on or off states. The relationship between the frame-sync pulse and the system clock (OSC_{out}) is shown in Figure 9.19. The data in the shift register of the LCD driver is loaded into the latch that drives the frontplane outputs when the frame-sync pulse is high. If new data is shifted in during this period, the display will flicker. To avoid this problem, new data should be shifted in when the frame-sync pulse is low. This can be easily achieved in the 68HC11 by

1. using the frame-sync pulse as the $\overline{IRQ}$ input to the 68HC11
2. selecting edge sensitive for the $\overline{IRQ}$ input to the 68HC11
3. updating the display data in the interrupt-handling routine

The binary data present in the latch determine the waveform signal to be generated by the frontplane drive circuits, whereas the backplane waveforms

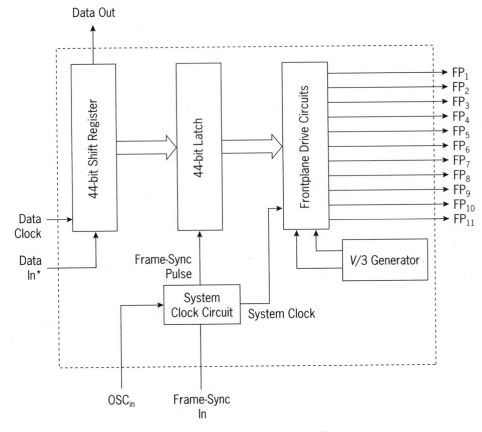

*Data-clock and data-in can be driven to voltages greater than V_{DD}.

Figure 9.18   Block diagram of the MC145001 (slave) LCD driver (Redrawn with permission of Motorola)

are invariant. The frontplane and backplane waveforms, FP_n and BP_n, are generated using the system clock (the oscillator output divided by 256) and voltages from the $V/3$ generator circuit (which divides V_{DD} into one-third increments). The backplane signal waveforms are shown in Figure 9.20, where you can see that the waveform of BP_{n+1} is shifted from the waveform of BP_n by one-fourth of a clock cycle. Each frontplane signal controls four segments. Because of the 1/4 multiplexing, either one segment or no segment is turned on at any time.

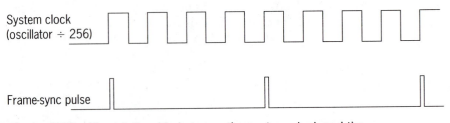

Figure 9.19   The relationship between the system clock and the frame-sync pulse

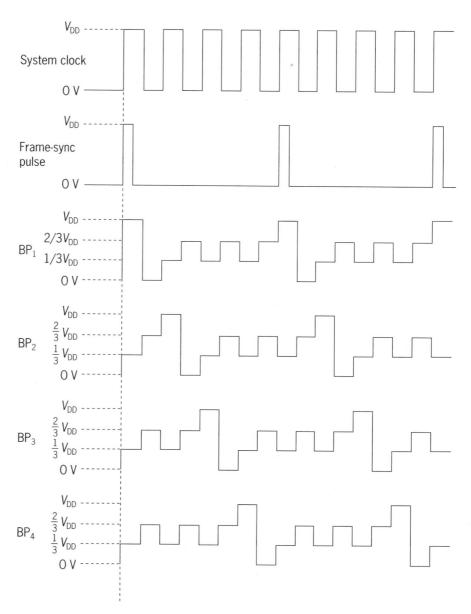

Figure 9.20 Backplane waveforms

The frontplane waveform of an on segment is shown in Figure 9.21a. The voltage across a segment is the voltage difference between the backplane and frontplane signals; an example of an on segment voltage waveform is shown in Figure 9.21b. The frontplane waveform of an off segment is shown in Figure 9.21c, and an example of an off segment voltage waveform is shown in Figure 9.21d. Twelve frontplane and four backplane drivers are available from the master unit. Since each digit needs two frontplane signals, a master LCD driver can drive six digits.

The slave unit consists of the same circuitry as the master unit, with two exceptions: it has no backplane driver circuitry, and its shift register and latch

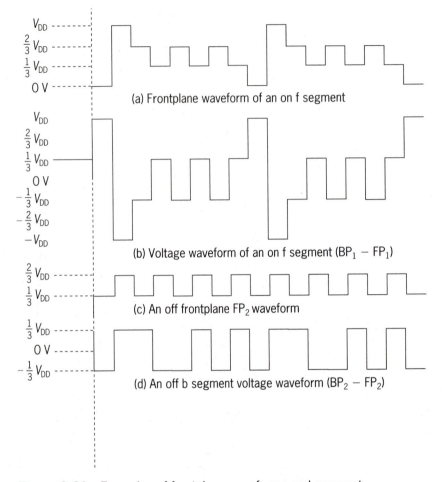

Figure 9.21 Examples of frontplane waveforms and segment voltage waveforms

hold 44 bits. Eleven frontplane and no backplane drivers are available from the slave unit. One slave unit can drive $5\frac{1}{2}$ digits.

PIN DESCRIPTIONS

The pin assignments of the MC145000 and MC145001 are shown in Figure 9.22. The functions and connections for most pins are self-explanatory. Only pins OSC_{in} and OSC_{out} need to be explained.

The OSC_{in} signal is the input to the system clock circuit. The oscillator frequency is either obtained from an external oscillator or generated in the master unit by connecting an external resistor between the OSC_{in} and OSC_{out} pins. Figure 9.23 shows the relationship between the resistor value and frequency.

The OSC_{out} signal is the system clock output generated by the master unit. This signal is connected to the OSC_{in} input of each slave unit to synchronize the updating of display data for the master and slave driver units.

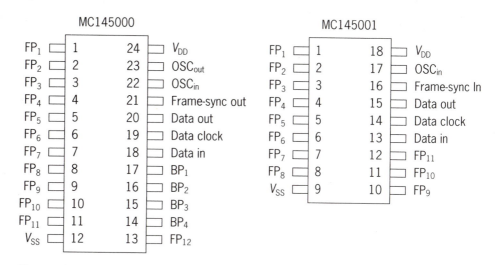

Figure 9.22 MC145000 and MC145001 pin assignments

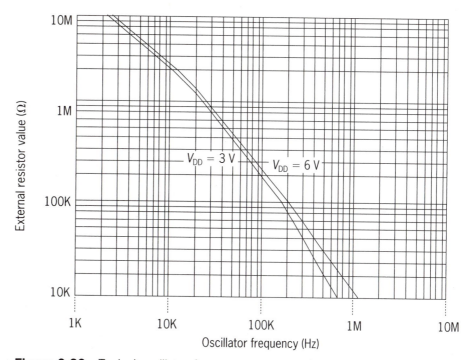

Figure 9.23 Typical oscillator frequency vs. external resistance value

DISPLAY PATTERNS

The data patterns to be displayed must be written into the LCD drivers in the right order. The LCD controller multiplexes four bits to drive the same frontplane output pin. The master unit also activates the corresponding backplane output to turn segments on or off. To turn on a segment, a 1 must be applied to it; to turn off a segment, a 0 must be applied. The bit locations (in the latch) that control the master unit LCD segments located at each frontplane-

Backplanes	Frontplanes											
	FP_1	FP_2	FP_3	FP_4	FP_5	FP_6	FP_7	FP_8	FP_9	FP_{10}	FP_{11}	FP_{12}
BP_1	4	8	12	16	20	24	28	32	36	40	44	48
BP_2	3	7	11	15	19	23	27	31	35	39	43	47
BP_3	2	6	10	14	18	22	26	30	34	38	42	46
BP_4	1	5	9	13	17	21	25	29	33	37	41	45

Table 9.3 ■ Bit locations for controlling the LCD segments

backplane intersection are shown in Table 9.3. To display a hexadecimal digit, eight bits must be shifted into the LCD driver. Since segment a is at location 8, it must be shifted into the LCD driver first. The order for shifting the segment patterns into the LCD driver is a, b, c, h, f, g, e, and d. For a slave LCD driver, column FP12 should be deleted. By combining Tables 9.2 and 9.3, the segment pattern for each BCD digit can be derived; these patterns are shown in Tables 9.4a and 9.4b.

LCD DRIVER SYSTEM CONFIGURATION

One master and several slave LCD driver units can be cascaded together to display more than six digits, as shown in Figure 9.24. The maximum number of slave units in a system is dictated by the maximum backplane drive capability of the device and by the system data update rate. Data is shifted serially first into the master unit and then into the following slave units on the falling

Digit	Segment								Hexadecimal representation
	a	b	c	h	f	g	e	d	
0	1	1	1	0	1	0	1	1	$EB
1	0	1	1	0	0	0	0	0	$60
2	1	1	0	0	0	1	1	1	$C7
3	1	1	1	0	0	1	0	1	$E5
4	0	1	1	0	1	1	0	0	$6C
5	1	0	1	0	1	1	0	1	$AD
6	0	0	1	0	1	1	1	1	$2F
7	1	1	1	0	0	0	0	0	$E0
8	1	1	1	0	1	1	1	1	$EF
9	1	1	1	0	1	1	0	0	$EC
A	1	1	1	0	1	1	1	0	$EE
B	0	0	0	0	1	0	1	0	$0A
C	0	1	1	0	1	0	1	0	$6A
D	0	1	1	0	1	0	1	1	$6B
E	0	0	0	0	0	1	0	0	$04
F	0	0	0	0	0	0	0	0	$00

Table 9.4a ■ BCD digits display patterns (without decimal point)

Digit	Segment								Hexadecimal representation
	a	b	c	h	f	g	e	d	
0	1	1	1	1	1	0	1	1	$FB
1	0	1	1	1	0	0	0	0	$70
2	1	1	0	1	0	1	1	1	$D7
3	1	1	1	1	0	1	0	1	$F5
4	0	1	1	1	1	1	0	0	$7C
5	1	0	1	1	1	1	0	1	$BD
6	0	0	1	1	1	1	1	1	$3F
7	1	1	1	1	0	0	0	0	$F0
8	1	1	1	1	1	1	1	1	$FF
9	1	1	1	1	1	1	0	0	$FC
A	1	1	1	1	1	1	1	0	$FE
B	0	0	0	1	1	0	1	0	$1A
C	0	1	1	1	1	0	1	0	$7A
D	0	1	1	1	1	0	1	1	$7B
E	0	0	0	1	0	1	0	0	$14
F	0	0	0	1	0	0	0	0	$10

Table 9.4b ■ BCD digits display patterns (with decimal point)

edge of the common data clock. The oscillator is common to the master unit and each of the slave units. At the beginning of each frame period, the master unit generates a frame-sync pulse to ensure that all slave unit frontplane drive circuits are synchronized to the master unit's backplane drive circuits. The master unit generates the backplane signals for all the LCD digits in the system.

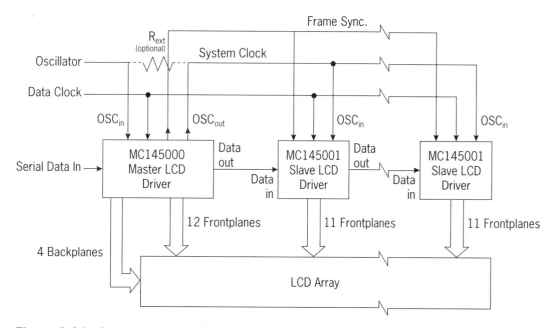

Figure 9.24 Frontplane and backplane connections to a multiplexed-by-four seven-segment (plus decimal point) LCD (Redrawn with permission of Motorola)

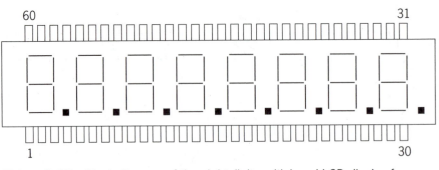

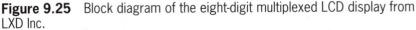

Figure 9.25   Block diagram of the eight-digit multiplexed LCD display from LXD Inc.

9.10.2 Interfacing the MC145000 and MC145001 to the 68HC11

Many LCDs can be driven by the MC145000 and MC145001 directly. One example is the 8-digit multiplexed LCD display (part #69) from LXD. Its block diagram and pin assignments are shown in Figures 9.25 and 9.26, respectively. This LCD is compatible with Motorola LCD drivers MC145000 and 145001, and it is a relatively large package in which many pins are not used. Figure 9.26 can be interpreted as follows:

- Backplane BP1 is assigned to pins 60 and 59.
- Backplane BP2 is assigned to pins 32 and 31.
- Backplane BP3 is assigned to pins 1 and 2.
- Backplane BP4 is assigned to pins 29 and 30.
- A_i, B_i, C_i, D_i, E_i, F_i, G_i, and DP stand for the segments and decimal point of the ith digit.
- A pin associated with segment letters is a frontplane pin.

All frontplanes and backplanes have two pins assigned to them, i.e., the user can use either pin (or both pins tied together) to drive a given front- or backplane pin.

The user can display from one to eight BCD digits with this LCD package. Since the MC145000 can drive only six digits, one slave LCD driver unit will be needed if more than six digits are to be displayed. The circuit connections between the 68HC11 SPI system, the MC145000, and an eight-digit multiplexed LCD display are shown in Figure 9.27. Frame-sync pulse is not used in this example. The $\overline{SS}$ pin should be configured as an output to avoid any problems. A 1-MΩ resistor connects the OSC_{in} and OSC_{out} pins to set the oscillator frequency to about 24 KHz (see Figure 9.23). The system clock is derived by dividing this frequency by 256 and is about 94 Hz. The LCD display is updated at this rate and should be fairly stable.

Example 9.5

Write a small program to display 123456 on the LCD in Figure 9.27.

Solution   The rated data clock frequency is higher than the highest operating frequency of the 68HC11, so we choose the highest SPI frequency—1 MHz. To

PAD	60	59	58	57	56	55	54	53	52	51	50	49	48	47	46	45	44	43	42	41	40	39	38	37	36	35	34	33	32	31
BP1	BP1	BP1	—	A8	A8	A7	A7	—	A6	A6	—	A5	A5	—	—	A4	A4	—	—	A3	A3	—	—	A2	A2	—	A1	A1	—	—
BP2	—	—	—	B8	B8	B7	B7	—	B6	B6	—	B5	B5	—	—	B4	B4	—	—	B3	B3	—	—	B2	B2	—	B1	B1	BP2	BP2
BP3	—	—	—	C8	C8	C7	C7	—	C6	C6	—	C5	C5	—	—	C4	C4	—	—	C3	C3	—	—	C2	C2	—	C1	C1	—	—
BP4	—	—	—	DP	DP	DP	DP	—	DP	DP	—	DP	DP	—	—	DP	DP	—	—	DP	DP	—	—	DP	DP	—	DP	DP	—	—
BP1	—	—	F8	F8	—	F7	F7	—	—	F6	F6	—	—	F5	F5	—	—	F4	F4	—	—	F3	F3	F2	F2	—	F1	F1	—	—
BP2	—	—	G8	G8	—	G7	G7	—	—	G6	G6	—	—	G5	G5	—	—	G4	G4	—	—	G3	G3	G2	G2	—	G1	G1	—	—
BP3	BP3	BP3	E8	E8	—	E7	E7	—	—	E6	E6	—	—	E5	E5	—	—	E4	E4	—	—	E3	E3	E2	E2	—	E1	E1	—	—
BP4	—	—	D8	D8	—	D7	D7	—	—	D6	D6	—	—	D5	D5	—	—	D4	D4	—	—	D3	D3	D2	D2	—	D1	D1	BP4	BP4
PAD	1	2	3	4	5	6	7	8	9	10	11	12	13	14	15	16	17	18	19	20	21	22	23	24	25	26	27	28	29	30

Figure 9.26 Pin assignments of the eight-digit multiplexed LCD from LXD Inc.

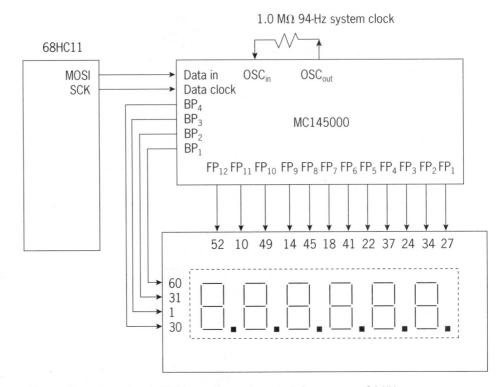

*The oscillator frequency is 256 times the system clock frequency or 24 KHz.

Figure 9.27   68HC11 SPI Drive 6-digit LCD Display

select this frequency, the SPR1 and SPR0 bits of the SPCR register must be set to 00. The SPI should be configured to operate in the master mode. Since data is clocked into the shift register on the falling edge of the data clock, the CPOL and CPHA bits in the SPCR register should be set to 01 (or 10). Port D pins should be set to normal CMOS output instead of open-drain. An interrupt is not needed for the SPI data transfer because the interrupt overhead is higher than polling for the SPI subsystem at 1 Mbits/sec. With these parameters, the value to be written into the SPCR register is $54.

The port D data direction register must also be set properly. Configure the $\overline{SS}$, SCK, and MOSI pins for output. The value $38 should be written into the DDRD register. The segment patterns are placed in a table in the order of the hexadecimal digits so that they can be accessed using the index addressing mode. The program is:

```
REGBAS    EQU    $1000        ; base address of I/O register block
DDRD      EQU    $09          ; offset of DDRD register from REGBAS
SPCR      EQU    $28          ; offset of SPCR register from REGBAS
SPSR      EQU    $29          ; offset of SPSR register from REGBAS
SPDR      EQU    $2A          ; offset of SPDR register from REGBAS
DD_INI    EQU    %00111000    ; value to set the DDRD register
SPCR_IN   EQU    %01010100    ; value to initialize the SPCR register

          ORG    $00
digits    FCB    1,2,3,4,5,6  ; digits to be displayed
```

```
lp_cnt      RMB    1                     ; loop count
            ORG    $C000
            LDS    #$FF                  ; set up the stack pointer (set to $3F for EVB)
            LDX    #REGBAS
            LDAA   #DD_INI
            STAA   DDRD,X                ; set up the SPI pin directions
            LDAA   #SPCR_INI
            STAA   SPCR,X                ; initialize the SPI system
```

* The following seven instructions send 48 zeros to clear the LCD. This action allows
* the user to see if the LCD circuit is connected properly

```
            LDAB   #6                    ; send out six zero bytes to initialize the LCD driver
loop        CLRA                         ;    "
            STAA   SPDR,X                ; send out eight zero bits to clear the LCD driver
*                                        ; registers and latches
            BRCLR  SPSR,X $80 *          ; wait SPI transfer is complete
            DECB                         ; decrement the loop count
            BNE    loop

            LDY    #digits               ; point Y to the start of the digits
            LDAB   #6
            STAB   lp_cnt                ; initialize the loop count
loop1       LDAB   0,Y                   ; get the digit to be displayed
            PSHY                         ; save the value of Y
            LDY    #lcdndp               ; place the base address of the LCD table in Y
            ABY                          ; index into the LCD digit pattern table
            LDAA   0,Y                   ; get the LCD pattern from the table
            STAA   SPDR,X                ; send out the digit pattern
            BRCLR  SPSR,X $80 *          ; wait until the digit has been shifted out
            PULY                         ; restore the digit pointer
            INY                          ; increment the digit pointer
            DEC    lp_cnt                ; decrement the loop count
            BNE    loop1
            .                            ; do something else
            .                            ;    "
            .                            ;    "
```

* The following lines contain the table of LCD display patterns for the hexdecimal digits

```
lcdndp      FCB    $EB,$60,$C7,$E5,$6C,$AD,$2F,$E0,$EF,$EC
            FCB    $EE,$0A,$6A,$6B,$04,$00
            END
```

When there are more than six digits to be displayed, the user will need one master and several slave drivers. The master LCD driver supplies frontplane signals for six digits, and the remaining frontplane signals are supplied by slave drivers. Backplane signals are supplied by the master driver.

9.11 The MC68HC68T1 Real-Time Clock with Serial Interface

The Motorola MC68HC68T1 is a dedicated clock chip that can keep track of seconds, minutes, hours, the day of the week, date, month, and year. It has a battery backup power supply input that can keep the device in operation during a power interruption, and its on-chip 32-byte RAM allows the user to save critical information when the power is interrupted. Its power control function can be used to sense the power transition, perform power-up or power-down, and reset the CPU when necessary. The 68HC68T1 is SPI-compatible and can be interfaced with the 68HC11 easily. The 68HC68T1 has a watchdog circuit similar to the COP system. When enabled, the watchdog circuit requires the microprocessor to toggle the slave select (SS) pin of the 68HC68T1 periodically without performing a serial transfer. If this condition is not sensed, the $\overline{\text{CPUR}}$ line resets the CPU (or interrupts it, depending on how this signal is connected to the 68HC11).

9.11.1 The MC68HC68T1 Signal Pins

The pin layout of the MC68HC68T1 is shown in Figure 9.28. The function of each signal is as follows:

CLK OUT: Clock output. This signal is the buffered clock output; it can provide one of seven selectable frequencies.

$\overline{\text{CPUR}}$: CPU reset (active low output). This pin provides an open-drain output and requires an external pull-up resistor. The active low output can be used to drive the reset of the microprocessor to allow orderly power-up/power-down. The $\overline{\text{CPUR}}$ signal is low for 15–40 ms when the watchdog circuit detects a CPU failure.

$\overline{\text{INT}}$: Interrupt (active low output). This active-low output is driven from a single N-channel transistor and must be tied to an external pull-up resistor. This interrupt is activated to a low level when any one of the following takes place:

1. Power sense operation is selected and a power failure occurs.
2. A previously set alarm time occurs. The alarm bit in the status register and the interrupt signal are delayed 30.5 ms when 32-KHz or 1-MHz operation is selected, 15.3 ms for 2-MHz operation, and 9.6 ms for 4-MHz operation.
3. A previously selected periodic interrupt signal activates.

The status register must be read to disable the interrupt output after the selected periodic interval occurs or when condition 1 or 2 activates the interrupt.

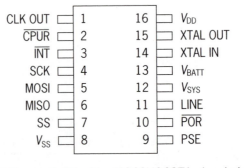

Figure 9.28   The MC68HC68T1 signal pins

SCK: Serial clock. This serial clock input is used to shift data into and out of the on-chip interface logic. SCK retains its previous state if the line driving it goes to a high-impedance state.

MOSI: Master out slave in. Data is shifted in from this pin, either on the rising or falling edge, with the most significant bit first. Data present at this pin is latched into the interface logic by SCK if the logic is enabled.

MISO: Master in slave out. The serial data present at this port is shifted out of the interface logic by SCK if the logic is enabled.

SS: Slave select. When high, the slave select input activates the interface logic; otherwise, the logic is in a reset state and the MISO pin is in a high-impedance state. The watchdog circuit is toggled at this pin.

V_{SS}: Negative power supply.

PSE: Power supply enable. This pin is used to control the system power supply and is enabled high under any one of the following conditions:

1. V_{sys} rises above the V_{batt} voltage after V_{sys} is reset low by a system failure.

2. An interrupt occurs (if the V_{sys} pin is powered up 0.7 V above V_{batt}).

3. A power-on reset occurs (if the V_{sys} pin is powered up 0.7 V above V_{batt}).

$\overline{POR}$: Power-on reset. This active low input generates an internal power-on reset using an external RC network.

Line: Line sense. The Line sense input can be used to drive one of two functions. The first function utilizes the input signal as the frequency source for the timekeeping counters. The second function enables the Line input to detect a power failure.

V_{sys}: System voltage. This input is connected to the system voltage.

V_{batt}: Battery voltage.

XTAL IN, XTAL OUT: Crystal input/output.

V_{dd}: Positive power supply.

9.11.2 On-Chip RAM and Registers

The MC68HC68T1 has 32 bytes of on-chip RAM and 13 registers. Its address map is shown in Figure 9.29. When accessing these RAM locations and/or registers, the microprocessor first sends in an 8-bit address using the SPI or port pins and then performs the actual access. Some registers are read/writable. Each read/writable register and each RAM byte has two addresses: one for read access and the other for write access. The separation of read addresses and write addresses makes the read and write accesses possible without an R/W signal.

All time counters and alarm registers (see Table 9.5) are in BCD format. The time-of-day can be displayed in 12-hour or 24-hour format. The selection of format is automatic: to use 24-hour format, the user enters the hour in the range 00–23; for 12-hour format, an hour in the range of 81–92 (for AM) or A1–B2 (for PM) is entered. The contents of the clock control, interrupt control, and status registers are described in the following subsections.

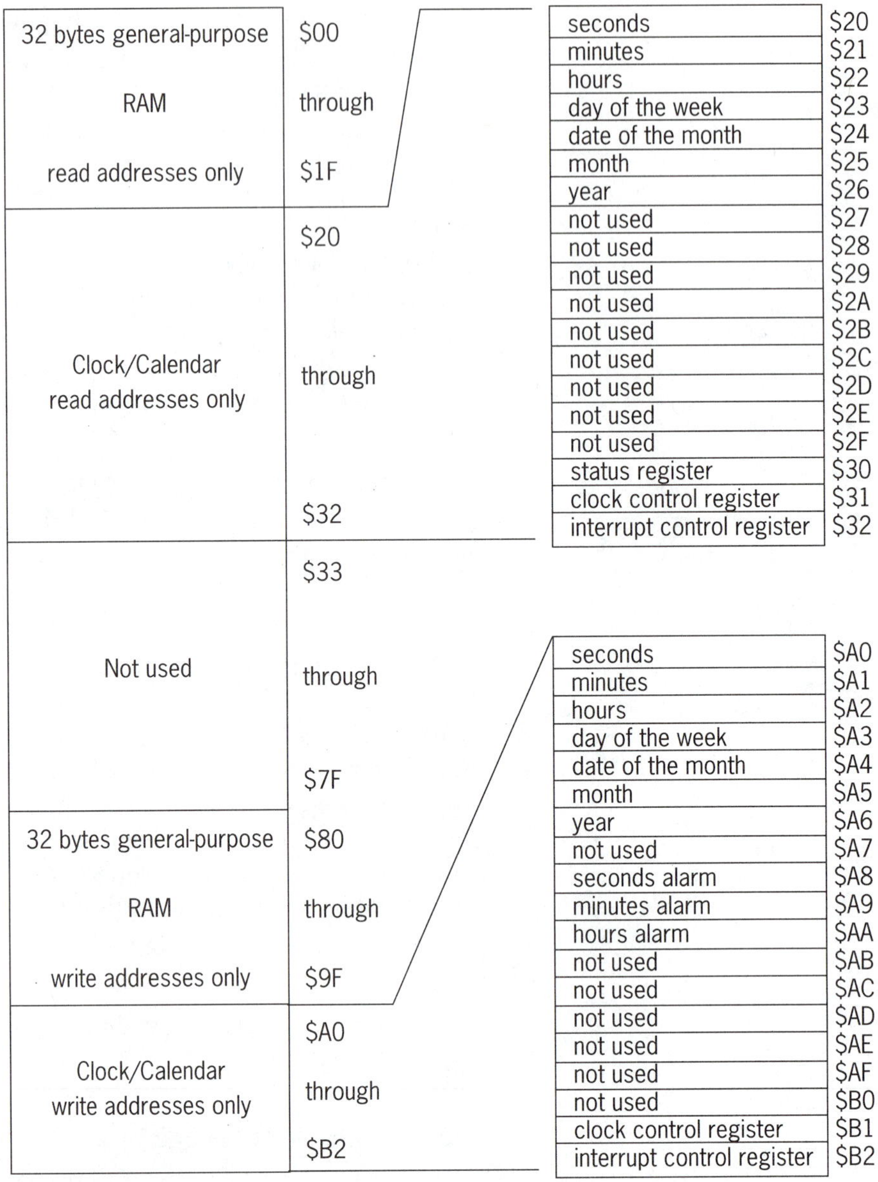

32 bytes general-purpose	$00
RAM	through
read addresses only	$1F

seconds	$20
minutes	$21
hours	$22
day of the week	$23
date of the month	$24
month	$25
year	$26
not used	$27
not used	$28
not used	$29
not used	$2A
not used	$2B
not used	$2C
not used	$2D
not used	$2E
not used	$2F
status register	$30
clock control register	$31
interrupt control register	$32

	$20
Clock/Calendar read addresses only	through
	$32

	$33
Not used	through
	$7F

seconds	$A0
minutes	$A1
hours	$A2
day of the week	$A3
date of the month	$A4
month	$A5
year	$A6
not used	$A7
seconds alarm	$A8
minutes alarm	$A9
hours alarm	$AA
not used	$AB
not used	$AC
not used	$AD
not used	$AE
not used	$AF
not used	$B0
clock control register	$B1
interrupt control register	$B2

32 bytes general-purpose	$80
RAM	through
write addresses only	$9F

	$A0
Clock/Calendar write addresses only	through
	$B2

Figure 9.29   MC68HC68T1 RAM and register address map (Redrawn with permission of Motorola)

CLOCK CONTROL REGISTER

The Clock Control Register has two addresses. The read address is at $31, and the write address is at $B1. The function of each bit is as follows:

D7	D6	D5	D4	D3	D2	D1	D0
Start	Line	Xtal Selct	Xtal Selct	50 Hz	CLK OUT	CLK OUT	CLK OUT
Stop	Xtal	1	0	60 Hz	2	1	0

| Address Location | | | Decimal | BCD data | BCD data |
Read	Write	Function	range	range	example
$20	$A0	seconds	0–59	00–59	21
$21	$A1	minutes	0–59	00–59	40
$22	$A2	hours (12-hour mode)	1–12	81–92 (AM) A1–B2 (PM)	90
		hours (24-hour mode)	0–23	00–23	10
$23	$A3	day of week (Sunday = 1)	1–7	01–07	03
$24	$A4	date of month	1–31	01–31	16
$25	$A5	month (January = 1)	1–12	01–12	06
$26	$A6	year	0–99	00–99	87
N/A	$A8	seconds alarm	0–59	00–59	21
N/A	$A9	minutes	0–59	00–59	40
N/A	$AA	hours alarm (12-hour mode)	1–12	01–12 (AM)	10
		hours alarm (24-hour mode)	0–23	00–23	10

Table 9.5 ■ Clock/calendar and alarm data modes
(Redrawn with permission of Motorola)

Start/$\overline{\text{Stop}}$. A 1 written into this bit enables the counter stages of the clock circuitry. A 0 holds all bits reset in the divider chain from 32 Hz to 1 Hz. The clock-out signal selected by bits D2, D1, and D0 is not affected by the stop function except for the 1- and 2-Hz outputs.

Line/$\overline{\text{Xtal}}$. When this bit is high, the clock operation uses a 50- or 60-cycle input at the Line input pin. When this bit is low, the XTAL IN pin is the source of the time update.

XTAL Select. Four possible crystal frequencies can be selected by the value in bits D5 and D4:

0 = 4.194304 MHz

1 = 2.097152 MHz

2 = 1.048576 MHz

3 = 32.768 KHz

At power-up, the device sets up for a 4-MHz oscillator. The MC68HC68T1 has an on-chip 150K resistor that is switched in series with the internal inverter when 32 KHz is selected through the clock control register. The recommended crystal connection is shown in Figure 9.30. Resistor R1 should be 22 MΩ for 32-KHz operation. Resistor R2 is used for 32-KHz operation only. Use a 100 K to 300 K range as specified by the crystal manufacturer.

50/$\overline{\text{60 Hz}}$. When this bit is 1, 50 Hz is the frequency at the Line input. Otherwise, 60 Hz is the Line frequency.

CLK OUT 2–CLK OUT 0. These three bits specify one of the seven frequencies to be used as the square wave output at the CLK OUT pin (see Table 9.6).

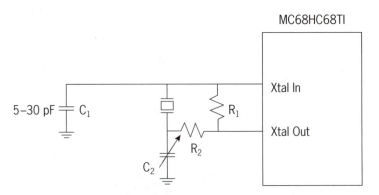

The values of C1 and C2 depend on the crystal frequency.

Figure 9.30   Recommended oscillator circuit

Periodic select	Clock output frequency
0	Xtal
1	Xtal/2
2	Xtal/4
3	Xtal/8
4	disable
5	1 Hz
6	2 Hz
7	50/60 Hz for Line operation
	64 Hz for Xtal operation

Table 9.6 ■ Clock output frequencies

INTERRUPT CONTROL REGISTER

This is a read/writable register. Its read address is $32, and its write address is $B2. The function of each bit is as follows:

Watchdog. When this bit is set to 1, the watchdog function is enabled.

Power down. A power-down operation is initiated when this bit is set to 1.

Power sense. When set to 1, this bit enables the Line input pin to sense a power failure. When power sense is selected, the input to the 50/60-Hz prescaler is disconnected and crystal operation is therefore required. An interrupt is generated when a power failure is sensed and the power sense and interrupt true bit in the status register are set. When power sense is activated, a logic low must be written to this location and followed by a high to reenable power sense.

D3–D0 value (hex)	Periodic interrupt output frequency	Frequency timebase	
		XTAL	Line
0	disable		
1	2048 Hz	x	
2	1024 Hz	x	
3	512 Hz	x	
4	256 Hz	x	
5	128 Hz	x	
6	64 Hz	x	
	50 or 60 Hz		x
7	32 Hz	x	
8	16 Hz	x	
9	8 Hz	x	
A	4 Hz	x	
B	2 Hz	x	x
C	1 Hz	x	x
D	1 cycle per minute	x	x
E	1 cycle per hour	x	x
F	1 cycle per day	x	x

Table 9.7 ■ Periodic interrupt output frequencies (at $\overline{\text{INT}}$ Pin)

Alarm. The output of the alarm comparator is enabled when this bit is set to 1. When a comparison of the seconds, minutes, and hours time counters and the alarm latches finds that they are equal, the interrupt output is activated.

Periodic select. The values in these four bits (D3, D2, D1, and D0) select the frequency of the periodic output. Table 9.7 lists all available frequencies.

STATUS REGISTER

This register is read-only and is located at $30. The meaning of each bit in this register is as follows:

D7	D6	D5	D4	D3	D2	D1	D0
0	Watch dog	0	First time up	Inter-rupt true	Power sense INT	Alarm INT	Clock INT

Watchdog. If this bit is high, the watchdog circuit has detected a CPU failure.

First time up. This bit is set to high by a power-on reset to indicate that the data in the RAM and Clock registers is not valid and should be initialized. After the status register is read, this bit is cleared if the POR pin is high. Otherwise, it remains at high.

Interrupt true. A high in this bit signifies that one of the three interrupts (power sense, alarm, or clock) is valid.

Power sense interrupt. This bit is set to high when the power sense circuit generates an interrupt. This bit is not cleared after a read of this register.

Alarm interrupt. When the contents of the seconds, minutes, and hours time counters and alarm latches are equal, this bit is set to high. The status register must be read before loading the interrupt control register for a valid alarm indication when the alarm activates.

Clock interrupt. A periodic interrupt sets this bit to high.

All bits except the first time up bit (which is set high) are reset low by a power-on reset. All bits except the power sense bits are reset after a read of the status register.

9.11.3 68HC68T1 Operations and Interfacing the 68HC68T1 to 68HC11

The operations of the MC68HC68T1 chip are explained in the following subsections.

REGISTERS AND RAM ADDRESSING

The 68HC68T1 has no address input. To select a particular RAM location or register, the microcontroller sends the corresponding address to the 68HC68T1 via the SPI interface. The address and data bytes are shifted, most significant bit first, into the serial data input (MOSI) and out of the serial data output (MISO). Any transfer of data requires the address of the byte, which specifies a write or read Clock/Calendar or RAM location; the address is followed by one or more bytes of data. Data transfer can occur one byte at a time or in multi-byte burst mode. Before sending the address, the SS pin should be set to high. The high level at the SS pin enables the 68HC68T1. The first byte sent to the 68HC68T1 is used to select either a location in RAM or a register. After that, the microcontroller can read or write single or multiple bytes from or into the 68HC68T1. Each read or write cycle causes the Clock/Calendar register or RAM address to automatically increment by 1. Incrementing continues after each byte transfer until the SS pin is set to low, which disables the serial interface logic. If the RAM is selected, the address increments to $1F or $9F and then wraps to $00 and continues. When the Clock/Calendar is selected, the address wraps to $20 after incrementing to $32 or $B2.

Interfacing the 68HC68T1 to the 68HC11 is fairly straightforward. Figure 9.31 shows the circuit connections for using the clock function of the 68HC68T1. A 32.760-KHz crystal oscillator supplies the clock signal to the 68HC68T1. An alarm transducer is added to inform the user when the preset alarm time is reached. The time of day and alarm time can be entered into the 68HC68T1 by using the keypad or switches. The alarm transducer is connected to the CLK OUT pin and will be driven by a square wave with a frequency of about 4 KHz. A battery consisting of three rechargeable NiCd cells is connected to the V_{batt} pin so that the time of day information can be maintained even without line power. The battery is charged through the resistor R_{charge} when line power is present.

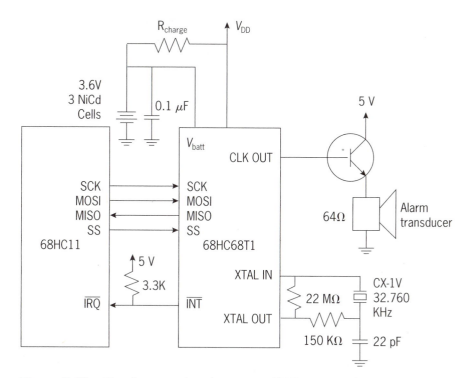

Figure 9.31 Circuit connections between a 68HC11 and a 68HC68T1 for the clock function

The following constant definitions will be used in all the programming examples related to the 68HC68T1:

```
REGBAS     EQU  $1000      ; I/O register block base address
SPDR       EQU  $2A        ; offset of the SPDR register from REGBAS
SPCR       EQU  $28        ; offset of the SPCR register from REGBAS
SPSR       EQU  $29        ; offset of the SPSR register from REGBAS
DDRD       EQU  $09        ; offset of the DDRD register from REGBAS
PORTD      EQU  $08        ; offset of the PORTD register from REGBAS
ram_addr   EQU  $80        ; starting address to write the RAM
second_r   EQU  $20        ; read address for seconds
minute_r   EQU  $21        ; read address for minutes
hours_r    EQU  $22        ; read address for hours
day_wk_r   EQU  $23        ; read address for day of week
date_m_r   EQU  $24        ; read address for date of month
month_r    EQU  $25        ; read address for month
year_r     EQU  $26        ; read address for year
stat_reg   EQU  $30        ; address of status register
clk_ctlr   EQU  $31        ; read address for clock control register
int_ctlr   EQU  $32        ; read address for interrupt control register
second_w   EQU  $A0        ; write address for seconds
minute_w   EQU  $A1        ; write address for minutes
hour_w     EQU  $A2        ; write address for hours
day_wk_w   EQU  $A3        ; write address for day of week
date_m_w   EQU  $A4        ; write address for date of month
```

```
month_w    EQU  $A5        ; write address for month
year_w     EQU  $A6        ; write address for year
s_alarm    EQU  $A8        ; write address for seconds alarm
m_alarm    EQU  $A9        ; write address for minutes alarm
h_alarm    EQU  $AA        ; write address for hours alarm
clk_ctlw   EQU  $AB        ; write address for clock control register
```

To enable data exchange between the 68HC11 and the 68HC68T1, the user needs to initialize the SPI system as follows:

1. Set the $\overline{SS}$, CLK, and MOSI pins as output and the MISO pin as input. The value %00111000 should be written into the DDRD register.

2. Disable the SPI interrupt—this requires bit 7 of the SPCR register to be set to 0.

3. Enable the SPI function. Bit 6 of the SPCR register must be set to 1.

4. Port D pins should be configured for normal CMOS pins—this requires bit 5 to be cleared to 0.

5. The clock (SCK) edge for shifting data in and out of the 68HC68T1 is unimportant. Let's choose the rising edge: set bits CPOL (bit 3) and CPHA (bit 2) of the SPCR register to 00.

6. Choose the highest usable data transfer rate because the 68HC68T1 can shift data in and out at a frequency higher than 2.1 MHz with a power supply of 5 V. To choose a 1-MHz frequency for the SCK clock, set bits SPR1 (bit 1) and SPR0 (bit 0) of the SPCR register to 00.

The following instruction sequence will set up these parameters:

```
LDX   #REGBAS     ; place the register block base address in X
LDAA  #$38        ; set up the SPI pin directions
STAA  DDRD,X      ;      "
LDAA  #$50        ; initialize the SPI system
STAA  SPCR,X      ;      "
```

The time-of-day, day of the week, date of the month, year, seconds alarm, minutes alarm, and hours alarm can be written into the 68HC68T1 using the multi-byte burst mode. To use burst mode transfer, the SS input to the 68HC68T1 must be set to high during the whole transfer process. At the end of transfer, the SS signal should be returned to low

Assume that the 24-hour mode is used in the 68HC68T1 and that clk_mgt contains the starting address of a memory block that stores seconds, minutes, hours, day of the week, date of the month, month, year, seconds alarm, minutes alarm, and hours alarm.

There are ten bytes in this block. Each piece of information occupies one byte and is stored in BCD format. The current time and the alarm can be entered interactively if the 68HC11 is connected to a PC using the SCI subsystem. The time-of-day and alarm time information can be defined by the following assembler directive:

```
clk_mgt   rmb 10    ; block of memory locations to store time and alarm info
```

This block of information can be written into the 68HC68T1 using the SPI function. Since the current times and alarm times are not stored in sequential addresses in the 68HC68T1, they must be sent to the 68HC68T1 in two bursts. The seconds write address and the current time are sent in the first burst, and the seconds alarm address and the alarm information are sent in the second burst, as shown in the following:

```
          ORG  $00
clk_mgt   rmb  10              ; block of memory to store time and alarm information

          ORG  $C000
          BSET PORTD,X $20     ; set the SS pin to high to enable SPI transfer
*                              ; to the 68HC68T1
          LDAA #second_w       ; load the seconds write address
          STAA SPDR,X          ; send the seconds write address to the 68HC68T1
LOOP      LDAA SPSR,X          ; wait until SPI transfer is complete
          BPL  LOOP            ;       "
          LDAB #7              ; set the byte count to 7
          LDY  #clk_mgt        ; load the base address of the current time into Y
LP        LDAA 0,Y             ; send one byte of time information to the
          STAA SPDR,X          ; 68HC68T1
here      LDAA SPSR,X          ; wait until SPI transfer is complete
          BPL  here            ;       "
          INY                  ; move the time pointer
          DECB
          BNE  LP
          BCLR PORTD,X $20     ; pull SS pin to low to disable transfer to 68HC68T1
```

* The following instruction sequence sends out the alarm information to the 68HC68T1

```
          BSET PORTD,X $20     ; set the SS pin to high to enable SPI transfer to the
*                              ; 68HC68T1
          LDAA #s_alarm        ; load the write address of the seconds alarm
          STAA SPDR,X          ; send the write address of the seconds alarm to the
*                              ; 68HC68T1
LOOP1     LDAA SPSR,X          ; wait until SPI transfer is complete
          BPL  LOOP1           ;       "
          LDAB #3              ; set the byte count to 3
          LDY  #clk_mgt+7      ; load the base address of the alarm into Y
LP1       LDAA 0,Y             ; send one byte of alarm information to the
          STAA SPDR,X          ; 68HC68T1
wait      LDAA SPSR,X          ; wait until SPI transfer is complete
          BPL  wait            ;       "
          INY                  ; move the time pointer
          DECB
          BNE  LP1
          BCLR PORTD,X $20     ; pull the SS pin to low to disable transfer to the
*                              ; 68HC68T1
```

To check the status register, use the following instruction sequence:

```
          BSET PORTD,X $20     ; enable SPI transfer to the 68HC68T1
          LDX  #REGBAS
          LDAA #stat_reg       ; send the status register address to the 68HC68T1
          STAA SPDR,X          ;       "
LP2       LDAA SPSR,X          ; wait until the address byte is shifted out
          BPL  LP2             ;       "
          STAA SPDR,X          ; start an SPI transfer to shift in the status register
```

```
LP3      LDAA  SPSR,X        ; wait until the contents of the status register is
         BPL   LP3           ; shifted in
         LDAA  SPDR,X        ; place the contents of the status register in A
         BCLR  PORTD,X $20   ;disable SPI transfer to the 68HC68T1
```

CLOCK/CALENDAR

The clock/calendar portion of this device consists of a long string of counters that are toggled by a 1-Hz input. The 1-Hz input can be derived from three different sources:

1. an external crystal oscillator applied between pins XTAL IN and XTAL OUT
2. an external frequency source applied to XTAL IN
3. a 50- or 60-Hz source connected to the LINE input

The example in Figure 9.31 uses an external 32.760-KHz crystal oscillator as the clock source. The time counters offer seconds, minutes, and hours data in 12- or 24-hour format. The values written into the time counters select the display format. An AM/PM indicator is available; once set, it toggles at 12:00 AM and 12:00 PM. The calendar counters contain day of the week, date of the month, month, and year information. The data in these counters are in BCD format. The hours counter uses BCD format for hours data plus bits for 12/24 hour and AM/PM modes. The seven time counters are readable at addresses $20 through $26 and can be written into at addresses $A0 through $A6.

ALARM

The alarm latches consist of seconds, minutes, and hours registers. When their outputs equal the values of the seconds, minutes, and hours time counters, an interrupt is generated. The interrupt output ($\overline{\text{INT}}$) goes low if the alarm bit in the status register is set and the interrupt output is activated. To preclude a false interrupt when loading the time counters, the alarm interrupt bit in the control register should be reset. This procedure is not needed when loading the alarm time.

The alarm function can be used to drive an alarm transducer to remind the user when the preset alarm time is reached. One method for doing this is as follows:

1. Use the CLK OUT pin to drive an alarm transducer (by means of a transistor).
2. Disable the CLK OUT (set to low) if the alarm time has not been reached.
3. Enable the alarm interrupt.
4. When the alarm time is reached, the alarm interrupt-handling routine enables a pulse output from the CLK OUT pin for some specific amount of time (say, 3 minutes).

In order to turn the alarm on for a specific amount of time, the user needs to enable the periodic interrupt of the 68HC68T1. This procedure is as follows:

1. Enable the 68HC68T1 to interrupt periodically (say, every second).

2. After setting up the alarm time, initialize an alarm counter. If the alarm is set for 3 minutes with a periodic interrupt every second, then the alarm count is 180 (3 × 60).

3. Stay in a wait loop while checking whether the alarm count has decremented to 0. If yes, turn off the alarm by setting the CLK OUT signal to low.

4. The interrupt-handling routine performs the following operations:
   a. Checks the interrupt source by reading the status register
   b. Returns if the interrupt is caused by the periodic interrupt and the alarm is not turned on
   c. Turns on the alarm if the interrupt is caused by the alarm function (i.e., the alarm time has been reached)
   d. Decrements the alarm count by 1 and returns from interrupt if the interrupt is caused by the periodic interrupt and the alarm has been turned on

Both the clock control and the interrupt-control registers must be set accordingly. The procedure for setting the interrupt control register is as follows:

1. Disable the watchdog interrupt (set bit 7 to 0)

2. Disable the power-down (set bit 6 to 0)

3. Disable the power sense (set bit 5 to 0)

4. Enable the alarm interrupt (set bit 4 to 1)

5. Select a periodic interrupt frequency of 1 Hz (set bits 3–0 to 1100)

Thus the value %00011100 should be written into the interrupt-control register. The procedure for setting the clock control register is as follows:

1. Set the start bit to start the clock (set bit 7 to 1)

2. Choose XTAL IN input to update the time (set bit 6 to 0)

3. Set the XTAL IN frequency to 32.760 KHz (set bits 5–4 to 11)

4. Bit 3 doesn't matter when XTAL IN is chosen to update the time (set it to 0 arbitrarily)

5. Disable CLK OUT output (set bits 2–0 to 100_2); to set the CLK OUT frequency to 32.760/8 ≈ 4 KHz, set these three bits to 011

Thus the value %10110100 should be written into the clock control register to turn off the CLK OUT output. The value %10110011 would be written into the clock control register to set an alarm device with a frequency of 4 KHz.

The following instruction sequence can set up the 68HC68T1 as desired:

```
        LDX   #REGBAS
        BSET  PORTD,X $20   ; set the SS pin to high to enable SPI transfer to
*                           ; the 68HC68T1

        LDAA  #clk_ctlw
        STAA  SPDR,X        ; send the write address of the clock control register
here    LDAA  SPSR,X        ; wait until the address has been shifted into
        BPL   here          ; the 68HC68T1
```

* The interrupt control and clock control registers are located at consecutive
* addresses and can be sent out in burst mode

```
            LDAA  #%10110100    ; send out the control byte to the clock
            STAA  SPDR,X        ; control register
first_bt    LDAA  SPSR,X        ; wait until the byte is shifted out
            BPL   first_bt      ;      "

            LDAA  #%00011100    ; send out the control byte to the interrupt control
            STAA  SPDR,X        ; register
sec_bt      LDAA  SPSR,X        ; wait until the byte is shifted out
            BPL   sec_bt        ;      "
            BCLR  PORTD,X $20   ; pull the SS pin to low to disable the SPI transfer
```

After initializing the 68HC68T1, the alarm count must be set up and updated to control the alarm interval. The 68HC68T1 will interrupt the 68HC11 every second once the alarm time is reached, so an interrupt-handling routine must be written to handle the interrupts. The following instruction sequence will set up and update the alarm count and clear the alarm flag when the alarm count is decremented to zero:

```
threemin    EQU   180          ; alarm count for three minutes
alarmcnt    rmb   1            ; memory location to store the alarm count
alarmflg    rmb   1            ; alarm flag indicating whether the alarm is on

set_alct    LDAA  #threemin
            STAA  alarmcnt      ; initialize the alarm count to 180
            CLR   alarmflg      ; initialize the alarm flag to 0
            .
            .
            .
```

With these settings, note that the periodic interrupt and the alarm interrupt may occur at the same time and that both are serviced by a single handling routine. The interrupt ($\overline{\text{IRQ}}$) handling routine must therefore check the status register of the 68HC68T1 in order to identify the cause of the interrupt. The alarm interrupt should be checked before the periodic interrupt. The interrupt status bit will be cleared when the status register is read. The interrupt-handling routine will perform the following functions:

1. Read the clock status register to identify the cause of the interrupt.

2. If the interrupt is caused by the alarm function, then turn on the alarm transducer and set the alarm flag to indicate that the alarm has been turned on.

3. If the interrupt is caused by the periodic interrupt (every second) and the alarm is not turned on, then exit.

4. If the interrupt is caused by the periodic interrupt and the alarm is set, then decrement the alarm count by 1. If the alarm count is decremented to 0 at this step, then clear the alarm flag, reinitialize the alarm count, and disable the CLK OUT signal. Return to the interrupted program.

Since the alarm device is driven by the CLK OUT pin, a proper frequency for the CLK OUT must be selected. To be an alarm, the signal frequency must be slightly higher but still audible. We will choose 4 KHz as the frequency of the CLK OUT signal. The value %011 should be set to the lowest three bits of the clock control register.

The clock interrupt handling routine is:

```
clk_hnd    LDX    #REGBAS
           BSET   PORTD,X $20      ; set the SS pin to high to enable SPI transfer
           LDAA   #stat_reg        ; send the status register address to the 68HC68T1
           STAA   SPDR,X           ;            "
           BRCLR  SPSR,X $80 *     ; wait until the status register address is shifted out
           STAA   SPDR,X           ; start SPI transfer to read the status register
           BRCLR  SPSR,X $80 *     ; wait until the contents of the status register is
    *                              ; shifted in
           BCLR   PORTD,X $20      ; disable SPI transfer to the 68HC68T1
           BSET   PORTD,X $20      ; enable SPI transfer to the 68HC68T1
    *                              ; new address is allowed to be transferred
    *                              ; to the 68HC68T1
           LDAA   SPDR,X           ; load the status register value into A
           ANDA   #$02             ; check the alarm interrupt bit
           BEQ    chkalarm         ; if interrupt is periodic, go and check the alarm flag
           LDAA   #clk_ctlw        ; send out the clock control register write address
           STAA   SPDR,X           ;            "
           BRCLR  SPSR,X $80 *     ; wait until the address is shifted out
           LDAA   #%10110011       ; send out a new control byte to the 68HC68T1 to
           STAA   SPDR,X           ; enable the CLK OUT signal to turn on the alarm
    *                              ; and set the frequency of CLK OUT to 4 KHz
           BRCLR  SPSR,X $80 *     ; wait until the control byte is shifted out
           LDAA   #1               ; set the alarm flag
           STAA   alarmflg         ;            "
           BRA    exit             ; prepare to return from interrupt
chkalarm   LDAA   alarmflg         ; check the alarm flag
           BEQ    exit             ; if the alarm is not turned on, then return
           DEC    alarmcnt         ; decrement the alarm count if the alarm has been
    *                              ; turned on
           BNE    exit             ; if the alarm count is not zero, return
```

* If the alarm count has been decremented to 0, then we need to disable the CLK OUT
* output, reinitialize the alarm count, and clear the alarm flag

```
           LDAA   #threemin
           STAA   alarmcnt         ; initialize the alarm count to 180
           CLR    alarmflg         ; initialize the alarm flag to 0
```

* To disable CLK OUT, write the value %10110100 into the clock control register

```
           LDAA   #clk_ctlw        ; send the write address of the clock control register
           STAA   SPDR,X           ; to the 68HC68T1
           BRCLR  SPSR,X $80 *     ; wait until the address is shifted out
           LDAA   #%10110100       ; send a value to the clock control register to disable
           STAA   SPDR,X           ; CLK OUT
           BRCLR  SPSR,X $80 *     ; wait until the new control byte is shifted out
exit       BCLR   PORTD,X $20      ; set the SS pin to low to disable SPI transfer
           RTI
```

The user can also add LCD drivers and display the time-of-day to make this clock function more useful.

POWER CONTROL

Power control consists of two operations: power sense and power-down/power-up. Two pins are involved in power sensing, the LINE input pin and the $\overline{\text{INT}}$ output pin. Two additional pins, PSE and V_{sys}, are utilized during a power-down/power-up operation.

Power Sensing   The power sense function is used mainly to detect power transitions. When power sensing is enabled, ac/dc transitions are sensed at the LINE input pin. Threshold detectors determine when transitions cease. After a delay of 2.68 to 4.64 ms plus the external input RC circuit time constant, an interrupt true bit in the status register is set to high. This bit can be sampled to see whether system power has turned back on.

The power sense circuitry operates by sensing the level of the voltage present at the LINE input pin. This voltage is centered around V_{DD}, and as long as the voltage is either plus or minus a certain threshold (approximately 0.7 V) from V_{DD}, a power sense failure is not indicated. With an ac signal, remaining in this V_{DD} window longer than a maximum of 4.64 ms activates the power sense circuit. The larger the amplitude of the signal, the less likely a power failure is to be detected. A 50- or 60-Hz, 10-V peak-to-peak sine-wave voltage is an acceptable signal to present at the LINE input pin to set up the power sense function.

Power-Down   Power-down is an operation initiated by the processor. The microcontroller sets the power-down bit in the interrupt-control register to initiate the power-down operation. During power-down, the PSE, CLK OUT, and the $\overline{\text{CPUR}}$ output are placed low. In addition, the serial interface (MOSI and MISO) is disabled.

In general, a power-down procedure consists of the following steps:

1. Set the power sense operation by setting bit 5 of the interrupt-control register to 1.

2. When an interrupt occurs, the CPU reads the status register to determine the interrupt source.

3. If a power failure is sensed, the CPU does the necessary housekeeping to prepare for shutdown.

4. The CPU reads the status register again after several milliseconds to determine the validity of the power failure.

5. The CPU sets power-down and disables all interrupts in the interrupt control register when power-down is verified. This causes the CPU reset ($\overline{\text{CPUR}}$ pin) and Clock Out pins to be held low and disconnects the serial interface.

6. When power returns and V_{sys} rises above $V_{\text{batt}} + 0.7$ V, power-up is initiated. The CPU reset is released and serial communication is established.

Power-Up   At the end of a power-on reset, the POR signal will go high. If the V_{sys} input also goes high, then the 68HC68T1 initiates the power-up operation by placing PSE, CLK OUT, and $\overline{\text{CPUR}}$ to high so that the CPU can start to boot.

	50 Hz		60 Hz		XTAL	
	Minimum	Maximum	Minimum	Maximum	Minimum	Maximum
Service time (ms)	—	10	—	8.3	—	7.8
Reset time (ms)	20	40	16.7	33.3	15.6	31.3

Table 9.8 ■ Watchdog service and reset times

THE WATCHDOG FUNCTION

When the Watchdog bit in the interrupt control register is set high, the SS pin must be toggled at regular intervals without a serial transfer. If the SS pin is not toggled, the 68HC68T1 supplies a CPU reset pulse at the $\overline{CPUR}$ pin and the watchdog bit in the status register is set to 1. The watchdog service time and the CPU reset duration are given in Table 9.8. This function is redundant for the 68HC11 because all the 68HC11 family members have a watchdog function. The 68HC11 COP timer may be better for some users because the time out period of the COP is programmable whereas it is fixed in the 68HC68T1.

9.12 Summary

Implementing the SPI function makes more of the limited number of signal pins in a microcontroller available for other functions. Two parties are involved in an SPI transfer: a master and one or more slaves. The SPI protocol uses a common clock signal supplied by the master to synchronize the shifting of data bits between the SPI master and slave units. Data bits are shifted in and out in the middle of a bit time. Both the master and the slave have master-out-slave-in (MOSI) and master-in-slave-out (MISO) pins for data transfer. An SPI slave usually has a slave select pin that, when asserted, enables data shifting. The SPI function is mainly used in low-to-medium speed I/O transfers. Examples in this chapter illustrated using the SPI subsystem to interface shift registers, seven-segment displays, LCD drivers, and the timer chip to the 68HC11.

9.13 Glossary

Frame sync   A signal that synchronizes the updating of the new display data between the LCD master (MC145000) and slave (MC145001) driver chips.

Mode fault   An SPI error that indicates that there may have been a multi-master conflict for system control. Mode fault is detected when the master SPI device has its $\overline{SS}$ pin pulled low.

Power-down   An operation that responds to the detection of low power supply; normally it performs some housekeeping function and then resets the system.

Power sense   A function available in several power management chips for detecting power source problems (for example, power dropping below the normal level).

Power-up   An operation that occurs in response to the detection of the power supply returning to normal. A power-up operation enables the microcomputer to boot.

Time-of-day chip   An integrated circuit that can keep track of the current year, month, day, hours, minutes, and seconds and may also provide the alarm function of the computer system. Like any other peripheral chip, the time-of-day chip must be initialized by the user before it can function properly.

Write collision   The SPI error that occurs when an attempt is made to write to the SPDR register while data transfer is taking place.

9.14 Exercises

E9.1. Is SPI a synchronous or an asynchronous serial data transfer protocol?

E9.2. Write an instruction sequence to set up the 68HC11 SPI with the following parameters: master mode, normal port D pins (no wired-or mode), disable the SPI interrupt, enable the SPI functions, use the falling edge of the SCK clock to shift data in and out, set the data rate to 1 Mbits/sec at the 4-MHz E clock, and use the SS pin as a general output pin.

E9.3. Write an instruction sequence to output the value $2B to a peripheral chip using the following pins and simulating the SPI operation:

D_{in}: serial data input

CLK: the falling edge of this clock signal is used to shift in the data

$\overline{EN}$: an active low signal to enable and disable the shifting of data

Assume the chip has a 6-bit register.

E9.4. Write a program to read a byte from HC589 #2 in Figure 9.7 into accumulator A.

E9.5. Write a program to output the byte $7F to HC595 #4 in Figure 9.10.

E9.6. Modify the program in Example 9.3 so that the circuit displays the value 12.34.

E9.7. Modify the program in Example 9.3 so that the circuit displays one digit at a time. Each digit should be displayed for one second and then turned off. The circuit displays a 1 in the first display for one second with the other three displays blank, then it displays a 2 in the second display for one second, etcetera. The operation is repeated forever. To blank a display, send in the binary value 1111_2.

E9.8. Use the circuit shown in Figure 9.13 but disable the SPI function. Write a program using the simulation method illustrated in section 9.5.3 to output the value 45.23 to four seven-segment displays. The data will be shifted out bit-by-bit using the MOSI (PD3) pin. The SCK should be toggled to simulate the SCK clock, and the $\overline{SS}$ pin should be used to load data into the latch of the MC14499.

E9.9. Use two 1/4-multiplexing LCDs, one MC145000, two MC145001s, and a 68HC68T1 to keep track of and display the current time. The fields of the current time are separated by decimal points. Enable the periodic and alarm interrupts. The periodic interrupt is to be generated every second, and the LCD display must be updated every second. Assume the current time and the alarm time are stored in a memory block that starts at $00 in an order that allows burst-mode transfer. Add an alarm transducer to the CLK OUT pin of the 68HC68T1. The alarm transducer should be turned on for five minutes when the preset alarm time is reached. Show the circuit connections and write a main program and an interrupt-handling routine to perform the specified functions.

E9.10. Use the SPI system to drive 2 MC14499s in cascade so that up to eight seven-segment displays can be driven. Draw the circuit connections and write a program to display the value 12345678 continuously.

9.15 Lab Exercises and Assignments

L9.1. Keyboard Scanning and Debouncing. Connect the 16-key membrane keypad as shown in Figure E6.1 and use the MC14499 to display the entered BCD digits. The entered digits should be shifted from left to right when more digits are entered. Use one MC14499 so that four BCD digits can be displayed at a time. Write a main program to initialize the SPI system and start to scan the keyboard. The display should be blank at the beginning. Whenever a key is pressed, the program should perform the debounce operation. If the debounce operation shows that the key was indeed pressed, then the main program should update the display. If the user enters more than four digits, then only the latest four digits should be displayed.

L9.2. Time-of-day display. Use the circuit developed in L9.1 to display the time of day. Use two MC14499s to control six seven-segment displays. The lab procedure is as follows:

1. Connect the circuit properly.
2. Write a program to
   a. Display six zeros as the prompt for the user to enter the time of day.
   b. Read two hours digits, then two minutes digits, then two seconds digits. Display the entered digits immediately and also check the validity of the entered time. If an invalid time is entered, the program should display six zeros again to prompt the user to reenter the time.
   c. Update the time of day once every second using the output-compare function.

10

68HC11

ANALOG

TO

DIGITAL

CONVERTER

10.1 Objectives

After completing this chapter, you should be able to:

- explain the A/D conversion process
- describe the precision, the various channels, and the operational modes of the 68HC11 A/D system
- interpret A/D conversion results
- describe the procedure for using the 68HC11 A/D system
- use the LM35 precision centigrade temperature sensors from National Semiconductor
- calculate the root-mean-square value of A/D conversion results
- use external A/D converters

10.2 Introduction

An analog signal quantity has a continuous set of values over a given range, in contrast to the discrete values of digital signals. Almost any measurable quantity, for example, current, voltage, temperature, speed, and time, is analog in nature. To be processed by a digital computer, analog signals must be represented in the digital form; thus an analog-to-digital (A/D) converter is required.

An A/D converter can deal only with electrical voltage. A nonelectric quantity must be converted into a voltage before A/D conversion can be performed. Conversion from a nonelectric quantity to a voltage requires the use of a *transducer*. For example, a temperature sensor is needed to convert a temperature into a voltage. To measure weight, a load cell is used to convert a weight into a voltage. However, the transducer output voltage may not be appropriate for processing by the A/D converter. A voltage level shifter and scaler are often needed to transform the voltage output into a range that can be handled by the A/D converter. Scaling and shifting of transducer outputs are beyond the scope of this text, but the overall process is illustrated in Figure 10.1.

The accuracy of an A/D converter is dictated by the number of bits used to represent the analog quantity. The greater the number of bits, the better the accuracy.

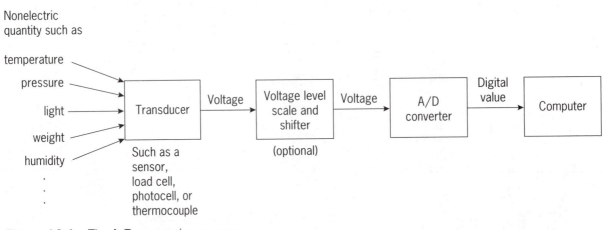

Figure 10.1   The A/D conversion process

10.3 An Overview of the 68HC11 A/D Converter

The 68HC11 has an eight-channel, 8-bit, multiplexed input, successive-approximation analog-to-digital converter with sample-and-hold circuitry to minimize conversion errors caused by rapidly changing input signals. Because it is equipped with a weighted array of capacitors, the 68HC11 uses an all-capacitive charge-redistribution technique to implement the successive-approximation A/D conversion method. This method will be explained in the next section. Implementing the successive-approximation A/D conversion method in the hardware circuit requires the use of a clock signal. The charge in the capacitors will leak away and will not be able to produce correct results if the clock frequency is below 750 KHz.

Two dedicated lines (V_{RL} and V_{RH}) provide the reference voltages for the A/D conversion. The V_{RH} line is used as the high reference voltage, and the V_{RL} line is used as the low reference voltage. V_{RH} cannot be higher than V_{DD} by more than 0.1 V, and V_{RL} cannot be lower than V_{SS}. The difference between V_{RH} and V_{RL} cannot be smaller than 2.5 V, but accuracy is tested and guaranteed only for $\Delta V_R = 5$ V $\pm$ 10% ($\Delta V_R = V_{RH} - V_{RL}$).

The A/D converter is ratiometric. An input voltage equal to V_{RL} converts to \$00, and an input voltage equal to V_{RH} converts to \$FF (full scale), with no overflow indication.

The A/D converter must be enabled before it can be used. It is enabled (or disabled) by setting (or clearing) the A/D power-up (ADPU) control bit in the OPTION register. Setting the ADPU bit starts the charge pump circuit in the A/D converter system. It takes about 100 μs for the charge pump circuit to stabilize.

10.4 The Successive Approximation A/D Conversion Method

The successive approximation method is a commonly used technique in data acquisition systems. A block diagram of a successive approximation A/D converter is shown in Figure 10.2.

A successive approximation A/D converter approximates the analog signal to n-bit code in n steps. It first initializes the SAR register to zero and then performs a series of guesses, starting with the most significant bit and proceeding toward the least significant bit. The algorithm of the successive approximation

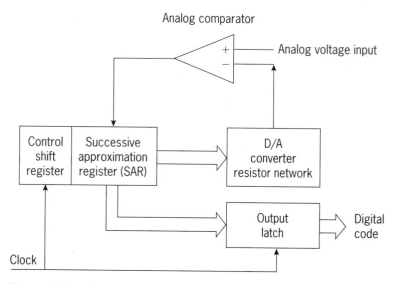

Figure 10.2 Block diagram of a successive approximation A/D converter

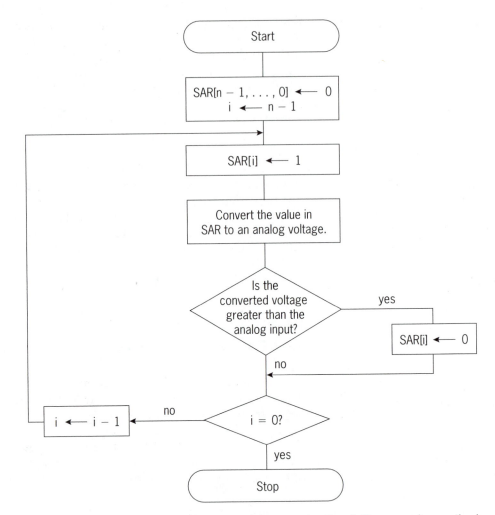

Figure 10.3   Flowchart for the successive approximation A/D conversion method

method is illustration in Figure 10.3. It assumes that the SAR register has n bits. For every bit of the SAR register, the algorithm

1. guesses the bit to be a 1
2. converts the value of the SAR register to an analog voltage
3. compares the D/A output with the analog input
4. clears the bit to 0 if the D/A output is larger (which indicates that the guess is wrong)

10.5 A/D System Options

The A/D converter has a multiplexer to select one of 16 analog signals. Eight of these channels correspond to port E input lines to the 68HC11; 4 of the channels are for internal reference points or test functions, and 4 channels are reserved for future use. Therefore, only 8 channels are available for normal use. The channels are selected by setting the channel select bits in the A/D control register (ADCTL).

The A/D system starts the conversion process one clock cycle after a control byte is written into the ADCTL register. The A/D system can be configured to perform conversion sequences four times on a selected channel in the single-channel mode or one time on each of four channels (either channels 1 to 4 or channels 5 to 8) in the multiple-channel mode.

There are two variations of single-channel operation. In the first variation (*nonscan mode*), four consecutive samples are taken from the single selected channel and are converted in turn, with the first result being stored in A/D result register 1 (ADR1), the second result being stored in register ADR2, etcetera. After the fourth conversion is completed, all conversion activity is halted until a new conversion command is written into the ADCTL register. In the second variation (*scan mode*), conversions are performed continually on the selected channel, with the fifth conversion being stored in register ADR1 (overwriting the first conversion result), the sixth conversion result overwriting ADR2, and so on.

There are also two variations of multiple-channel operation. In the first variation (nonscan mode), the selected four channels are converted, one at a time, with the first result being stored in register ADR1, the second result being stored in register ADR2, and so on. After the fourth conversion is complete, all conversion activity is halted until a new conversion command is written into the ADCTL register. In the second variation (scan mode), conversions are performed continually on the selected group of four channels, with the fifth conversion being stored in register ADR1 (replacing the earlier conversion result for the first channel in the group), the sixth conversion overwriting ADR2, and so on.

10.6 The Clock Frequency Issue

An A/D converter requires a clock signal to operate. The A/D system of the 68HC11 can select either the E clock signal or the internal RC circuit output as the clock source. The RC clock signal runs at about 1.5 MHz. The clock source is selected by setting or clearing the CSEL bit (bit 6) in the OPTION register. When this bit is 0, the E clock signal is selected. Otherwise, the RC clock source is selected. The RC clock requires 10 ms to start and settle. The RC clock source should be selected only if the E clock frequency is lower than 750 KHz; otherwise, the capacitor array charge leakage will become too large to obtain correct A/D conversion results.

10.7 A/D System Registers

Six registers are involved in the A/D conversion process: the A/D control/status register (ADCTL), the OPTION register, and four A/D result registers (ADR1, ADR2, ADR3, and ADR4).

10.7.1 The A/D Control/Status Register

The ADCTL register is located at $1030. All bits in this register can be read or written, except for bit 7, which is a read-only status indicator, and bit 6, which always reads as a zero. The contents of the ADCTL register are as follows:

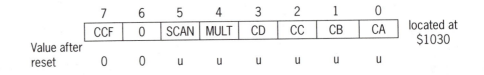

CCF: Conversion complete flag. This read-only status indicator is set to 1 when all four A/D result registers contain valid conversion results. Each time the ADCTL register is written, this bit is automatically cleared to zero and a conversion sequence is started.

SCAN: Continuous scan control. When this control bit is 0, the four requested conversions are performed once to fill the four result registers. When this control bit is 1, conversions continue in a round-robin fashion with the result registers being updated as data becomes available.

MULT: Multiple-channel/single-channel control. When this bit is 0, the A/D system performs four consecutive conversions on the single channel specified by the four channel select bits (CD through CA, bits 3–0 of the ADCTL register). When this bit is 1, the A/D system is configured to perform a conversion on each of four channels, with each result register corresponding to one channel.

CD: Channel select D.

CC: Channel select C.

CB: Channel select B.

CA: Channel select A.

These four bits are used to select one of the 16 A/D channels. When a multiple-channel mode is selected, the two least significant bits (CB and CA) have no meaning, and the CD and CC bits specify which group of four channels is to be converted. The channels selected by the four channel select bits are shown in Table 10.1.

10.7.2 A/D Result Registers 1, 2, 3, and 4 (ADR1, ADR2, ADR3, and ADR4)

The A/D result registers are read-only registers that hold an 8-bit conversion result. Writes to these registers have no effect. Data in the A/D result registers are valid when the CCF flag bit in the ADCTL register is 1, indicating that a conversion sequence is complete.

The A/D conversion process is started one clock cycle after the ADCTL register is written into. The conversion of one sample takes 32 clock cycles. The CCF bit in the ADCTL register will be set to 1 when all four A/D result

CD	CC	CB	CA	Channel signal	Result in ADRx if MULT = 1
0	0	0	0	AN0	ADR1
0	0	0	1	AN1	ADR2
0	0	1	0	AN2	ADR3
0	0	1	1	AN3	ADR4
0	1	0	0	AN4*	ADR1
0	1	0	1	AN5*	ADR2
0	1	1	0	AN6*	ADR3
0	1	1	1	AN7*	ADR4
1	0	0	0	reserved	ADR1
1	0	0	1	reserved	ADR2
1	0	1	0	reserved	ADR3
1	0	1	1	reserved	ADR4
1	1	0	0	V_{RH} pin**	ADR1
1	1	0	1	V_{RL} pin**	ADR2
1	1	1	0	$(V_{RH})/2$**	ADR3
1	1	1	1	reserved**	ADR4

* Not available in 48-pin package.
** These channels are intended for factory testing.

Table 10.1 ■ Analog-to-digital channel assignments

registers contain valid conversion results, thus it will be set 129 clock cycles after the ADCTL register is written into.

The addresses of A/D result registers are as follows:

ADR1 is at $1031.

ADR2 is at $1032.

ADR3 is at $1033.

ADR4 is at $1034.

10.7.3 The OPTION Register

The OPTION register is located at $1039. Only bits 7 and 6 are related to the operation of the A/D converter. Bit 7 (ADPU) enables the A/D converter when it is set to 1. After setting this bit, the user must wait for at least 100 μs before using the A/D converter system to allow the charge pump and comparator circuit to stabilize. Bit 6 (CSEL) selects the clock source for the A/D converter. When CSEL = 1, the internal RC clock is selected. Otherwise, the E clock is selected. The contents of the OPTION register are as follows:

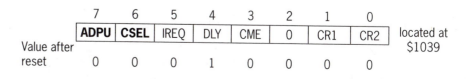

	7	6	5	4	3	2	1	0	
	ADPU	CSEL	IREQ	DLY	CME	0	CR1	CR2	located at $1039
Value after reset	0	0	0	1	0	0	0	0	

10.8 The Interpretation of Conversion Results

Since the 68HC11 A/D converter is ratiometric, the voltage corresponding to an A/D conversion can be computed using the following method:

1. Let x be the A/D conversion result.
2. Let $range = (V_{RH} - V_{RL})$.
3. A conversion result of 255 corresponds to the voltage V_{RH}. A conversion result of 0 corresponds to the voltage V_{RL}. Thus the voltage (V_x) corresponding to x can be computed using the following formula:

$$V_x = V_{RL} + (range \times x)/255 \qquad \qquad \textbf{10.1}$$

■

Example 10.1

Assume that $V_{RH} = 4$ V and $V_{RL} = 1$ V. Find out the corresponding voltage values for A/D conversion results of 10, 40, 60, 127, and 210.

Solution

$$range = V_{RH} - V_{RL} = 4\,V - 1\,V = 3\,V$$

The voltages corresponding to the results of 10, 40, 60, 127, and 210 are:

$1 + (10 \times 3)/255 = 1.12\,V$
$1 + (40 \times 3)/255 = 1.47\,V$
$1 + (60 \times 3)/255 = 1.71\,V$
$1 + (127 \times 3)/255 = 2.49\,V$
$1 + (210 \times 3)/255 = 3.47\,V$

10.9 The Procedure for Using the A/D System

The procedure for using the 68HC11 A/D system is as follows:

Step 1
Connect the hardware properly. The high reference voltage V_{RH} should not be higher than V_{DD} (5 volts). The low reference voltage V_{RL} should not be lower than V_{SS} (0 volts). The analog signal to be converted should fall between V_{RL} and V_{RH}. If the value of the analog signal is not between V_{RL} and V_{RH}, then the user must scale and shift it to the desired range. Connect the analog signal(s) to the appropriate A/D input pin(s).

Step 2
Set the ADPU bit in the OPTION register to enable the A/D system. Set the CSEL bit to 0 or 1, depending on the E clock frequency. A 0 selects the E clock as the clock signal for the A/D conversion process.

Step 3

Wait for the charge pump to stabilize. The user can set up a program loop to create a delay of at least 100 μs. If the internal RC oscillator output is selected as the clock source, then a 10-ms delay is needed for the RC clock output to settle.

Step 4

Select the appropriate channel(s) and operation modes by programming the ADCTL register. The A/D conversion will be started one clock cycle after the ADCTL register is written into.

Step 5

Wait until the CCF bit in the ADCTL register is set, then collect the A/D conversion results and store them in memory. When the CCF bit is set, the value of the ADCTL register becomes negative and allows the user to decide whether the A/D conversion is completed.

Example 10.2

Write an instruction sequence to set up the following A/D conversion environment:

Nonscan mode.

Single-channel mode.

Select channel AN0.

Choose the E clock as the clock source for the A/D converter.

Enable the A/D converter.

Solution To set up the first three A/D parameters, we need to write a byte containing the following bits into the ADCTL register:

■ to select nonscan mode, set bit 5 of the ADCTL register to 0

■ to select single-channel mode, set bit 4 of the ADCTL register to 0.

■ to select channel AN0, set the bits 3–0 of the ADCTL register to 0000.

Since the msb of the ADCTL register is a status bit, we can simply write a 0 into it. Therefore, we must write the byte %00000000 into the ADCTL register.

To choose the E clock as the clock source for the A/D converter, we must set the CSEL bit (bit 6) of the OPTION register to 0. To enable the A/D converter, we need to set the bit 7 of the ADCTL register to 1 and wait for 100 μs. The following instruction sequence will set up the specified A/D conversion environment:

```
REGBAS    EQU  $1000        ; base address of I/O register block
ADCTL     EQU  $30          ; offset of ADCTL from REGBAS
OPTION    EQU  $39          ; offset of OPTION from REGBAS

          LDX  #REGBAS
          BCLR OPTION,X $40  ; select the E clock as the clock source for the A/D
*                            ; converter
```

```
                    BSET OPTION,X $80      ; enable the A/D converter
                    LDY  #30              ; delay 105 μs for the charge pump to stabilize
        delay       DEY                  ;      "
                    BNE  delay           ;      "
                    LDAB #%00000000      ; select nonscan, single-channel mode and channel
        *                                ; AN0

                    STAB ADCTL,X         ; start the A/D conversion
                    END
```

Example 10.3

Write an instruction sequence to set up the following A/D conversion environment:

    Nonscan mode.

    Multiple-channel mode.

    Select channels AN4–AN7.

    Choose the E clock as the clock source for the A/D converter.

    Enable the A/D converter.

Solution To set up the first three A/D parameters, we need to write a byte containing the following bits into the ADCTL register:

- to select non-scan mode, set bit 5 of the ADCTL register to 0.

- to select multiple-channel mode, set bit 4 of the ADCTL register to 1.

- to select channels AN4–AN7, set bits 3–0 of the ADCTL register to 0100 (bits 1 and 0 can be any value).

Since the msb of the ADCTL register is a status bit, we can simply write a 0 into it. Therefore, we must write the byte %00010100 into the ADCTL register.

To choose the E clock as the clock source for the A/D converter, we must set the CSEL bit (bit 6) of the OPTION register to 0. To enable the A/D converter, we need to set bit 7 of the ADCTL register to 1 and wait for 100 μs. The following instruction sequence will set up the specified A/D conversion environment:

```
        REGBAS   EQU  $1000            ; base address of I/O register block
        ADCTL    EQU  $30              ; offset of ADCTL from REGBAS
        OPTION   EQU  $39              ; offset of OPTION from REGBAS

                 LDX  #REGBAS
                 BCLR OPTION,X $40      ; select the E clock as the clock source for the A/D
        *                              ; converter
                 BSET OPTION,X $80      ; enable the A/D converter
                 LDY  #30              ; delay 105 μs for the charge pump to stabilize
        delay    DEY                  ;      "
                 BNE  delay           ;      "
                 LDAB #%00010100      ; select non-scan, multiple-channel mode and
        *                              ; channel AN4-AN7
                 STAB ADCTL,X         : start the A/D conversion
                 END
```

Example 10.4

Write an instruction sequence to convert the analog signal connected to channel AN0 into digital form. Perform four conversions and stop. Assume the frequency of the E clock is 2 MHz.

Solution   The circuit connections for this example are shown in Figure 10.4. The user should notice that the high reference voltage V_{RH} and the low reference voltage V_{RL} have been set to 5 V and 0 V, respectively.

Since only four A/D conversions are to be performed, the user needs to write only one byte into the ADCTL register. The program is:

```
REGBAS    EQU   $1000       ; I/O registers base address
OPTION    EQU   $39         ; offset of OPTION from REGBAS
ADCTL     EQU   $30         ; offset of ADCTL from REGBAS
ADR1      EQU   $31         ; offset of ADR1 from REGBAS
ADR2      EQU   $32         ; offset of ADR2 from REGBAS
ADR3      EQU   $33         ; offset of ADR3 from REGBAS
ADR4      EQU   $34         ; offset of ADR4 from REGBAS
ADPU      EQU   $80         ; mask to set the ADPU bit of the OPTION register
ADCLK     EQU   $40         ; mask to clear bit 6 of the OPTION register to select
*                           ; the E clock as the A/D clock signal
A2D_INI   EQU   $00         ; value to be written into ADCTL to select
*                           ; single-channel, nonscan mode, and channel AN0
          ORG   $00
result    RMB   4           ; reserve four bytes to store the results

          ORG   $C000       ; starting address of the program
          LDX   #REGBAS
          BSET  OPTION,X ADPU    ; enable the charge pump to start A/D conversion
          BCLR  OPTION,X ADCLK   ; select the E clock for A/D conversion
          LDY   #30         ; delay 105 µs to wait for the charge pump to stabilize
delay     DEY               ;    "
          BNE   delay       ;    "
          LDAA  #A2D_INI    ; initialize the ADCTL register and start A/D conversion
          STAA  ADCTL,X     ;    "
again     LDAA  ADCTL,X     ; check CCF bit
          BPL   again       ; wait until the CCF bit is set
          LDAA  ADR1,X      ; get the first result
```

68HC11

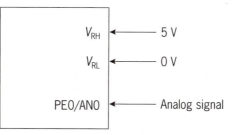

Figure 10.4   Circuit connections for Example 10.4

```
STAA  result      ; save it
LDAA  ADR2,X      ; get the second result
STAA  result+1    ; save it
LDAA  ADR3,X      ; get the third result
STAA  result+2    ; save it
LDAA  ADR4,X      ; get the fourth result
STAA  result+3    ; save it
END
```

Example 10.5

Take 20 samples from each of the A/D channels AN0 to AN3, convert them to digital form, and store them in memory locations from $00 to $4F.

Solution   The circuit connections are shown in Figure 10.5. The samples must be taken at as regular intervals as possible. This goal can be achieved by starting the next conversion (by writing a new byte into the ADCTL register) immediately after the previous conversion has been completed but before collecting the result.

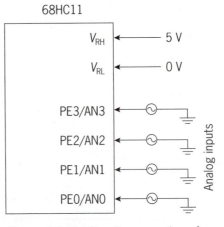

Figure 10.5   Circuit connections for Example 10.5

The following program performs the conversion with this objective in mind:

```
N       EQU  20       ; number of samples
REGBAS  EQU  $1000    ; I/O registers base address
OPTION  EQU  $39      ; offset of OPTION from REGBAS
ADCTL   EQU  $30      ; offset of ADCTL from REGBAS
ADR1    EQU  $31      ; offset of ADR1 from REGBAS
ADR2    EQU  $32      ; offset of ADR2 from REGBAS
ADR3    EQU  $33      ; offset of ADR3 from REGBAS
ADR4    EQU  $34      ; offset of ADR4 from REGBAS
ADPU    EQU  $80      ; mask to set or clear the ADPU bit of the OPTION
*                     ; register
ADCLK   EQU  $40      ; mask to clear or set the CSEL bit of the OPTION
*                     ; register
```

```
A2D_INI    EQU    $10              ; value to be written into the ADCTL register to select
*                                  ; multi-channel, nonscan mode, and channels AN0–AN3

           ORG    $00
RESULT     RMB    80               ; reserved 80 bytes to store the results

           ORG    $C000            ; starting address of the program
           LDX    #REGBAS
           BSET   OPTION,X ADPU    ; enable the charge pump to start the A/D
*                                  ; conversion
           BCLR   OPTION,X ADCLK   ; select the E clock for A/D conversion
           LDY    #30              ; delay for 105 μs to wait for the charge pump to
*                                  ; stabilize
delay      DEY                     ;         "
           BNE    delay            ;         "
           LDAA   #A2D_INI         ; initialize the ADCTL register
           STAA   ADCTL,X          ;         "
           LDAB   #N               ; number of samples to be collected from each channel
           LDY    #RESULT          ; point Y to the result area
wait       LDAA   ADCTL,X          ; wait until the conversion of four samples is completed
           BPL    wait             ;         "
```

* start the next conversion immediately so that samples can be taken more uniformly in
* time

```
           LDAA   #A2D_INI         ; start the next conversion
           STAA   ADCTL,X          ;         "
```

* The following eight instructions collect the previous conversion results

```
           LDAA   ADR1,X           ; fetch the result from channel 1
           STAA   0,Y
           LDAA   ADR2,X           ; fetch the result from channel 2
           STAA   1,Y
           LDAA   ADR3,X           ; fetch the result from channel 3
           STAA   2,Y
           LDAA   ADR4,X           ; fetch the result from channel 4
           STAA   3,Y
           INY                     ; move the result pointer
           INY                     ;     "
           INY                     ;     "
           INY                     ;     "
           DECB                    ; decrement the loop count
           BNE    LOOP
           END
```

10.10 Using the LM35 Precision Centigrade Temperature Sensors

National Semiconductor produces an LM35 series of precision integrated-circuit temperature sensors whose output voltage is linearly proportional to Celsius (centigrade) temperature. The LM35 does not require any external calibration or trimming to provide typical accuracies of $\pm\frac{1}{4}$ °C at room temperature and $\pm\frac{3}{4}$ °C over its complete −55 to +150 °C temperature range. It can be used with single power supplies or with plus-and-minus supplies. As it draws

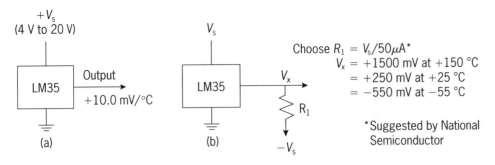

(a) Basic centigrade temperature sensor (12 °C to 150 °C)
(b) Full-range centigrade temperature sensor

Figure 10.6   Circuit connections for the LM35 temperature sensor

only 60 μA current from its power supply, it has very low self-heating (less than 0.1 °C in still air).

The LM35 is available in two packages: the TO-46 metal can and the TO-92 plastic package. It has three pins: V_{out} for voltage output, V_s for supply, and GND for ground. Two typical circuit connections for the LM35 are shown in Figure 10.6. These two circuits will sense the surface temperature of the LM35 and provide very accurate readings.

Example 10.6   Digital Thermometer

Use the circuit shown in Figure 10.6b as a building block in a system to measure room temperature. Display the result in two integral digits and one fractional digit using seven-segment displays. Assume that room temperature never goes below 0 °C and never goes above 42.5 °C so that the A/D converter of the 68HC11 can be used to perform the conversion and drive the seven-segment displays.

Solution   The voltage output from the circuit shown in Figure 10.6b will be 0 V at 0 °C and 425 mV at 42.5 °C. Higher precision can be obtained from the 68HC11 A/D converter if the voltage output corresponding to 42.5 °C is scaled to 5 V. The circuit shown in Figure 10.7 will scale V_x from the range of

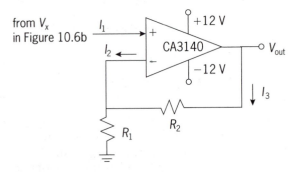

Figure 10.7   Circuit to scale V_x

Characteristic	Value
maximum ($V_{s+} \sim V_{s-}$)	36 V
minimum ($V_{s+} \sim V_{s-}$)	4 V
maximum (V_+ or V_-)	$V_{s+} + 8$
minimum (V_+ or V_-)	$V_{s-} - 0.5$
max common mode input	$V_{s+} - 2.5$
min common mode input	$V_{s-} - 0.5$
input resistance	1 TΩ
input capacitance	4 pF
input current	2 pA
input offset voltage	5 mV
input offset current	0.1 pA
output resistance	60 Ω
maximum output voltage	$V_{s+} - 3V$
minimum output voltage	$V_{s-} + .13V$
maximum sourcing current	10 mA
maximum sinking current	1 mA
amplification	100,000
slew rate	7 V/μs
gain bandwidth product	3.7 MHz
transient response	80 ns
supply current	1.6 mA
device dissipation	8 mW

Table 10.2 ■ Characteristics of the CA3140

0–425 mV to the range of 0–5 V. The CA3140 op amp shown in Figure 10.7 is particularly suitable for microcontrollers and microprocessors because it has CMOS inputs, which use almost no current, and bipolar transistor outputs, which can supply plenty of current. The characteristics of the CA3140 op amp are listed in Table 10.2.

Because of the high input impedance of the op amp, both I_1 and I_2 are zeros. The voltage drop across the resistor R_1 is equal to V_x. On the other hand, $V_{R_1} = I_3 \times R_1$, and $V_{out} = I_3 (R_1 + R_2)$. The voltage gain of the op amp circuit in Figure 10.7 is given by the following equation:

$$A_v = \frac{V_{out}}{V_x} = \frac{R_1 + R_2}{R_1} = 1 + \frac{R_2}{R_1} \qquad \textbf{10.2}$$

By setting the ratio of R_1 to R_2 to 10.765, the range of V_{out} will be from 0 V to 5 V for temperatures from 0 °C to 42.5 °C. To obtain this ratio using standard resistors, the user can choose 4.7 KΩ for R_1 and 51 KΩ for R_2. The error caused by these resistance values will be within 0.1 °C. The circuit connections between the LM35 and the 68HC11 are shown in Figure 10.8.

The circuit in Figure 9.13 can be used to display the temperature. However, only three digits are needed for this digital thermometer. The first seven-segment display in Figure 9.13 will be left unconnected, as shown in Figure 10.9. The decimal point will be displayed in this example. Since the 68HC11 A/D converter has 8-bit precision, the conversion result 255 corre-

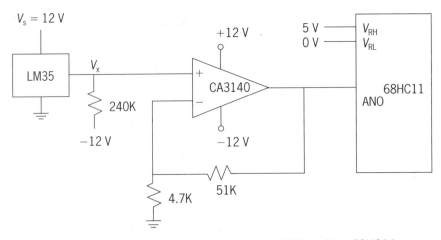

Figure 10.8 Circuit connections between the LM35 and the 68HC11

sponds to 42.5 °C and 0 corresponds to 0 °C. To obtain the temperature reading, the user must divide the A/D conversion result by 6 (255 ÷ 42.5 = 6). In the 68HC11, the integer quotient of this division becomes the two integral digits of the temperature. The first fractional digit can be derived by multiplying the remainder by 10 and then dividing the product by 6. The E clock frequency is

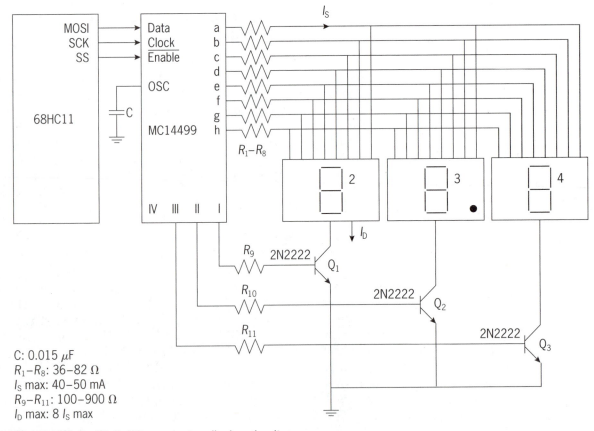

C: 0.015 μF
R_1–R_8: 36–82 Ω
I_S max: 40–50 mA
R_9–R_{11}: 100–900 Ω
I_D max: 8 I_S max

Figure 10.9 Digital thermometer display circuit

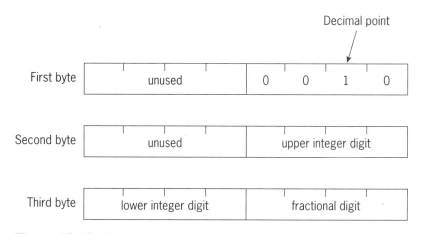

Figure 10.10 Displaying the temperature digits

assumed to be 2 MHz. We will use +12 V and −12 V as the power supply of the temperature sensor in Figure 10.6b because these two voltages are available in EVB and EVM boards.

Since room temperature does not change very fast, it should be adequate to measure the temperature once every second. Three bytes are needed to hold the value to be sent to the MC14499 circuit. The first byte will be used to display the decimal point. The value to be stored in this byte is a constant %00000010. The binary digit 1 allows the decimal point associated with display 3 in Figure 10.9 to be lighted. The upper four bits of this byte are not needed—they are included simply to fill up eight bits so that the decimal digits pattern can be sent in one SPI transfer. The upper four bits of the second byte are not needed and will be arbitrarily set to %0000. The lower four bits of the second byte will hold the digit for tens degree. The upper four bits of the third byte will hold the value of the lower digit of the integer part, and the lower four bits will hold the fractional digit. The contents of these three bytes are illustrated in Figure 10.10.

The following program will start the A/D converter and display the temperature using three digits:

```
REGBAS   BRCLR $1000        ; I/O register block base address
SPCR     EQU   $28          ; offset of SPCR from REGBAS
SPDR     EQU   $2A          ; offset of SPDR from REGBAS
SPSR     EQU   $29          ; offset of SPSR from REGBAS
DDRD     EQU   $09          ; offset of DDRD from REGBAS
ADR1     EQU   $31          ; offset of ADR1 from REGBAS
OPTION   EQU   $39          ; offset of OPTION from REGBAS
ADCTL    EQU   $30          ; offset of ADCTL from REGBAS
PORTB    EQU   $04          ; offset of PORTB from REGBAS
PORTD    EQU   $08          ; offset of PORTD from REGBAS
SP_DIR   EQU   %00111000    ; value to set SPI pin directions (set SS, MOSI, and
*                           ; SCK pins to be outputs, and MISO to be an input)
SPCR_IN  EQU   %01010110    ; value to be written into SPCR to enable SPI,
*                           ; choose falling edge to shift data, master mode,
*                           ; normal port D output, and 128 Kbits/sec data
*                           ; rate
ADPU     EQU   $80          ; mask to set the ADPU bit of the OPTION register
```

```
ADCLK        EQU    $40              ; mask to clear the CSEL bit of the OPTION
*                                    ; register
A2D_INI      EQU    $00              ; value to be written into ADCTL to set nonscan
*                                    ; mode select single-channel mode, and select
*                                    ; channel AN0
TCNT         EQU    $0E              ; offset of TCNT from REGBAS
TOC2         EQU    $18              ; offset of TOC2 from REGBAS
TFLG1        EQU    $23              ; offset of TFLG1 from REGBAS
OC2          EQU    $40              ; mask to check the OC2F of TFLG1
OC2M         EQU    $BF              ; mask to clear the OC2F bit of TFLG1 for BCLR
*                                    ; instruction


             ORG    $00
BYTE1        RMB    1                ; storage for results
BYTE2        RMB    1                ;        "
BYTE3        RMB    1                ;        "
remain       RMB    2                ; to hold the remainder of division
oc2cnt       RMB    1                ; output comparison count of OC2


             ORG    $C000
             LDX    #REGBAS
             LDAA   #%00000010       ; store the value %00000010 in BYTE1
             STAA   BYTE1            ;        "
```

* The following four instructions initialize the SPI system to prepare to shift data out
* to the MC14499

```
             LDAA   #SP_DIR          ; set SPI pin directions
             STAA   DDRD,X           ;        "
             LDAA   #SPCR_INI        ; initialize the SPI system
             STAA   SPCR,X           ;        "
```

* The following five instructions enable the A/D converter, select the E clock to control
* the A/D conversion process, and wait for the charge pump to stabilize

```
             BSET   OPTION,X ADPU    ; enable charge pump to start A/D conversion
             BCLR   OPTION,X ADCLK   ; select E clock for A/D conversion
             LDY    #30              ; delay 105 μs so that the charge pump is
*                                    ; stabilized
delay        DEY                     ;        "
             BNE    delay            ;        "
forever      LDAA   #A2D_INI         ; initialize the ADCTL and start an A/D conversion
             STAA   ADCTL,X          ;        "
HERE         LDAA   ADCTL,X          ; check the CCF bit
             BPL    HERE             ; wait until the A/D conversion is complete
             LDAB   ADR1,X           ; read the temperature
```

* The following three instructions convert the A/D result into a temperature reading

```
             CLRA                    ; the value 255 corresponds to 42.5 °C
             LDX    #6               ; 1 °C is equivalent to 6
             IDIV                    ; convert to temperature in Celsius
             STD    remain           ; save the remainder
             XGDX                    ; leave the integer part of the temperature in D
```

* The following two instructions separate the upper and lower integer digits. The upper
* integer digit stays in accumulator B and the upper integer digit stays in X

```
            LDX     #10                 ;
            IDIV                        ;

            LSLB                        ; shift the lower integer digit to the upper 4 bits of B
            LSLB                        ;        "
            LSLB                        ;        "
            LSLB                        ;        "
            STAB    BYTE3               ; the lower four bits in BYTE3 are 0000
            XGDX                        ; swap X and D so that the upper integer digit is
*                                       ; in B
            STAB    BYTE2               ; save the upper integer digit in memory
            LDD     remain              ; get back the remainder in the division. The
*                                       ; remainder only stays in accumulator B because
*                                       ; it is less than 6
            LDAA    #10                 ; multiply the remainder by 10
            MUL                         ;        "
            LDX     #6                  ; set the divisor to 6
            IDIV                        ; compute remainder × 10 ÷ 6 to derive the first
*                                       ; fractional digit
```

* The quotient will be less than 10 and will stay in the lowest four bits of accumulator B
* after execution of the next instruction

```
            XGDX                        ; swap the quotient into A and B
            ADDB    BYTE3               ; combine the lower integer and the fractional
*                                       ; digits
            STAB    BYTE3               ; prepare to transfer to SPI for display
            BCLR    PORTD,X $20         ; pull SS pin to low to enable SPI transfer to
*                                       ; the MC14499
            LDAA    BYTE1               ; get BYTE1 and send it out
            STAA    SPDR,X              ;        "
            BRCLR   SPSR,X $80 *        ; wait until 8 bits have been shifted out
            LDAA    BYTE2               ; get BYTE2 and send it out
            STAA    SPDR,X              ;        "
            BRCLR   SPSR,X $80 *        ; wait until 8 bits have been shifted out
            LDAA    BYTE3               ; get BYTE3 and send it out
            STAA    SPDR,X              ;        "
            BRCLR   SPSR,X $80 *        ; wait until 8 bits have been shifted out
            BSET    PORTD,X $20         ; pull the SS pin to high so that the value shifted to
*                                       ; the MC14499 can be loaded into the latch for
*                                       ; display
```

* The following instruction sequence uses the OC2 function 100 times to create a delay
* of 1 second. Each output compare using OC2 creates a delay of 10 ms.

```
            LDAB    #100
            STAB    oc2cnt              ; initialize OC2 count to 100 to create a 1-second
*                                       ; delay
            LDX     #REGBAS
            BCLR    TFLG1,X OC2M        ; clear the OC2F bit in TFLG1 before using the
*                                       ; OC2 function
```

```
          LDD    TCNT,X
repeat    ADDD   #20000              ; create a delay of 10 ms
          STD    TOC2,X              ;     "
wait      BRCLR  TFLG1,X OC2 *       ; check whether the OC2F flag of the TFLG1
    *                                ; register is set
          BCLR   TFLG1,X OC2M        ; clear the OC2F flag of the TFLG1 register
          LDD    TOC2,X
          DEC    oc2cnt
          BNE    repeat
          BRA    forever             ; repeat forever
          END
```

10.11 Processing the Results of A/D Conversions

The results of the A/D conversion often need to be processed. One of the most useful measurements of an AC signal is its root-mean-square (RMS) value. The root-mean-square value of an AC signal is defined as:

$$V_{RMS} = \sqrt{\frac{1}{T} \int_0^T V^2(t)\, dt} \qquad\qquad \textbf{10.3}$$

Since the 68HC11 cannot calculate the integral, the following equation will be used to approximate the RMS value:

$$V_{RMS} \cong \sqrt{\frac{1}{N} \sum_{i=0}^{N} V_i^2} \qquad\qquad \textbf{10.4}$$

To make this equation a good approximation of the real RMS value, samples must be as equally spaced in time as possible. It is obvious that the more samples collected over the time period, the better the result.

Example 10.7

Write a short program to compute the average of the squared values of 64 samples stored at $00–$3F and save the result at $61 and $62.

Solution   The square of an 8-bit value requires two bytes to hold it, and three bytes (at locations $60, $61, and $62) are needed to store the sum of sixty-four 16-bit values. The average of these 64 values will need to be stored in two bytes. Division by 64 can be performed by shifting the sum to the right by 6 positions. To shift a 24-bit value (in $60, $61, and $62), execute the following instruction sequence six times:

```
      LSR  $60    ; bit 0 of the memory location at $60 is shifted into the C flag
      ROR  $61    ; the C flag is shifted into bit 7 of the memory location at $61 and
   *              ; bit 0 of the memory location at $61 is shifted into the C flag
      ROR  $62    ; the C flag is shifted into bit 7 of the memory location at $62, and
   *              ; bit 0 of the memory location at $62 is shifted into the C flag
```

The program to perform the squaring, summing, and averaging is as follows:

```
              ORG   $C000   ; starting address of the program
              LDX   #00     ; load the base address of the array of samples
              CLR   $60     ; initialize the sum to 0
              CLR   $61     ;     "
              CLR   $62     ;     "
              LDY   #64     ; initialize the loop count
      LOOP    LDAA  0,X     ; get a sample
              LDAB  0,X
              MUL           ; compute the square of the sample
              ADDD  $61     ; add to the temporary sum
              STD   $61     ; save the temporary sum
              LDAB  $60     ; add the carry to the upper byte
              ADCB  #0      ;     "
              STAB  $60     ;     "
              INX           ; move to the next sample
              DEY           ; decrement the loop count
              BNE   LOOP    ; repeat if the loop count is not yet 0
              LDAB  #6      ; prepare to shift right six times
```

* The next five instructions divide the sum by 64

```
      LOOP1   LSR   $60
              ROR   $61
              ROR   $62
              DECB
              BNE   LOOP1
              END
```

The next step in computing an RMS value is to compute the square root of the sum. Integer arithmetic can be used to perform this computation. The technique is based on the following equation:

$$\sum_{i=0}^{n-1} i = \frac{n(n-1)}{2} \qquad\qquad \textbf{10.5}$$

This equation can be transformed into:

$$n^2 = \sum_{i=0}^{n-1} (2i + 1) \qquad\qquad \textbf{10.6}$$

Equation 10.6 gives us a clue about how to compute the square root of a value. Suppose we want to compute the square root of p, and n is the integer value that is closest to the true square root. Then one of the following three relationships is satisfied:

$n^2 < p$
$n^2 = p$
$n^2 > p$

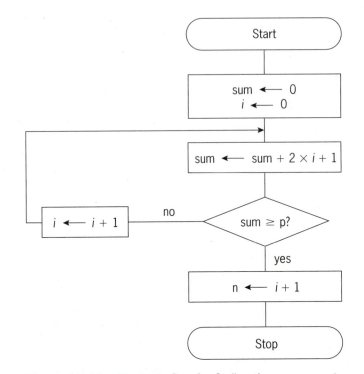

Figure 10.11   The logic flow for finding the square root

The logic shown in Figure 10.11 can be used to find a value for n, and this algorithm is translated into the program in Example 10.8. The flowchart in Figure 10.11 stops the iteration when sum $\geq p$. However, it may not find the closest square root when sum $> p$. This problem can be corrected by making a minor modification to the program, as will be illustrated in Example 10.9.

Example 10.8

Write a program to compute the square root of the average value computed in Example 10.7 (stored at $61 and $62) and save the square root at $63.

Solution   The program is as follows:

```
          ORG   $64
sq_root   RMB   2              ; to hold the value of Σ(2i + 1)
i         RMB   1              ; index i
          ORG   $C200          ; starting address of the program
          CLR   sq_root        ; initialize Σ(2i + 1) to 0
          CLR   sq_root+1      ;      "
          LDAB  #-1
          STAB  i              ; place a − 1 into i
REPEAT    INC   i              ; i starts from 0 because of the previous instruction
          LDAB  i
          CLRA
          LSLD                 ; compute 2i
          ADDD  #1             ; compute 2i + 1
```

```
ADDD  sq_root     ; add 2i + 1 to the accumulating sum
STD   sq_root     ; update the accumulating sum
CPD   $61         ; compare to n²
BLO   REPEAT      ; repeat when Σ(2i + 1) < n²
INC   i           ; add 1 to i to obtain the square root
END               ; square root is in memory location i
```

Example 10.9

Add a sequence of instructions to the program in Example 10.8 to find the closest square root of the square sum.

Solution If n is the true square root of p (i.e., $n^2 = p$), then there is no problem. If n is not the true square root of p, then either n or $n - 1$ is the integer closest to the true square root of p. The choice between n and $n - 1$ can be made by comparing the following two expressions:

$$n^2 - p \qquad\qquad (1)$$
$$p - (n - 1)^2 \qquad (2)$$

If the value of expression (1) is smaller, choose n; otherwise, choose $n - 1$. Note that both values are non-negative.

This idea can be translated into the following instruction sequence:

```
LDAA  i       ; place n in A
TAB           ; also place n in B
MUL           ; compute n²
CPD   $61     ; compare to p
BEQ   exit    ; n is the true square root of p
```

* Expressions (1) and (2) are compared in the following sequence of instructions:

```
        SUBD $61     ; compute n² - p
        STD  $64     ; save it in the memory
        LDAA i
        DECA         ; compute n - 1
        TAB          ; place n - 1 in B
        MUL          ; compute (n - 1)²
        STD  $66     ; save (n - 1)² in the memory
        LDD  $61     ; place p in D
        SUBD $66     ; compute p - (n - 1)²
        CPD  $64     ; compare p - (n - 1)² to n² - p
        BHI  exit    ; n is the closest to the true square root
        DEC  i       ; n - 1 is closest to the true square root
  exit  .            ; do something else
        .
        .
```

This instruction sequence should be added after the last instruction (INC i) of the program in Example 10.8.

10.12 Using the MC145050 External A/D Converter

Most members of the 68HC11 family have only an 8-bit A/D converter, but it is sometimes desirable to have higher resolution. Using an external A/D converter or choosing a member of the 68HC11 family that has higher conversion resolution are two possible alternatives. This section will discuss the use of an external 10-bit A/D converter with a serial interface.

The MC145050 is a 10-bit A/D converter manufactured by Motorola. The block diagram of this device is shown in Figure 10.12. The MC145050 has the following features:

- 11 analog input channels with a sample-and-hold circuit
- uses a successive approximation method to perform conversions and completes one conversion in 22 μs at 2 MHz
- direct interface to the SPI interface
- analog input range of 0 to 5 V with a 5-volt supply

10.12.1 The MC145050 Signal Pins

The MC145050 has 20 pins. The function of each signal is listed below:

$\overline{CS}$: Chip select. Being active low, this signal initializes the chip to perform conversions and provides tristate control of the data output pin.

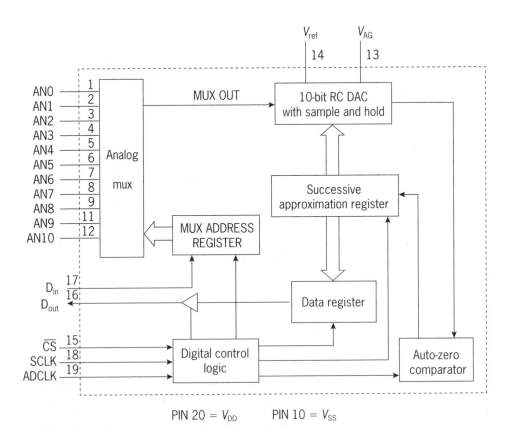

PIN 20 = V_{DD} PIN 10 = V_{SS}

Figure 10.12   Block diagram of the MC145050 (Redrawn with permission of Motorola)

A high-to-low transition on $\overline{CS}$ resets the serial data port and synchronizes it to the microprocessor data stream. $\overline{CS}$ can remain active during the conversion cycle for multiple serial transfers, or $\overline{CS}$ can be inactive high after each transfer. If $\overline{CS}$ is kept active low between transfers, the length of each transfer is limited to either 10 or 16 SCLK cycles. If $\overline{CS}$ is inactive high between transfers, each transfer can be anywhere from 10 to 16 SCLK cycles long.

D_{out}: Serial data output of the A/D conversion result. D_{out} is in a high-impedance state when $\overline{CS}$ is inactive high. When the chip recognizes a valid active low on $\overline{CS}$, D_{out} is taken out of the high-impedance state and is driven with the most significant bit (MSB) of the previous conversion result. The value on D_{out} changes to the second-most significant result bit upon the first falling edge of SCLK. The remaining result bits are shifted out in order.

D_{in}: Serial data input. The MC145050 accepts the first four bits input from this pin as the select signals (address) for the 11 analog input sources. The address is shifted MSB first.

SCLK: Serial data clock. This clock input drives the internal I/O state machine to perform three major functions: (1) drive the data shift registers to simultaneously shift in the next mux address from the D_{in} pin and shift out the previous conversion result on the D_{out} pin, (2) begin sampling the analog voltage onto the RC DAC as soon as the new mux address is available, and (3) transfer control to the A/D conversion state machine (driven by ADCLK) after the last bit of the previous conversion result has been shifted out on the D_{out} pin.

ADCLK: A/D conversion clock. This pin clocks the dynamic A/D conversion sequence and may be asynchronous to SCLK. Control of the chip passes to ADCLK after the tenth falling edge of SCLK. Control of the chip passes back to SCLK after the successive approximation conversion sequence is complete or after a valid chip select is recognized.

AN0 through AN10: Analog multiplexer inputs. Inputs AN0 through AN10 are analog voltage inputs to be converted into digital values. The inputs AN0, AN1, . . . , AN10 are addressed by $0, $1, . . . , $A, respectively.

V_{SS} and V_{DD}: Device supply pins. V_{SS} is normally connected to ground, and V_{DD} is normally connected to a positive supply voltage.

V_{AG} and V_{ref}: Analog reference voltage. V_{AG} and V_{ref} are analog reference low and high voltages, respectively. The MC145050 is a ratiometric A/D converter. Analog input voltages $\geq V_{ref}$ produce a full-scale output equal to 1023, and input voltages $\leq V_{AG}$ produce an output zero. Caution: The analog input voltage must be no lower than V_{SS} and no higher than V_{DD}.

10.12.2 Chip Functioning

The MC145050 can be interfaced to the 68HC11 using the SPI interface. The circuit connections are shown in Figure 10.13. The 68HC11 sends the analog voltage channel address to the converter via the MOSI pin, and the previous conversion result is returned from the MISO pin. The total conversion

68HC11   MC145050

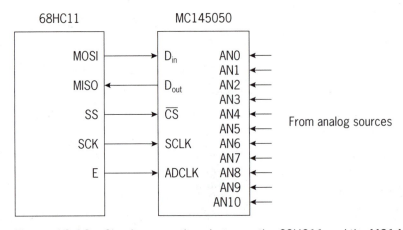

Figure 10.13 Circuit connections between the 68HC11 and the MC145050

time (the time from sending the analog input address until the result becomes available) is given by the following formula:

$$t_{conv} = 10 \text{ SCLK} + 44 \text{ ADCLK}$$

If the period of SCLK is set to be twice that of the ADCLK clock, then the total conversion time is equal to 64 ADCLK clock cycles.

When the microcontroller sends in the analog channel address for the first time, no previous conversion result is available, and hence the value returned from the MISO pin should be ignored.

The procedure for using the MC145050 is as follows:

Step 1
Enable the SPI subsystem for serial data transfer.
Step 2
Set $\overline{CS}$ to low to initialize the MC145050 and enable the data transfer and the A/D conversion process.
Step 3
Send in the address in two bytes to allow the previous conversion result to be sent back in 16-bit format.
Step 4
Wait for 44 ADCLK cycles to collect the conversion result, then start the next conversion.
Step 5
Repeat steps 2 to 4 as many times as needed.
Step 6
Set $\overline{CS}$ to high to end the process.

Example 10.10

An MC145050 is interfaced to a 68HC11 using the circuit connections shown in Figure 10.13. Write a program to start A/D conversion from channels AN0 to AN10 and store the results starting at location $00. Take one sample from each channel.

Solution The channel address must be sent to the MC145050 in order to start the A/D conversion and receive the previous conversion result. The channel address will be sent in two bytes using the SPI function, although only the first four bits will be used by the MC145050 to address the channel. The upper byte of the previous conversion result is returned to the SPDR register when the first byte of the channel address is sent, and the lower byte of the previous conversion result is returned after the second byte of the channel address is sent. A buffer of 22 bytes will be allocated to hold the A/D conversion results; index register Y will be used as a pointer to this area. This program will perform A/D conversion using the following procedure:

Step 1
Initialize the SPI subsystem for SPI transfer. This step includes programming the DDRD and SPCR registers. Also initialize index register Y to point to two bytes before the result area.

Step 2
Pull the $\overline{\text{CS}}$ pin to low to initialize the MC145050 and enable the data transfer and A/D conversion.

Step 3
Initialize the channel address upper byte ch_adr to $00.

Step 4
Send ch_adr to MC145050 and wait until transfer is complete.

Step 5
Read the upper byte of the previous conversion result and save it in the memory location pointed to by Y. Increment Y by 1.

Step 6
Send in a dummy byte as the lower byte of the channel address and wait until transfer is complete.

Step 7
Read the lower byte of the previous conversion result and save it in the memory location pointed to by Y. Increment Y by 1.

Step 8
Wait until the current A/D conversion is complete.

Step 9
Is ch_adr equal to $A0? If not, then set ch_adr equal to ch_adr + $10 and go to Step 4.

Step 10
Send $00 to the MC145050 to read back the upper byte of the last conversion result. Wait until the SPI transfer is complete. Save the result at the location pointed to by Y. Increment Y by 1.

Step 11
Send $00 to the MC145050 to read back the lower byte of the last conversion result. Wait until the SPI transfer is complete, and save the result at the location pointed to by Y.

The program is:

```
REGBAS    EQU  $1000        ; I/O block base address
DDRD      EQU  $09          ; offset of DDRD register from the base address
SPCR      EQU  $28          ; offset of SPCR register from the base address
SPDR      EQU  $2A          ; offset of SPDR register from the base address
SPSR      EQU  $29          ; offset of SPSR register from the base address
```

```
            PORTD    EQU   $08            ; offset of PORTD register from the base address
            BRKPT    EQU   $F000          ; breakpoint address
            CR_INI   EQU   %01011000      ; value to be written into SPCR to enable SPI transfer,
  *                                       ; disable SPI interrupt, choose falling edge as active
  *                                       ; edge, set period of SCK to be 2 E cycles
            DD_INI   EQU   %00111000      ; value to be written into DDRD to set port D pin
  *                                       ; directions

                     ORG   $00
            dummy    RMB   2              ; these two bytes are used to hold two bytes of invalid
  *                                       ; conversion results when the address of channel AN0
  *                                       ; is sent. This is done to simplify the program loop
            RESULT   RMB   22             ; reserve two bytes for each result

                     ORG   $C000          ; starting address of the program
                     LDX   #REGBAS
                     LDY   #dummy         ; point Y to $00
                     LDAA  #DD_INI
                     STAA  DDRD,X         ; set port D pin direction
                     LDAA  #CR_INI
                     STAA  SPCR,X         ; initialize SPI
                     BCLR  PORTD,X $20    ; pull CS to low
                     CLRA                 ; accumulator A holds the channel address
            loop1    STAA  SPDR,X         ; send analog channel upper address byte
            LP0      LDAB  SPSR,X
                     BPL   LP0            ; wait for SPI transfer to be complete
                     LDAB  SPDR,X         ; get the upper byte of the previous result
                     STAB  0,Y            ; save the upper byte of the previous result
                     INY                  ; move the pointer to the A/D result area
                     STAA  SPDR,X         ; send dummy lower address byte
            LPX      LDAB  SPSR,X         ; wait until the SPI transfer is complete
                     BPL   LPX
                     LDAB  SPDR,X         ; get the lower byte of the previous result
                     STAB  0,Y            ; save the lower byte of the previous result
                     INY                  ; move the pointer to the result area
```

* The last three instructions take 13 E clock cycles to execute. We need to wait
* another 31 E clock cycles for the A/D conversion to be complete. The following five
* instructions create a delay of 32 E clock cycles.

```
                     LDAB  #3             ; this instruction takes 2 E clock cycles to execute.
            LP2      NOP                  ; this instruction takes 2 E clock cycles
                     LDX   #REGBAS        ; this instruction takes 3 E clock cycles
                     DECB                 ; this instruction takes 2 E clock cycles
                     BNE   LP2            ; this instruction takes 3 E clock cycles
                     ADDA  #$10           ; select the next channel by incrementing the upper
  *                                       ; 4 bits of A because these 4 bits are used as the
  *                                       ; address
                     CMPA  #$B0           ; when the address sent to MC145050 is $B0
  *                                       ; the sampling of channels AN0–AN10 is complete
                     BNE   loop1
```

* The following 12 instructions send out address $00 to read back the last conversion
* result

```
                CLRA
                STAA  SPDR,X        ; send AN0 upper address byte
        LP1     LDAB  SPSR,X        ; wait until the SPI transfer is completed
                BPL   LP1           ;     "
                LDAB  SPDR,X        ; get the upper byte of the last A/D result
                STAB  0,Y           ; save it
                INY                 ; move the pointer to the A/D result area
                STAA  SPDR,X        ; send dummy lower address byte
        LPY     LDAB  SPSR,X        ; wait until the SPI transfer is completed
                BPL   LPY           ;     "
                LDAB  SPDR,X        ; get the lower byte of the last A/D result
                STAB  0,Y           ; save it
                BSET  PORTD,X $20   ; pull CS to high to disable the A/D converter
                END
```

Example 10.11

Assume that $V_{ref} = 4.5$ V and $V_{AG} = 1.5$ V for the circuit in Figure 10.13. Find the conversion results for the analog voltages of 1.0 V, 2.0 V, 2.5 V, 3.0 V, 3.5 V, and 4.0 V.

Solution

$$range = 4.5 - 1.5 = 3.0 \text{ V}$$

The conversion result for 1.0 V is 0. The conversion results for remaining voltages are:

$(2.0 - 1.5)/3 \times 1023 = 170$
$(2.5 - 1.5)/3 \times 1023 = 341$
$(3.0 - 1.5)/3 \times 1023 = 511$
$(3.5 - 1.5)/3 \times 1023 = 682$
$(4.0 - 1.5)/3 \times 1023 = 852$

10.13 Summary

This chapter explains the features and operation of the 68HC11 A/D converter in detail. The 68HC11 has an eight-channel, 8-bit, multiplexed input A/D converter. The highest and lowest analog input voltages are limited to +5 V and 0 V, respectively. The A/D converter needs a clock signal to perform the conversion; the user can choose either the E clock or the on-chip RC circuit output as the clock source. The RC-circuit output should be chosen when the E clock frequency is lower than 750 KHz. The conversion of one sample takes 32 clock cycles to complete. Four registers are provided to hold the conversion results.

Applications of A/D conversion normally involve the use of a transducer to convert a nonelectric quantity into volts. The voltage output from the transducer may not be suitable for the A/D converter. Thus a scaling and level-shifting circuit is often needed to translate the transducer output to an appropriate range. An example is given to demonstrate this technique. An example using a temperature sensor demonstrates the measurement of temperature using the 68HC11 A/D converter. The output of an A/D converter often needs further processing and manipulation, such as calculation of the root-mean-square value.

10.14 Glossary

Charge pump   A circuit technique that can raise a low voltage to a level above the power supply. A charge pump is often used in an A/D converter, in EEPROM and EPROM programming, etcetera.

Load cell   A transducer that can convert weight into a voltage.

Thermocouple   A transducer that converts a high temperature into a voltage.

Temperature sensor   A transducer that can convert temperature into a voltage.

Transducer   A device that can convert a nonelectric quantity into a voltage.

10.15 Exercises

E10.1. Explain the difference between digital and analog signals.

E10.2. Survey the available D/A and A/D methods.

E10.3. Survey methods and devices for converting temperature, pressure, and weight into electric voltage.

E10.4. How long does it take to complete the A/D conversion of four samples if the E clock frequency is 3 MHz and the E clock is used as the clock signal to the 68HC11 A/D converter?

E10.5. Design an op amp circuit that can scale and shift the voltage range of $0-+10$ mV to $0-5$ V.

E10.6. What value should be written into the ADCTL register so that the 68HC11 A/D converter will convert signals from channel 3 in the scan mode continually?

E10.7. Assume the V_{RH} and V_{RL} inputs to the 68HC11 A/D converter are 5 V and 0 V, respectively. What are the corresponding voltages for conversion results of 25, 64, 100, 150, and 200?

E10.8. Use the 68HC11 A/D converter to sample an unknown signal applied at channel AN0 (pin PE0). Measure the signal every 10 ms and compute the average value at the end of one second. Display the average value using two seven-segment displays. Use port B to display the average voltage. Repeat the same measurement forever. The following components are used in this experiment:

Two common cathode (or anode) seven-segment displays

Two seven-segment LED drivers. Use a 74LS47 for common-anode displays or a 74LS48 for common-cathode displays.

Fourteen 100-ohm resistors.

V_{RL} and V_{RH} are set to 0 V and 5 V respectively.

E10.9. Design a digital thermometer so that it can measure a temperature range from $-10\,°C$ to $+50\,°C$. The requirements are:

1. Display the temperature in a three-digit format: two integer digits and one fractional digit. The sign and the decimal point must be displayed.

2. Use four seven-segment displays to display the temperature. Segment h of the left-most display is used for the sign. When the temperature is negative, light the segment h; otherwise, turn off the left-most display.

3. Use a 10-bit A/D external converter such as the MC145050 to achieve accuracy.

4. Build an op amp circuit that can scale and shift the voltage output to the 0–5 V range so that it can be properly handled by the external converter.

5. Measure the temperature ten times every second and display the average. Update the display once every second.

6. Write a program to perform the required control functions.

10.16 Lab Exercises and Assignments

L10.1. Simple A/D conversion experiment. Using either the EVM, EVB, or EVBU, perform the following steps:

Step 1
Adjust the function generator output to between 0 and 5 V. Set the frequency to about 1 KHz.

Step 2
Connect the AN0 (PE0) pin to the function generator output.

Step 3
Set the V_{RH} and V_{RL} outputs to the single-board 68HC11 computer to 5 V and 0 V, respectively.

Step 4
Write a program that does the following operations:

1. starts the A/D converter

2. takes 32 samples, converts them, and stores the A/D conversion results at $00–$1F

3. computes the average value and saves the result at $20

Step 5
Assemble the program and download it to the single-board computer for execution.

Step 6
Run the program for sine, triangular, and square waves. Compare the results with the results of manual computations.

L10.2. Digital thermometer. Use the LM35 temperature sensor, the CA3140 op amp, and an MC145050 10-bit external A/D converter to construct a digital thermometer. The temperature should be displayed in three integral digits and one fractional digit. Use the MC14499 to drive seven-segment displays. The requirements are:

1. The temperature range is from 0 °C to 100 °C. However, the temperature must be displayed in Fahrenheit.
2. Use four seven-segment displays to display the temperature reading, which must include one fractional digit.
3. Use a water bath to change the ambient temperature to the LM35. Remember to insulate the LM35 so that it does not get wet.
4. Measure the temperature 10 times every second and display the average of the measurements.
5. Update the temperature once every second.

APPENDIX

NUMBER

SYSTEMS

A.1 Objectives

After completing this appendix, you should be able to:

- convert a decimal number to a binary, octal, or hexadecimal number
- convert a binary, octal, or hexadecimal number to a decimal number
- convert numbers to their two's complement representation
- add, subtract, and negate binary, octal, and hexadecimal numbers
- describe the BCD, excess-3, 2-out-of-5, and Gray codes

A.2 An Introduction to Number Systems

For most of our lives, we have been exposed to the base 10 number system. Our preference for a 10-digit number system is probably based on our ten fingers. However, base 10 is not a natural system for digital hardware, where arithmetic is based on the binary digits 0 and 1. Binary numbers are used in digital electronics and computer and data communication applications to represent the logic high and low states of digital circuits. All information in a digital system is encoded and transmitted as binary numbers. This appendix will cover the main positional number systems related to digital hardware: the decimal, binary, octal, and hexadecimal systems. The concepts covered in this section are applied in many of the logic circuit examples throughout this text. Conversion between number systems and arithmetic in other systems will be discussed in the following sections.

In this appendix, the base of a number is written as a subscript on its lower right. For example, the binary number 1101 is written as 1101_2.

A.3 Binary, Octal, and Hexadecimal Numbers

In any system, a number is written from its most significant digit on the left to its least significant digit on the right. In general, an $(n + m)$-digit number in any number system can be represented by the mathematical equation

$$N = A_{n-1}r^{n-1} + A_{n-2}r^{n-2} + \ldots + A_0 r^0 + A_{-1}r^{-1} + A_{-2}r^{-2} + \ldots + A_{-m}r^{-m} \qquad \textbf{A-1}$$

where the As can be any of the digits allowed in the number system and r is the *radix* or *base* of the system. The digits allowed in a number system range from 0 to $r - 1$. The allowable digits in the decimal, binary, octal, and hexadecimal number systems are listed in Table A.1. Equation A-1 also provides a method for converting a number in any number system to decimal representation.

The primary reason for introducing hexadecimal and octal numbers is their natural correspondence to binary numbers. They permit compact repre-

Number system	Allowable digits
decimal	0, 1, 2, 3, 4, 5, 6, 7, 8, 9
binary	0, 1
octal	0, 1, 2, 3, 4, 5, 6, 7
hexadecimal	0, 1, 2, 3, 4, 5, 6, 7, 8, 9, A, B, C, D, E, F*

*A, B, C, D, E, and F correspond to the decimal numbers 10, 11, 12, 13, 14, and 15.

Table A.1 ■ Allowable digits in the decimal, binary, octal, and hexadecimal systems

sentation of machine code instructions, data, and memory addresses. Hexadecimal numbers are used more often than octal numbers.

A.3.1 Binary-to-Decimal Conversion

A binary number can be easily converted to its decimal equivalent using equation A-1. For example, the decimal equivalent of the binary number 10010111_2 can be computed as follows:

$$10010111_2 = 1 \times 2^7 + 0 \times 2^6 + 0 \times 2^5 + 1 \times 2^4 + 0 \times 2^3 + 1 \times 2^2 + 1 \times 2^1$$
$$+ 1 \times 2^0 = 151_{10}$$

Example A.1

Convert the binary number 100100110.101_2 to its decimal equivalent.

Solution: Simply apply equation A-1:

$$100100110.101_2 = 1 \times 2^8 + 0 \times 2^7 + 0 \times 2^6 + 1 \times 2^5 + 0 \times 2^4 + 0 \times 2^3$$
$$+ 1 \times 2^2 + 1 \times 2^1 + 0 \times 2^0 + 1 \times 2^{-1} + 0 \times 2^{-2} + 1 \times 2^{-3}$$
$$= 294.625_{10}$$

A.3.2 Decimal-to-Binary Conversion

Converting a number from the decimal number system to the binary number system requires two separate methods: one to convert integers and one to convert fractional numbers. Decimal numbers containing both integer and fractional parts are converted to binary by applying those techniques separately to the integer and fractional parts of the number. Any value that is an integer in one number system will remain so in another number system, and the same is true for fractions.

Decimal integers are converted to binary integers by a method of repeated division by 2. The integer is divided by 2 and the remainder, either a 0 or a 1, is the binary value for the bit position of the least significant bit (LSB). The quotient resulting from the first division is the next value to be divided by 2. The remainder from the second division process becomes the second-least significant bit of the binary number, and so on. Successive divisions produce bit values occupying positions further and further from the binary point until the most significant bit (MSB) is reached. Division continues until the quotient is 0 and no further division can take place.

Example A.2

Convert the decimal integer 53_{10} to its binary equivalent.

Solution:

Division	Quotient	Remainder	Binary equivalent	
53 ÷ 2	26	1	1	LSB
26 ÷ 2	13	0	0	
13 ÷ 2	6	1	1	
6 ÷ 2	3	0	0	
3 ÷ 2	1	1	1	
1 ÷ 2	0	1	1	MSB

We can multiply the resulting binary number by bit weights to verify our answer:

$$110101_2 = 1 \times 2^5 + 1 \times 2^4 + 0 \times 2^3 + 1 \times 2^2 + 0 \times 2^1 + 1 \times 2^0 = 53_{10}$$

The result is verified as correct.

■

Decimal fractions are converted to binary fractions by a method of repeated multiplication by 2. The fraction is multiplied by 2, and the carry-out into the first integer position, either a 0 or 1, is the bit value for the MSB of the binary fraction. The fractional result from the first multiplication is then multiplied by 2, and its carry-out into the integer portion of the number is the next bit value. Each successive multiplication produces a binary bit value to occupy the next position, moving from the binary point toward the LSB. The multiplication process continues until the fractional part of the number is 0 or until the desired level of precision has been met if exact conversion is impossible.

It is often necessary to decide how many multiplication steps must be performed in order to obtain the same degree of precision that the original decimal number had. Note that in scientific representation, the right-most fractional digit is often inaccurate due to rounding. The number of multiplication steps to be performed can be determined as follows:

Suppose the decimal number has m fractional digits. Then the number has an inaccuracy of 10^{-m}. The number of multiplication steps to be performed (k) should make the resulting binary as precise as 10^{-m}. k can be computed by the following equation:

$$2^k \times 10^{-m} \geq 1 \qquad \text{A-2}$$

For example, if a decimal number has a fractional part with 3 digits, then the number of multiplication steps to be performed is computed as follows:

$$2^k \times 10^{-3} \geq 1 \Rightarrow 2^k \geq 1000$$
$$\therefore k \geq 10$$

Therefore, at least ten multiplication steps must be performed in order to obtain a binary equivalent with the same degree of precision as the original decimal number.

Example A.3

Convert the decimal fraction $.24_{10}$ to its binary equivalent with the same degree of precision.

Solution: Multiply the fraction by 2. The carry-out is the bit value. Since there are two fractional digits in the decimal representation, the number of multiplication steps (k) to be performed should be such that $2^k \times 10^{-2} \geq 1$ if exact conversion is impossible. Therefore, $k \geq 7$.

Multiplication	Product	Carry-out	
0.24×2	0.48	0	MSB
0.48×2	0.96	0	
0.96×2	1.92	1	
0.92×2	1.84	1	
0.84×2	1.68	1	
0.68×2	1.36	1	
0.36×2	0.72	0	LSB

$$0.24_{10} = 0.001111_2$$

Example A.4

Convert the decimal value 53.375_{10} to binary.

Solution: The integer and fraction must be converted separately.

$$53.375_{10} = 53_{10} + .375_{10}$$

Integer:

$$53_{10} = 110101_2$$

Division	Quotient	Remainder	
$53 \div 2$	26	1	LSB
$26 \div 2$	13	0	
$13 \div 2$	6	1	
$6 \div 2$	3	0	
$3 \div 2$	1	1	
$1 \div 2$	0	1	MSB

Fractional part:

$$.375_{10} = .011_2$$

Multiplication	Product	Carry-out	
0.375 × 2	0.75	0	MSB
0.75 × 2	1.50	1	↓
0.50 × 2	1.00	1	LSB

Result: $53.375_{10} = 110101.011_2$

A.3.3 Octal to Decimal Conversion

An octal number can be converted to its decimal equivalent using equation A-1.

Example A.5

Convert the octal number 127.75_8 to its decimal equivalent.

Solution: Applying equation A-1,

$$127.75_8 = 1 \times 8^2 + 2 \times 8^1 + 7 \times 8^0 + 7 \times 8^{-1} + 5 \times 8^{-2} = 87.953125_{10}$$

Example A.6

Convert the octal number 376.444 to its decimal equivalent.

Solution: Applying equation A-1,

$$376.444_8 = 3 \times 8^2 + 7 \times 8^1 + 6 \times 8^0 + 4 \times 8^{-1} + 4 \times 8^{-2} + 4 \times 8^{-3}$$
$$= 254.5703125_{10}$$

A.3.4 Decimal to Octal Conversion

As in decimal to binary conversion, a decimal integer is converted to its octal equivalent by a method of repeated division by 8, while a decimal fraction is converted to its octal equivalent by a method of repeated multiplication by 8. The decimal fraction to be converted is multiplied by 8, and the carry-out into the first integer position is the octal value for the MSB of the octal fraction. The fractional result from the first multiplication is then multiplied by 8, and its carry-out into the integer portion of the number is the next octal value. Each successive multiplication produces an octal digit value to occupy the next position, moving from the fractional point toward the LSB. The multiplication process continues until the fractional part of the number is 0 or until the desired degree of precision has been met if exact conversion is impossible.

Also as in decimal to binary conversion, it is often necessary to decide how many multiplication steps should be performed to obtain the same degree of precision as were found in the original decimal number, since the right-most fractional digit is often inaccurate due to a rounding operation. The number of multiplication steps to carry out can be determined as follows:

> If the decimal number has m fractional digits, it has an inaccuracy of 10^{-m}. The number of multiplication steps to be performed (k) should give the resulting octal the same degree of precision as 10^{-m}. k can be computed by the following equation:

$$8^k \times 10^{-m} \geq 1$$

A-3

Example A.7

Convert the decimal number 100.625_{10} to its octal equivalent.

Solution: The integer and fractional parts must be converted separately.

$$100.625_{10} = 100_{10} + .625_{10}$$

Integer: $100_{10} = 144_8$

Division	Quotient	Remainder	
100 ÷ 8	12	4	LSB
12 ÷ 8	1	4	↓
1 ÷ 8	0	1	MSB

If an exact conversion is impossible for the fractional part, the minimal number of multiplication steps to be performed in order to maintain the same degree of precision as in the original decimal number is computed using equation A-3. In this example, it is found to be four.

Fractional part: $625_{10} = .5_8$

Multiplication	Product	Carry-out
0.625 × 8	5.00	5

Result: $100.625_{10} = 144.5_8$

■

Example A.8

Convert the decimal number 246.245 to its octal equivalent. If an exact conversion for the fractional part cannot be done, perform as many multiplication steps as necessary to maintain the same degree of precision as in the decimal representation.

Solution: The integer and fractional parts must be converted separately.

$$246.245_{10} = 246_{10} + .245_{10}$$

Integer part: $246_{10} = 366_8$

Division	Quotient	Remainder	
246 ÷ 8	30	6	LSB
30 ÷ 8	3	6	↓
3 ÷ 8	0	3	MSB

There are three fractional digits in the decimal number, so the minimal number of multiplication steps needed to maintain the same accuracy as the decimal representation if the exact conversion is impossible is found to be four (using the equation A-3).

Fractional part: $.245_{10} = .1753_8$

Multiplication	Product	Carry-out	
0.245 × 8	1.96	1	MSB
0.96 × 8	7.68	7	↓
0.68 × 8	5.44	5	
0.44 × 8	3.52	3	LSB

Result: $246.245_{10} = 366.1753_8$

■

A.3.5 Hexadecimal to Decimal Conversion

A hexadecimal number can be converted to its decimal equivalent by using equation A-1.

Example A.9

Convert the hexadecimal number $100.4A_{16}$ to its decimal equivalent.

Solution: Applying equation A-1,

$$100.4A_{16} = 1 \times 16^2 + 0 \times 16^1 + 0 \times 16^0 + 4 \times 16^{-1} + 10 \times 16^{-2}$$
$$= 256.2890625_{10}$$

Example A.10

Convert the hexadecimal number $2FF.88_{16}$ to its decimal equivalent.

Solution: Applying equation A-1,

$$2FF.88_{16} = 2 \times 16^2 + 15 \times 16^1 + 15 \times 16^0 + 8 \times 16^{-1} + 8 \times 16^{-2}$$
$$= 767.53125_{10}$$

A.3.6 Decimal to Hexadecimal Conversion

A decimal integer can be converted to its hexadecimal equivalent by a method of repeated division by 16, while the decimal fraction can be converted to its hexadecimal equivalent by a method of repeated multiplication by 16. The conversion from the decimal fraction to the hexadecimal fraction may not be exact, and again the user must decide how many digits are necessary. The number of multiplication steps to perform can be determined as follows:

If the decimal number has m fractional digits, it has an inaccuracy of 10^{-m}. The number of multiplication steps to be performed (k) should give the resulting hexadecimal the same degree of precision as 10^{-m}. k can be computed using the following equation:

$$16^k \times 10^{-m} \geq 1 \qquad \text{A-4}$$

Example A.11

Convert the decimal number 420.625_{10} to its hexadecimal equivalent and maintain the same degree of precision as in the decimal representation.

Solution: $420.625_{10} = 420_{10} + .625_{10}$

Integer part: $420_{10} = 1A4_{16}$

Division	Quotient	Remainder	
$420 \div 16$	26	4	LSB
$26 \div 16$	1	10 (or A)	↓
$1 \div 16$	0	1	MSB

Fractional part: $625_{10} = .A_{16}$

Multiplication	Product	Carry-out
0.625×16	10.00	10 (or A)

Result: $420.625_{10} = 1A4.A_{16}$

Example A.12

Convert the decimal number 980.475_{10} to its hexadecimal equivalent. Maintain the same degree of precision as in the decimal representation if exact conversion is impossible.

Solution: $980.475_{10} = 980_{10} + .475_{10}$

Integer part: $980_{10} = 3D4_{16}$

Division	Quotient	Remainder	
$980 \div 16$	61	4	LSB
$61 \div 16$	3	13 (or D)	↓
$3 \div 16$	0	3	MSB

Fractional part: $.475_{10} = .799_{16}$

The minimal number of multiplication steps to be performed to obtain the same degree of precision is computed using equation A-4 and is found to be three.

Multiplication	Product	Carry-out	
0.475×16	7.6	7	MSB
0.6×16	9.6	9	↓
0.6×16	9.6	9	LSB

Result: $980.475_{10} = 3D4.799_{16}$

Octal digit	Binary code
0	000
1	001
2	010
3	011
4	100
5	101
6	110
7	111

Table A.2 ■ Binary code for octal digit

A.3.7 Binary to Octal Conversion

Since an octal digit is in the range from 0–7, three bits are needed to represent an octal digit. The binary code of each octal digit is shown in Table A.2. The following procedure converts a binary number to its octal equivalent:

1. Start from the fraction point.
2. Partition the integer part from right to left into groups of three bits.
3. Partition the fractional part from left to right into groups of three bits.
4. Write each group of three bits as an octal digit.
5. If the left-most group in the integer part does not have three bits, then add one or two leading 0s to make it a three-bit group.
6. If the right-most group in the fractional part does not have three bits, then append one or two 0s to its right to make it a three-bit group.

Example A.13

Convert the binary number 101011001.110011_2 to its octal equivalent.

Solution: The integer part of this binary number is 101011001. Partition it from right to left into groups of three bits, and write each group of three bits as an octal digit as follows:

$$101,011,001_2 = 531_8$$

Partition the fractional part from left to right into groups of three bits, and write each group of three bits as an octal digit as follows:

$$.110011_2 = .110,011_2 = .63_8$$

Result: $101011001.110011_2 = 531.63_8$

Example A.14

Convert the binary number 1010101010.0010011_2 to its octal equivalent.

Solution: The integer part of this number is 1010101010. Partition it from right to left into groups of three bits, and write each group as an octal digit as follows:

$$1010101010_2 = 1,010,101,010_2 = 001,010,101,010_2 = 1252_8$$

Partition the fractional part from left to right into groups of three bits, and write each group of three bits as an octal digit as follows:

$$.0010011_2 = .001,001,1_2 = .001,001,100_2 = .114_8$$

Result: $1010101010.0010011_2 = 1252.114_8$

A.3.8 Octal to Binary Conversion

To convert an octal number to its binary equivalent, simply convert each octal digit to its corresponding binary combination and then delete the leading and trailing 0s.

Example A.15

Convert the octal number 1375.426_8 to its binary equivalent.

Solution: Convert each octal digit to its corresponding binary code:

$$1375.426_8 = 001011111101.100010110_2 = 1011111101.10001011_2$$

Example A.16

Convert the octal number 364.231_8 to its binary equivalent.

Solution: Convert each octal digit to its corresponding binary code:

$$364.231_8 = 011110100.010011001_2 = 11110100.010011001_2$$

A.3.9 Binary to Hexadecimal Conversion

Since a hexadecimal digit can range from 0 to 15, four bits are needed to encode it. The binary code of each hexadecimal digit is shown in Table A.3.

Hexadecimal digit	Binary code
0	0000
1	0001
2	0010
3	0011
4	0100
5	0101
6	0110
7	0111
8	1000
9	1001
A	1010
B	1011
C	1100
D	1101
E	1110
F	1111

Table A.3 ■ Binary code for hexadecimal digits

The following procedure can be used to convert a binary number to its hexadecimal equivalent:

1. Start from the binary point.
2. Partition the integer part from right to left into groups of four bits.
3. Partition the fractional part from left to right into groups of four bits.
4. Write each group of four bits as a hexadecimal digit.
5. If the left-most group in the integer part does not have four bits, then add 0s to its left to make it a four-bit group.
6. If the right-most group in the fractional part does not have four bits, then append 0s to its right to make it a four-bit group.

Example A.17

Convert the binary number $1000111000100100.101000010011_2$ to its hexadecimal equivalent.

Solution: The integer part of the given binary number is 1000111000100100_2. Partition it from right to left into four-bit groups, and write each group of four bits as its corresponding hexadecimal digit:

$$1000111000100100_2 = 1000,1110,0010,0100_2 = 8E24_{16}$$

The fractional part is $.101000010011_2$. Partition the fraction from left to right into four-bit groups and write each group of four bits into its corresponding hexadecimal digit:

$$.101000010011_2 = .1010,0001,0011 = .A13_{16}$$

Therefore, the equivalent hexadecimal number is $8E24.A13_{16}$.

Example A.18

Convert the binary number $100011100010.00101001001_2$ to its hexadecimal equivalent.

Solution: The integer part of the given number is 100011100010_2. Partition this number from right to left into four-bit groups, and write each group of four bits as its corresponding hexadecimal digit:

$$100011100010_2 = 1000,1110,0010_2 = 8E2_{16}$$

The fractional part is $.00101001001_2$. Partition the fraction from left to right into four-bit groups and write each group of four bits as its corresponding hexadecimal digit:

$$.00101001001_2 = .0010,1001,001_2 = .0010,1001,0010_2 = .292_{16}$$

Therefore, the hexadecimal equivalent of the given number is $8E2.292_{16}$.

A.3.10 Hexadecimal to Binary Conversion

To convert from hexadecimal to binary, simply convert each hexadecimal digit to its binary code and then delete the leading and trailing 0's.

Example A.19

Convert the hexadecimal number $6AB2.04_{16}$ to its binary equivalent.

Solution: Convert each hexadecimal digit to its binary code.

$$6AB2.04_{16} = 0110,1010,1011,0010.0000,0100_2 = 110101010110010.000001_2$$

A.4 Binary Addition and Subtraction

Now that you are familiar with the basics of binary numbers, we will consider addition and subtraction. Addition in binary follows the familiar rules of decimal addition. When adding two numbers, add the successive bits and any carry. You will need only a few addition facts:

$$0 + 0 = 0$$
$$0 + 1 = 1$$
$$1 + 1 = 0 \quad \text{carry} = 1$$
$$1 + 1 + 1 = 1 \quad \text{carry} = 1$$

In this section, we will deal with positive binary numbers only, using as many bits as required to represent a number. The carry generated in any bit position is added to the next higher bit.

Binary numbers are subtracted using the following rules:

$$0 - 0 = 0$$
$$0 - 1 = 1 \quad \text{with a borrow}$$
$$1 - 0 = 1$$
$$1 - 1 = 0$$

Example A.20

Add the following pairs of positive binary numbers:

    a. 1110110 and 1011101

    b. 01101 and 10001

Solution:

a.
```
carry    1 1 1 1
         1 1 1 0 1 1 0₂
       + 1 0 1 1 1 0 1₂
       1 1 0 1 0 0 1 1₂
```

b.
```
carry            1
           0 1 1 0 1₂
         + 1 0 0 0 1₂
           1 1 1 1 0₂
```

a. carry $1\,1\,1\,1$
$$1\,1\,1\,0\,1\,1\,0_2$$
$$+\ 1\,0\,1\,1\,1\,0\,1_2$$
$$1\,1\,0\,1\,0\,0\,1\,1_2$$

b. carry 1
$$0\,1\,1\,0\,1_2$$
$$+\ 1\,0\,0\,0\,1_2$$
$$1\,1\,1\,1\,0_2$$

Example A.21

Perform the following binary subtractions.

    a. $11001_2 - 110_2$

    b. $110011_2 - 11101_2$

Solution: The subtraction is performed step by step as described below (the format may look a little bit confusing). Subtraction starts in column 0 and works toward the higher columns. A column with a 1 in the borrow row indicates that the column to its right has borrowed a 1 from it.

a.
4 3 2 1 0	column
0 1 1 0 0	borrow
1 1 0 0 1₂	minuend
− 1 1 0₂	subtrahend
1 0 0 1 1₂	difference

Column 0
$$1 - 0 = 1$$

Column 1

$0 - 1 = 1$. A borrow is generated in this step. Since the number in column 2 is also a 0, we need to borrow from column 3, which is not a 0. The 0 in column 2 should be changed to 1.

Column 2

Because there was a borrow from column 3, this column becomes $1 - 1 = 0$.

Column 3

After the borrow, this column becomes $0 - 0 = 0$.

Column 4

$1 - 0 = 1$

```
b.   5 4 3 2 1 0     column
     1 1 1 0 0 0     borrow
     1 1 0 0 1 1     minuend
  −    1 1 1 0 1     subtrahend
     0 1 0 1 1 0     difference
```

Column 0

$1 - 1 = 0$

Column 1

$1 - 0 = 1$

Column 2

$0 - 1 = 1$. A borrow is generated. Since the number in column 3 is also a 0, we need to borrow from column 4, which is not a 0. The 0 in column 3 is changed to 1.

Column 3

Because this column has a borrow from column 4, it becomes $1 - 1 = 0$.

Column 4

After the borrow, this column becomes $0 - 1 = 1$, and a borrow is generated.

Column 5

Because of the borrow, this column becomes $0 - 0 = 0$.

A.5 Two's Complement Numbers

In a computer, the number of bits that can be used to express a number is fixed, so computer arithmetic is performed on data stored in fixed-length memory locations. The size of the location is determined by the number of bits. The memory locations of most 8-bit computers, including the 68HC11, are 8 bits in length.

Another restriction on number representation in a computer is that both positive and negative numbers must be expressed. However, a computer does not include the plus and minus signs in a number. Instead, all modern computers use the two's complement number system to represent positive and negative numbers. In the two's complement system, all numbers that have a most significant bit (MSB) of 0 are positive, and all numbers with an MSB of 1 are negative. Positive two's complement numbers are identical to binary numbers, except that the MSB must be a 0. If N is a positive number, then its two's complement N_C is given by the following expression:

$$N_C = 2^n - N \qquad \textbf{A-5}$$

The two's complement of N is used to represent $-N$. Machines that use the two's complement number system can represent integers in the range

$$-2^{n-1} \le N \le 2^{n-1} - 1 \qquad \textbf{A-6}$$

where n is the number of bits available for representing N.

Example A.22

Find the range of integers that can be represented by an 8-bit two's complement number system.

Solution: The range of integers that can be represented by the 8-bit two's complement number system is:

$$-2^7 \le N \le 2^n - 1$$
$$-128_{10} \le N \le 127_{10}$$

Example A.23

Represent the negative binary number -11001_2 in 8-bit two's complement format.

Solution: Use equation A-5 and the subtraction method in section A.4.

$$N_C = 2^8 - 11001 = 11100111_2$$

Example A.24

Represent the negative binary number -1100101_2 in two's complement format.

Solution: Use equation A-5 and the subtraction method in section A.4.

$$N_C = 2^8 - 1100101_2 = 10011011_2$$

It is obvious that a negative two's complement binary number does not represent the magnitude of the given number. However, the magnitude of the negative number can be found using the following equation:

$$N = 2^8 - N_C \qquad \textbf{A-7}$$

where N is the magnitude of the negative number and N_C is the negative two's complement number.

Example A.25

Find the magnitude of the two's complement numbers 11001100_2 and 10010010_2.

Solution: Using equation A-7, the magnitude of 11001100_2 is

$$2^8 - 11001100_2 = 110100_2 = 52_{10}$$

and the magnitude of 10010010_2 is

$$2^8 - 10010010_2 = 1101110_2 = 110_{10}$$

A.6 Two's Complement Negation, Addition, and Subtraction

In this section we will discuss negation, addition, and subtraction of two's complement numbers.

A.6.1 Negating Two's Complement Numbers

To negate a number in the two's complement number system,

1. Invert all bits of the number.
2. Add 1.

Example A.26

Negate the numbers 00110101_2 and 10100011_2.

Solution:
To negate 00110101_2,

1. Invert all bits of $00110101_2 \rightarrow 11001010_2$
2. Add 1 $\rightarrow 11001010_2 + 1_2 = 11001011_2$

To negate 10100011_2,

1. Invert all bits of $10100011_2 \rightarrow 01011100_2$
2. Add 1 $\rightarrow 01011100_2 + 1_2 = 01011101_2$

A.6.2 Two's Complement Addition

In decimal addition, we must be concerned about the signs of two operands. If two numbers are of the same sign, then we simply perform the addition. If they are of different signs, however, they must be subtracted. Two's complement addition is much easier since the signs of the numbers do not have to be considered. As long as overflow does not occur, the sign of the result will always be correct. The carry-out, if it occurs, can simply be discarded. Overflow will be discussed shortly.

Example A.27

Add the decimal numbers 63_{10} and 27_{10} using two's complement arithmetic.

Solution: The 8-bit binary equivalent of 63_{10} is 00111111_2, and the 8-bit binary equivalent of 27_{10} is 00011011_2. Adding these two numbers together,

$$
\begin{array}{r}
0\ 0\ 1\ 1\ 1\ 1\ 1\ 1_2 \\
+\ 0\ 0\ 0\ 1\ 1\ 0\ 1\ 1_2 \\
\hline
0\ 1\ 0\ 1\ 1\ 0\ 1\ 0_2 = 90_{10} = 63_{10} + 27_{10}
\end{array}
$$

Example A.28

Add the decimal numbers 97_{10} and -13_{10} using two's complement arithmetic.

Solution: The 8-bit binary equivalent of the decimal value 97_{10} is 01100001_2, and the 8-bit two's complement equivalent of 13_{10} is 00001101_2. The 8-bit two's complement equivalent of the decimal value -13 is obtained by negating 00001101_2, which yields 11110011_2. Adding these two numbers together,

$$
\begin{array}{r}
0\ 1\ 1\ 0\ 0\ 0\ 0\ 1_2 \\
+\ 1\ 1\ 1\ 1\ 0\ 0\ 1\ 1_2 \\
\hline
1\ 0\ 1\ 0\ 1\ 0\ 1\ 0\ 0_2
\end{array}
$$

carry-out to be discarded

The result is decimal value 84 (01010100_2), which is correct ($97_{10} - 13_{10} = 84_{10}$).

A.6.3 Two's Complement Subtraction

Subtraction in the two's complement number system is performed by negating the subtrahend and then adding it to the minuend. The sign will be correct. The carry, if it occurs, is simply discarded.

Example A.29

Subtract the decimal number 8 from the decimal number 15.

Solution: The two's complement equivalent of 15_{10} is represented in 8-bit format as 00001111_2. The two's complement equivalent of 8_{10} is represented in 8-bit format as 00001000_2; negation of this value yields 11111000_2. Adding these two numbers together,

$$
\begin{array}{r}
0\,0\,0\,0\,1\,1\,1\,1_2 \\
+\ 1\,1\,1\,1\,1\,0\,0\,0_2 \\
\hline
1\ 0\,0\,0\,0\,0\,1\,1\,1_2
\end{array}
$$

carry-out to be discarded

After discarding the carry-out, the result is the decimal value 7_{10}, which is correct.

A.6.4 Overflow

Overflow can occur with either addition or subtraction in two's complement representation. During addition, overflow occurs when the sum of two numbers with like signs have a result with the opposite sign. Overflow never occurs when adding two numbers with unlike signs. In subtraction, overflow can occur when subtracting two numbers with unlike signs. If

negative − positive = positive

or

positive − negative = negative

then overflow has occurred.
Overflow never occurs when subtracting two numbers with like signs.

Example A.30

Does overflow occur in the following 8-bit operations?

    a. $01111111_2 - 00000111_2$
    b. $01100101_2 + 01100000_2$
    c. $10010001_2 - 01110000_2$

Solution:
    a. Negation of 00000111_2 yields 11111001_2. $01111111 - 111$ is performed as follows:

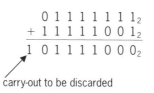

$$0\ 1\ 1\ 1\ 1\ 1\ 1\ 1_2$$
$$+\ 1\ 1\ 1\ 1\ 1\ 0\ 0\ 1_2$$
$$1\ 0\ 1\ 1\ 1\ 1\ 0\ 0\ 0_2$$

carry-out to be discarded

The difference is $01111000_2 = 120_{10}$. There is no overflow.

b. The sum of these two numbers is as follows:

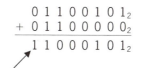

$$0\ 1\ 1\ 0\ 0\ 1\ 0\ 1_2$$
$$+\ 0\ 1\ 1\ 0\ 0\ 0\ 0\ 0_2$$
$$1\ 1\ 0\ 0\ 0\ 1\ 0\ 1_2$$

The sign has changed from positive to negative.

Overflow has occurred.

c. The negation of 01110000 is 10010000. $10010001 - 01110000$ is performed as follows:

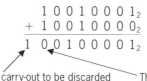

$$1\ 0\ 0\ 1\ 0\ 0\ 0\ 1_2$$
$$+\ 1\ 0\ 0\ 1\ 0\ 0\ 0\ 0_2$$
$$1\ 0\ 0\ 1\ 0\ 0\ 0\ 0\ 1_2$$

carry-out to be discarded   The sign bit is 0, which indicates that the difference is positive. This is impossible because we are subtracting a positive number from a negative number.

Overflow has occurred.

A.7 Octal and Hexadecimal Arithmetic

Although we don't do octal and hexadecimal arithmetic in our daily life, we occasionally need to add, subtract, or negate hexadecimal numbers when studying microprocessors and microcontrollers. The arithmetic in these two number systems is not very different from arithmetic in the decimal number system.

Octal numbers are added and subtracted in the same way as in the decimal number system, except when carries and borrows are generated. During addition, any sum that is equal to or larger than 8 is expressed by subtracting 8 from it and sending a carry to the next higher place. A borrow occurs when the subtrahend is larger than the minuend in any place of subtraction. A borrow has a value of 8, and the minuend in the next more significant place is decremented by 1. If this place is a 0, then it is changed to a 7 and the next higher place is decremented.

Hexadecimal numbers are added and subtracted by mentally replacing any letters (A, B, and so on) with their numeric equivalents. During addition, any sum that is equal to or larger than the decimal value 16 is expressed by subtracting 16 from it and sending a carry to the next more significant digit. During subtraction, a borrow has a decimal value of 16, and the minuend in the next more significant digit is decremented by 1. If this digit is a zero, then it is changed to F and the next higher digit is decremented by 1.

Example A.31

Add the octal numbers 237_8 and 425_8.

Solution: Octal addition is similar to decimal addition.

```
2 1 0      column
0 1 0      carry
2 3 7₈
+ 4 2 5₈
─────────
6 6 4₈     sum
```

Column 0
$5 + 7 = ?$ In decimal, the sum is 12. Because this number is larger than 8, there is a carry into the next higher digit. The sum is 8 less than 12 or 4 in octal.
Column 1
$3 + 2 = ?$ Here $3 + 2 = 5$, but the carry-in from the previous column increments this number to 6.
Column 2
$2 + 4 = ?$ Here $2 + 4 = 6$.

The sum is the octal value 664_8.

Example A.32

Subtract the octal value 236_8 from 432_8.

Solution: The difference is computed as follows:

```
2 1 0      column
1 1 0      borrow
4 3 2₈     minuend
− 2 3 6₈   subtrahend
─────────
1 7 4₈     difference
```

Column 0
$2 − 6 = ?$ Here the subtrahend is larger than the minuend, so a borrow occurs. The borrow allows us to add 8 to the minuend, so column 0 becomes $2 + 8 − 6 = 4$.
Column 1
The borrow from column 0 has changed the minuend in column 1 from 3 to 2. Hence the subtrahend is larger and a borrow occurs.

The borrow allows us to add 8 to the minuend, so column 1 becomes $2 + 8 - 3 = 7$.
Column 2
The borrow from column 1 has changed the minuend in column 2 from 4 to 3. Hence, the subtraction is simply $3 - 2 = 1$.

The difference is the octal value 174_8.

Example A.33

Add the hexadecimal numbers 3215_{16} and $A07B_{16}$.

Solution: The sum is computed as follows:

```
  3 2 1 0       column
  0 0 1 0       carry
  3 2 1 5₁₆
+ A 0 7 B₁₆
  ─────────
  D 2 9 0₁₆     sum
```

Column 0
$5 + B = ?$ In decimal, the answer is 16, so a carry is generated. The sum is 16 less than 16 and is thus 0.
Column 1
$1 + 7 = ?$ $1 + 7 = 8$, and the carry-in from the previous column increments this number to 9.
Column 2
$2 + 0 = ?$ The sum is 2.
Column 3
$3 + A = ?$ $3 + 10 = 13_{10} = D_{16}$.

The sum is the hexadecimal value $D290_{16}$.

Example A.34

Subtract $3CF3_{16}$ from $663F_{16}$.

Solution: The difference is computed as follows:

```
  3 2 1 0       column
  1 1 0 0       borrow
  6 6 3 F₁₆     minuend
− 3 C F 3₁₆     subtrahend
  ─────────
  2 9 4 C₁₆     difference
```

Column 0
$F - 3 = ?$ The hexadecimal number F is 15 in decimal. Therefore, $F - 3 = 15 - 3 = 12_{10} = C_{16}$.

Column 1

$3 - F = ?$ Here the subtrahend is larger than the minuend, so a borrow occurs. The borrow allows us to add 16 to the minuend, so column 1 becomes $3 + 16 - 15 = 4$.

Column 2

The borrow from column 1 has changed the minuend in column 2 from 6 to 5. Hence the subtrahend is larger and a borrow occurs. The borrow allows us to add 16 to the minuend, so column 2 becomes $5 + 16 - 12 = 9$.

Column 3

The borrow in column 2 has changed the minuend in column 3 from 6 to 5. Hence the difference is $5 - 3 = 2$.

The difference is the hexadecimal value $294C_{16}$.

A.8 Negating Octal and Hexadecimal Numbers

Negating octal numbers or hexadecimal numbers means finding the negative equivalent of a given octal or hexadecimal number. Therefore, if an octal or hexadecimal number is positive (has an MSB of 0 in its binary equivalent), the result of the negation will be a negative number, and vice versa. In this section, we will consider only fixed-length numbers (either 8 bits or 16 bits). The 68HC11 can work only on 8-bit data, and a memory address is represented in 16 bits.

Since every octal digit is encoded into 3 binary bits and neither 8 nor 16 is a multiple of 3, we will use the following steps to negate an octal number:

Step 1

Convert the octal number to its binary equivalent with an appropriate number of bits (either 8 or 16).

Step 2

Negate the binary equivalent.

Step 3

Convert the two's complement number to its octal equivalent.

Example A.35

Negate the following octal numbers.

a. 144_8

b. 06624_8

c. 120006_8

Solution: Convert these octal numbers to binary and then negate them:

a. Convert the octal number 144 to an 8-bit binary number: 01100100_2
   Negate the binary equivalent: 10011100
   Convert to octal: 234_8

b. Convert the octal number 06624 to a 16-bit binary number: 0000110110010100_2.
Negate the binary equivalent: 1111001001101100_2
Convert to octal: 171154_8

c. Convert the octal number 120006 to a 16-bit binary number: 1010000000000110_2.
Negate the binary equivalent: 0101111111111010_2
Convert to octal: 057772_8

We can use the same algorithm to negate hexadecimal numbers. Since each hexadecimal digit can be encoded in 4 bits, each 8-bit and 16-bit binary number can be converted to 2 and 4 hexadecimal digits. The following procedure will speed up the negation of hexadecimal numbers:

Step 1
Start with the least significant nonzero digit. Subtract the digit from the hexadecimal number 10 and write down the difference.
Step 2
Subtract each higher digit from the hexadecimal value F and write down the difference.

Example A.36

Negate the following hexadecimal numbers:

a. $A107_{16}$
b. 2349_{16}
c. $44A0_{16}$

Solution:

a. The least significant nonzero digit is 7. Subtract it from the hexadecimal value 10. The difference is 9. Subtract each higher digit from F as follows:

$$F - 0 = F$$
$$F - 1 = E$$
$$F - A = 5$$

Therefore

$$-A107_{16} = 5EF9_{16}$$

b. The least significant nonzero digit is 9. Subtract it from the hexadecimal value 10.

$$10_{16} - 9 = 7$$

Subtract each higher digit from F as follows:

$$F - 4 = B$$
$$F - 3 = C$$
$$F - 2 = D$$

Therefore

$$-2349_{16} = DCB7_{16}$$

c. The least significant nonzero digit is A. Subtract it from the hexadecimal value 10.

$$10_{16} - A = 6$$

Subtract each higher digit from F as follows:

F − 4 = B
F − 4 = B

Therefore,

$$-44A0_{16} = BB60_{16}$$

A.9 Coding Decimal Numbers

Although virtually all digital systems are binary in the sense that all internal signals can take on only two values, some nevertheless perform arithmetic in the decimal system. In some circumstances, the identity of decimal numbers is retained to the extent that each decimal digit is individually represented by a binary code. There are ten decimal digits, so four binary bits are required for each code element. The most obvious choice is to let each decimal digit be represented by its corresponding four-bit binary number, as shown in Table A.4. This form of representation is known as the *BCD* (binary-coded decimal) representation or code.

Most input to and output from computer systems is in the decimal system because this system is most convenient for human users. On input, decimal

Decimal digit	Binary representation
0	0000
1	0001
2	0010
3	0011
4	0100
5	0101
6	0110
7	0111
8	1000
9	1001

Table A.4 ■   Binary-coded decimal digits

Decimal digit	Excess-3 d_3 d_2 d_1 d_0				2-out-of-5 d_4 d_3 d_2 d_1 d_0					Gray code d_3 d_2 d_1 d_0			
0	0	0	1	1	0	0	0	1	1	0	0	1	0
1	0	1	0	0	0	0	1	0	1	0	1	1	0
2	0	1	0	1	0	0	1	1	0	0	1	1	1
3	0	1	1	0	0	1	0	0	1	0	1	0	1
4	0	1	1	1	0	1	0	1	0	0	1	0	0
5	1	0	0	0	0	1	1	0	0	1	1	0	0
6	1	0	0	1	1	0	0	0	1	1	1	0	1
7	1	0	1	0	1	0	0	1	0	1	1	1	1
8	1	0	1	1	1	0	1	0	0	1	1	1	0
9	1	1	0	0	1	1	0	0	0	1	0	1	0

Table A.5 ■ Other coding methods for decimal digits

numbers are converted into some binary form for processing, and this conversion is reversed on output. In a "straight binary" computer, a decimal number such as 46 is converted into its binary equivalent, that is, 101110. In a computer using the BCD system, 46 would be converted into 0100 0110. If the BCD format is used, it must be preserved during arithmetic processing. For example, addition in a BCD machine would be carried out as shown below.

```
decimal        BCD
   46          0100 0110
 + 38        + 0011 1000
 ----          ---------
   84          1000 0100
```

The 68HC11 handles BCD addition with the DAA instruction. DAA stands for *decimal adjust accumulator*; this instruction must be executed immediately following a BCD addition. The interested reader should refer to section 2.8.4.

The principal advantage of the BCD system is the simplicity of input/output conversion; its principal disadvantage is the complexity of arithmetic processing. The choice (between binary and BCD) depends on the type of problems the system will be handling.

The BCD code is not the only possible coding method for decimal digits. The excess-3 code is larger than its corresponding decimal digit by 3. The 2-out-of-5 code represents each decimal digit by one of the ten possible combinations of two 1s and three 0s. The distinguishing feature of the Gray code is that successive coded characters never differ in more than one bit. The excess-3, 2-out-of-5, and Gray codes are shown in Table A.5.

A.10 Representing Character Data

Computers use numbers to represent characters, but there is no natural correspondence between the characters and the numbers, so the assignment of number codes to characters must be defined. Two character code sets are in widespread use today: ASCII (American Standard Code for Information Interchange) and EBCDIC (Extended Binary Coded Decimal Interchange Code). Several types of mainframe computers use EBCDIC for internal storage and processing of characters. In EBCDIC, each character is represented by a unique 8-bit number, and a total of 256 different characters can be represented. Most microcomputer and minicomputer systems use the ASCII character set. A computer does not depend on any particular set, but most I/O devices that display characters require the use of ASCII codes. Only ASCII codes will be discussed in this text.

A.10.1 ASCII Codes

ASCII characters have a 7-bit code. They are usually stored in a fixed-length 8-bit number. The $2^7 = 128$ different codes are partitioned into 95 print characters and 33 control characters. The control characters define communication protocols and special operations on peripheral devices. The printable characters consist of the following:

> 26 uppercase letters (A–Z)
>
> 26 lowercase letters (a–z)
>
> 10 digits (0–9)
>
> 1 blank space
>
> 32 special-character symbols, including !@#$%^&*()-_=+'[];:'",<.>/?{}

The complete ASCII character set is shown in Table A.6. The binary number corresponding to each character is given in hexadecimal.

A.10.2 Control Characters

For efficient code conversion, the alphanumeric characters occupy contiguous binary ranges. This property is important for I/O code conversion and the design of I/O devices. Table A.7 shows the ranges of the control characters, the decimal digits, and the alphabetic characters.

BEL is one of the control characters in the ASCII set. It has a binary value of 00000111 (7), and its transmission to the terminal causes the bell to ring. Used infrequently, this character can remind a typist of a "near-end-of-line" condition in a word processing program or indicate an invalid editing stroke. Like other control characters, it has special terminal or data transmission applications. Table A.8 lists some noteworthy control characters, along with descriptions of their actions on typical peripheral devices.

Many of the other 27 control characters have special applications in conjunction with intelligent terminals and printers (for example, clearing the screen, inserting a line, or specifying cursor motion or special character fonts). Some of these characters are used as control codes in data communications for message switching and telegrams.

Seven-bit hexadecimal code	Character	Seven-bit hexadecimal code	Character	Seven-bit hexadecimal code	Character	Seven-bit hexadecimal code	Character
00	NUL	20	SP	40	@	60	`
01	SOH	21	!	41	A	61	a
02	STX	22	"	42	B	62	b
03	ETX	23	#	43	C	63	c
04	EOT	24	$	44	D	64	d
05	ENQ	25	%	45	E	65	e
06	ACK	26	&	46	F	66	f
07	BEL	27	'	47	G	67	g
08	BS	28	(	48	H	68	h
09	HT	29	)	49	I	69	i
0A	LF	2A	*	4A	J	6A	j
0B	VT	2B	+	4B	K	6B	k
0C	FF	2C	,	4C	L	6C	l
0D	CR	2D	-	4D	M	6D	m
0E	SO	2E	.	4E	N	6E	n
0F	SI	2F	/	4F	O	6F	o
10	DLE	30	0	50	P	70	p
11	DC1	31	1	51	Q	71	q
12	DC2	32	2	52	R	72	r
13	DC3	33	3	53	S	73	s
14	DC4	34	4	54	T	74	t
15	NAK	35	5	55	U	75	u
16	SYN	36	6	56	V	76	v
17	ETB	37	7	57	W	77	w
18	CAN	38	8	58	X	78	x
19	EM	39	9	59	Y	79	y
1A	SUB	3A	:	5A	Z	7A	z
1B	ESC	3B	;	5B	[	7B	{
1C	FS	3C	<	5C	\	7C		
1D	GS	3D	=	5D	]	7D	}
1E	RS	3E	>	5E	^	7E	~
1F	US	3F	?	5F	_	7F	DEL

Table A.6 ■ ASCII code chart

ASCII characters	Eight-bit code range		
	Decimal	Hexadecimal	Binary
control characters	0–31	00–1F	00000000–00011111
blank space	32	20	00100000
decimal digits (0–9)	48–57	30–39	00110000–00111001
uppercase letters (A–Z)	65–90	41–5A	01000001–01011010
lowercase letters (a–z)	97–122	61–7A	01100001–01101010
special characters	unused codes in the range 33–126		
delete character	127		01111111

Table A.7 ■ ASCII character ranges

| Symbol | Value | | Meaning |
	Hexadecimal	Decimal	
CR	0D	13	carriage return—move the screen cursor to the beginning of the current line
LF	0A	10	line feed—advance the screen cursor down one line
BS	08	8	backspace—backspace the screen cursor
FF	0C	12	form feed—advance the hard copy unit to the top of the next page
BEL	07	7	ring the bell on the terminal
HT	09	9	horizontal tab—advance the screen cursor to the next tab stop (as on a typewriter)

Table A.8 ■ Selected ASCII control characters

A.11 Summary

Binary numbers are used to represent data in digital systems. The binary values 1 and 0 correspond to the two allowed states in digital electronics. The binary number system is a weighted base-2 system in which n bits can represent 2^n numbers.

The octal and hexadecimal codes are mainly used as shorthands for the binary code. Three binary bits can be represented by one octal digit, and four binary bits can be represented by one hexadecimal digit. Numbers can be converted between the binary, octal, decimal, and hexadecimal formats.

The binary number system is used in all arithmetic operations (addition, subtraction, multiplication, and division). Signed-number arithmetic operations can be accomplished entirely through addition by representing numbers in their two's complement form.

A.12 Glossary

ASCII (American Standard Code for Information Interchange) code   A code that uses seven bits to encode all printable and control characters.

BCD (Binary Coded Decimal)   A code in which each decimal digit is represented by its equivalent 4-bit binary number.

EBCDIC (Extended Binary Coded Decimal Interchange Code)   A code used mainly in IBM mainframe computers; it uses 8 bits to represent each character.

Excess-3   A coding method for decimal digits; each excess-3 code is larger than its corresponding decimal digit by 3.

Gray code   A coding method for decimal digits in which successive coded characters never differ in more than one bit.

Overflow   A condition that occurs when the result of an arithmetic operation cannot be accommodated by the preset number of bits (say, 8 or 16 bits); it occurs fairly often when numbers are represented by fixed numbers of bits.

A.13 Exercises

EA.1. Convert the following binary numbers to their decimal equivalents:

    a. 1001011000_2

    b. 100011110_2

    c. 100000001_2

    d. 111111111_2

EA.2. Convert the following octal numbers to their decimal equivalents:

    a. 2335_8

    b. 167_8

    c. 4005.246_8

    d. 625.575_8

EA.3. Convert the following hexadecimal numbers to their decimal equivalents:

    a. 250_{16}

    b. 1200_{16}

    c. $4AE0_{16}$

    d. $FFFE.428_{16}$

    e. $245.62B_{16}$

EA.4. Convert the following decimal numbers to their binary, octal, and hexadecimal equivalents:

    a. 4135_{10}

    b. 625.625_{10}

    c. 910.125_{10}

    d. 1360.275_{10}

    e. 415_{10}

EA.5. Convert the following binary numbers to their octal and hexadecimal equivalents:

    a. 100100111110_2

    b. 101010111100_2

    c. 1000001110110_2

    d. 1001101100001_2

EA.6. Convert the following octal numbers to their binary equivalents:

    a. 423_8

    b. 1246_8

    c. 3314_8

    d. 5425_8

    e. 2400_8

EA.7. Convert the following hexadecimal numbers to their binary equivalents:

    a. 38_{16}

    b. $4AC0_{16}$

    c. $FF02_{16}$

    d. 1240_{16}

    e. $8B0_{16}$

EA.8. Compute the sums of the following pairs of binary numbers:

    a. 10010001_2 and $10110110 0_2$

    b. 110010_2 and 100011_2

EA.9. Negate the following 8-bit binary numbers:

    a. 00100001_2

    b. 10101010_2

    c. 01100001_2

    d. 00010011_2

    e. 10011100_2

EA.10. Use 8-bit two's complement addition to perform the indicated operations:

    a. $01001011_2 - 00010111_2$

    b. $01000001_2 - 00010001_2$

    c. $10000001_2 - 00100000_2$

Does overflow occur in any of these operations?

EA.11. What range of integers can be represented by the 16-bit two's complement number system?

EA.12. What range of integers can be represented by the 32-bit two's complement number system?

EA.13. Perform the following octal operations:

    a. $75_8 - 12_8$

    b. $420_8 + 365_8$

    c. $330_8 - 225_8$

    d. $471_8 - 335_8$

EA.14. Perform the following hexadecimal operations:

    a. $A7_{16} - 66_{16}$

    b. $58_{16} - 3F_{16}$

    c. $459_{16} + 1FA_{16}$

    d. $330_{16} + 8FB_{16}$

    e. $12F4_{16} - 25F9_{16}$

EA.15. Negate the following hexadecimal numbers:

    a. $357A_{16}$

    b. $A03B_{16}$

    c. 2488_{16}

    d. 1239_{16}

    e. $F0A7_{16}$

APPENDIX

B

DATA

SHEETS

B.1

68HC11

INSTRUCTION

SET

APPENDIX A
INSTRUCTION SET DETAILS

A.1 INTRODUCTION

This appendix contains complete detailed information for all M68HC11 instructions. The instructions are arranged in alphabetical order with the instruction mnemonic set in larger type for easy reference.

A.2 NOMENCLATURE

The following nomenclature is used in the subsequent defintions.

(a) Operators

( )	= Contents of Register Shown Inside Parentheses
◂	= Is Transferred to
▲	= Is Pulled from Stack
▼	= Is Pushed onto Stack
•	= Boolean AND
+	= Arithmetic Addition Symbol Except Where Used as Inclusive-OR Symbol in Boolean Formula
⊕	= Exclusive-OR
×	= Multiply
:	= Concatenation
−	= Arithmetic Subtraction Symbol or Negation Symbol (Twos Complement)

(b) Registers in the MPU

ACCA	= Accumulator A
ACCB	= Accumulator B
ACCX	= Accumulator ACCA or ACCB
ACCD	= Double Accumulator — Accumulator A Concatenated with Accumulator B Where A is the Most Significant Byte
CCR	= Condition Code Register
IX	= Index Register X, 16 Bits
IXH	= Index Register X, Higher Order 8 Bits
IXL	= Index Register X, Lower Order 8 Bits
IY	= Index Register Y, 16 Bits
IYH	= Index Register Y, Higher Order 8 Bits
IYL	= Index Register Y, Lower Order 8 Bits
PC	= Program Counter, 16 Bits
PCH	= Program Counter, Higher Order (Most Significant) 8 Bits
PCL	= Program Counter, Lower Order (Least Significant) 8 Bits
SP	= Stack Pointer, 16 Bits
SPH	= Stack Pointer, Higher Order 8 Bits
SPL	= Stack Pointer, Lower Order 8 Bits

(c) Memory and Addressing
- M   = A Memory Location (One Byte)
- M + 1   = The Byte of Memory at $0001 Plus the Address of the Memory Location Indicated by "M"
- Rel   = Relative Offset (i.e., the Twos Complement Number Stored in the Last Byte of Machine Code Corresponding to a Branch Instruction)
- (opr)   = Operand
- (msk)   = Mask Used in Bit Manipulation Instructions
- (rel)   = Relative Offset Used in Branch Instructions

(d) Bits 7–0 of the Condition Code Register
- S   = Stop Disable, Bit 7
- X   = X Interrupt Mask, Bit 6
- H   = Half Carry, Bit 5
- I   = I Interrupt Mask, Bit 4
- N   = Negative Indicator, Bit 3
- Z   = Zero Indicator, Bit 2
- V   = Twos Complement Overflow Indicator, Bit 1
- C   = Carry/Borrow, Bit 0

(e) Status of Individual Bits BEFORE Execution of an Instruction
- An   = Bit n of ACCA (n = 7, 6, 5 . . . 0)
- Bn   = Bit n of ACCB (n = 7, 6, 5 . . . 0)
- Dn   = Bit n of ACCD (n = 15, 14, 13 . . . 0)
         Where Bits 15–8 Refer to ACCA and Bit 7–0 Refer to ACCB
- IXn   = Bit n of IX (n = 15, 14, 13 . . . 0)
- IXHn   = Bit n of IXH (n = 7, 6, 5 . . . 0)
- IXLn   = Bit n of IXL (n = 7, 6, 5 . . . 0)
- IYn   = Bit n of IY (n = 15, 14, 13 . . . 0)
- IYHn   = Bit n of IYH (n = 7, 6, 5 . . . 0)
- IYLn   = Bit n of IYL (n = 7, 6, 5 . . . 0)
- Mn   = Bit n of M (n = 7, 6, 5 . . . 0)
- SPHn   = Bit n of SPH (n = 7, 6, 5 . . . 0)
- SPLn   = Bit n of SPL (n = 7, 6, 5 . . . 0)
- Xn   = Bit n of ACCX (n = 7, 6, 5 . . . 0)

(f) Status of Individual Bits of RESULT of Execution of an Instruction
  (i) For 8-Bit Results
   - Rn   = Bit n of the Result (n = 7, 6, 5 . . . 0)
          This applies to instructions which provide a result contained in a single byte of memory or in an 8-bit register.
  (ii) For 16-Bit Results
   - RHn = Bit n of the Most Significant Byte of the Result (n = 7, 6, 5 . . . 0)
   - RLn = Bit n of the Least Significant Byte of the Result (n = 7, 6, 5 . . . 0)
          This applies to instructions which provide a result contained in two consecutive bytes of memory or in a 16-bit register.
   - Rn   = Bit n of the Result (n = 15, 14, 13 . . . 0)

(g) Notation Used in CCR Activity Summary Figures
  —   = Bit Not Affected
  0   = Bit Forced to Zero
  1   = Bit Forced to One
  ♦   = Bit Set or Cleared According to Results of Operation
  ♦   = Bit may change from one to zero, remain zero, or remain one as a result
            of this operation, but cannot change from zero to one.

(h) Notation Used in Cycle-by-Cycle Execution Tables
  —   = Irrelevant Data
  ii   = One Byte of Immediate Data
  jj   = High-Order Byte of 16-Bit Immediate Data
  kk   = Low-Order Byte of 16-Bit Immediate Data
  hh   = High-Order Byte of 16-Bit Extended Address
  ll   = Low-Order Byte of 16-Bit Extended Address
  dd   = Low-Order 8 Bits of Direct Address $0000–$00FF
            (High Byte Assumed to be $00)
  mm   = 8-Bit Mask (Set Bits Correspond to Operand Bits Which Will Be Affected)
  ff   = 8-Bit Forward Offset $00 (0) to $FF (255) (Is Added to Index)
  rr   = Signed Relative Offset $80 (− 128) to $7F (+ 127)
            (Offset Relative to Address Following Machine Code Offset Byte)
  OP   = Address of Opcode Byte
  OP + n   = Address of nth Location after Opcode Byte
  SP   = Address Pointed to by Stack Pointer Value (at the Start of an Instruction)
  SP + n   = Address of nth Higher Address Past That Pointed to by Stack Pointer
  SP − n   = Address of nth Lower Address Before That Pointed to by Stack Pointer
  Sub   = Address of Called Subroutine
  Nxt op   = Opcode of Next Instruction
  Rtn hi   = High-Order Byte of Return Address
  Rtn lo   = Low-Order Byte of Return Address
  Svc hi   = High-Order Byte of Address for Service Routine
  Svc lo   = Low-Order Byte of Address for Service Routine
  Vec hi   = High-Order Byte of Interrupt Vector
  Vec lo   = Low-Order Byte of Interrupt Vector

ABA   Add Accumulator B to Accumulator A   ABA

Operation: ACCA ⬢ (ACCA) + (ACCB)

Description: Adds the contents of accumulator B to the contents of accumulator A and places the result in accumulator A. Accumulator B is not changed. This instruction affects the H condition code bit so it is suitable for use in BCD arithmetic operations (see DAA instruction for additional information).

Condition Codes and Boolean Formulae:

S	X	H	I	N	Z	V	C
—	—	⬣	—	⬣	⬣	⬣	⬣

H $A3 \cdot B3 + B3 \cdot \overline{R3} + \overline{R3} \cdot A3$
Set if there was a carry from bit 3; cleared otherwise.

N   R7
Set if MSB of result is set; cleared otherwise.

Z $\overline{R7} \cdot \overline{R6} \cdot \overline{R5} \cdot \overline{R4} \cdot \overline{R3} \cdot \overline{R2} \cdot \overline{R1} \cdot \overline{R0}$
Set if result is $00; cleared otherwise.

V $A7 \cdot B7 \cdot \overline{R7} + \overline{A7} \cdot \overline{B7} \cdot R7$
Set if a twos complement overflow resulted from the operation; cleared otherwise.

C $A7 \cdot B7 + B7 \cdot \overline{R7} + \overline{R7} \cdot A7$
Set if there was a carry from the MSB of the result; cleared otherwise.

Source Forms: ABA

Addressing Modes, Machine Code, and Cycle-by-Cycle Execution:

Cycle	ABA (INH)		
	Addr	Data	R/$\overline{W}$
1	OP	1B	1
2	OP + 1	—	1

ABX   Add Accumulator B to Index Register X   **ABX**

Operation: IX ◀ (IX) + (ACCB)

Description: Adds the 8-bit unsigned contents of accumulator B to the contents of index register X (IX) considering the possible carry out of the low-order byte of the index register X; places the result in index register X (IX). Accumulator B is not changed. There is no equivalent instruction to add accumulator A to an index register.

Condition Codes and Boolean Formulae:

S	X	H	I	N	Z	V	C
—	—	—	—	—	—	—	—

None affected

Source Forms: ABX

Addressing Modes, Machine Code, and Cycle-by-Cycle Execution:

Cycle	ABX (INH)		
	Addr	Data	R/W̄
1	OP	3A	1
2	OP + 1	—	1
3	FFFF	—	1

ABY   Add Accumulator B to Index Register Y   **ABY**

Operation: IY ⬦ (IY) + (ACCB)

Description: Adds the 8-bit unsigned contents of accumulator B to the contents of index register Y (IY) considering the possible carry out of the low-order byte of index register Y; places the result in index register Y (IY). Accumulator B is not changed. There is no equivalent instruction to add accumulator A to an index register.

Condition Codes and Boolean Formulae:

S	X	H	I	N	Z	V	C
—	—	—	—	—	—	—	—

None affected

Source Forms: ABY

Addressing Modes, Machine Code, and Cycle-by-Cycle Execution:

Cycle	ABY (INH)		
	Addr	Data	R/W̄
1	OP	18	1
2	OP+1	3A	1
3	OP+2	—	1
4	FFFF	—	1

ADC   Add with Carry   ADC

Operation: ACCX ◂ (ACCX) + (M) + (C)

Description: Adds the contents of the C bit to the sum of the contents of ACCX and M and places the result in ACCX. This instruction affects the H condition code bit so it is suitable for use in BCD arithmetic operations (see DAA instruction for additional information).

Condition Codes and Boolean Formulae:

S	X	H	I	N	Z	V	C
—	—	$\updownarrow$	—	$\updownarrow$	$\updownarrow$	$\updownarrow$	$\updownarrow$

H $X3 \cdot M3 + M3 \cdot \overline{R3} + \overline{R3} \cdot X3$
Set if there was a carry from bit 3; cleared otherwise.

N $R7$
Set if MSB of result is set; cleared otherwise.

Z $\overline{R7} \cdot \overline{R6} \cdot \overline{R5} \cdot \overline{R4} \cdot \overline{R3} \cdot \overline{R2} \cdot \overline{R1} \cdot \overline{R0}$
Set if result is $00; cleared otherwise.

V $X7 \cdot M7 \cdot \overline{R7} + \overline{X7} \cdot \overline{M7} \cdot R7$
Set if a twos complement overflow resulted from the operation; cleared otherwise.

C $X7 \cdot M7 + M7 \cdot \overline{R7} + \overline{R7} \cdot X7$
Set if there was a carry from the MSB of the result; cleared otherwise.

Source Forms: ADCA (opr); ADCB (opr)

Addressing Modes, Machine Code, and Cycle-by-Cycle Execution:

Cycle	ADCA (IMM)			ADCA (DIR)			ADCA (EXT)			ADCA (IND, X)			ADCA (IND, Y)		
	Addr	Data	R/$\overline{W}$	Addr	Data	R/$\overline{W}$	Addr	Data	R/$\overline{W}$	Addr	Data	R/$\overline{W}$	Addr	Data	R/$\overline{W}$
1	OP	89	1	OP	99	1	OP	B9	1	OP	A9	1	OP	18	1
2	OP+1	ii	1	OP+1	dd	1	OP+1	hh	1	OP+1	ff	1	OP+1	A9	1
3				00dd	(00dd)	1	OP+2	ll	1	FFFF	—	1	OP+2	ff	1
4							hhll	(hhll)	1	X+ff	(X+ff)	1	FFFF	—	1
5													Y+ff	(Y+ff)	1

Cycle	ADCB (IMM)			ADCB (DIR)			ADCB (EXT)			ADCB (IND, X)			ADCB (IND, Y)		
	Addr	Data	R/$\overline{W}$	Addr	Data	R/$\overline{W}$	Addr	Data	R/$\overline{W}$	Addr	Data	R/$\overline{W}$	Addr	Data	R/$\overline{W}$
1	OP	C9	1	OP	D9	1	OP	F9	1	OP	E9	1	OP	18	1
2	OP+1	ii	1	OP+1	dd	1	OP+1	hh	1	OP+1	ff	1	OP+1	E9	1
3				00dd	(00dd)	1	OP+2	ll	1	FFFF	—	1	OP+2	ff	1
4							hhll	(hhll)	1	X+ff	(X+ff)	1	FFFF	—	1
5													Y+ff	(Y+ff)	1

ADD

Add without Carry

ADD

Operation: ACCX ◀ (ACCX) + (M)

Description: Adds the contents of M to the contents of ACCX and places the result in ACCX. This instruction affects the H condition code bit so it is suitable for use in the BCD arithmetic operations (see DAA instruction for additional information).

Condition Codes and Boolean Formulae:

S	X	H	I	N	Z	V	C
—	—	$\updownarrow$	—	$\updownarrow$	$\updownarrow$	$\updownarrow$	$\updownarrow$

H $X3 \cdot M3 + M3 \cdot \overline{R3} + \overline{R3} \cdot X3$
Set if there was a carry from bit 3; cleared otherwise.

N $R7$
Set if MSB of result is set; cleared otherwise.

Z $\overline{R7} \cdot \overline{R6} \cdot \overline{R5} \cdot \overline{R4} \cdot \overline{R3} \cdot \overline{R2} \cdot \overline{R1} \cdot \overline{R0}$
Set if result is $00; cleared otherwise.

V $X7 \cdot M7 \cdot \overline{R7} + \overline{X7} \cdot \overline{M7} \cdot R7$
Set if a twos complement overflow resulted from the operation; cleared otherwise.

C $X7 \cdot M7 + M7 \cdot \overline{R7} + \overline{R7} \cdot X7$
Set if there was a carry from the MSB of the result; cleared otherwise.

Source Forms: ADDA (opr); ADDB (opr)

Addressing Modes, Machine Code, and Cycle-by-Cycle Execution:

Cycle	ADDA (IMM)			ADDA (DIR)			ADDA (EXT)			ADDA (IND, X)			ADDA (IND, Y)		
	Addr	Data	R/$\overline{W}$	Addr	Data	R/$\overline{W}$	Addr	Data	R/$\overline{W}$	Addr	Data	R/$\overline{W}$	Addr	Data	R/$\overline{W}$
1	OP	8B	1	OP	9B	1	OP	BB	1	OP	AB	1	OP	18	1
2	OP+1	ii	1	OP+1	dd	1	OP+1	hh	1	OP+1	ff	1	OP+1	AB	1
3				00dd	(00dd)	1	OP+2	ll	1	FFFF	—	1	OP+2	ff	1
4							hhll	(hhll)	1	X+ff	(X+ff)	1	FFFF	—	1
5													Y+ff	(Y+ff)	1

Cycle	ADDB (IMM)			ADDB (DIR)			ADDB (EXT)			ADDB (IND, X)			ADDB (IND, Y)		
	Addr	Data	R/$\overline{W}$	Addr	Data	R/$\overline{W}$	Addr	Data	R/$\overline{W}$	Addr	Data	R/$\overline{W}$	Addr	Data	R/$\overline{W}$
1	OP	CB	1	OP	DB	1	OP	FB	1	OP	EB	1	OP	18	1
2	OP+1	ii	1	OP+1	dd	1	OP+1	hh	1	OP+1	ff	1	OP+1	EB	1
3				00dd	(00dd)	1	OP+2	ll	1	FFFF	—	1	OP+2	ff	1
4							hhll	(hhll)	1	X+ff	(X+ff)	1	FFFF	—	1
5													Y+ff	(Y+ff)	1

ADDD   Add Double Accumulator   ADDD

Operation: ACCD ◀ (ACCD) + (M:M + 1)

Description: Adds the contents of M concatenated with M + 1 to the contents of ACCD and places the results in ACCD. Accumulator A corresponds to the high-order half of the 16-bit double accumulator D.

Condition Codes and Boolean Formulae:

S	X	H	I	N	Z	V	C
—	—	—	—	↕	↕	↕	↕

N   R15
Set if MSB of result is set; cleared otherwise.

Z $\overline{R15} \cdot \overline{R14} \cdot \overline{R13} \cdot \overline{R12} \cdot \overline{R11} \cdot \overline{R10} \cdot \overline{R9} \cdot \overline{R8} \cdot \overline{R7} \cdot \overline{R6} \cdot \overline{R5} \cdot \overline{R4} \cdot \overline{R3} \cdot \overline{R2} \cdot \overline{R1} \cdot \overline{R0}$
Set if result is $0000; cleared otherwise.

V $D15 \cdot M15 \cdot \overline{R15} + \overline{D15} \cdot \overline{M15} \cdot R15$
Set if a twos complement overflow resulted from the operation; cleared otherwise.

C $D15 \cdot M15 + M15 \cdot \overline{R15} + \overline{R15} \cdot D15$
Set if there was a carry from the MSB of the result; cleared otherwise.

Source Form: ADDD (opr)

Addressing Modes, Machine Code, and Cycle-by-Cycle Execution:

Cycle	ADDD (IMM) Addr	Data	R/W̄	ADDD (DIR) Addr	Data	R/W̄	ADDD (EXT) Addr	Data	R/W̄	ADDD (IND, X) Addr	Data	R/W̄	ADDD (IND, Y) Addr	Data	R/W̄
1	OP	C3	1	OP	D3	1	OP	F3	1	OP	E3	1	OP	18	1
2	OP + 1	jj	1	OP + 1	dd	1	OP + 1	hh	1	OP + 1	ff	1	OP + 1	E3	1
3	OP + 2	kk	1	00dd	(00dd)	1	OP + 2	ll	1	FFFF	—	1	OP + 2	ff	1
4	FFFF	—	1	00dd + 1	(00dd + 1)	1	hhll	(hhll)	1	X + ff	(X + ff)	1	FFFF	—	1
5				FFFF	—	1	hhll + 1	(hhll + 1)	1	X + ff + 1	(X + ff + 1)	1	Y + ff	(Y + ff)	1
6							FFFF	—	1	FFFF	—	1	Y + ff + 1	(Y + ff + 1)	1
7													FFFF	—	1

AND AND

Logical AND

Operation: ACCX ⬦ (ACCX) • (M)

Description: Performs the logical AND between the contents of ACCX and the contents of M and places the result in ACCX. (Each bit of ACCX after the operation will be the logical AND of the corresponding bits of M and of ACCX before the operation.)

Condition Codes and Boolean Formulae:

S	X	H	I	N	Z	V	C
—	—	—	—	⬍	⬍	0	—

N R7
   Set if MSB of result is set; cleared otherwise.

Z $\overline{R7} \cdot \overline{R6} \cdot \overline{R5} \cdot \overline{R4} \cdot \overline{R3} \cdot \overline{R2} \cdot \overline{R1} \cdot \overline{R0}$
   Set if result is $00; cleared otherwise.

V 0
   Cleared

Source Forms: ANDA (opr); ANDB (opr)

Addressing Modes, Machine Code, and Cycle-by-Cycle Execution:

Cycle	ANDA (IMM) Addr	Data	R/W̄	ANDA (DIR) Addr	Data	R/W̄	ANDA (EXT) Addr	Data	R/W̄	ANDA (IND, X) Addr	Data	R/W̄	ANDA (IND, Y) Addr	Data	R/W̄
1	OP	84	1	OP	94	1	OP	B4	1	OP	A4	1	OP	18	1
2	OP+1	ii	1	OP+1	dd	1	OP+1	hh	1	OP+1	ff	1	OP+1	A4	1
3				00dd	(00dd)	1	OP+2	ll	1	FFFF	—	1	OP+2	ff	1
4							hhll	(hhll)	1	X+ff	(X+ff)	1	FFFF	—	1
5													Y+ff	(Y+ff)	1

Cycle	ANDB (IMM) Addr	Data	R/W̄	ANDB (DIR) Addr	Data	R/W̄	ANDB (EXT) Addr	Data	R/W̄	ANDB (IND, X) Addr	Data	R/W̄	ANDB (IND, Y) Addr	Data	R/W̄
1	OP	C4	1	OP	D4	1	OP	F4	1	OP	E4	1	OP	18	1
2	OP+1	ii	1	OP+1	dd	1	OP+1	hh	1	OP+1	ff	1	OP+1	E4	1
3				00dd	(00dd)	1	OP+2	ll	1	FFFF	—	1	OP+2	ff	1
4							hhll	(hhll)	1	X+ff	(X+ff)	1	FFFF	—	1
5													Y+ff	(Y+ff)	1

ASL

Arithmetic Shift Left
(Same as LSL)

ASL

Operation:

$$C \leftarrow \boxed{b7 - - - - - - b0} \leftarrow 0$$

Description: Shifts all bits of the ACCX or M one place to the left. Bit 0 is loaded with a zero. The C bit in the CCR is loaded from the most significant bit of ACCX or M.

Condition Codes and Boolean Formulae:

S	X	H	I	N	Z	V	C
—	—	—	—	↕	↕	↕	↕

N R7
   Set if MSB of result is set; cleared otherwise.

Z $\overline{R7} \cdot \overline{R6} \cdot \overline{R5} \cdot \overline{R4} \cdot \overline{R3} \cdot \overline{R2} \cdot \overline{R1} \cdot \overline{R0}$
   Set if result is $00; cleared otherwise.

V $N \oplus C = [N \cdot \overline{C}] + [\overline{N} \cdot C]$ (for N and C after the shift)
   Set if (N is set and C is clear) or (N is clear and C is set); cleared otherwise (for values of N and C after the shift).

C M7
   Set if, before the shift, the MSB of ACCX or M was set; cleared otherwise.

Source Forms: ASLA; ASLB; ASL (opr)

Addressing Modes, Machine Code, and Cycle-by-Cycle Execution:

Cycle	ASLA (INH)			ASLB (INH)			ASL (EXT)			ASL (IND, X)			ASL (IND, Y)		
	Addr	Data	R/$\overline{W}$	Addr	Data	R/$\overline{W}$	Addr	Data	R/$\overline{W}$	Addr	Data	R/$\overline{W}$	Addr	Data	R/$\overline{W}$
1	OP	48	1	OP	58	1	OP	78	1	OP	68	1	OP	18	1
2	OP+1	—	1	OP+1	—	1	OP+1	hh	1	OP+1	ff	1	OP+1	68	1
3							OP+2	ll	1	FFFF	—	1	OP+2	ff	1
4							hhll	(hhll)	1	X+ff	(X+ff)	1	FFFF	—	1
5							FFFF	—	1	FFFF	—	1	Y+ff	(Y+ff)	1
6							hhll	result	0	X+ff	result	0	FFFF	—	1
7													Y+ff	result	0

ASLD

Arithmetic Shift Left Double Accumulator
(Same as LSLD)

ASLD

Operation:

Description: Shifts all bits of ACCD one place to the left. Bit 0 is loaded with a zero. The C bit in the CCR is loaded from the most significant bit of ACCD.

Condition Codes and Boolean Formulae:

S	X	H	I	N	Z	V	C
—	—	—	—	↕	↕	↕	↕

N   R15
  Set if MSB of result is set; cleared otherwise.

Z $\overline{R15} \cdot \overline{R14} \cdot \overline{R13} \cdot \overline{R12} \cdot \overline{R11} \cdot \overline{R10} \cdot \overline{R9} \cdot \overline{R8} \cdot \overline{R7} \cdot \overline{R6} \cdot \overline{R5} \cdot \overline{R4} \cdot \overline{R3} \cdot \overline{R2} \cdot \overline{R1} \cdot \overline{R0}$
  Set if result is $0000; cleared otherwise.

V $N \oplus C = [N \cdot \overline{C}] + [\overline{N} \cdot C]$ (for N and C after the shift)
  Set if (N is set and C is clear) or (N is clear and C is set); cleared otherwise (for values of N and C after the shift).

C   D15
  Set if, before the shift, the MSB of ACCD was set; cleared otherwise.

Source Form: ASLD

Addressing Modes, Machine Code, and Cycle-by-Cycle Execution:

Cycle	ASLD (INH)		
	Addr	Data	R/$\overline{W}$
1	OP	05	1
2	OP + 1	—	1
3	FFFF	—	1

ASR   **Arithmetic Shift Right**   # ASR

Operation:

$$\boxed{} \rightarrow \boxed{b7 ----- b0} \rightarrow \boxed{C}$$

Description: Shifts all of ACCX or M one place to the right. Bit 7 is held constant. Bit 0 is loaded into the C bit of the CCR. This operation effectively divides a twos complement value by two without changing its sign. The carry bit can be used to round the result.

Condition Codes and Boolean Formulae:

S	X	H	I	N	Z	V	C
—	—	—	—	↕	↕	↕	↕

N   R7
   Set if MSB of result is set; cleared otherwise.

Z $\overline{R7} \cdot \overline{R6} \cdot \overline{R5} \cdot \overline{R4} \cdot \overline{R3} \cdot \overline{R2} \cdot \overline{R1} \cdot \overline{R0}$
   Set if result is $00; cleared otherwise.

V $N \oplus C = [N \cdot \overline{C}] + [\overline{N} \cdot C]$ (for N and C after the shift)
   Set if (N is set and C is clear) or (N is clear and C is set); cleared otherwise (for values of N and C after the shift).

C   M0
   Set if, before the shift, the LSB of ACCX or M was set; cleared otherwise.

Source Forms: ASRA; ASRB; ASR (opr)

Addressing Modes, Machine Code, and Cycle-by-Cycle Execution:

Cycle	ASRA (INH) Addr	Data	R/$\overline{W}$	ASRB (INH) Addr	Data	R/$\overline{W}$	ASR (EXT) Addr	Data	R/$\overline{W}$	ASR (IND, X) Addr	Data	R/$\overline{W}$	ASR (IND, Y) Addr	Data	R/$\overline{W}$
1	OP	47	1	OP	57	1	OP	77	1	OP	67	1	OP	18	1
2	OP+1	—	1	OP+1	—	1	OP+1	hh	1	OP+1	ff	1	OP+1	67	1
3							OP+2	ll	1	FFFF	—	1	OP+2	ff	1
4							hhll	(hhll)	1	X+ff	(X+ff)	1	FFFF	—	1
5							FFFF	—	1	FFFF	—	1	Y+ff	(Y+ff)	1
6							hhll	result	0	X+ff	result	0	FFFF	—	1
7													Y+ff	result	0

BCC

Branch if Carry Clear
(Same as BHS)

BCC

Operation: PC ◀ (PC) + $0002 + Rel   if (C) = 0

Description: Tests the state of the C bit in the CCR and causes a branch if C is clear.

See BRA instruction for further details of the execution of the branch.

Condition Codes and Boolean Formulae:

S	X	H	I	N	Z	V	C
—	—	—	—	—	—	—	—

None affected

Source Form: BCC (rel)

Addressing Modes, Machine Code, and Cycle-by-Cycle Execution:

Cycle	BCC (REL)		
	Addr	Data	R/$\overline{W}$
1	OP	24	1
2	OP + 1	rr	1
3	FFFF	—	1

The following table is a summary of all branch instructions.

Test	Boolean	Mnemonic	Opcode	Complementary		Branch	Comment
r > m	Z + (N ⊕ V) = 0	BGT	2E	r ≤ m	BLE	2F	Signed
r ≥ m	N ⊕ V = 0	BGE	2C	r < m	BLT	2D	Signed
r = m	Z = 1	BEQ	27	r ≠ m	BNE	26	Signed
r ≤ m	Z + (N ⊕ V) = 1	BLE	2F	r > m	BGT	2E	Signed
r < m	N ⊕ V = 1	BLT	2D	r ≥ m	BGE	2C	Signed
r > m	C + Z = 0	BHI	22	r ≤ m	BLS	23	Unsigned
r ≥ m	C = 0	BHS/BCC	24	r < m	BLO/BCS	25	Unsigned
r = m	Z = 1	BEQ	27	r ≠ m	BNE	26	Unsigned
r ≤ m	C + Z = 1	BLS	23	r > m	BHI	22	Unsigned
r < m	C = 1	BLO/BCS	25	r ≥ m	BHS/BCC	24	Unsigned
Carry	C = 1	BCS	25	No Carry	BCC	24	Simple
Negative	N = 1	BMI	2B	Plus	BPL	2A	Simple
Overflow	V = 1	BVS	29	No Overflow	BVC	28	Simple
r = 0	Z = 1	BEQ	27	r ≠ 0	BNE	26	Simple
Always	—	BRA	20	Never	BRN	21	Unconditional

BCLR   Clear Bit(s) in Memory   BCLR

Operation: M $\Leftarrow$ (M)•$\overline{(PC+2)}$
         M $\Leftarrow$ (M)•$\overline{(PC+3)}$ (for IND, Y address mode only)

Description: Clear multiple bits in location M. The bit(s) to be cleared are specified by ones in the mask byte. All other bits in M are rewritten to their current state.

Condition Codes and Boolean Formulae:

S	X	H	I	N	Z	V	C
—	—	—	—	$\updownarrow$	$\updownarrow$	0	—

N R7
   Set if MSB of result is set; cleared otherwise.

Z $\overline{R7} \cdot \overline{R6} \cdot \overline{R5} \cdot \overline{R4} \cdot \overline{R3} \cdot \overline{R2} \cdot \overline{R1} \cdot \overline{R0}$
   Set if result is $00; cleared otherwise.

V 0
   Cleared

Source Forms: BCLR (opr) (msk)

Addressing Modes, Machine Code, and Cycle-by-Cycle Execution:

Cycle	BCLR (DIR)			BCLR (IND, X)			BCLR (IND, Y)		
	Addr	Data	R/$\overline{W}$	Addr	Data	R/$\overline{W}$	Addr	Data	R/$\overline{W}$
1	OP	15	1	OP	1D	1	OP	18	1
2	OP+1	dd	1	OP+1	ff	1	OP+1	1D	1
3	00dd	(00dd)	1	FFFF	—	1	OP+2	ff	1
4	OP+2	mm	1	X+ff	(X+ff)	1	FFFF	—	1
5	FFFF	—	1	OP+2	mm	1	(IY)+ff	(Y+ff)	1
6	00dd	result	0	FFFF	—	1	OP+3	mm	1
7				X+ff	result	0	FFFF	—	1
8							Y+ff	result	0

BCS

**Branch if Carry Set
(Same as BLO)**

BCS

Operation: PC ◀ (PC) + $0002 + Rel   if (C) = 1

Description: Tests the state of the C bit in the CCR and causes a branch if C is set.

See BRA instruction for further details of the execution of the branch.

Condition Codes and Boolean Formulae:

S	X	H	I	N	Z	V	C
—	—	—	—	—	—	—	—

None affected

Source Form: BCS (rel)

Addressing Modes, Machine Code, and Cycle-by-Cycle Execution:

Cycle	BCS (REL)		
	Addr	Data	R/$\overline{W}$
1	OP	25	1
2	OP + 1	rr	1
3	FFFF	—	1

The following table is a summary of all branch instructions.

Test	Boolean	Mnemonic	Opcode	Complementary	Branch	Comment
r > m	Z + (N ⊕ V) = 0	BGT	2E	r ≤ m	BLE	2F	Signed
r ≥ m	N ⊕ V = 0	BGE	2C	r < m	BLT	2D	Signed
r = m	Z = 1	BEQ	27	r ≠ m	BNE	26	Signed
r ≤ m	Z + (N ⊕ V) = 1	BLE	2F	r > m	BGT	2E	Signed
r < m	N ⊕ V = 1	BLT	2D	r ≥ m	BGE	2C	Signed
r > m	C + Z = 0	BHI	22	r ≤ m	BLS	23	Unsigned
r ≥ m	C = 0	BHS/BCC	24	r < m	BLO/BCS	25	Unsigned
r = m	Z = 1	BEQ	27	r ≠ m	BNE	26	Unsigned
r ≤ m	C + Z = 1	BLS	23	r > m ·	BHI	22	Unsigned
r < m	C = 1	BLO/BCS	25	r ≥ m	BHS/BCC	24	Unsigned
Carry	C = 1	BCS	25	No Carry	BCC	24	Simple
Negative	N = 1	BMI	2B	Plus	BPL	2A	Simple
Overflow	V = 1	BVS	29	No Overflow	BVC	28	Simple
r = 0	Z = 1	BEQ	27	r ≠ 0	BNE	26	Simple
Always	—	BRA	20	Never	BRN	21	Unconditional

BEQ

Branch if Equal

Operation: PC ◀ (PC) + $0002 + Rel   if (Z) = 1

Description: Tests the state of the Z bit in the CCR and causes a branch if Z is set.

See BRA instruction for further details of the execution of the branch.

Condition Codes and Boolean Formulae:

S	X	H	I	N	Z	V	C
—	—	—	—	—	—	—	—

None affected

Source Form: BEQ (rel)

Addressing Modes, Machine Code, and Cycle-by-Cycle Execution:

Cycle	BEQ (REL)		
	Addr	Data	R/$\overline{W}$
1	OP	27	1
2	OP + 1	rr	1
3	FFFF	—	1

The following table is a summary of all branch instructions.

Test	Boolean	Mnemonic	Opcode	Complementary		Branch	Comment
r>m	Z + (N ⊕ V) = 0	BGT	2E	r≤m	BLE	2F	Signed
r≥m	N ⊕ V = 0	BGE	2C	r<m	BLT	2D	Signed
r = m	Z = 1	BEQ	27	r ≠ m	BNE	26	Signed
r≤m	Z + (N ⊕ V) = 1	BLE	2F	r>m	BGT	2E	Signed
r<m	N ⊕ V = 1	BLT	2D	r≥m	BGE	2C	Signed
r>m	C + Z = 0	BHI	22	r≤m	BLS	23	Unsigned
r≥m	C = 0	BHS/BCC	24	r<m	BLO/BCS	25	Unsigned
r = m	Z = 1	BEQ	27	r ≠ m	BNE	26	Unsigned
r≤m	C + Z = 1	BLS	23	r>m	BHI	22	Unsigned
r<m	C = 1	BLO/BCS	25	r≥m	BHS/BCC	24	Unsigned
Carry	C = 1	BCS	25	No Carry	BCC	24	Simple
Negative	N = 1	BMI	2B	Plus	BPL	2A	Simple
Overflow	V = 1	BVS	29	No Overflow	BVC	28	Simple
r = 0	Z = 1	BEQ	27	r ≠ 0	BNE	26	Simple
Always	—	BRA	20	Never	BRN	21	Unconditional

BGE

Branch if Greater than or Equal to Zero

BGE

Operation: PC ◆ (PC) + $0002 + Rel   if $(N) \oplus (V) = 0$
i.e., if $(ACCX) \geq (M)$ (twos-complement "signed" numbers)

Description: If the BGE instruction is executed immediately after execution of any of the instructions, CBA, CMP(A, B, or D), CP(X or Y), SBA, SUB(A, B, or D), the branch will occur if and only if the twos-complement number represented by the ACCX was greater than or equal to the two-complement number represented by M.

See BRA instruction for further details of the execution of the branch.

Condition Codes and Boolean Formulae:

S	X	H	I	N	Z	V	C
—	—	—	—	—	—	—	—

None affected

Source Form: BGE (rel)

Addressing Modes, Machine Code, and Cycle-by-Cycle Execution:

Cycle	BGE (REL)		
	Addr	Data	R/W̄
1	OP	2C	1
2	OP + 1	rr	1
3	FFFF	—	1

The following table is a summary of all branch instructions.

Test	Boolean	Mnemonic	Opcode	Complementary		Branch	Comment
r>m	$Z + (N \oplus V) = 0$	BGT	2E	r≤m		BLE	2F	Signed
r≥m	$N \oplus V = 0$	BGE	2C	r<m		BLT	2D	Signed
r=m	$Z = 1$	BEQ	27	r≠m		BNE	26	Signed
r≤m	$Z + (N \oplus V) = 1$	BLE	2F	r>m		BGT	2E	Signed
r<m	$N \oplus V = 1$	BLT	2D	r≥m		BGE	2C	Signed
r>m	$C + Z = 0$	BHI	22	r≤m		BLS	23	Unsigned
r≥m	$C = 0$	BHS/BCC	24	r<m		BLO/BCS	25	Unsigned
r=m	$Z = 1$	BEQ	27	r≠m		BNE	26	Unsigned
r≤m	$C + Z = 1$	BLS	23	r>m		BHI	22	Unsigned
r<m	$C = 1$	BLO/BCS	25	r≥m		BHS/BCC	24	Unsigned
Carry	$C = 1$	BCS	25	No Carry		BCC	24	Simple
Negative	$N = 1$	BMI	2B	Plus		BPL	2A	Simple
Overflow	$V = 1$	BVS	29	No Overflow		BVC	28	Simple
r=0	$Z = 1$	BEQ	27	r≠0		BNE	26	Simple
Always	—	BRA	20	Never		BRN	21	Unconditional

BGT

Branch if Greater than Zero

BGT

Operation: PC ◀ (PC) + $0002 + Rel   if (Z) + [(N) ⊕ (V)] = 0
i.e., if (ACCX)>(M)   (twos-complement signed numbers)

Description: If the BGT instruction is executed immediately after execution of any of the instructions, CBA, CMP(A, B, or D), CP(X or Y), SBA, SUB(A, B, or D), the branch will occur if and only if the twos-complement number represented by ACCX was greater than the twos-complement number represented by M.

See BRA instruction for further details of the execution of the branch.

Condition Codes and Boolean Formulae:

S	X	H	I	N	Z	V	C
—	—	—	—	—	—	—	—

None affected

Source Form: BGT (rel)

Addressing Modes, Machine Code, and Cycle-by-Cycle Execution:

Cycle	BGT (REL)		
	Addr	Data	R/W̄
1	OP	2E	1
2	OP + 1	rr	1
3	FFFF	—	1

The following table is a summary of all branch instructions.

Test	Boolean	Mnemonic	Opcode	Complementary		Branch	Comment
r>m	Z + (N ⊕ V) = 0	BGT	2E	r≤m	BLE	2F	Signed
r≥m	N ⊕ V = 0	BGE	2C	r<m	BLT	2D	Signed
r = m	Z = 1	BEQ	27	r ≠ m	BNE	26	Signed
r≤m	Z + (N ⊕ V) = 1	BLE	2F	r>m	BGT	2E	Signed
r<m	N ⊕ V = 1	BLT	2D	r≥m	BGE	2C	Signed
r>m	C + Z = 0	BHI	22	r≤m	BLS	23	Unsigned
r≥m	C = 0	BHS/BCC	24	r<m	BLO/BCS	25	Unsigned
r = m	Z = 1	BEQ	27	r ≠ m	BNE	26	Unsigned
r≤m	C + Z = 1	BLS	23	r>m	BHI	22	Unsigned
r<m	C = 1	BLO/BCS	25	r≥m	BHS/BCC	24	Unsigned
Carry	C = 1	BCS	25	No Carry	BCC	24	Simple
Negative	N = 1	BMI	2B	Plus	BPL	2A	Simple
Overflow	V = 1	BVS	29	No Overflow	BVC	28	Simple
r = 0	Z = 1	BEQ	27	r ≠ 0	BNE	26	Simple
Always	—	BRA	20	Never	BRN	21	Unconditional

BHI   Branch if Higher   BHI

Operation: PC ◀ (PC) + $0002 + Rel   if (C) + (Z) = 0
                 i.e., if (ACCX) > (M)   (unsigned binary numbers)

Description: If the BHI instruction is executed immediately after execution of any of the instructions, CBA, CMP(A, B, or D), CP(X or Y), SBA, SUB(A, B, or D), the branch will occur if and only if the unsigned binary number represented by ACCX was greater than the unsigned binary number represented by M. Generally not useful after INC/DEC, LD/ST, TST/CLR/COM because these instructions do not affect the C bit in the CCR.

See BRA instruction for further details of the execution of the branch.

Condition Codes and Boolean Formulae:

S	X	H	I	N	Z	V	C
—	—	—	—	—	—	—	—

None affected

Source Form: BHI (rel)

Addressing Modes, Machine Code, and Cycle-by-Cycle Execution:

Cycle	BHI (REL)		
	Addr	Data	R/W̄
1	OP	22	1
2	OP + 1	rr	1
3	FFFF	—	1

The following table is a summary of all branch instructions.

Test	Boolean	Mnemonic	Opcode	Complementary		Branch	Comment
r>m	Z + (N ⊕ V) = 0	BGT	2E	r≤m	BLE	2F	Signed
r≥m	N ⊕ V = 0	BGE	2C	r<m	BLT	2D	Signed
r = m	Z = 1	BEQ	27	r ≠ m	BNE	26	Signed
r≤m	Z + (N ⊕ V) = 1	BLE	2F	r>m	BGT	2E	Signed
r<m	N ⊕ V = 1	BLT	2D	r≥m	BGE	2C	Signed
r>m	C + Z = 0	BHI	22	r≤m	BLS	23	Unsigned
r≥m	C = 0	BHS/BCC	24	r<m	BLO/BCS	25	Unsigned
r = m	Z = 1	BEQ	27	r ≠ m	BNE	26	Unsigned
r≤m	C + Z = 1	BLS	23	r>m	BHI	22	Unsigned
r<m	C = 1	BLO/BCS	25	r≥m	BHS/BCC	24	Unsigned
Carry	C = 1	BCS	25	No Carry	BCC	24	Simple
Negative	N = 1	BMI	2B	Plus	BPL	2A	Simple
Overflow	V = 1	BVS	29	No Overflow	BVC	28	Simple
r = 0	Z = 1	BEQ	27	r ≠ 0	BNE	26	Simple
Always	—	BRA	20	Never	BRN	21	Unconditional

BHS

Branch if Higher or Same
(Same as BCC)

BHS

Operation: PC ◀ (PC) + $0002 + Rel   if (C) = 0
i.e., if (ACCX) ≥ (M)   (unsigned binary numbers)

Description: If the BHS instruction is executed immediately after execution of any of the instructions, CBA, CMP(A, B, or D), CP(X or Y), SBA, SUB(A, B, or D), the branch will occur if and only if the unsigned binary number represented by ACCX was greater than or equal to the unsigned binary number represented by M. Generally not useful after INC/DEC, LD/ST, TST/CLR/COM because these instructions do not affect the C bit in the CCR.

See BRA instruction for further details of the execution of the branch.

Condition Codes and Boolean Formulae:

S	X	H	I	N	Z	V	C
—	—	—	—	—	—	—	—

None affected

Source Form: BHS (rel)

Addressing Modes, Machine Code, and Cycle-by-Cycle Execution:

Cycle	BHS (REL)		
	Addr	Data	R/W̅
1	OP	24	1
2	OP + 1	rr	1
3	FFFF	—	1

The following table is a summary of all branch instructons.

Test	Boolean	Mnemonic	Opcode	Complementary		Branch	Comment
r>m	Z + (N ⊕ V) = 0	BGT	2E	r≤m	BLE	2F	Signed
r≥m	N ⊕ V = 0	BGE	2C	r<m	BLT	2D	Signed
r = m	Z = 1	BEQ	27	r ≠ m	BNE	26	Signed
r≤m	Z + (N ⊕ V) = 1	BLE	2F	r>m	BGT	2E	Signed
r<m	N ⊕ V = 1	BLT	2D	r≥m	BGE	2C	Signed
r>m	C + Z = 0	BHI	22	r≤m	BLS	23	Unsigned
r≥m	C = 0	BHS/BCC	24	r<m	BLO/BCS	25	Unsigned
r = m	Z = 1	BEQ	27	r ≠ m	BNE	26	Unsigned
r≤m	C + Z = 1	BLS	23	r>m	BHI	22	Unsigned
r<m	C = 1	BLO/BCS	25	r≥m	BHS/BCC	24	Unsigned
Carry	C = 1	BCS	25	No Carry	BCC	24	Simple
Negative	N = 1	BMI	2B	Plus	BPL	2A	Simple
Overflow	V = 1	BVS	29	No Overflow	BVC	28	Simple
r = 0	Z = 1	BEQ	27	r ≠ 0	BNE	26	Simple
Always	—	BRA	20	Never	BRN	21	Unconditional

BIT

Bit Test

BIT

Operation: (ACCX)•(M)

Description: Performs the logical AND operation between the contents of ACCX and the contents of M and modifies the condition codes accordingly. Neither the contents of ACCX or M operands are affected. (Each bit of the result of the AND would be the logical AND of the corresponding bits of ACCX and M.)

Condition Codes and Boolean Formulae:

S	X	H	I	N	Z	V	C
—	—	—	—	↕	↕	0	—

N R7
  Set if MSB of result is set; cleared otherwise.

Z $\overline{R7} \cdot \overline{R6} \cdot \overline{R5} \cdot \overline{R4} \cdot \overline{R3} \cdot \overline{R2} \cdot \overline{R1} \cdot \overline{R0}$
  Set if result is $00; cleared otherwise.

V 0
  Cleared

Source Forms: BITA (opr); BITB (opr)

Addressing Modes, Machine Code, and Cycle-by-Cycle Execution:

Cycle	BITA (IMM) Addr	Data	R/W̄	BITA (DIR) Addr	Data	R/W̄	BITA (EXT) Addr	Data	R/W̄	BITA (IND, X) Addr	Data	R/W̄	BITA (IND, Y) Addr	Data	R/W̄
1	OP	85	1	OP	95	1	OP	B5	1	OP	A5	1	OP	18	1
2	OP+1	ii	1	OP+1	dd	1	OP+1	hh	1	OP+1	ff	1	OP+1	A5	1
3				00dd	(00dd)	1	OP+2	ll	1	FFFF	—	1	OP+2	ff	1
4							hhll	(hhll)	1	X+ff	(X+ff)	1	FFFF	—	1
5													Y+ff	(Y+ff)	1

Cycle	BITB (IMM) Addr	Data	R/W̄	BITB (DIR) Addr	Data	R/W̄	BITB (EXT) Addr	Data	R/W̄	BITB (IND, X) Addr	Data	R/W̄	BITB (IND, Y) Addr	Data	R/W̄
1	OP	C5	1	OP	D5	1	OP	F5	1	OP	E5	1	OP	18	1
2	OP+1	ii	1	OP+1	dd	1	OP+1	hh	1	OP+1	ff	1	OP+1	E5	1
3				00dd	(00dd)	1	OP+2	ll	1	FFFF	—	1	OP+2	ff	1
4							hhll	(hhll)	1	X+ff	(X+ff)	1	FFFF	—	1
5													Y+ff	(Y+ff)	1

BLE

Branch if Less than or Equal to Zero

BLE

Operation: $\text{PC} \blacklozenge (\text{PC}) + \$0002 + \text{Rel}$ if $(Z) + [(N) \oplus (V)] = 1$
i.e., if $(\text{ACCX}) \leq (\text{M})$ (twos-complement signed numbers)

Description: If the BLE instruction is executed immediately after execution of any of the instructions, CBA, CMP(A, B, or D), CP(X, or Y), SBA, SUB(A, B, or D), the branch will occur if and only if the twos-complement number represented by ACCX was less than or equal to the twos-complement number represented by M.

See BRA instruction for further details of the execution of the branch.

Condition Codes and Boolean Formulae:

S	X	H	I	N	Z	V	C
—	—	—	—	—	—	—	—

None affected

Source Form: BLE (rel)

Addressing Modes, Machine Code, and Cycle-by-Cycle Execution:

Cycle	BLE (REL)		
	Addr	Data	R/$\overline{\text{W}}$
1	OP	2F	1
2	OP + 1	rr	1
3	FFFF	—	1

The following table is a summary of all branch instructions.

Test	Boolean	Mnemonic	Opcode	Complementary		Branch	Comment
r>m	Z + (N ⊕ V) = 0	BGT	2E	r≤m	BLE	2F	Signed
r≥m	N ⊕ V = 0	BGE	2C	r<m	BLT	2D	Signed
r = m	Z = 1	BEQ	27	r ≠ m	BNE	26	Signed
r≤m	Z + (N ⊕ V) = 1	BLE	2F	r>m	BGT	2E	Signed
r<m	N ⊕ V = 1	BLT	2D	r≥m	BGE	2C	Signed
r>m	C + Z = 0	BHI	22	r≤m	BLS	23	Unsigned
r≥m	C = 0	BHS/BCC	24	r<m	BLO/BCS	25	Unsigned
r = m	Z = 1	BEQ	27	r ≠ m	BNE	26	Unsigned
r≤m	C + Z = 1	BLS	23	r>m	BHI	22	Unsigned
r<m	C = 1	BLO/BCS	25	r≥m	BHS/BCC	24	Unsigned
Carry	C = 1	BCS	25	No Carry	BCC	24	Simple
Negative	N = 1	BMI	2B	Plus	BPL	2A	Simple
Overflow	V = 1	BVS	29	No Overflow	BVC	28	Simple
r = 0	Z = 1	BEQ	27	r ≠ 0	BNE	26	Simple
Always	—	BRA	20	Never	BRN	21	Unconditional

BLO

**Branch if Lower
(Same as BCS)**

BLO

Operation: PC ◀ (PC) + $0002 + Rel   if (C) = 1
                   i.e., if (ACCX)<(M)   (unsigned binary numbers)

Description: If the BLO instruction is executed immediately after execution of any of
the instructions, CBA, CMP(A, B, or D), CP(X or Y), SBA, SUB(A, B, or D), the branch
will occur if and only if the unsigned binary number represented by ACCX was less
than the unsigned binary number represented by M. Generally not useful after INC/
DEC, LD/ST, TST/CLR/COM because these instructions do not affect the C bit in the
CCR.

See BRA instruction for further details of the execution of the branch.

Condition Codes and Boolean Formulae:

S	X	H	I	N	Z	V	C
—	—	—	—	—	—	—	—

None affected

Source Form: BLO (rel)

Addressing Modes, Machine Code, and Cycle-by-Cycle Execution:

Cycle	BLO (REL)		
	Addr	Data	R/W̄
1	OP	25	1
2	OP + 1	rr	1
3	FFFF	—	1

The following table is a summary of all branch instructions.

Test	Boolean	Mnemonic	Opcode	Complementary		Branch	Comment
r>m	Z + (N ⊕ V) = 0	BGT	2E	r≤m	BLE	2F	Signed
r≥m	N ⊕ V = 0	BGE	2C	r<m	BLT	2D	Signed
r = m	Z = 1	BEQ	27	r ≠ m	BNE	26	Signed
r≤m	Z + (N ⊕ V) = 1	BLE	2F	r>m	BGT	2E	Signed
r<m	N ⊕ V = 1	BLT	2D	r≥m	BGE	2C	Signed
r>m	C + Z = 0	BHI	22	r≤m	BLS	23	Unsigned
r≥m	C = 0	BHS/BCC	24	r<m	BLO/BCS	25	Unsigned
r = m	Z = 1	BEQ	27	r ≠ m	BNE	26	Unsigned
r≤m	C + Z = 1	BLS	23	r>m	BHI	22	Unsigned
r<m	C = 1	BLO/BCS	25	r≥m	BHS/BCC	24	Unsigned
Carry	C = 1	BCS	25	No Carry	BCC	24	Simple
Negative	N = 1	BMI	2B	Plus	BPL	2A	Simple
Overflow	V = 1	BVS	29	No Overflow	BVC	28	Simple
r = 0	Z = 1	BEQ	27	r ≠ 0	BNE	26	Simple
Always	—	BRA	20	Never	BRN	21	Unconditional

BLS

Branch if Lower or Same

BLS

Operation: PC ◄ (PC) + $0002 + Rel   if (C) + (Z) = 1
                  i.e., if (ACCX)≤(M)   (unsigned binary numbers)

Description: If the BLS instruction is executed immediately after execution of any of the instructions, CBA, CMP(A, B, or D), CP(X or Y), SBA, SUB(A, B, or D), the branch will occur if and only if the unsigned binary number represented by ACCX was less than or equal to the unsigned binary number represented by M. Generally not useful after INC/DEC, LD/ST, TST/CLR/COM because these instructions do not affect the C bit in the CCR.

     See BRA instruction for further details of the execution of the branch.

Condition Codes and Boolean Formulae:

S	X	H	I	N	Z	V	C
—	—	—	—	—	—	—	—

     None affected

Source Form: BLS (rel)

Addressing Modes, Machine Code, and Cycle-by-Cycle Execution:

Cycle	BLS (REL)		
	Addr	Data	R/W̄
1	OP	23	1
2	OP + 1	rr	1
3	FFFF	—	1

The following table is a summary of all branch instructions.

Test	Boolean	Mnemonic	Opcode	Complementary		Branch	Comment
r>m	Z + (N ⊕ V) = 0	BGT	2E	r≤m	BLE	2F	Signed
r≥m	N ⊕ V = 0	BGE	2C	r<m	BLT	2D	Signed
r = m	Z = 1	BEQ	27	r ≠ m	BNE	26	Signed
r≤m	Z + (N ⊕ V) = 1	BLE	2F	r>m	BGT	2E	Signed
r<m	N ⊕ V = 1	BLT	2D	r≥m	BGE	2C	Signed
r>m	C + Z = 0	BHI	22	r≤m	BLS	23	Unsigned
r≥m	C = 0	BHS/BCC	24	r<m	BLO/BCS	25	Unsigned
r = m	Z = 1	BEQ	27	r ≠ m	BNE	26	Unsigned
r≤m	C + Z = 1	BLS	23	r>m	BHI	22	Unsigned
r<m	C = 1	BLO/BCS	25	r≥m	BHS/BCC	24	Unsigned
Carry	C = 1	BCS	25	No Carry	BCC	24	Simple
Negative	N = 1	BMI	2B	Plus	BPL	2A	Simple
Overflow	V = 1	BVS	29	No Overflow	BVC	28	Simple
r = 0	Z = 1	BEQ	27	r ≠ 0	BNE	26	Simple
Always	—	BRA	20	Never	BRN	21	Unconditional

BLT
Branch if Less than Zero
BLT

Operation:
PC ◀ (PC) + $0002 + Rel   if (N)⊕(V) = 1
i.e., if (ACCX)<(M)   (twos-complement signed numbers)

Description: If the BLT instruction is executed immediately after execution of any of the instructons, CBA, CMP(A, B, or D), CP(X or Y), SBA, SUB(A, B, or D), the branch will occur if and only if the twos-complement number represented by ACCX was less than the twos-complement number represented by M.

See BRA instruction for further details of the execution of the branch.

Condition Codes and Boolean Formulae:

S	X	H	I	N	Z	V	C
—	—	—	—	—	—	—	—

None affected

Source Form: BLT (rel)

Addressing Modes, Machine Code, and Cycle-by-Cycle Execution:

Cycle	BLT (REL)		
	Addr	Data	R/W̄
1	OP	2D	1
2	OP + 1	rr	1
3	FFFF	—	1

The following table is a summary of all branch instructions.

Test	Boolean	Mnemonic	Opcode	Complementary		Branch	Comment
r>m	Z+(N ⊕ V) = 0	BGT	2E	r≤m	BLE	2F	Signed
r≥m	N ⊕ V = 0	BGE	2C	r<m	BLT	2D	Signed
r = m	Z = 1	BEQ	27	r ≠ m	BNE	26	Signed
r≤m	Z+(N ⊕ V) = 1	BLE	2F	r>m	BGT	2E	Signed
r<m	N ⊕ V = 1	BLT	2D	r≥m	BGE	2C	Signed
r>m	C+Z = 0	BHI	22	r≤m	BLS	23	Unsigned
r≥m	C = 0	BHS/BCC	24	r<m	BLO/BCS	25	Unsigned
r = m	Z = 1	BEQ	27	r ≠ m	BNE	26	Unsigned
r≤m	C+Z = 1	BLS	23	r>m	BHI	22	Unsigned
r<m	C = 1	BLO/BCS	25	r≥m	BHS/BCC	24	Unsigned
Carry	C = 1	BCS	25	No Carry	BCC	24	Simple
Negative	N = 1	BMI	2B	Plus	BPL	2A	Simple
Overflow	V = 1	BVS	29	No Overflow	BVC	28	Simple
r = 0	Z = 1	BEQ	27	r ≠ 0	BNE	26	Simple
Always	—	BRA	20	Never	BRN	21	Unconditional

BMI

Branch if Minus

BMI

Operation: PC ◀ (PC) + $0002 + Rel   if (N) = 1

Description: Tests the state of the N bit in the CCR and causes a branch if N is set.

See BRA instruction for further details of the execution of the branch.

Condition Codes and Boolean Formulae:

S	X	H	I	N	Z	V	C
—	—	—	—	—	—	—	—

None affected

Source Form: BMI (rel)

Addressing Modes, Machine Code, and Cycle-by-Cycle Execution:

Cycle	BMI (REL)		
	Addr	Data	R/W̄
1	OP	2B	1
2	OP + 1	rr	1
3	FFFF	—	1

The following table is a summary of all branch instructions.

Test	Boolean	Mnemonic	Opcode	Complementary		Branch	Comment
r>m	Z + (N ⊕ V) = 0	BGT	2E	r≤m	BLE	2F	Signed
r≥m	N ⊕ V = 0	BGE	2C	r<m	BLT	2D	Signed
r=m	Z = 1	BEQ	27	r≠m	BNE	26	Signed
r≤m	Z + (N ⊕ V) = 1	BLE	2F	r>m	BGT	2E	Signed
r<m	N ⊕ V = 1	BLT	2D	r≥m	BGE	2C	Signed
r>m	C + Z = 0	BHI	22	r≤m	BLS	23	Unsigned
r≥m	C = 0	BHS/BCC	24	r<m	BLO/BCS	25	Unsigned
r=m	Z = 1	BEQ	27	r≠m	BNE	26	Unsigned
r≤m	C + Z = 1	BLS	23	r>m	BHI	22	Unsigned
r<m	C = 1	BLO/BCS	25	r≥m	BHS/BCC	24	Unsigned
Carry	C = 1	BCS	25	No Carry	BCC	24	Simple
Negative	N = 1	BMI	2B	Plus	BPL	2A	Simple
Overflow	V = 1	BVS	29	No Overflow	BVC	28	Simple
r=0	Z = 1	BEQ	27	r≠0	BNE	26	Simple
Always	—	BRA	20	Never	BRN	21	Unconditional

BNE
Branch if Not Equal to Zero
BNE

Operation: PC ◄ (PC) + $0002 + Rel   if (Z) = 0

Description: Tests the state of the Z bit in the CCR and causes a branch if Z is clear.

See BRA instruction for further details of the execution of the branch.

Condition Codes and Boolean Formulae:

S	X	H	I	N	Z	V	C
—	—	—	—	—	—	—	—

None affected

Source Form: BNE (rel)

Addressing Modes, Machine Code, and Cycle-by-Cycle Execution:

Cycle	BNE (REL)		
	Addr	Data	R/$\overline{W}$
1	OP	26	1
2	OP + 1	rr	1
3	FFFF	—	1

The following table is a summary of all branch instructions.

Test	Boolean	Mnemonic	Opcode	Complementary		Branch	Comment
r>m	Z+(N ⊕ V)=0	BGT	2E	r≤m	BLE	2F	Signed
r≥m	N ⊕ V=0	BGE	2C	r<m	BLT	2D	Signed
r=m	Z=1	BEQ	27	r≠m	BNE	26	Signed
r≤m	Z+(N ⊕ V)=1	BLE	2F	r>m	BGT	2E	Signed
r<m	N ⊕ V=1	BLT	2D	r≥m	BGE	2C	Signed
r>m	C+Z=0	BHI	22	r≤m	BLS	23	Unsigned
r≥m	C=0	BHS/BCC	24	r<m	BLO/BCS	25	Unsigned
r=m	Z=1	BEQ	27	r≠m	BNE	26	Unsigned
r≤m	C+Z=1	BLS	23	r>m	BHI	22	Unsigned
r<m	C=1	BLO/BCS	25	r≥m	BHS/BCC	24	Unsigned
Carry	C=1	BCS	25	No Carry	BCC	24	Simple
Negative	N=1	BMI	2B	Plus	BPL	2A	Simple
Overflow	V=1	BVS	29	No Overflow	BVC	28	Simple
r=0	Z=1	BEQ	27	r≠0	BNE	26	Simple
Always	—	BRA	20	Never	BRN	21	Unconditional

BPL   Branch if Plus   BPL

Operation: PC ◄ (PC) + $0002 + Rel   if (N) = 0

Description: Tests the state of the N bit in the CCR and causes a branch if N is clear.

See BRA instruction for details of the execution of the branch.

Condition Codes and Boolean Formulae:

S	X	H	I	N	Z	V	C
—	—	—	—	—	—	—	—

None affected

Source Form: BPL (rel)

Addressing Modes, Machine Code, and Cycle-by-Cycle Execution:

Cycle	BPL (REL) Addr	Data	R/$\overline{W}$
1	OP	2A	1
2	OP + 1	rr	1
3	FFFF	—	1

The following table is a summary of all branch instructions.

Test	Boolean	Mnemonic	Opcode	Complementary		Branch	Comment
r>m	Z + (N ⊕ V) = 0	BGT	2E	r≤m	BLE	2F	Signed
r≥m	N ⊕ V = 0	BGE	2C	r<m	BLT	2D	Signed
r = m	Z = 1	BEQ	27	r≠m	BNE	26	Signed
r≤m	Z + (N ⊕ V) = 1	BLE	2F	r>m	BGT	2E	Signed
r<m	N ⊕ V = 1	BLT	2D	r≥m	BGE	2C	Signed
r>m	C + Z = 0	BHI	22	r≤m	BLS	23	Unsigned
r≥m	C = 0	BHS/BCC	24	r<m	BLO/BCS	25	Unsigned
r = m	Z = 1	BEQ	27	r≠m	BNE	26	Unsigned
r≤m	C + Z = 1	BLS	23	r>m	BHI	22	Unsigned
r<m	C = 1	BLO/BCS	25	r≥m	BHS/BCC	24	Unsigned
Carry	C = 1	BCS	25	No Carry	BCC	24	Simple
Negative	N = 1	BMI	2B	Plus	BPL	2A	Simple
Overflow	V = 1	BVS	29	No Overflow	BVC	28	Simple
r = 0	Z = 1	BEQ	27	r≠0	BNE	26	Simple
Always	—	BRA	20	Never	BRN	21	Unconditional

BRA

Branch Always

BRA

Operation: PC ◀ (PC) + $0002 + Rel

Description: Unconditional branch to the address given by the foregoing formula, in which Rel is the relative offset stored as a twos complement number in the second byte of machine code corresponding to the branch instruction.

The source program specifies the destination of any branch instruction by its absolute address, either as a numerical value or as a symbol or expression, that can be numerically evaluated by the assembler. The assembler obtains the relative address, Rel, from the absolute address and the current value of the location counter.

Condition Codes and Boolean Formulae:

S	X	H	I	N	Z	V	C
—	—	—	—	—	—	—	—

None affected

Source Form: BRA (rel)

Addressing Modes, Machine Code, and Cycle-by-Cycle Execution:

Cycle	BRA (REL)		
	Addr	Data	R/$\overline{W}$
1	OP	20	1
2	OP + 1	rr	1
3	FFFF	—	1

The following table is a summary of all branch instructions.

Test	Boolean	Mnemonic	Opcode	Complementary	Branch		Comment
r>m	Z + (N ⊕ V) = 0	BGT	2E	r≤m	BLE	2F	Signed
r≥m	N ⊕ V = 0	BGE	2C	r<m	BLT	2D	Signed
r = m	Z = 1	BEQ	27	r ≠ m	BNE	26	Signed
r≤m	Z + (N ⊕ V) = 1	BLE	2F	r>m	BGT	2E	Signed
r<m	N ⊕ V = 1	BLT	2D	r≥m	BGE	2C	Signed
r>m	C + Z = 0	BHI	22	r≤m	BLS	23	Unsigned
r≥m	C = 0	BHS/BCC	24	r<m	BLO/BCS	25	Unsigned
r = m	Z = 1	BEQ	27	r ≠ m	BNE	26	Unsigned
r≤m	C + Z = 1	BLS	23	r>m	BHI	22	Unsigned
r<m	C = 1	BLO/BCS	25	r≥m	BHS/BCC	24	Unsigned
Carry	C = 1	BCS	25	No Carry	BCC	24	Simple
Negative	N = 1	BMI	2B	Plus	BPL	2A	Simple
Overflow	V = 1	BVS	29	No Overflow	BVC	28	Simple
r = 0	Z = 1	BEQ	27	r ≠ 0	BNE	26	Simple
Always	—	BRA	20	Never	BRN	21	Unconditional

BRCLR

Branch if Bit(s) Clear

BRCLR

Operation: PC ◄ (PC) + $0004 + Rel   if (M)•(PC + 2) = 0
                   PC ◄ (PC) + $0005 + Rel   if (M)•(PC + 3) = 0 (for IND, Y address mode only)

Description: Performs the logical AND of location M and the mask supplied with the instruction, then branches if the result is zero (only if all bits corresponding to ones in the mask byte are zeros in the tested byte).

Condition Codes and Boolean Formulae:

S	X	H	I	N	Z	V	C
—	—	—	—	—	—	—	—

None affected

Source Form: BRCLR (opr) (msk) (rel)

Addressing Modes, Machine Code, and Cycle-by-Cycle Execution:

Cycle	BRCLR (DIR)			BRCLR (IND, X)			BRCLR (IND, Y)		
	Addr	Data	R/W̄	Addr	Data	R/W̄	Addr	Data	R/W̄
1	OP	13	1	OP	1F	1	OP	18	1
2	OP + 1	dd	1	OP + 1	ff	1	OP + 1	1F	1
3	00dd	(00dd)	1	FFFF	—	1	OP + 2	ff	1
4	OP + 2	mm	1	X + ff	(X + ff)	1	FFFF	—	1
5	OP + 3	rr	1	OP + 2	mm	1	(IY) + ff	(Y + ff)	1
6	FFFF	—	1	OP + 3	rr	1	OP + 3	mm	1
7				FFFF	—	1	OP + 4	rr	1
8							FFFF	—	1

BRN

Branch Never

BRN

Operation: PC ◀ (PC) + $0002

Description: Never branches. In effect, this instruction can be considered as a two-byte NOP (no operation) requiring three cycles for execution. Its inclusion in the instruction set is to provide a complement for the BRA instruction. The instruction is useful during program debug to negate the effect of another branch instruction without disturbing the offset byte. Having a complement for BRA is also useful in compiler implementations.

Condition Codes and Boolean Formulae:

S	X	H	I	N	Z	V	C
—	—	—	—	—	—	—	—

None affected

Source Form: BRN (rel)

Addressing Modes, Machine Code, and Cycle-by-Cycle Execution:

Cycle	BRN (REL)		
	Addr	Data	R/$\overline{W}$
1	OP	21	1
2	OP + 1	rr	1
3	FFFF	—	1

The following table is a summary of all branch instructions.

Test	Boolean	Mnemonic	Opcode	Complementary		Branch	Comment
r > m	Z + (N ⊕ V) = 0	BGT	2E	r ≤ m	BLE	2F	Signed
r ≥ m	N ⊕ V = 0	BGE	2C	r < m	BLT	2D	Signed
r = m	Z = 1	BEQ	27	r ≠ m	BNE	26	Signed
r ≤ m	Z + (N ⊕ V) = 1	BLE	2F	r > m	BGT	2E	Signed
r < m	N ⊕ V = 1	BLT	2D	r ≥ m	BGE	2C	Signed
r > m	C + Z = 0	BHI	22	r ≤ m	BLS	23	Unsigned
r ≥ m	C = 0	BHS/BCC	24	r < m	BLO/BCS	25	Unsigned
r = m	Z = 1	BEQ	27	r ≠ m	BNE	26	Unsigned
r ≤ m	C + Z = 1	BLS	23	r > m	BHI	22	Unsigned
r < m	C = 1	BLO/BCS	25	r ≥ m	BHS/BCC	24	Unsigned
Carry	C = 1	BCS	25	No Carry	BCC	24	Simple
Negative	N = 1	BMI	2B	Plus	BPL	2A	Simple
Overflow	V = 1	BVS	29	No Overflow	BVC	28	Simple
r = 0	Z = 1	BEQ	27	r ≠ 0	BNE	26	Simple
Always	—	BRA	20	Never	BRN	21	Unconditional

BRSET

Branch if Bit(s) Set

BRSET

Operation: PC ◀ (PC) + $0004 + Rel   if $\overline{(M)} \cdot (PC + 2) = 0$
PC ◀ (PC) + $0005 + Rel   if $\overline{(M)} \cdot (PC + 3) = 0$ (for IND, Y address mode only)

Description: Performs the logical AND of location M inverted and the mask supplied with the instruction, then branches if the result is zero (only if all bits corresponding to ones in the mask byte are ones in the tested byte).

Condition Codes and Boolean Formulae:

S	X	H	I	N	Z	V	C
—	—	—	—	—	—	—	—

None affected

Source Form: BRSET (opr) (msk) (rel)

Addressing Modes, Machine Code, and Cycle-by-Cycle Execution:

Cycle	BRSET (DIR)			BRSET (IND, X)			BRSET (IND, Y)		
	Addr	Data	R/$\overline{W}$	Addr	Data	R/$\overline{W}$	Addr	Data	R/$\overline{W}$
1	OP	12	1	OP	1E	1	OP	18	1
2	OP + 1	dd	1	OP + 1	ff	1	OP + 1	1E	1
3	00dd	(00dd)	1	FFFF	—	1	OP + 2	ff	1
4	OP + 2	mm	1	X + ff	(X + ff)	1	FFFF	—	1
5	OP + 3	rr	1	OP + 2	mm	1	(IY) + ff	(Y + ff)	1
6	FFFF	—	1	OP + 3	rr	1	OP + 3	mm	1
7				FFFF	—	1	OP + 4	rr	1
8							FFFF	—	1

BSET

Set Bit(s) in Memory

 **BSET**

Operation: $M \Leftarrow (M) + (PC + 2)$
$M \Leftarrow (M) + (PC + 3)$ (for IND, Y address mode only)

Description: Set multiple bits in location M. The bit(s) to be set are specified by ones in the mask byte (last machine code byte of the instruction). All other bits in M are unaffected.

Condition Codes and Boolean Formulae:

S	X	H	I	N	Z	V	C
—	—	—	—	$\updownarrow$	$\updownarrow$	0	—

N R7
    Set if MSB of result is set; cleared otherwise.

Z $\overline{R7} \cdot \overline{R6} \cdot \overline{R5} \cdot \overline{R4} \cdot \overline{R3} \cdot \overline{R2} \cdot \overline{R1} \cdot \overline{R0}$
    Set if result is $00; cleared otherwise.

V 0
    Cleared

Source Form: BSET (opr) (msk)

Addressing Modes, Machine Code, and Cycle-by-Cycle Execution:

Cycle	BSET (DIR)			BSET (IND, X)			BSET (IND, Y)		
	Addr	Data	R/$\overline{W}$	Addr	Data	R/$\overline{W}$	Addr	Data	R/$\overline{W}$
1	OP	14	1	OP	1C	1	OP	18	1
2	OP + 1	dd	1	OP + 1	ff	1	OP + 1	1C	1
3	00dd	(00dd)	1	FFFF	—	1	OP + 2	ff	1
4	OP + 2	mm	1	X + ff	(X + ff)	1	FFFF	—	1
5	FFFF	—	1	OP + 2	mm	1	(IY) + ff	(Y + ff)	1
6	00dd	result	0	FFFF	—	1	OP + 3	mm	1
7				X + ff	result	0	FFFF	—	1
8							Y + ff	result	0

BSR

Branch to Subroutine

BSR

Operation:

PC ◀ (PC) + $0002	Advance PC to return address
◀ (PCL)	Push low-order return onto stack
SP ◀ (SP) − 0001	
◀ (PCH)	Push high-order return onto stack
SP ◀ (SP) − $0001	
PC ◀ (PC) + Rel	Load start address of requested subroutine

Description: The program counter is incremented by two (this will be the return address). The least significant byte of the contents of the program counter (low-order return address) is pushed onto the stack. The stack pointer is then decremented by one. The most significant byte of the contents of the program counter (high-order return address) is pushed onto the stack. The stack pointer is then decremented by one. A branch then occurs to the location specified by the branch offset.

See BRA instruction for further details of the execution of the branch.

Condition Codes and Boolean Formulae:

S	X	H	I	N	Z	V	C
—	—	—	—	—	—	—	—

None affected

Source Form: BSR (rel)

Addressing Modes, Machine Code, and Cycle-by-Cycle Execution:

Cycle	BSR (REL)		
	Addr	Data	R/W̄
1	OP	8D	1
2	OP + 1	rr	1
3	FFFF	—	1
4	Sub	Nxt op	1
5	SP	Rtn lo	0
6	SP − 1	Rtn hi	0

BVC   **Branch if Overflow Clear**   # BVC

Operation: $PC \blacklozenge (PC) + \$0002 + Rel$ if $(V) = 0$

Description: Tests the state of the V bit in the CCR and causes a branch if V is clear.

Used after an operation on twos-complement binary values, this instruction will cause a branch if there was NO overflow. That is, branch if the twos-complement result was valid.

See BRA instruction for further details of the execution of the branch.

Condition Codes and Boolean Formulae:

S	X	H	I	N	Z	V	C
—	—	—	—	—	—	—	—

None affected

Source Form: BVC (rel)

Addressing Modes, Machine Code, and Cycle-by-Cycle Execution:

Cycle	BVC (REL)		
	Addr	Data	R/W̄
1	OP	28	1
2	OP+1	rr	1
3	FFFF	—	1

The following table is a summary of all branch instructions.

Test	Boolean	Mnemonic	Opcode	Complementary		Branch	Comment
r>m	$Z + (N \oplus V) = 0$	BGT	2E	r≤m	BLE	2F	Signed
r≥m	$N \oplus V = 0$	BGE	2C	r<m	BLT	2D	Signed
r=m	$Z = 1$	BEQ	27	r≠m	BNE	26	Signed
r≤m	$Z + (N \oplus V) = 1$	BLE	2F	r>m	BGT	2E	Signed
r<m	$N \oplus V = 1$	BLT	2D	r≥m	BGE	2C	Signed
r>m	$C + Z = 0$	BHI	22	r≤m	BLS	23	Unsigned
r≥m	$C = 0$	BHS/BCC	24	r<m	BLO/BCS	25	Unsigned
r=m	$Z = 1$	BEQ	27	r≠m	BNE	26	Unsigned
r≤m	$C + Z = 1$	BLS	23	r>m	BHI	22	Unsigned
r<m	$C = 1$	BLO/BCS	25	r≥m	BHS/BCC	24	Unsigned
Carry	$C = 1$	BCS	25	No Carry	BCC	24	Simple
Negative	$N = 1$	BMI	2B	Plus	BPL	2A	Simple
Overflow	$V = 1$	BVS	29	No Overflow	BVC	28	Simple
r=0	$Z = 1$	BEQ	27	r≠0	BNE	26	Simple
Always	—	BRA	20	Never	BRN	21	Unconditional

BVS   Branch if Overflow Set   BVS

Operation: PC ◀ (PC) + $0002 + Rel   if (V) = 1

Description: Tests the state of the V bit in the CCR and causes a branch if V is set.

Used after an operation on twos-complement binary values, this instruction will cause a branch if there was an overflow. That is, branch if the twos-complement result was invalid.

See BRA instruction for details of the execution of the branch.

Condition Codes and Boolean Formulae:

S	X	H	I	N	Z	V	C
—	—	—	—	—	—	—	—

None affected

Source Form: BVS (rel)

Addressing Modes, Machine Code, and Cycle-by-Cycle Execution:

Cycle	BVS (REL)		
	Addr	Data	R/W̄
1	OP	29	1
2	OP + 1	rr	1
3	FFFF	—	1

The following table is a summary of all branch instructions.

Test	Boolean	Mnemonic	Opcode	Complementary		Branch	Comment
r>m	Z + (N ⊕ V) = 0	BGT	2E	r≤m	BLE	2F	Signed
r≥m	N ⊕ V = 0	BGE	2C	r<m	BLT	2D	Signed
r = m	Z = 1	BEQ	27	r ≠ m	BNE	26	Signed
r≤m	Z + (N ⊕ V) = 1	BLE	2F	r>m	BGT	2E	Signed
r<m	N ⊕ V = 1	BLT	2D	r≥m	BGE	2C	Signed
r>m	C + Z = 0	BHI	22	r≤m	BLS	23	Unsigned
r≥m	C = 0	BHS/BCC	24	r<m	BLO/BCS	25	Unsigned
r = m	Z = 1	BEQ	27	r ≠ m	BNE	26	Unsigned
r≤m	C + Z = 1	BLS	23	r>m	BHI	22	Unsigned
r<m	C = 1	BLO/BCS	25	r≥m	BHS/BCC	24	Unsigned
Carry	C = 1	BCS	25	No Carry	BCC	24	Simple
Negative	N = 1	BMI	2B	Plus	BPL	2A	Simple
Overflow	V = 1	BVS	29	No Overflow	BVC	28	Simple
r = 0	Z = 1	BEQ	27	r ≠ 0	BNE	26	Simple
Always	—	BRA	20	Never	BRN	21	Unconditional

CBA

Compare Accumulators

CBA

Operation: (ACCA) − (ACCB)

Description: Compares the contents of ACCA to the contents of ACCB and sets the condition codes, which may be used for arithmetic and logical conditional branches. Both operands are unaffected.

Condition Codes and Boolean Formulae:

S	X	H	I	N	Z	V	C
—	—	—	—	↕	↕	↕	↕

N R7
   Set if MSB of result is set; cleared otherwise.

Z $\overline{R7} \cdot \overline{R6} \cdot \overline{R5} \cdot \overline{R4} \cdot \overline{R3} \cdot \overline{R2} \cdot \overline{R1} \cdot \overline{R0}$
   Set if result is $00; cleared otherwise.

V $A7 \cdot \overline{B7} \cdot \overline{R7} + \overline{A7} \cdot B7 \cdot R7$
   Set if a twos complement overflow resulted from the operation; cleared otherwise.

C $\overline{A7} \cdot B7 + B7 \cdot R7 + R7 \cdot \overline{A7}$
   Set if there was a borrow from the MSB of the result; cleared otherwise.

Source Form: CBA

Addressing Modes, Machine Code, and Cycle-by-Cycle Execution:

Cycle	CBA (INH)		
	Addr	Data	R/$\overline{W}$
1	OP	11	1
2	OP + 1	—	1

CLC

Clear Carry

CLC

Operation: C bit ◄ 0

Description: Clears the C bit in the CCR.

CLC may be used to set up the C bit prior to a shift or rotate instruction involving the C bit.

Condition Codes and Boolean Formulae:

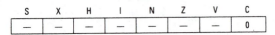

S	X	H	I	N	Z	V	C
—	—	—	—	—	—	—	0

C   0
  Cleared

Source Form: CLC

Addressing Modes, Machine Code, and Cycle-by-Cycle Execution:

Cycle	CLC (INH)		
	Addr	Data	R/W̄
1	OP	0C	1
2	OP+1	—	1

CLI

Clear Interrupt Mask

CLI

Operation: I bit ← 0

Description: Clears the interrupt mask bit in the CCR. When the I bit is clear, interrupts are enabled. There is a one E-clock cycle delay in the clearing mechanism for the I bit so that, if interrupts were previously disabled, the next instruction after a CLI will always be executed, even if there was an interrupt pending prior to execution of the CLI instruction.

Condition Codes and Boolean Formulae:

S	X	H	I	N	Z	V	C
—	—	—	0	—	—	—	—

I 0
  Cleared

Source Form: CLI

Addressing Modes, Machine Code, and Cycle-by-Cycle Execution:

Cycle	CLI (INH)		
	Addr	Data	R/W̄
1	OP	0E	1
2	OP+1	—	1

CLR

Clear

CLR

Operation: ACCX ◀ 0   or: M ◀ 0

Description;   The contents of ACCX or M are replaced with zeros.

Condition Codes and Boolean Formulae:

S	X	H	I	N	Z	V	C
—	—	—	—	0	1	0	0

N 0
  Cleared

Z 1
  Set

V 0
  Cleared

C 0
  Cleared

Source Forms: CLRA; CLRB; CLR (opr)

Addressing Modes, Machine Code, and Cycle-by-Cycle Execution:

Cycle	CLRA (INH)			CLRB (INH)			CLR (EXT)			CLR (IND, X)			CLR (IND, Y)		
	Addr	Data	R/W̄	Addr	Data	R/W̄	Addr	Data	R/W̄	Addr	Data	R/W̄	Addr	Data	R/W̄
1	OP	4F	1	OP	5F	1	OP	7F	1	OP	6F	1	OP	18	1
2	OP + 1	—	1	OP + 1	—	1	OP + 1	hh	1	OP + 1	ff	1	OP + 1	6F	1
3							OP + 2	ll	1	FFFF	—	1	OP + 2	ff	1
4							hhll	(hhll)	1	X + ff	(X + ff)	1	FFFF	—	1
5							FFFF	—	1	FFFF	—	1	Y + ff	(Y + ff)	1
6							hhll	00	0	X + ff	00	0	FFFF	—	1
7													Y + ff	00	0

CLV   Clear Twos-Complement Overflow Bit   CLV

Operation: V bit 0

Description: Clears the twos complement overflow bit in the CCR.

Condition Codes and Boolean Formulae:

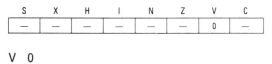

S	X	H	I	N	Z	V	C
—	—	—	—	—	—	0	—

V 0
  Cleared

Source Form: CLV

Addressing Modes, Machine Code, and Cycle-by-Cycle Execution:

Cycle	CLV (INH)		
	Addr	Data	R/W̄
1	OP	0A	1
2	OP + 1	—	1

CMP

Compare

CMP

Operation: (ACCX) − (M)

Description: Compares the contents of ACCX to the contents of M and sets the condition codes, which may be used for arithmetic and logical conditional branching. Both operands are unaffected.

Condition Codes and Boolean Formulae:

S	X	H	I	N	Z	V	C
—	—	—	—	$\updownarrow$	$\updownarrow$	$\updownarrow$	$\updownarrow$

N   R7
     Set if MSB of result is set; cleared otherwise.

Z $\overline{R7} \cdot \overline{R6} \cdot \overline{R5} \cdot \overline{R4} \cdot \overline{R3} \cdot \overline{R2} \cdot \overline{R1} \cdot \overline{R0}$
     Set if result is $00; cleared otherwise.

V $X7 \cdot \overline{M7} \cdot \overline{R7} + \overline{X7} \cdot M7 \cdot R7$
     Set if a twos complement overflow resulted from the operation; cleared otherwise.

C $\overline{X7} \cdot M7 + M7 \cdot R7 + R7 \cdot \overline{X7}$
     Set if there was a borrow from the MSB of the result; cleared otherwise.

Source Forms: CMPA (opr); CMPB (opr)

Addressing Modes, Machine Code, and Cycle-by-Cycle Execution:

Cycle	CMPA (IMM)			CMPA (DIR)			CMPA (EXT)			CMPA (IND, X)			CMPA (IND, Y)		
	Addr	Data	R/$\overline{W}$	Addr	Data	R/$\overline{W}$	Addr	Data	R/$\overline{W}$	Addr	Data	R/$\overline{W}$	Addr	Data	R/$\overline{W}$
1	OP	81	1	OP	91	1	OP	B1	1	OP	A1	1	OP	18	1
2	OP+1	ii	1	OP+1	dd	1	OP+1	hh	1	OP+1	ff	1	OP+1	A1	1
3				00dd	(00dd)	1	OP+2	ll	1	FFFF	—	1	OP+2	ff	1
4							hhll	(hhll)	1	X+ff	(X+ff)	1	FFFF	—	1
5													Y+ff	(Y+ff)	1

Cycle	CMPB (IMM)			CMPB (DIR)			CMPB (EXT)			CMPB (IND, X)			CMPB (IND, Y)		
	Addr	Data	R/$\overline{W}$	Addr	Data	R/$\overline{W}$	Addr	Data	R/$\overline{W}$	Addr	Data	R/$\overline{W}$	Addr	Data	R/$\overline{W}$
1	OP	C1	1	OP	D1	1	OP	F1	1	OP	E1	1	OP	18	1
2	OP+1	ii	1	OP+1	dd	1	OP+1	hh	1	OP+1	ff	1	OP+1	E1	1
3				00dd	(00dd)	1	OP+2	ll	1	FFFF	—	1	OP+2	ff	1
4							hhll	(hhll)	1	X+ff	(X+ff)	1	FFFF	—	1
5													Y+ff	(Y+ff)	1

COM

Complement

COM

Operation: ACCX ◀ ($\overline{ACCX}$) = $FF − (ACCX) **or:** M ◀ ($\overline{M}$) = $FF − (M)

Description: Replaces the contents of ACCX or M with its ones complement. (Each bit of the contents of ACCX or M is replaced with the complement of that bit.) To complement a value without affecting the C-bit, EXclusive-OR the value with $FF.

Condition Codes and Boolean Formulae:

S	X	H	I	N	Z	V	C
—	—	—	—	✱	✱	0	1

N R7
  Set if MSB of result is set; cleared otherwise.

Z $\overline{R7} \cdot \overline{R6} \cdot \overline{R5} \cdot \overline{R4} \cdot \overline{R3} \cdot \overline{R2} \cdot \overline{R1} \cdot \overline{R0}$
  Set if result is $00; cleared otherwise.

V 0
  Cleared

C 1
  Set (For compatibility with M6800)

Source Forms: COMA; COMB; COM (opr)

Addressing Modes, Machine Code, and Cycle-by-Cycle Execution:

Cycle	COMA (INH) Addr	Data	R/$\overline{W}$	COMB (INH) Addr	Data	R/$\overline{W}$	COM (EXT) Addr	Data	R/$\overline{W}$	COM (IND, X) Addr	Data	R/$\overline{W}$	COM (IND, Y) Addr	Data	R/$\overline{W}$
1	OP	43	1	OP	53	1	OP	73	1	OP	63	1	OP	18	1
2	OP + 1	—	1	OP + 1	—	1	OP + 1	hh	1	OP + 1	ff	1	OP + 1	63	1
3							OP + 2	ll	1	FFFF	—	1	OP + 2	ff	1
4							hhll	(hhll)	1	X + ff	(X + ff)	1	FFFF	—	1
5							FFFF	—	1	FFFF	—	1	Y + ff	(Y + ff)	1
6							hhll	result	0	X + ff	result	0	FFFF	—	1
7													Y + ff	result	0

CPD   **Compare Double Accumulator**   CPD

Operation: $(ACCD) - (M:M+1)$

Description: Compares the contents of accumulator D with a 16-bit value at the address specified and sets the condition codes accordingly. The compare is accomplished internally by doing a 16-bit subtract of $(M:M+1)$ from accumulator D without modifying either accumulator D or $(M:M+1)$.

Condition Codes and Boolean Formulae:

S	X	H	I	N	Z	V	C
—	—	—	—	↕	↕	↕	↕

N   R15
    Set if MSB of result is set; cleared otherwise.

Z $\overline{R15} \cdot \overline{R14} \cdot \overline{R13} \cdot \overline{R12} \cdot \overline{R11} \cdot \overline{R10} \cdot \overline{R9} \cdot \overline{R8} \cdot \overline{R7} \cdot \overline{R6} \cdot \overline{R5} \cdot \overline{R4} \cdot \overline{R3} \cdot \overline{R2} \cdot \overline{R1} \cdot \overline{R0}$
    Set if result is $0000; cleared otherwise.

V $D15 \cdot \overline{M15} \cdot \overline{R15} + \overline{D15} \cdot M15 \cdot R15$
    Set if a twos complement overflow resulted from the operation; cleared otherwise.

C $\overline{D15} \cdot M15 + M15 \cdot R15 + R15 \cdot \overline{D15}$
    Set if the absolute value of the contents of memory is larger than the absolute value of the accumulator; cleared otherwise.

Source Form: CPD (opr)

Addressing Modes, Machine Code, and Cycle-by-Cycle Execution:

Cycle	CPD (IMM) Addr	Data	R/W̅	CPD (DIR) Addr	Data	R/W̅	CPD (EXT) Addr	Data	R/W̅	CPD (IND, X) Addr	Data	R/W̅	CPD (IND, Y) Addr	Data	R/W̅
1	OP	1A	1	OP	1A	1	OP	1A	1	OP	1A	1	OP	CD	1
2	OP+1	83	1	OP+1	93	1	OP+1	B3	1	OP+1	A3	1	OP+1	A3	1
3	OP+2	jj	1	OP+2	dd	1	OP+2	hh	1	OP+2	ff	1	OP+2	ff	1
4	OP+3	kk	1	00dd	(00dd)	1	OP+3	ll	1	FFFF	—	1	FFFF	—	1
5	FFFF	—	1	00dd+1	(00dd+1)	1	hhll	(hhll)	1	X+ff	(X+ff)	1	Y+ff	(Y+ff)	1
6				FFFF	—	1	hhll+1	(hhll+1)	1	X+ff+1	(X+ff+1)	1	Y+ff+1	(Y+ff+1)	1
7							FFFF	—	1	FFFF	—	1	FFFF	—	1

CPX

Compare Index Register X

CPX

Operation: (IX) − (M:M + 1)

Description: Compares the contents of the index register X with a 16-bit value at the address specified and sets the condition codes accordingly. The compare is accomplished internally by doing a 16-bit subtract of (M:M + 1) from index register X without modifying either index register X or (M:M + 1).

Condition Codes and Boolean Formulae:

S	X	H	I	N	Z	V	C
—	—	—	—	↕	↕	↕	↕

N R15
Set if MSB of result is set; cleared otherwise.

Z $\overline{R15} \cdot \overline{R14} \cdot \overline{R13} \cdot \overline{R12} \cdot \overline{R11} \cdot \overline{R10} \cdot \overline{R9} \cdot \overline{R8} \cdot \overline{R7} \cdot \overline{R6} \cdot \overline{R5} \cdot \overline{R4} \cdot \overline{R3} \cdot \overline{R2} \cdot \overline{R1} \cdot \overline{R0}$
Set if result is $0000; cleared otherwise.

V $IX15 \cdot \overline{M15} \cdot \overline{R15} + \overline{IX15} \cdot M15 \cdot R15$
Set if a twos complement overflow resulted from the operation; cleared otherwise.

C $\overline{IX15} \cdot M15 + M15 \cdot R15 + R15 \cdot \overline{IX15}$
Set if the absolute value of the contents of memory is larger than the absolute value of the index register; cleared otherwise.

Source Form: CPX (opr)

Addressing Modes, Machine Code, and Cycle-by-Cycle Execution:

Cycle	CPX (IMM)			CPX (DIR)			CPX (EXT)			CPX (IND, X)			CPX (IND, Y)		
	Addr	Data	R/W̄	Addr	Data	R/W̄	Addr	Data	R/W̄	Addr	Data	R/W̄	Addr	Data	R/W̄
1	OP	8C	1	OP	9C	1	OP	BC	1	OP	AC	1	OP	CD	1
2	OP + 1	jj	1	OP + 1	dd	1	OP + 1	hh	1	OP + 1	ff	1	OP + 1	AC	1
3	OP + 2	kk	1	00dd	(00dd)	1	OP + 2	ll	1	FFFF	—	1	OP + 2	ff	1
4	FFFF	—	1	00dd + 1	(00dd + 1)	1	hhll	(hhll)	1	X + ff	(X + ff)	1	FFFF	—	1
5				FFFF	—	1	hhll + 1	(hhll + 1)	1	X + ff + 1	(X + ff + 1)	1	Y + ff	(Y + ff)	1
6							FFFF	—	1	FFFF	—	1	Y + ff + 1	(Y + ff + 1)	1
7													FFFF	—	1

CPY   **Compare Index Register Y**   CPY

Operation: $(IY) - (M:M + 1)$

Description: Compares the contents of the index register Y with a 16-bit value at the address specified and sets the condition codes accordingly. The compare is accomplished internally by doing a 16-bit subtract of $(M:M + 1)$ from index register Y without modifying either index register Y or $(M:M + 1)$.

Condition Codes and Boolean Formulae:

S	X	H	I	N	Z	V	C
—	—	—	—	↕	↕	↕	↕

N R15
   Set if MSB of result is set; cleared otherwise.

Z $\overline{R15} \cdot \overline{R14} \cdot \overline{R13} \cdot \overline{R12} \cdot \overline{R11} \cdot \overline{R10} \cdot \overline{R9} \cdot \overline{R8} \cdot \overline{R7} \cdot \overline{R6} \cdot \overline{R5} \cdot \overline{R4} \cdot \overline{R3} \cdot \overline{R2} \cdot \overline{R1} \cdot \overline{R0}$
   Set if result is $0000; cleared otherwise.

V $IY15 \cdot \overline{M15} \cdot \overline{R15} + \overline{IY15} \cdot M15 \cdot R15$
   Set if a twos complement overflow resulted from the operation; cleared otherwise.

C $\overline{IY15} \cdot M15 + M15 \cdot R15 + R15 \cdot \overline{IY15}$
   Set if the absolute value of the contents of memory is larger than the absolute value of the index register; cleared otherwise.

Source Form: CPY (opr)

Addressing Modes, Machine Code, and Cycle-by-Cycle Execution:

Cycle	CPY (IMM)			CPY (DIR)			CPY (EXT)			CPY (IND, X)			CPY (IND, Y)		
	Addr	Data	R/W̄	Addr	Data	R/W̄	Addr	Data	R/W̄	Addr	Data	R/W̄	Addr	Data	R/W̄
1	OP	18	1	OP	18	1	OP	18	1	OP	1A	1	OP	18	1
2	OP+1	8C	1	OP+1	9C	1	OP+1	BC	1	OP+1	AC	1	OP+1	AC	1
3	OP+2	jj	1	OP+2	dd	1	OP+2	hh	1	OP+2	ff	1	OP+2	ff	1
4	OP+3	kk	1	00dd	(00dd)	1	OP+3	ll	1	FFFF	—	1	FFFF	—	1
5	FFFF	—	1	00dd+1	(00dd+1)	1	hhll	(hhll)	1	X+ff	(X+ff)	1	Y+ff	(Y+ff)	1
6				FFFF	—	1	hhll+1	(hhll+1)	1	X+ff+1	(X+ff+1)	1	Y+ff+1	(Y+ff+1)	1
7							FFFF	—	1	FFFF	—	1	FFFF	—	1

DAA   Decimal Adjust ACCA   DAA

Operation: The following table summarizes the operation of the DAA instruction for all legal combinations of input operands. A correction factor (column 5 in the following table) is added to ACCA to restore the result of an addition of two BCD operands to a valid BCD value and set or clear the carry bit.

State of C Bit Before DAA (Column 1)	Upper Half-Byte of ACCA (Bits 7–4) (Column 2)	Initial Half-Carry H Bit from CCR (Column 3)	Lower Half-Byte of ACCA (Bits 3–0) (Column 4)	Number Added to ACCA by DAA (Column 5)	State of C Bit After DAA (Column 6)
0	0-9	0	0-9	00	0
0	0-8	0	A-F	06	0
0	0-9	1	0-3	06	0
0	A-F	0	0-9	60	1
0	9-F	0	A-F	66	1
0	A-F	1	0-3	66	1
1	0-2	0	0-9	60	1
1	0-2	0	A-F	66	1
1	0-3	1	0-3	66	1

NOTE

Columns (1) through (4) of the above table represent all possible cases which can result from any of the operations ABA, ADD, or ADC, with initial carry either set or clear, applied to two binary-coded-decimal operands. The table shows hexadecimal values.

Description: If the contents of ACCA and the state of the carry/borrow bit C and the state of the half-carry bit H are all the result of applying any of the operations ABA, ADD, or ADC to binary-coded-decimal operands, with or without an initial carry, the DAA operation will adjust the contents of ACCA and the carry bit C in the CCR to represent the correct binary-coded-decimal sum and the correct state of the C bit.

Condition Codes and Boolean Formulae:

S	X	H	I	N	Z	V	C
—	—	—	—	⬢	⬢	?	⬢

N R7
   Set if MSB of result is set; cleared otherwise.

Z $\overline{R7} \cdot \overline{R6} \cdot \overline{R5} \cdot \overline{R4} \cdot \overline{R3} \cdot \overline{R2} \cdot \overline{R1} \cdot \overline{R0}$
   Set if result is $00; cleared otherwise.

V ?
   Not defined

C See table above.

DAA

Decimal Adjust ACCA
(Continued)

DAA

Source Form: DAA

Addressing Modes, Machine Code, and Cycle-by-Cycle Execution:

Cycle	DAA (INH)		
	Addr	Data	R/W̄
1	OP	19	1
2	OP+1	—	1

For the purpose of illustration, consider the case where the BCD value $99 was just added to the BCD value $22. The add instruction is a binary operation, which yields the result $BB with no carry (C) or half carry (H). This corresponds to the fifth row of the table on the previous page. The DAA instruction will therefore add the correction factor $66 to the result of the addition, giving a result of $21 with the carry bit set. This result corresponds to the BCD value $121, which is the expected BCD result.

DEC

Decrement

DEC

Operation: ACCX ⬧ (ACCX) − $01 **or:** M ⬧ (M) − $01

Description: Subtract one from the contents of ACCX or M.

The N, Z, and V bits in the CCR are set or cleared according to the results of the operation. The C bit in the CCR is not affected by the operation, thus allowing the DEC instruction to be used as a loop counter in multiple-precision computations.

When operating on unsigned values, only BEQ and BNE branches can be expected to perform consistently. When operating on twos-complement values, all signed branches are available.

Condition Codes and Boolean Formulae:

S	X	H	I	N	Z	V	C
—	—	—	—	⬧	⬧	⬧	—

N   R7
   Set if MSB of result is set; cleared otherwise.

Z $\overline{R7} \cdot \overline{R6} \cdot \overline{R5} \cdot \overline{R4} \cdot \overline{R3} \cdot \overline{R2} \cdot \overline{R1} \cdot \overline{R0}$
   Set if result is $00; cleared otherwise

V $X7 \cdot \overline{X6} \cdot \overline{X5} \cdot \overline{X4} \cdot X3 \cdot \overline{X2} \cdot \overline{X1} \cdot \overline{X0} = \overline{R7} \cdot R6 \cdot R5 \cdot R4 \cdot R3 \cdot R2 \cdot R1 \cdot R0$
   Set if there was a twos complement overflow as a result of the operation; cleared otherwise. Twos complement overflow occurs if and only if (ACCX) or (M) was $80 before the operation.

Source Forms: DECA; DECB; DEC (opr)

Addressing Modes, Machine Code, and Cycle-by-Cycle Execution:

Cycle	DECA (INH)			DECB (INH)			DEC (EXT)			DEC (IND, X)			DEC (IND, Y)		
	Addr	Data	R/W̄	Addr	Data	R/W̄	Addr	Data	R/W̄	Addr	Data	R/W̄	Addr	Data	R/W̄
1	OP	4A	1	OP	5A	1	OP	7A	1	OP	6A	1	OP	18	1
2	OP+1	—	1	OP+1	—	1	OP+1	hh	1	OP+1	ff	1	OP+1	6A	1
3							OP+2	ll	1	FFFF	—	1	OP+2	ff	1
4							hhll	(hhll)	1	X+ff	(X+ff)	1	FFFF	—	1
5							FFFF	—	1	FFFF	—	1	Y+ff	(Y+ff)	1
6							hhll	result	0	X+ff	result	0	FFFF	—	1
7													Y+ff	result	0

DES

Decrement Stack Pointer

DES

Operation: SP ◄ (SP) – $0001

Description: Subtract one from the stack pointer.

Condition Codes and Boolean Formulae:

S	X	H	I	N	Z	V	C
—	—	—	—	—	—	—	—

None affected

Source Form: DES

Addressing Modes, Machine Code, and Cycle-by-Cycle Execution:

Cycle	DES (INH)		
	Addr	Data	R/W̄
1	OP	34	1
2	OP+1	—	1
3	SP	—	1

DEX
DEX   Decrement Index Register X   DEX

Operation: IX ⬦ (IX) − $0001

Description: Subtract one from the index register X.

Only the Z bit is set or cleared according to the result of this operation.

Condition Codes and Boolean Formulae:

S	X	H	I	N	Z	V	C
—	—	—	—	—	⬤	—	—

Z $\overline{R15} \cdot \overline{R14} \cdot \overline{R13} \cdot \overline{R12} \cdot \overline{R11} \cdot \overline{R10} \cdot \overline{R9} \cdot \overline{R8} \cdot \overline{R7} \cdot \overline{R6} \cdot \overline{R5} \cdot \overline{R4} \cdot \overline{R3} \cdot \overline{R2} \cdot \overline{R1} \cdot \overline{R0}$
Set if result is $0000; cleared otherwise.

Source Form: DEX

Addressing Modes, Machine Code, and Cycle-by-Cycle Execution:

Cycle	DEX (INH)		
	Addr	Data	R/W̄
1	OP	09	1
2	OP+1	—	1
3	FFFF	—	1

DEY

Decrement Index Register Y

DEY

Operation: IY ⬩ (IY) − $0001

Description: Subtract one from the index register Y.

Only the Z bit is set or cleared according to the result of this operation.

Condition Codes and Boolean Formulae:

S	X	H	I	N	Z	V	C
—	—	—	—	—	✦	—	—

Z $\overline{R15} \cdot \overline{R14} \cdot \overline{R13} \cdot \overline{R12} \cdot \overline{R11} \cdot \overline{R10} \cdot \overline{R9} \cdot \overline{R8} \cdot \overline{R7} \cdot \overline{R6} \cdot \overline{R5} \cdot \overline{R4} \cdot \overline{R3} \cdot \overline{R2} \cdot \overline{R1} \cdot \overline{R0}$
Set if result is $0000; cleared otherwise.

Source Form: DEY

Addressing Modes, Machine Code, and Cycle-by-Cycle Execution:

Cycle	DEY (INH)		
	Addr	Data	R/$\overline{W}$
1	OP	18	1
2	OP + 1	09	1
3	OP + 2	—	1
4	FFFF	—	1

EOR EOR

Exclusive-OR

Operation: ACCX ◀ (ACCX) ⊕ (M)

Description: Performs the logical exclusive-OR between the contents of ACCX and the contents of M and places the result in ACCX. (Each bit of ACCX after the operation will be the logical exclusive-OR of the corresponding bits of M and ACCX before the operation.)

Condition Codes and Boolean Formulae:

S	X	H	I	N	Z	V	C
—	—	—	—	↕	↕	0	—

N R7
   Set if MSB of result is set; cleared otherwise.

Z $\overline{R7} \cdot \overline{R6} \cdot \overline{R5} \cdot \overline{R4} \cdot \overline{R3} \cdot \overline{R2} \cdot \overline{R1} \cdot \overline{R0}$
   Set if result is $00; cleared otherwise

V 0
   Cleared

Source Forms: EORA (opr); EORB (opr)

Addressing Modes, Machine Code, and Cycle-by-Cycle Execution:

Cycle	EORA (IMM)			EORA (DIR)			EORA (EXT)			EORA (IND, X)			EORA (IND, Y)		
	Addr	Data	R/W̄	Addr	Data	R/W̄	Addr	Data	R/W̄	Addr	Data	R/W̄	Addr	Data	R/W̄
1	OP	88	1	OP	98	1	OP	B8	1	OP	A8	1	OP	18	1
2	OP+1	ii	1	OP+1	dd	1	OP+1	hh	1	OP+1	ff	1	OP+1	A8	1
3				00dd	(00dd)	1	OP+2	ll	1	FFFF	—	1	OP+2	ff	1
4							hhll	(hhll)	1	X+ff	(X+ff)	1	FFFF	—	1
5													Y+ff	(Y+ff)	1

Cycle	EORB (IMM)			EORB (DIR)			EORB (EXT)			EORB (IND, X)			EORB (IND, Y)		
	Addr	Data	R/W̄	Addr	Data	R/W̄	Addr	Data	R/W̄	Addr	Data	R/W̄	Addr	Data	R/W̄
1	OP	C8	1	OP	D8	1	OP	F8	1	OP	E8	1	OP	18	1
2	OP+1	ii	1	OP+1	dd	1	OP+1	hh	1	OP+1	ff	1	OP+1	E8	1
3				00dd	(00dd)	1	OP+2	ll	1	FFFF	—	1	OP+2	ff	1
4							hhll	(hhll)	1	X+ff	(X+ff)	1	FFFF	—	1
5													Y+ff	(Y+ff)	1

FDIV

Fractional Divide

FDIV

Operation: (ACCD)/(IX); IX ◆ Quotient, ACCD ◆ Remainder

Description: Performs an usigned fractional divide of the 16-bit numerator in the D accumulator by the 16-bit denominator in the index register X and sets the condition codes accordingly. The quotient is placed in the index register X, and the remainder is placed in the D accumulator. The radix point is assumed to be in the same place for both the numerator and the denominator. The radix point is to the left of bit 15 for the quotient. The numerator is assumed to be less than the denominator. In the case of overflow (denominator is less than or equal to the numerator) or divide by zero, the quotient is set to $FFFF, and the remainder is indeterminate.

FDIV is equivalent to multiplying the numerator by 2^{16} and then performing a 32×16-bit integer divide. The result is interpreted as a binary-weighted fraction, which resulted from the division of a 16-bit integer by a larger 16-bit integer. A result of $0001 corresponds to 0.000015, and $FFFF corresponds to 0.99998. The remainder of an IDIV instruction can be resolved into a binary-weighted fraction by an FDIV instruction. The remainder of an FDIV instruction can be resolved into the next 16-bits of binary-weighted fraction by another FDIV instruction.

Condition Codes and Boolean Formulae:

S	X	H	I	N	Z	V	C
—	—	—	—	—	✸	✸	✸

Z $\overline{R15} \cdot \overline{R14} \cdot \overline{R13} \cdot \overline{R12} \cdot \overline{R11} \cdot \overline{R10} \cdot \overline{R9} \cdot \overline{R8} \cdot \overline{R7} \cdot \overline{R6} \cdot \overline{R5} \cdot \overline{R4} \cdot \overline{R3} \cdot \overline{R2} \cdot \overline{R1} \cdot \overline{R0}$
Set if quotient is $0000; cleared otherwise.

V   1 if IX≤D
Set if denominator was less than or equal to the numerator; cleared otherwise.

C $\overline{IX15} \cdot \overline{IX14} \cdot \overline{IX13} \cdot \overline{IX12} \cdot \overline{IX11} \cdot \overline{IX10} \cdot \overline{IX9} \cdot \overline{IX8} \cdot$
$\overline{IX7} \cdot \overline{IX6} \cdot \overline{IX5} \cdot \overline{IX4} \cdot \overline{IX3} \cdot \overline{IX2} \cdot \overline{IX1} \cdot \overline{IX0}$
Set if denominator was $0000; cleared otherwise.

Source Form: FDIV

Addressing Modes, Machine Code, and Cycle-by-Cycle Execution:

Cycle	FDIV (INH)		
	Addr	Data	R/$\overline{W}$
1	OP	03	1
2	OP+1	—	1
3–41	FFFF	—	1

IDIV

Integer Divide

IDIV

Operation: (ACCD)/(IX); IX ◀ Quotient, ACCD ◀ Remainder

Description: Performs an unsigned integer divide of the 16-bit numerator in D accumulator by the 16-bit denominator in index register X and sets the condition codes accordingly. The quotient is placed in index register X, and the remainder is placed in accumulator D. The radix point is assumed to be in the same place for both the numerator and the denominator. The radix point is to the right of bit zero for the quotient. In the case of divide by zero, the quotient is set to $FFFF, and the remainder is indeterminate.

Condition Codes and Boolean Formulae:

S	X	H	I	N	Z	V	C
—	—	—	—	—	✦	0	✦

Z $\overline{R15} \cdot \overline{R14} \cdot \overline{R13} \cdot \overline{R12} \cdot \overline{R11} \cdot \overline{R10} \cdot \overline{R9} \cdot \overline{R8} \cdot \overline{R7} \cdot \overline{R6} \cdot \overline{R5} \cdot \overline{R4} \cdot \overline{R3} \cdot \overline{R2} \cdot \overline{R1} \cdot \overline{R0}$
Set if result is $0000; cleared otherwise.

V   0
Cleared.

C $\overline{IX15} \cdot \overline{IX14} \cdot \overline{IX13} \cdot \overline{IX12} \cdot \overline{IX11} \cdot \overline{IX10} \cdot \overline{IX9} \cdot \overline{IX8} \cdot$
$\overline{IX7} \cdot \overline{IX6} \cdot \overline{IX5} \cdot \overline{IX4} \cdot \overline{IX3} \cdot \overline{IX2} \cdot \overline{IX1} \cdot \overline{IX0}$
Set if denominator was $0000; cleared otherwise.

Source Form: IDIV

Addressing Modes, Machine Code, and Cycle-by-Cycle Execution:

Cycle	IDIV (INH)		
	Addr	Data	R/W̄
1	OP	02	1
2	OP+1	—	1
3–41	FFFF	—	1

INC

Increment

INC

Operation: ACCX ⇐ (ACCX) + $01   **or:**   M ⇐ (M) + $01

Description: Add one to the contents of ACCX or M.

The N, Z, and V bits in the CCR are set or cleared according to the results of the operation. The C bit in the CCR is not affected by the operation, thus allowing the INC instruction to be used as a loop counter in multiple-precision computations.

When operating on unsigned values, only BEQ and BNE branches can be expected to perform consistently. When operating on twos-complement values, all signed branches are available.

Condition Codes and Boolean Formulae:

S	X	H	I	N	Z	V	C
—	—	—	—	↕	↕	↕	—

N R7
Set if MSB of result is set; cleared otherwise.

Z $\overline{R7} \cdot \overline{R6} \cdot \overline{R5} \cdot \overline{R4} \cdot \overline{R3} \cdot \overline{R2} \cdot \overline{R1} \cdot \overline{R0}$
Set if result is $00; cleared otherwise.

V $\overline{X7} \cdot X6 \cdot X5 \cdot X4 \cdot X3 \cdot X2 \cdot X1 \cdot X0$
Set if there is a twos complement overflow as a result of the operation; cleared otherwise. Twos complement overflow occurs if and only if (ACCX) or (M) was $7F before the operation.

Source Forms: INCA; INCB; INC (opr)

Addressing Modes, Machine Code, and Cycle-by-Cycle Execution:

Cycle	INCA (INH)			INCB (INH)			INC (EXT)			INC (IND, X)			INC (IND, Y)		
	Addr	Data	R/W̄	Addr	Data	R/W̄	Addr	Data	R/W̄	Addr	Data	R/W̄	Addr	Data	R/W̄
1	OP	4C	1	OP	5C	1	OP	7C	1	OP	6C	1	OP	18	1
2	OP+1	—	1	OP+1	—	1	OP+1	hh	1	OP+1	ff	1	OP+1	6C	1
3							OP+2	ll	1	FFFF	—	1	OP+2	ff	1
4							hhll	(hhll)	1	X+ff	(X+ff)	1	FFFF	—	1
5							FFFF	—	1	FFFF	—	1	Y+ff	(Y+ff)	1
6							hhll	result	0	X+ff	result	0	FFFF	—	1
7													Y+ff	result	0

INS

Increment Stack Pointer

INS

Operation: SP ◀ (SP) + $0001

Description: Add one to the stack pointer.

Condition Codes and Boolean Formulae:

S	X	H	I	N	Z	V	C
—	—	—	—	—	—	—	—

None affected

Source Form: INS

Addressing Modes, Machine Code, and Cycle-by-Cycle Execution:

Cycle	INS (INH)		
	Addr	Data	R/W̄
1	OP	31	1
2	OP + 1	—	1
3	SP	—	1

INX
INX

Increment Index Register X

Operation: IX ◀ (IX) + $0001

Description: Add one to index register X.

Only the Z bit is set or cleared according to the result of this operation.

Condition Codes and Boolean Formulae:

S	X	H	I	N	Z	V	C
—	—	—	—	—	⬥	—	—

Z $\overline{R15} \cdot \overline{R14} \cdot \overline{R13} \cdot \overline{R12} \cdot \overline{R11} \cdot \overline{R10} \cdot \overline{R9} \cdot \overline{R8} \cdot \overline{R7} \cdot \overline{R6} \cdot \overline{R5} \cdot \overline{R4} \cdot \overline{R3} \cdot \overline{R2} \cdot \overline{R1} \cdot \overline{R0}$
Set if result is $0000; cleared otherwise.

Source Form: INX

Addressing Modes, Machine Code, and Cycle-by-Cycle Execution:

Cycle	INX (INH)		
	Addr	Data	R/$\overline{W}$
1	OP	08	1
2	OP + 1	—	1
3	FFFF	—	1

INY

Increment Index Register Y

INY

Operation:　　IY ⬑ (IY) + $0001

Description:　　Add one to index register Y.

Only the Z bit is set or cleared according to the result of this operation.

Condition Codes and Boolean Formulae:

S	X	H	I	N	Z	V	C
—	—	—	—	—	⬦	—	—

Z　$\overline{R15} \cdot \overline{R14} \cdot \overline{R13} \cdot \overline{R12} \cdot \overline{R11} \cdot \overline{R10} \cdot \overline{R9} \cdot \overline{R8} \cdot \overline{R7} \cdot \overline{R6} \cdot \overline{R5} \cdot \overline{R4} \cdot \overline{R3} \cdot \overline{R2} \cdot \overline{R1} \cdot \overline{R0}$
Set if result is $0000; cleared otherwise.

Source Form:　　INY

Addressing Modes, Machine Code, and Cycle-by-Cycle Execution:

Cycle	INY (INH)		
	Addr	Data	R/W̄
1	OP	18	1
2	OP + 1	08	1
3	OP + 2	—	1
4	FFFF	—	1

JMP   Jump   JMP

Operation: PC ⬅ Effective Address

Description: A jump occurs to the instruction stored at the effective address. The effective address is obtained according to the rules for EXTended or INDexed addressing.

Condition Codes and Boolean Formulae:

S	X	H	I	N	Z	V	C
—	—	—	—	—	—	—	—

None affected

Source Form: JMP (opr)

Addressing Modes, Machine Code, and Cycle-by-Cycle Execution:

Cycle	JMP (EXT)			JMP (IND, X)			JMP (IND, Y)		
	Addr	Data	R/W̄	Addr	Data	R/W̄	Addr	Data	R/W̄
1	OP	7E	1	OP	6E	1	OP	18	1
2	OP+1	hh	1	OP+1	ff	1	OP+1	6E	1
3	OP+2	ll	1	FFFF	—	1	OP+2	ff	1
4							FFFF	—	1

JSR   Jump to Subroutine   JSR

Operation: PC ◄ (PC) + $0003   (for EXTended or INDexed, Y addressing) **or:**
PC ◄ (PC) + $0002   (for DIRect or INDexed, X addressing)
↜(PCL)   Push low-order return address onto stack
SP ◄ (SP) − $0001
↜(PCH)   Push high-order return address onto stack
SP ◄ (SP) − $0001
PC ◄ Effective Addr   Load start address of requested subroutine

Description: The program counter is incremented by three or by two, depending on
the addressing mode, and is then pushed onto the stack, eight bits at a time, least
significant byte first. The stack pointer points to the next empty location in the stack.
A jump occurs to the instruction stored at the effective address. The effective address
is obtained according to the rules for EXTended, DIRect, or INDexed addressing.

Condition Codes and Boolean Formulae:

S	X	H	I	N	Z	V	C
—	—	—	—	—	—	—	—

None affected

Source Form: JSR (opr)

Addressing Modes, Machine Code, and Cycle-by-Cycle Execution:

Cycle	JSR (DIR)			JSR (EXT)			JSR (IND, X)			JSR (IND, Y)		
	Addr	Data	R/W̄	Addr	Data	R/W̄	Addr	Data	R/W̄	Addr	Data	R/W̄
1	OP	9D	1	OP	BD	1	OP	AD	1	OP	18	1
2	OP + 1	dd	1	OP + 1	hh	1	OP + 1	ff	1	OP + 1	AD	1
3	00dd	(00dd)	1	OP + 2	ll	1	FFFF	—	1	OP + 2	ff	1
4	SP	Rtn lo	0	hhll	(hhll)	1	X + ff	(X + ff)	1	FFFF	—	1
5	SP − 1	Rtn hi	0	SP	Rtn lo	0	SP	Rtn lo	0	Y + ff	(Y + ff)	1
6				SP − 1	Rtn hi	0	SP − 1	Rtn hi	0	SP	Rtn lo	0
7										SP − 1	Rtn hi	0

LDA   Load Accumulator   LDA

Operation: ACCX ◀ (M)

Description: Loads the contents of memory into the 8-bit accumulator. The condition codes are set according to the data.

Condition Codes and Boolean Formulae:

S	X	H	I	N	Z	V	C
—	—	—	—	↕	↕	0	—

N   R7
     Set if MSB of result is set; cleared otherwise.

Z $\overline{R7} \cdot \overline{R6} \cdot \overline{R5} \cdot \overline{R4} \cdot \overline{R3} \cdot \overline{R2} \cdot \overline{R1} \cdot \overline{R0}$
     Set if result is $00; cleared otherwise

V   0
     Cleared

Source Form: LDAA (opr); LDAB (opr)

Addressing Modes, Machine Code, and Cycle-by-Cycle Execution:

Cycle	LDAA (IMM) Addr	Data	R/W̄	LDAA (DIR) Addr	Data	R/W̄	LDAA (EXT) Addr	Data	R/W̄	LDAA (IND, X) Addr	Data	R/W̄	LDAA (IND, Y) Addr	Data	R/W̄
1	OP	86	1	OP	96	1	OP	B6	1	OP	A6	1	OP	18	1
2	OP+1	ii	1	OP+1	dd	1	OP+1	hh	1	OP+1	ff	1	OP+1	A6	1
3				00dd	(00dd)	1	OP+2	ll	1	FFFF	—	1	OP+2	ff	1
4							hhll	(hhll)	1	X+ff	(X+ff)	1	FFFF	—	1
5													Y+ff	(Y+ff)	1

Cycle	LDAB (IMM) Addr	Data	R/W̄	LDAB (DIR) Addr	Data	R/W̄	LDAB (EXT) Addr	Data	R/W̄	LDAB (IND, X) Addr	Data	R/W̄	LDAB (IND, Y) Addr	Data	R/W̄
1	OP	C6	1	OP	D6	1	OP	F6	1	OP	E6	1	OP	18	1
2	OP+1	ii	1	OP+1	dd	1	OP+1	hh	1	OP+1	ff	1	OP+1	E6	1
3				00dd	(00dd)	1	OP+2	ll	1	FFFF	—	1	OP+2	ff	1
4							hhll	(hhll)	1	X+ff	(X+ff)	1	FFFF	—	1
5													Y+ff	(Y+ff)	1

LDD   Load Double Accumulator   LDD

Opeation: ACCD $\Leftarrow$ (M:M + 1); ACCA $\Leftarrow$ (M), ACCB $\Leftarrow$ (M + 1)

Description: Loads the contents of memory locations M and M + 1 into the double accumulator D. The condition codes are set according to the data. The information from location M is loaded into accumulator A, and the information from location M + 1 is loaded into accumulator B.

Condition Codes and Boolean Formulae:

S	X	H	I	N	Z	V	C
—	—	—	—	$\updownarrow$	$\updownarrow$	0	—

N R15
   Set if MSB of result is set; cleared otherwise.

Z $\overline{R15} \cdot \overline{R14} \cdot \overline{R13} \cdot \overline{R12} \cdot \overline{R11} \cdot \overline{R10} \cdot \overline{R9} \cdot \overline{R8} \cdot \overline{R7} \cdot \overline{R6} \cdot \overline{R5} \cdot \overline{R4} \cdot \overline{R3} \cdot \overline{R2} \cdot \overline{R1} \cdot \overline{R0}$
   Set if result is $0000; cleared otherwise.

V 0
   Cleared

Source Form: LDD (opr)

Addressing Modes, Machine Code, and Cycle-by-Cycle Execution:

Cycle	LDD (IMM)			LDD (DIR)			LDD (EXT)			LDD (IND, X)			LDD (IND, Y)		
	Addr	Data	R/$\overline{W}$	Addr	Data	R/$\overline{W}$	Addr	Data	R/$\overline{W}$	Addr	Data	R/$\overline{W}$	Addr	Data	R/$\overline{W}$
1	OP	CC	1	OP	DC	1	OP	FC	1	OP	EC	1	OP	18	1
2	OP + 1	jj	1	OP + 1	dd	1	OP + 1	hh	1	OP + 1	ff	1	OP + 1	EC	1
3	OP + 2	kk	1	00dd	(00dd)	1	OP + 2	ll	1	FFFF	—	1	OP + 2	ff	1
4				00dd + 1	(00dd + 1)	1	hhll	(hhll)	1	X + ff	(X + ff)	1	FFFF	—	1
5							hhll + 1	(hhll + 1)	1	X + ff + 1	(X + ff + 1)	1	Y + ff	(Y + ff)	1
6													Y + ff + 1	(Y + ff + 1)	1

LDS   Load Stack Pointer   LDS

Operation: SPH ◀ (M), SPL ◀ (M + 1)

Description: Loads the most significant byte of the stack pointer from the byte of memory at the address specified by the program, and loads the least significant byte of the stack pointer from the next byte of memory at one plus the address specified by the program.

Condition Codes and Boolean Formulae:

S	X	H	I	N	Z	V	C
—	—	—	—	⬍	⬍	0	—

N   R15
    Set if MSB of result is set; cleared otherwise.

Z $\overline{R15} \cdot \overline{R14} \cdot \overline{R13} \cdot \overline{R12} \cdot \overline{R11} \cdot \overline{R10} \cdot \overline{R9} \cdot \overline{R8} \cdot \overline{R7} \cdot \overline{R6} \cdot \overline{R5} \cdot \overline{R4} \cdot \overline{R3} \cdot \overline{R2} \cdot \overline{R1} \cdot \overline{R0}$
    Set if result is $0000; cleared otherwise.

V   0
    Cleared

Source Form: LDS (opr)

Addressing Modes, Machine Code, and Cycle-by-Cycle Execution:

Cycle	LDS (IMM)			LDS (DIR)			LDS (EXT)			LDS (IND, X)			LDS (IND, Y)		
	Addr	Data	R/W̄	Addr	Data	R/W̄	Addr	Data	R/W̄	Addr	Data	R/W̄	Addr	Data	R/W̄
1	OP	8E	1	OP	9E	1	OP	BE	1	OP	AE	1	OP	18	1
2	OP+1	jj	1	OP+1	dd	1	OP+1	hh	1	OP+1	ff	1	OP+1	AE	1
3	OP+2	kk	1	00dd	(00dd)	1	OP+2	ll	1	FFFF	—	1	OP+2	ff	1
4				00dd+1	(00dd+1)	1	hhll	(hhll)	1	X+ff	(X+ff)	1	FFFF	—	1
5							hhll+1	(hhll+1)	1	X+ff+1	(X+ff+1)	1	Y+ff	(Y+ff)	1
6													Y+ff+1	(Y+ff+1)	1

LDX

Load Index Register X

LDX

Operation;   IXH ⬦ (M), IXL ⬦ (M + 1)

Description: Loads the most significant byte of index register X from the byte of memory at the address specified by the program, and loads the least significant byte of index register X from the next byte of memory at one plus the address specified by the program.

Condition Codes and Boolean Formulae:

S	X	H	I	N	Z	V	C
—	—	—	—	⬦	⬦	0	—

N R15
   Set if MSB of result is set; cleared otherwise.

Z $\overline{R15} \cdot \overline{R14} \cdot \overline{R13} \cdot \overline{R12} \cdot \overline{R11} \cdot \overline{R10} \cdot \overline{R9} \cdot \overline{R8} \cdot \overline{R7} \cdot \overline{R6} \cdot \overline{R5} \cdot \overline{R4} \cdot \overline{R3} \cdot \overline{R2} \cdot \overline{R1} \cdot \overline{R0}$
   Set if result is $0000; cleared otherwise.

V 0
   Cleared

Source Form: LDX (opr)

Addressing Modes, Machine Code, and Cycle-by-Cycle Execution:

Cycle	LDX (IMM)			LDX (DIR)			LDX (EXT)			LDX (IND, X)			LDX (IND, Y)		
	Addr	Data	R/W̄	Addr	Data	R/W̄	Addr	Data	R/W̄	Addr	Data	R/W̄	Addr	Data	R/W̄
1	OP	CE	1	OP	DE	1	OP	FE	1	OP	EE	1	OP	CD	1
2	OP + 1	jj	1	OP + 1	dd	1	OP + 1	hh	1	OP + 1	ff	1	OP + 1	EE	1
3	OP + 2	kk	1	00dd	(00dd)	1	OP + 2	ll	1	FFFF	—	1	OP + 2	ff	1
4				00dd + 1	(00dd + 1)	1	hhll	(hhll)	1	X + ff	(X + ff)	1	FFFF	—	1
5							hhll + 1	(hhll + 1)	1	X + ff + 1	(X + ff + 1)	1	Y + ff	(Y + ff)	1
6													Y + ff + 1	(Y + ff + 1)	1

LDY

Load Index Register Y

LDY

Operation: IYH ◄ (M), IYL ◄ (M + 1)

Description: Loads the most significant byte of index register Y from the byte of memory at the address specified by the program, and loads the least significant byte of index register Y from the next byte of memory at one plus the address specified by the program.

Condition Codes and Boolean Formulae:

S	X	H	I	N	Z	V	C
—	—	—	—	↕	↕	0	—

N R15
Set if MSB of result is set; cleared otherwise.

Z $\overline{R15} \cdot \overline{R14} \cdot \overline{R13} \cdot \overline{R12} \cdot \overline{R11} \cdot \overline{R10} \cdot \overline{R9} \cdot \overline{R8} \cdot \overline{R7} \cdot \overline{R6} \cdot \overline{R5} \cdot \overline{R4} \cdot \overline{R3} \cdot \overline{R2} \cdot \overline{R1} \cdot \overline{R0}$
Set if result is $0000; cleared otherwise.

V 0
Cleared

Source Form: LDY (opr)

Addressing Modes, Machine Code, and Cycle-by-Cycle Execution:

Cycle	LDY (IMM) Addr	Data	R/W̄	LDY (DIR) Addr	Data	R/W̄	LDY (EXT) Addr	Data	R/W̄	LDY (IND, X) Addr	Data	R/W̄	LDY (IND, Y) Addr	Data	R/W̄
1	OP	18	1	OP	18	1	OP	18	1	OP	1A	1	OP	18	1
2	OP+1	CE	1	OP+1	DE	1	OP+1	FE	1	OP+1	EE	1	OP+1	EE	1
3	OP+2	jj	1	OP+2	dd	1	OP+2	hh	1	OP+2	ff	1	OP+2	ff	1
4	OP+3	kk	1	00dd	(00dd)	1	OP+3	ll	1	FFFF	—	1	FFFF	—	1
5				00dd+1	(00dd+1)	1	hhll	(hhll)	1	X+ff	(X+ff)	1	Y+ff	(Y+ff)	1
6							hhll+1	(hhll+1)	1	X+ff+1	(X+ff+1)	1	Y+ff+1	(Y+ff+1)	1

LSL

Logical Shift Left
(Same as ASL)

LSL

Operation:

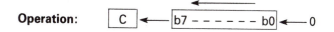

$$C \longleftarrow \boxed{b7 - - - - - - b0} \longleftarrow 0$$

Description: Shifts all bits of the ACCX or M one place to the left. Bit 0 is loaded with zero. The C bit is loaded from the most significant bit of ACCX or M.

Condition Codes and Boolean Formulae:

S	X	H	I	N	Z	V	C
—	—	—	—	↕	↕	↕	↕

N R7
   Set if MSB of result is set; cleared otherwise.

Z $\overline{R7} \cdot \overline{R6} \cdot \overline{R5} \cdot \overline{R4} \cdot \overline{R3} \cdot \overline{R2} \cdot \overline{R1} \cdot \overline{R0}$
   Set if result is $00; cleared otherwise.

V $N \oplus C = [N \cdot \overline{C}] + [\overline{N} \cdot C]$ (for N and C after the shift)
   Set if (N is set and C is clear) or (N is clear and C is set); cleared otherwise (for values of N and C after the shift).

C M7
   Set if, before the shift, the MSB of ACCX or M was set; cleared otherwise.

Source Forms: LSLA; LSLB; LSL (opr)

Addressing Modes, Machine Code, and Cycle-by-Cycle Execution:

Cycle	LSLA (INH)			LSLB (INH)			LSL (EXT)			LSL (IND, X)			LSL (IND, Y)		
	Addr	Data	R/W̄	Addr	Data	R/W̄	Addr	Data	R/W̄	Addr	Data	R/W̄	Addr	Data	R/W̄
1	OP	48	1	OP	58	1	OP	78	1	OP	68	1	OP	18	1
2	OP+1	—	1	OP+1	—	1	OP+1	hh	1	OP+1	ff	1	OP+1	68	1
3							OP+2	ll	1	FFFF	—	1	OP+2	ff	1
4							hhll	(hhll)	1	X+ff	(X+ff)	1	FFFF	—	1
5							FFFF	—	1	FFFF	—	1	Y+ff	(Y+ff)	1
6							hhll	result	0	X+ff	result	0	FFFF	—	1
7													Y+ff	result	0

LSLD

Logical Shift Left Double
(Same as ASLD)

LSLD

Operation:

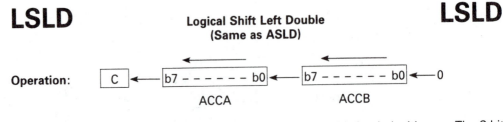

ACCA   ACCB

Description: Shifts all of ACCD one place to the left. Bit 0 is loaded with zero. The C bit is loaded from the most significant bit of ACCD.

Condition Codes and Boolean Formulae:

S	X	H	I	N	Z	V	C
—	—	—	—	↕	↕	↕	↕

N   R15
    Set if MSB of result is set; cleared otherwise.

Z $\overline{R15} \cdot \overline{R14} \cdot \overline{R13} \cdot \overline{R12} \cdot \overline{R11} \cdot \overline{R10} \cdot \overline{R9} \cdot \overline{R8} \cdot \overline{R7} \cdot \overline{R6} \cdot \overline{R5} \cdot \overline{R4} \cdot \overline{R3} \cdot \overline{R2} \cdot \overline{R1} \cdot \overline{R0}$
    Set if result is $0000; cleared otherwise.

V $N \oplus C = [N \cdot \overline{C}] + [\overline{N} \cdot C]$ (for N and C after the shift)
    Set if (N is set and C is clear) or (N is clear and C is set); cleared otherwise (for values of N and C after the shift).

C   D15
    Set if, before the shift, the MSB of ACCD was set; cleared otherwise.

Source Form: LSLD

Addressing Modes, Machine Code, and Cycle-by-Cycle Execution:

Cycle	LSLD (INH)		
	Addr	Data	R/$\overline{W}$
1	OP	05	1
2	OP + 1	—	1
3	FFFF	—	1

LSR

Logical Shift Right

LSR

Operation:

$$0 \rightarrow \boxed{b7 - - - - - - b0} \rightarrow \boxed{C}$$

Description: Shifts all bits of ACCX or M one place to the right. Bit 7 is loaded with zero. The C bit is loaded from the least significant bit of ACCX or M.

Condition Codes and Boolean Formulae:

S	X	H	I	N	Z	V	C
—	—	—	—	0	↕	↕	↕

N 0
Cleared.

Z $\overline{R7} \cdot \overline{R6} \cdot \overline{R5} \cdot \overline{R4} \cdot \overline{R3} \cdot \overline{R2} \cdot \overline{R1} \cdot \overline{R0}$
Set if result is $00; cleared otherwise.

V $N \oplus C = [N \cdot \overline{C}] + [\overline{N} \cdot C]$ (for N and C after the shift)
Since N = 0, this simplifies to C (after the shift).

C M0
Set if, before the shift, the LSB of ACCX or M was set; cleared otherwise.

Source Forms: LSRA; LSRB; LSR (opr)

Addressing Modes, Machine Code, and Cycle-by-Cycle Execution:

Cycle	LSRA (INH)			LSRB (INH)			LSR (EXT)			LSR (IND, X)			LSR (IND, Y)		
	Addr	Data	R/W̄	Addr	Data	R/W̄	Addr	Data	R/W̄	Addr	Data	R/W̄	Addr	Data	R/W̄
1	OP	44	1	OP	54	1	OP	74	1	OP	64	1	OP	18	1
2	OP+1	—	1	OP+1	—	1	OP+1	hh	1	OP+1	ff	1	OP+1	64	1
3							OP+2	ll	1	FFFF	—	1	OP+2	ff	1
4							hhll	(hhll)	1	X+ff	(X+ff)	1	FFFF	—	1
5							FFFF	—	1	FFFF	—	1	Y+ff	(Y+ff)	1
6							hhll	result	0	X+ff	result	0	FFFF	—	1
7													Y+ff	result	0

LSRD   Logical Shift Right Double Accumulator   LSRD

Operation: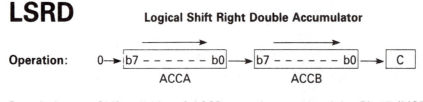

Description: Shifts all bits of ACCD one place to the right. Bit 15 (MSB of ACCA) is loaded with zero. The C bit is loaded from the least significant bit of ACCD (LSB of ACCB).

Condition Codes and Boolean Formulae:

S	X	H	I	N	Z	V	C
—	—	—	—	0	⇕	⇕	⇕

N   0
      Cleared

Z $\overline{R15} \cdot \overline{R14} \cdot \overline{R13} \cdot \overline{R12} \cdot \overline{R11} \cdot \overline{R10} \cdot \overline{R9} \cdot \overline{R8} \cdot \overline{R7} \cdot \overline{R6} \cdot \overline{R5} \cdot \overline{R4} \cdot \overline{R3} \cdot \overline{R2} \cdot \overline{R1} \cdot \overline{R0}$
      Set if result is $0000; cleared otherwise.

V   D0
      Set if, after the shift operaton, C is set; cleared otherwise.

C   D0
      Set if, before the shift, the least significant bit of ACCD was set; cleared otherwise.

Source Form: LSRD

Addressing Modes, Machine Code, and Cycle-by-Cycle Execution:

Cycle	LSRD (INH)		
	Addr	Data	R/$\overline{W}$
1	OP	04	1
2	OP + 1	—	1
3	FFFF	—	1

MUL   Multiply Unsigned   MUL

Operation: ACCD ◄ (ACCA) × (ACCB)

Description: Multiplies the 8-bit unsigned binary value in accumulator A by the 8-bit unsigned binary value in accumulator B to obtain a 16-bit unsigned result in the double accumulator D. Unsigned multiply allows multiple-precision operations. The carry flag allows rounding the most significant byte of the result through the sequence: MUL, ADCA #0.

Condition Codes and Boolean Formulae:

S	X	H	I	N	Z	V	C
—	—	—	—	—	—	—	↕

C   R7
    Set if bit 7 of the result (ACCB bit 7) is set; cleared otherwise.

Source Form: MUL

Addressing Modes, Machine Code, and Cycle-by-Cycle Execution:

Cycle	MUL (INH)		
	Addr	Data	R/W̄
1	OP	3D	1
2	OP + 1	—	1
3–10	FFFF	—	1

NEG   Negate   NEG

Operation: $(ACCX) \leftarrow -(ACCX) = \$00 - (ACCX)$ **or:** $(M) \leftarrow -(M) = \$00 - (M)$

Description: Replaces the contents of ACCX or M with its twos complement; the value $80 is left unchanged.

Condition Codes and Boolean Formulae:

S	X	H	I	N	Z	V	C
—	—	—	—	↕	↕	↕	↕

N   R7
   Set if MSB of result is set; cleared otherwise.

Z $\overline{R7} \cdot \overline{R6} \cdot \overline{R5} \cdot \overline{R4} \cdot \overline{R3} \cdot \overline{R2} \cdot \overline{R1} \cdot \overline{R0}$
   Set if result is $00; cleared otherwise.

V $R7 \cdot \overline{R6} \cdot \overline{R5} \cdot \overline{R4} \cdot \overline{R3} \cdot \overline{R2} \cdot \overline{R1} \cdot \overline{R0}$
   Set if there is a twos complement overflow from the implied subtraction from zero; cleared otherwise. A twos complement overflow will occur if and only if the contents of ACCX or M is $80.

C $R7 + R6 + R5 + R4 + R3 + R2 + R1 + R0$
   Set if there is a borrow in the implied subtraction from zero; cleared otherwise. The C bit will be set in all cases except when the contents of ACCX or M is $00.

Source Forms: NEGA; NEGB; NEG (opr)

Addressing Modes, Machine Code, and Cycle-by-Cycle Execution:

Cycle	NEGA (INH)			NEGB (INH)			NEG (EXT)			NEG (IND, X)			NEG (IND, Y)		
	Addr	Data	R/W̄	Addr	Data	R/W̄	Addr	Data	R/W̄	Addr	Data	R/W̄	Addr	Data	R/W̄
1	OP	40	1	OP	50	1	OP	70	1	OP	60	1	OP	18	1
2	OP+1	—	1	OP+1	—	1	OP+1	hh	1	OP+1	ff	1	OP+1	60	1
3							OP+2	ll	1	FFFF	—	1	OP+2	ff	1
4							hhll	(hhll)	1	X+ff	(X+ff)	1	FFFF	—	1
5							FFFF	—	1	FFFF	—	1	Y+ff	(Y+ff)	1
6							hhll	result	0	X+ff	result	0	FFFF	—	1
7													Y+ff	result	0

NOP

No Operation

NOP

Description: This is a single-byte instruction that causes only the program counter to be incremented. No other registers are affected. This instruction is typically used to produce a time delay although some software disciplines discourage CPU frequency-based time delays. During debug, NOP instructions are sometimes used to temporarily replace other machine code instructions, thus disabling the replaced instruction(s).

Condition Codes and Boolean Formulae:

S	X	H	I	N	Z	V	C
—	—	—	—	—	—	—	—

None affected

Source Form: NOP

Addressing Modes, Machine Code, and Cycle-by-Cycle Execution:

Cycle	NOP (INH)		
	Addr	Data	R/W̄
1	OP	01	1
2	OP + 1	—	1

Reprinted with permission of Motorola

ORA

Inclusive-OR

ORA

Operation: ACCX ◀ (ACCX) + (M)

Description: Performs the logical inclusive-OR between the contents of ACCX and the contents of M and places the result in ACCX. (Each bit of ACCX after the operation will be the logical inclusive-OR of the corresponding bits of M and of ACCX before the operation.)

Condition Codes and Boolean Formulae:

S	X	H	I	N	Z	V	C
—	—	—	—	↕	↕	0	—

N   R7
   Set if MSB of result is set; cleared otherwise.

Z $\overline{R7} \cdot \overline{R6} \cdot \overline{R5} \cdot \overline{R4} \cdot \overline{R3} \cdot \overline{R2} \cdot \overline{R1} \cdot \overline{R0}$
   Set if result is $00; cleared otherwise

V   0
   Cleared

Source Forms: ORAA (opr); ORAB (opr)

Addressing Modes, Machine Code, and Cycle-by-Cycle Execution:

Cycle	ORAA (IMM) Addr	Data	R/W̄	ORAA (DIR) Addr	Data	R/W̄	ORAA (EXT) Addr	Data	R/W̄	ORAA (IND, X) Addr	Data	R/W̄	ORAA (IND, Y) Addr	Data	R/W̄
1	OP	8A	1	OP	9A	1	OP	BA	1	OP	AA	1	OP	18	1
2	OP+1	ii	1	OP+1	dd	1	OP+1	hh	1	OP+1	ff	1	OP+1	AA	1
3				00dd	(00dd)	1	OP+2	ll	1	FFFF	—	1	OP+2	ff	1
4							hhll	(hhll)	1	X+ff	(X+ff)	1	FFFF	—	1
5													Y+ff	(Y+ff)	1

Cycle	ORAB (IMM) Addr	Data	R/W̄	ORAB (DIR) Addr	Data	R/W̄	ORAB (EXT) Addr	Data	R/W̄	ORAB (IND, X) Addr	Data	R/W̄	ORAB (IND, Y) Addr	Data	R/W̄
1	OP	CA	1	OP	DA	1	OP	FA	1	OP	EA	1	OP	18	1
2	OP+1	ii	1	OP+1	dd	1	OP+1	hh	1	OP+1	ff	1	OP+1	EA	1
3				00dd	(00dd)	1	OP+2	ll	1	FFFF	—	1	OP+2	ff	1
4							hhll	(hhll)	1	X+ff	(X+ff)	1	FFFF	—	1
5													Y+ff	(Y+ff)	1

PSH   Push Data onto Stack   **PSH**

Operation: ➡ACCX, SP ⬍ (SP) − $0001

Description: The contents of ACCX are stored on the stack at the address contained in the stack pointer. The stack pointer is then decremented.

Push instructions are commonly used to save the contents of one or more CPU registers at the start of a subroutine. Just before returning from the subroutine, corresponding pull instructions are used to restore the saved CPU registers so the subroutine will appear not to have affected these registers.

Condition Codes and Boolean Formulae:

S	X	H	I	N	Z	V	C
—	—	—	—	—	—	—	—

None affected

Source Forms: PSHA; PSHB

Addressing Modes, Machine Code, and Cycle-by-Cycle Execution:

Cycle	PSHA (INH)			PSHB (INH)		
	Addr	Data	R/W̄	Addr	Data	R/W̄
1	OP	36	1	OP	37	1
2	OP+1	—	1	OP+1	—	1
3	SP	(A)	0	SP	(B)	0

PSHX   Push Index Register X onto Stack   PSHX

Operation: ⬅(IXL), SP ⬇ (SP) − $0001
⬅(IXH), SP ⬇ (SP) − $0001

Description: The contents of the index register X are pushed onto the stack (low-order byte first) at the address contained in the stack pointer. The stack pointer is then decremented by two.

Push instructions are commonly used to save the contents of one or more CPU registers at the start of a subroutine. Just before returning from the subroutine, corresponding pull instructions are used to restore the saved CPU registers so the subroutine will appear not to have affected these registers.

Condition Codes and Boolean Formulae:

S	X	H	I	N	Z	V	C
—	—	—	—	—	—	—	—

None affected

Source Form: PSHX

Addressing Modes, Machine Code, and Cycle-by-Cycle Execution:

Cycle	PSHX (INH)		
	Addr	Data	R/$\overline{W}$
1	OP	3C	1
2	OP + 1	—	1
3	SP	(IXL)	0
4	SP − 1	(IXH)	0

PSHY

Push Index Register Y onto Stack

PSHY

Operation: ➡(IYL), SP ⬧ (SP) − $0001
➡(IYH), SP ⬧ (SP) − $0001

Description: The contents of the index register Y are pushed onto the stack (low-order byte first) at the address contained in the stack pointer. The stack pointer is then decremented by two.

Push instructions are commonly used to save the contents of one or more CPU registers at the start of a subroutine. Just before returning from the subroutine, corresponding pull instructions are used to restore the saved CPU registers so the subroutine will appear not to have affected these registers.

Condition Codes and Boolean Formulae:

S	X	H	I	N	Z	V	C
—	—	—	—	—	—	—	—

None affected

Source Form: PSHY

Addressing Modes, Machine Code, and Cycle-by-Cycle Execution:

Cycle	PSHY (INH)		
	Addr	Data	R/W̄
1	OP	18	1
2	OP + 1	3C	1
3	OP + 2	—	1
4	SP	(IYL)	0
5	SP − 1	(IYH)	0

PUL   **Pull Data from Stack**   PUL

Operation: SP ⬦ (SP) + $0001, ◆(ACCX)

Description: The stack pointer is incremented. The ACCX is then loaded from the stack at the address contained in the stack pointer.

Push instructions are commonly used to save the contents of one or more CPU registers at the start of a subroutine. Just before returning from the subroutine, corresponding pull instructions are used to restore the saved CPU registers so the subroutine will appear not to have affected these registers.

Condition Codes and Boolean Formulae:

S	X	H	I	N	Z	V	C
—	—	—	—	—	—	—	—

None affected

Source Forms: PULA; PULB

Addressing Modes, Machine Code, and Cycle-by-Cycle Execution:

Cycle	PULA (INH)			PULB (INH)		
	Addr	Data	R/W̄	Addr	Data	R/W̄
1	OP	32	1	OP	33	1
2	OP+1	—	1	OP+1	—	1
3	SP	—	1	SP	—	1
4	SP+1	get A	1	SP+1	get B	1

PULX   Pull Index Register X from Stack   PULX

Operation: SP ◄ (SP) + $0001; ◄(IXH)
SP ◄ (SP) + $0001; ◄(IXL)

Description: The index register X is pulled from the stack (high-order byte first), beginning at the address contained in the stack pointer plus one. The stack pointer is incremented by two in total.

Push instructions are commonly used to save the contents of one or more CPU registers at the start of a subroutine. Just before returning from the subroutine, corresponding pull instructions are used to restore the saved CPU registers so the subroutine will appear not to have affected these registers.

Condition Codes and Boolean Formulae:

S	X	H	I	N	Z	V	C
—	—	—	—	—	—	—	—

None affected

Source Form: PULX

Addressing Modes, Machine Code, and Cycle-by-Cycle Execution:

Cycle	PULX (INH)		
	Addr	Data	R/W̄
1	OP	38	1
2	OP + 1	—	1
3	SP	—	1
4	SP + 1	get IXH	1
5	SP + 2	get IXL	1

PULY
Pull Index Register Y from Stack
PULY

Operation: SP ◀ (SP) + $0001; ◆(IYH)
SP ◀ (SP) + $0001; ◆(IYL)

Description: The index register Y is pulled from the stack (high-order byte first) beginning at the address contained in the stack pointer plus one. The stack pointer is incremented by two in total.

Push instructions are commonly used to save the contents of one or more CPU registers at the start of a subroutine. Just before returning from the subroutine, corresponding pull instructions are used to restore the saved CPU registers so the subroutine will appear not to have affected these registers.

Condition Codes and Boolean Formulae:

S	X	H	I	N	Z	V	C
—	—	—	—	—	—	—	—

None affected

Source Form: PULY

Addressing Modes, Machine Code, and Cycle-by-Cycle Execution:

Cycle	PULY (INH)		
	Addr	Data	R/$\overline{\text{W}}$
1	OP	18	1
2	OP + 1	38	1
3	OP + 2	—	1
4	SP	—	1
5	SP + 1	get IYH	1
6	SP + 2	get IYL	1

ROL

Rotate Left

ROL

Operation:

$$C \leftarrow \boxed{b7 \text{ -- -- -- -- -- } b0} \leftarrow \boxed{C}$$

Description: Shifts all bits of ACCX or M one place to the left. Bit 0 is loaded from the C bit. The C bit is loaded from the most significant bit of ACCX or M. The rotate operations include the carry bit to allow extension of the shift and rotate operations to multiple bytes. For example, to shift a 24-bit vaue left one bit, the sequence ASL LOW, ROL MID, ROL HIGH could be used where LOW, MID, and HIGH refer to the low-order, middle, and high-order bytes of the 24-bit value, respectively.

Condition Codes and Boolean Formulae:

S	X	H	I	N	Z	V	C
—	—	—	—	⬍	⬍	⬍	⬍

N R7
   Set if MSB of result is set; cleared otherwise.

Z $\overline{R7} \cdot \overline{R6} \cdot \overline{R5} \cdot \overline{R4} \cdot \overline{R3} \cdot \overline{R2} \cdot \overline{R1} \cdot \overline{R0}$
   Set if result is $00; cleared otherwise.

V $N \oplus C = [N \cdot \overline{C}] + [\overline{N} \cdot C]$ (for N and C after the rotate)
   Set if (N is set and C is clear) or (N is clear and C is set); cleared otherwise (for values of N and C after the rotate).

C M7
   Set if, before the rotate, the MSB of ACCX or M was set; cleared otherwise.

Source Forms: ROLA; ROLB; ROL (opr)

Addressing Modes, Machine Code, and Cycle-by-Cycle Execution:

Cycle	ROLA (INH)			ROLB (INH)			ROL (EXT)			ROL (IND, X)			ROL (IND, Y)		
	Addr	Data	R/$\overline{W}$	Addr	Data	R/$\overline{W}$	Addr	Data	R/$\overline{W}$	Addr	Data	R/$\overline{W}$	Addr	Data	R/$\overline{W}$
1	OP	49	1	OP	59	1	OP	79	1	OP	69	1	OP	18	1
2	OP+1	—	1	OP+1	—	1	OP+1	hh	1	OP+1	ff	1	OP+1	69	1
3							OP+2	ll	1	FFFF	—	1	OP+2	ff	1
4							hhll	(hhll)	1	X+ff	(X+ff)	1	FFFF	—	1
5							FFFF	—	1	FFFF	—	1	Y+ff	(Y+ff)	1
6							hhll	result	0	X+ff	result	0	FFFF	—	1
7													Y+ff	result	0

ROR   Rotate Right   **ROR**

Operation: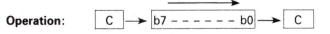

Description: Shift all bits of ACCX or M one place to the right. Bit 7 is loaded from the C bit. The C bit is loaded from the least significant bit of ACCX or M. The rotate operations include the carry bit to allow extension of the shift and rotate operations to multiple bytes. For example, to shift a 24-bit value right one bit, the sequence LSR HIGH, ROR MID, ROR LOW could be used where LOW, MID, and HIGH refer to the low-order, middle, and high-order bytes of the 24-bit value, respectively. The first LSR could be replaced by ASR to maintain the original value of the sign bit (MSB of high-order byte) of the 24-bit value.

Condition Codes and Boolean Formulae:

S	X	H	I	N	Z	V	C
—	—	—	—	$\updownarrow$	$\updownarrow$	$\updownarrow$	$\updownarrow$

N R7
Set if MSB of result is set; cleared otherwise.

Z $\overline{R7} \cdot \overline{R6} \cdot \overline{R5} \cdot \overline{R4} \cdot \overline{R3} \cdot \overline{R2} \cdot \overline{R1} \cdot \overline{R0}$
Set if result is $00; cleared otherwise.

V $N \oplus C = [N \cdot \overline{C}] + [\overline{N} \cdot C]$ (for N and C after the rotate)
Set if (N is set and C is clear) or (N is clear and C is set); cleared otherwise (for values of N and C after the rotate).

C M0
Set if, before the rotate, the LSB of ACCX or M was set; cleared otherwise.

Source Forms: RORA; RORB; ROR (opr)

Addressing Modes, Machine Code, and Cycle-by-Cycle Execution:

Cycle	RORA (INH)			RORB (INH)			ROR (EXT)			ROR (IND, X)			ROR (IND, Y)		
	Addr	Data	R/$\overline{W}$	Addr	Data	R/$\overline{W}$	Addr	Data	R/$\overline{W}$	Addr	Data	R/$\overline{W}$	Addr	Data	R/$\overline{W}$
1	OP	46	1	OP	56	1	OP	76	1	OP	66	1	OP	18	1
2	OP+1	—	1	OP+1	—	1	OP+1	hh	1	OP+1	ff	1	OP+1	66	1
3							OP+2	ll	1	FFFF	—	1	OP+2	ff	1
4							hhll	(hhll)	1	X+ff	(X+ff)	1	FFFF	—	1
5							FFFF	—	1	FFFF	—	1	Y+ff	(Y+ff)	1
6							hhll	result	0	X+ff	result	0	FFFF	—	1
7													Y+ff	result	0

RTI

Return from Interrupt

RTI

Operation: SP ⬦ (SP) + $0001, ⬥(CCR)
SP ⬦ (SP) + $0001, ⬥(ACCB)
SP ⬦ (SP) + $0001, ⬥(ACCA)
SP ⬦ (SP) + $0001, ⬥(IXH)
SP ⬦ (SP) + $0001, ⬥(IXL)
SP ⬦ (SP) + $0001, ⬥(IYH)
SP ⬦ (SP) + $0001, ⬥(IYL)
SP ⬦ (SP) + $0001, ⬥(PCH)
SP ⬦ (SP) + $0001, ⬥(PCL)

Description: The condition code, accumulators B and A, index registers X and Y, and the program counter will be restored to a state pulled from the stack. The X bit in the CCR may be cleared as a result of an RTI instruction but may not be set if it was cleared prior to execution of the RTI instruction.

Condition Codes and Boolean Formulae:

S	X	H	I	N	Z	V	C
⬍	⬩	⬍	⬍	⬍	⬍	⬍	⬍

Condition code bits take on the value of the corresponding bit of the unstacked CCR except that the X bit may not change from a zero to a one. Software can leave X set, leave X clear, or change X from one to zero. The XIRQ interrupt mask can only become set as a $\overline{RESET}$ of a reset or recognition of an XIRQ interrupt.

Source Form: RTI

Addressing Modes, Machine Code, and Cycle-by-Cycle Execution:

Cycle	RTI (INH)		
	Addr	Data	R/$\overline{W}$
1	OP	3B	1
2	OP + 1	—	1
3	SP	—	1
4	SP + 1	get CC	1
5	SP + 2	get B	1
6	SP + 3	get A	1
7	SP + 4	get IXH	1
8	SP + 5	get IXL	1
9	SP + 6	get IXH	1
10	SP + 7	get IXL	1
11	SP + 8	Rtn hi	1
12	SP + 9	Rtn lo	1

RTS

Return from Subroutine

RTS

Operation: SP ◄ (SP) + $0001, ►(PCH)
SP ◄ (SP) + $0001, ►(PCL)

Description: The stack pointer is incremented by one. The contents of the byte of memory, at the address now contained in the stack pointer, are loaded into the high-order eight bits of the program counter. The stack pointer is again incremented by one. The contents of the byte of memory, at the address now contained in the stack pointer, are loaded into the low-order eight bits of the program counter.

Condition Codes and Boolean Formulae:

S	X	H	I	N	Z	V	C
—	—	—	—	—	—	—	—

None affected

Source Form: RTS

Addressing Modes, Machine Code, and Cycle-by-Cycle Execution:

Cycle	RTS (INH)		
	Addr	Data	R/W̄
1	OP	39	1
2	OP+1	—	1
3	SP	—	1
4	SP+1	Rtn hi	1
5	SP+2	Rtn lo	1

SBA   **Subtract Accumulators**   # SBA

Operation: ACCA ◀ (ACCA) − (ACCB)

Description: Subtracts the contents of ACCB from the contents of ACCA and places the result in ACCA. The contents of ACCB are not affected. For subtract instructions, the C bit in the CCR represents a borrow.

Condition Codes and Boolean Formulae:

S	X	H	I	N	Z	V	C
—	—	—	—	↕	↕	↕	↕

N R7
   Set if MSB of result is set; cleared otherwise.

Z $\overline{R7} \cdot \overline{R6} \cdot \overline{R5} \cdot \overline{R4} \cdot \overline{R3} \cdot \overline{R2} \cdot \overline{R1} \cdot \overline{R0}$
   Set if result is $00; cleared otherwise.

V $A7 \cdot \overline{B7} \cdot \overline{R7} + \overline{A7} \cdot B7 \cdot R7$
   Set if a twos complement overflow resulted from the operation; cleared otherwise.

C $\overline{A7} \cdot B7 + B7 \cdot R7 + R7 \cdot \overline{A7}$
   Set if the absolute value of ACCB is larger than the absolute value of ACCA; cleared otherwise.

Source Form: SBA

Addressing Modes, Machine Code, and Cycle-by-Cycle Execution:

Cycle	SBA (INH)		
	Addr	Data	R/$\overline{W}$
1	OP	10	1
2	OP + 1	—	1

SBC   Subtract with Carry   SBC

Operation: ACCX ◀ (ACCX) − (M) − (C)

Description: Subtracts the contents of M and the contents of C from the contents of ACCX and places the result in ACCX. For subtract instructions the C bit in the CCR represents a borrow.

Condition Codes and Boolean Formulae:

S	X	H	I	N	Z	V	C
—	—	—	—	⬍	⬍	⬍	⬍

N   R7
    Set if MSB of result is set; cleared otherwise.

Z $\overline{R7} \cdot \overline{R6} \cdot \overline{R5} \cdot \overline{R4} \cdot \overline{R3} \cdot \overline{R2} \cdot \overline{R1} \cdot \overline{R0}$
    Set if result is $00; cleared otherwise.

V $X7 \cdot \overline{M7} \cdot \overline{R7} + \overline{X7} \cdot M7 \cdot R7$
    Set if a twos complement overflow resulted from the operation; cleared otherwise.

C $\overline{X7} \cdot M7 + M7 \cdot R7 + R7 \cdot \overline{X7}$
    Set if the absolute value of the contents of memory plus previous carry is larger than the absolute value of the accumulator; cleared otherwise.

Source Forms: SBCA (opr); SBCB (opr)

Addressing Modes, Machine Code, and Cycle-by-Cycle Execution:

Cycle	SBCA (IMM) Addr	Data	R/W̄	SBCA (DIR) Addr	Data	R/W̄	SBCA (EXT) Addr	Data	R/W̄	SBCA (IND, X) Addr	Data	R/W̄	SBCA (IND, Y) Addr	Data	R/W̄
1	OP	82	1	OP	92	1	OP	B2	1	OP	A2	1	OP	18	1
2	OP+1	ii	1	OP+1	dd	1	OP+1	hh	1	OP+1	ff	1	OP+1	A2	1
3				00dd	(00dd)	1	OP+2	ll	1	FFFF	—	1	OP+2	ff	1
4							hhll	(hhll)	1	X+ff	(X+ff)	1	FFFF	—	1
5													Y+ff	(Y+ff)	1

Cycle	SBCB (IMM) Addr	Data	R/W̄	SBCB (DIR) Addr	Data	R/W̄	SBCB (EXT) Addr	Data	R/W̄	SBCB (IND, X) Addr	Data	R/W̄	SBCB (IND, Y) Addr	Data	R/W̄
1	OP	C2	1	OP	D2	1	OP	F2	1	OP	E2	1	OP	18	1
2	OP+1	ii	1	OP+1	dd	1	OP+1	hh	1	OP+1	ff	1	OP+1	E2	1
3				00dd	(00dd)	1	OP+2	ll	1	FFFF	—	1	OP+2	ff	1
4							hhll	(hhll)	1	X+ff	(X+ff)	1	FFFF	—	1
5													Y+ff	(Y+ff)	1

SEC

Set Carry

SEC

Operation: C bit ◀ 1

Description: Sets the C bit in the CCR.

Condition Codes and Boolean Formulae:

S	X	H	I	N	Z	V	C
—	—	—	—	—	—	—	1

C   1
    Set

Source Form: SEC

Addressing Modes, Machine Code, and Cycle-by-Cycle Execution:

Cycle	SEC (INH)		
	Addr	Data	R/W̄
1	OP	0D	1
2	OP + 1	—	1

SEI   **Set Interrupt Mask**   # SEI

Operation: I bit ◀ 1

Description: Sets the interrupt mask bit in the CCR. When the I bit is set, all maskable interrupts are inhibited, and the MPU will recognize only non-maskable interrupt sources or an SWI.

Condition Codes and Boolean Formulae:

S	X	H	I	N	Z	V	C
—	—	—	1	—	—	—	—

I   1
   Set

Source Form: SEI

Addressing Modes, Machine Code, and Cycle-by-Cycle Execution:

Cycle	SEI (INH)		
	Addr	Data	R/W̄
1	OP	0F	1
2	OP+1	—	1

SEV   Set Twos Complement Overflow Bit   SEV

Operation: V bit ↟ 1

Description: Sets the twos complement overflow bit in the CCR.

Condition Codes and Boolean Formulae:

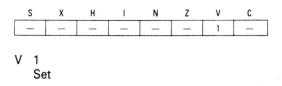

S	X	H	I	N	Z	V	C
—	—	—	—	—	—	1	—

V 1
    Set

Source Form: SEV

Addressing Modes, Machine Code, and Cycle-by-Cycle Execution:

Cycle	SEV (INH)		
	Addr	Data	R/W̄
1	OP	0B	1
2	OP + 1	—	1

STA   Store Accumulator   STA

Operation: M ◀ (ACCX)

Description: Stores the contents of ACCX in memory. The contents of ACCX remains unchanged.

Condition Codes and Boolean Formulae:

S	X	H	I	N	Z	V	C
—	—	—	—	↕	↕	0	—

N X7
   Set if MSB of result is set; cleared otherwise.

Z $\overline{X7} \cdot \overline{X6} \cdot \overline{X5} \cdot \overline{X4} \cdot \overline{X3} \cdot \overline{X2} \cdot \overline{X1} \cdot \overline{X0}$
   Set if result is $00; cleared otherwise.

V 0
   Cleared

Source Forms: STAA (opr); STAB (opr)

Addressing Modes, Machine Code, and Cycle-by-Cycle Execution:

Cycle	STAA (DIR) Addr	Data	R/W̄	STAA (EXT) Addr	Data	R/W̄	STAA (IND, X) Addr	Data	R/W̄	STAA (IND, Y) Addr	Data	R/W̄
1	OP	97	1	OP	B7	1	OP	A7	1	OP	18	1
2	OP+1	dd	1	OP+1	hh	1	OP+1	ff	1	OP+1	A7	1
3	00dd	(A)	0	OP+2	ll	1	FFFF	—	1	OP+2	ff	1
4				hhll	(A)	0	X+ff	(A)	0	FFFF	—	1
5										Y+ff	(A)	0

Cycle	STAB (DIR) Addr	Data	R/W̄	STAB (EXT) Addr	Data	R/W̄	STAB (IND, X) Addr	Data	R/W̄	STAB (IND, Y) Addr	Data	R/W̄
1	OP	D7	1	OP	F7	1	OP	E7	1	OP	18	1
2	OP+1	dd	1	OP+1	hh	1	OP+1	ff	1	OP+1	E7	1
3	00dd	(B)	0	OP+2	ll	1	FFFF	—	1	OP+2	ff	1
4				hhll	(B)	0	X+ff	(B)	0	FFFF	—	1
5										Y+ff	(B)	0

STD   Store Double Accumulator   STD

Operation: M:M + 1 ◀ (ACCD); M ◀ (ACCA), M + 1 ◀ (ACCB)

Description: Stores the contents of double accumulator ACCD in memory. The contents of ACCD remain unchanged.

Condition Codes and Boolean Formulae:

S	X	H	I	N	Z	V	C
—	—	—	—	✿	✿	0	—

N   D15
Set if MSB of result is set; cleared otherwise.

Z $\overline{D15} \cdot \overline{D14} \cdot \overline{D13} \cdot \overline{D12} \cdot \overline{D11} \cdot \overline{D10} \cdot \overline{D9} \cdot \overline{D8} \cdot \overline{D7} \cdot \overline{D6} \cdot \overline{D5} \cdot \overline{D4} \cdot \overline{D3} \cdot \overline{D2} \cdot \overline{D1} \cdot \overline{D0}$
Set if result is \$0000; cleared otherwise.

V   0
Cleared

Source Form: STD (opr)

Addressing Modes, Machine Code, and Cycle-by-Cycle Execution:

Cycle	STD (DIR)			STD (EXT)			STD (IND, X)			STD (IND, Y)		
	Addr	Data	R/W̅	Addr	Data	R/W̅	Addr	Data	R/W̅	Addr	Data	R/W̅
1	OP	DD	1	OP	FD	1	OP	ED	1	OP	18	1
2	OP + 1	dd	1	OP + 1	hh	1	OP + 1	ff	1	OP + 1	ED	1
3	00dd	(A)	0	OP + 2	ll	1	FFFF	—	1	OP + 2	ff	1
4	00dd + 1	(B)	0	hhll	(A)	0	X + ff	(A)	0	FFFF	—	1
5				hhll + 1	(B)	0	X + ff + 1	(B)	0	Y + ff	(A)	0
6										Y + ff + 1	(B)	0

STOP   **Stop Processing**   # STOP

Description: If the S bit in the CCR is set, then the STOP instruction is disabled and operates like the NOP instruction. If the S bit in the CCR is clear, the STOP instruction causes all system clocks to halt, and the system is placed in a minimum-power standby mode. All CPU registers remain unchanged. I/O pins also remain unaffected.

Recovery from STOP may be accomplished by $\overline{RESET}$, $\overline{XIRQ}$, or an unmasked $\overline{IRQ}$. When recovering from STOP with $\overline{XIRQ}$, if the X bit in the CCR is clear, execution will resume with the stacking operations for the $\overline{XIRQ}$ interrupt. If the X bit in the CCR is set, masking $\overline{XIRQ}$ interrupts, execution will resume with the opcode fetch for the instruction which follows the STOP instruction (continue).

An error in some mask sets of the M68HC11 caused incorrect recover from STOP under very specific unusual conditions. If the opcode of the instruction before the STOP instruction came from column 4 or 5 of the opcode map, the STOP instruction was incorrectly interpreted as a two-byte instruction. A simple way to avoid this potential problem is to put a NOP instruction (which is a column 0 opcode) immediately before any STOP instruction.

Condition Codes and Boolean Formulae:

S	X	H	I	N	Z	V	C
—	—	—	—	—	—	—	—

None affected

Source Form: STOP

Addressing Modes, Machine Code, and Cycle-by-Cycle Execution:

Cycle	STOP (INH)		
	Addr	Data	R/$\overline{W}$
1	OP	CF	1
2	OP + 1	—	1

STS
Store Stack Pointer
STS

Operation: M ⭫ (SPH), M + 1 ⭫ (SPL)

Description: Stores the most significant byte of the stack pointer in memory at the address specified by the program and stores the least significant byte of the stack pointer at the next location in memory, at one plus the address specified by the program.

Condition Codes and Boolean Formulae:

S	X	H	I	N	Z	V	C
—	—	—	—	⭫	⭫	0	—

N SP15
Set if MSB of result is set; cleared otherwise.

Z $\overline{SP15} \cdot \overline{SP14} \cdot \overline{SP13} \cdot \overline{SP12} \cdot \overline{SP11} \cdot \overline{SP10} \cdot \overline{SP9} \cdot \overline{SP8} \cdot$
$\overline{SP7} \cdot \overline{SP6} \cdot \overline{SP5} \cdot \overline{SP4} \cdot \overline{SP3} \cdot \overline{SP2} \cdot \overline{SP1} \cdot \overline{SP0}$
Set if result is $0000; cleared otherwise.

V 0
Cleared

Source Form: STS (opr)

Addressing Modes, Machine Code, and Cycle-by-Cycle Execution:

Cycle	STS (DIR)			STS (EXT)			STS (IND, X)			STS (IND, Y)		
	Addr	Data	R/W̄	Addr	Data	R/W̄	Addr	Data	R/W̄	Addr	Data	R/W̄
1	OP	9F	1	OP	BF	1	OP	AF	1	OP	18	1
2	OP + 1	dd	1	OP + 1	hh	1	OP + 1	ff	1	OP + 1	AF	1
3	00dd	(SPH)	0	OP + 2	ll	1	FFFF	—	1	OP + 2	ff	1
4	oodd + 1	(SPL)	0	hhll	(SPH)	0	X + ff	(SPH)	0	FFFF	—	1
5				hhll + 1	(SPL)	0	X + ff + 1	(SPL)	0	Y + ff	(SPH)	0
6										Y + ff + 1	(SPL)	0

STX   Store Index Register X   STX

Operation: M ⬦ (IXH), M + 1 ⬦ (IXL)

Description: Stores the most significant byte of index register X in memory at the address specified by the program and stores the least significant byte of index register X at the next location in memory, at one plus the address specified by the program.

Condition Codes and Boolean Formulae:

S	X	H	I	N	Z	V	C
—	—	—	—	⬧	⬧	0	—

N IX15
   Set if MSB of result is set; cleared otherwise.

Z $\overline{IX15} \cdot \overline{IX14} \cdot \overline{IX13} \cdot \overline{IX12} \cdot \overline{IX11} \cdot \overline{IX10} \cdot \overline{IX9} \cdot \overline{IX8} \cdot$
   $\overline{IX7} \cdot \overline{IX6} \cdot \overline{IX5} \cdot \overline{IX4} \cdot \overline{IX3} \cdot \overline{IX2} \cdot \overline{IX1} \cdot \overline{IX0}$
   Set if result is $0000; cleared otherwise.

V 0
   Cleared

Source Form: STX (opr)

Addressing Modes, Machine Code, and Cycle-by-Cycle Execution:

Cycle	STX (DIR)			STX (EXT)			STX (IND, X)			STX (IND, Y)		
	Addr	Data	R/$\overline{W}$	Addr	Data	R/$\overline{W}$	Addr	Data	R/$\overline{W}$	Addr	Data	R/$\overline{W}$
1	OP	DF	1	OP	FF	1	OP	EF	1	OP	CD	1
2	OP + 1	dd	1	OP + 1	hh	1	OP + 1	ff	1	OP + 1	EF	1
3	00dd	(IXH)	0	OP + 2	ll	1	FFFF	—	1	OP + 2	ff	1
4	oodd + 1	(IXL)	0	hhll	(IXH)	0	X + ff	(IXH)	0	FFFF	—	1
5				hhll + 1	(IXL)	0	X + ff + 1	(IXL)	0	Y + ff	(IXH)	0
6										Y + ff + 1	(IXL)	0

STY   **Store Index Register Y**   # STY

Operation: M ◀ (IYH), M + 1 ◀ (IYL)

Description: Stores the most significant byte of index register Y in memory at the address specified by the program and stores the least significant byte of index register Y at the next location in memory, at one plus the address specified by the program.

Condition Codes and Boolean Formulae:

S	X	H	I	N	Z	V	C
—	—	—	—	↕	↕	0	—

N   IY15
    Set if MSB of result is set; cleared otherwise.

Z $\overline{IY15} \cdot \overline{IY14} \cdot \overline{IY13} \cdot \overline{IY12} \cdot \overline{IY11} \cdot \overline{IY10} \cdot \overline{IY9} \cdot \overline{IY8} \cdot$
    $\overline{IY7} \cdot \overline{IY6} \cdot \overline{IY5} \cdot \overline{IY4} \cdot \overline{IY3} \cdot \overline{IY2} \cdot \overline{IY1} \cdot \overline{IY0}$
    Set if result is \$0000; cleared otherwise.

V   0
    Cleared

Source Form: STY (opr)

Addressing Modes, Machine Code, and Cycle-by-Cycle Execution:

Cycle	STY (DIR)			STY (EXT)			STY (IND, X)			STY (IND, Y)		
	Addr	Data	R/W̄	Addr	Data	R/W̄	Addr	Data	R/W̄	Addr	Data	R/W̄
1	OP	18	1	OP	18	1	OP	1A	1	OP	18	1
2	OP + 1	DF	1	OP + 1	FF	1	OP + 1	EF	1	OP + 1	EF	1
3	OP + 2	dd	1	OP + 2	hh	1	OP + 2	ff	1	OP + 2	ff	1
4	00dd	(IYH)	0	OP + 3	ll	1	FFFF	—	1	FFFF	—	1
5	00dd + 1	(IYL)	0	hhll	(IYH)	0	X + ff	(IYH)	0	Y + ff	(IYH)	0
6				hhll + 1	(IYL)	0	X + ff + 1	(IYL)	0	Y + ff + 1	(IYL)	0

SUB

Subtract

SUB

Operation: ACCX ◀ (ACCX) − (M)

Description: Subtracts the contents of M from the contents of ACCX and places the result in ACCX. For subtract instructions, the C bit in the CCR represents a borrow.

Condition Codes and Boolean Formulae:

S	X	H	I	N	Z	V	C
—	—	—	—	↕	↕	↕	↕

N R7
  Set if MSB of result is set; cleared otherwise.

Z $\overline{R7} \cdot \overline{R6} \cdot \overline{R5} \cdot \overline{R4} \cdot \overline{R3} \cdot \overline{R2} \cdot \overline{R1} \cdot \overline{R0}$
  Set if result is $00; cleared otherwise.

V $X7 \cdot \overline{M7} \cdot \overline{R7} + \overline{X7} \cdot M7 \cdot R7$
  Set if a twos complement overflow resulted from the operation; cleared otherwise.

C $\overline{X7} \cdot M7 + M7 \cdot R7 + R7 \cdot \overline{X7}$
  Set if the absolute value of the contents of memory are larger than the absolute value of the contents of the accumulator; cleared otherwise.

Source Forms: SUBA (opr); SUBB (opr)

Addressing Modes, Machine Code, and Cycle-by-Cycle Execution:

Cycle	SUBA (IMM)			SUBA (DIR)			SUBA (EXT)			SUBA (IND, X)			SUBA (IND, Y)		
	Addr	Data	R/W̄	Addr	Data	R/W̄	Addr	Data	R/W̄	Addr	Data	R/W̄	Addr	Data	R/W̄
1	OP	80	1	OP	90	1	OP	B0	1	OP	A0	1	OP	18	1
2	OP+1	ii	1	OP+1	dd	1	OP+1	hh	1	OP+1	ff	1	OP+1	A0	1
3				00dd	(00dd)	1	OP+2	ll	1	FFFF	—	1	OP+2	ff	1
4							hhll	(hhll)	1	X+ff	(X+ff)	1	FFFF	—	1
5													Y+ff	(Y+ff)	1

Cycle	SUBB (IMM)			SUBB (DIR)			SUBB (EXT)			SUBB (IND, X)			SUBB (IND, Y)		
	Addr	Data	R/W̄	Addr	Data	R/W̄	Addr	Data	R/W̄	Addr	Data	R/W̄	Addr	Data	R/W̄
1	OP	C0	1	OP	D0	1	OP	F0	1	OP	E0	1	OP	18	1
2	OP+1	ii	1	OP+1	dd	1	OP+1	hh	1	OP+1	ff	1	OP+1	E0	1
3				00dd	(00dd)	1	OP+2	ll	1	FFFF	—	1	OP+2	ff	1
4							hhll	(hhll)	1	X+ff	(X+ff)	1	FFFF	—	1
5													Y+ff	(Y+ff)	1

SUBD

Subtract Double Accumulator

SUBD

Operation: ACCD ⬥ (ACCD) − (M:M + 1)

Description: Subtracts the contents of M:M + 1 from the contents of double accumulator D and places the result in ACCD. For subtract instructions, the C bit in the CCR represents a borrow.

Condition Codes and Boolean Formulae:

S	X	H	I	N	Z	V	C
—	—	—	—	⇕	⇕	⇕	⇕

N R15
   Set if MSB of result is set; cleared otherwise.

Z $\overline{R15} \cdot \overline{R14} \cdot \overline{R13} \cdot \overline{R12} \cdot \overline{R11} \cdot \overline{R10} \cdot \overline{R9} \cdot \overline{R8} \cdot \overline{R7} \cdot \overline{R6} \cdot \overline{R5} \cdot \overline{R4} \cdot \overline{R3} \cdot \overline{R2} \cdot \overline{R1} \cdot \overline{R0}$
   Set if result is $0000; cleared otherwise.

V $D15 \cdot \overline{M15} \cdot \overline{R15} + \overline{D15} \cdot M15 \cdot R15$
   Set if a twos complement overflow resulted from the operation; cleared otherwise.

C $\overline{D15} \cdot M15 + M15 \cdot R15 + R15 \cdot \overline{D15}$
   Set if the absolute value of the contents of memory is larger than the absolute value of the accumulator; cleared otherwise.

Source Form: SUBD (opr)

Addressing Modes, Machine Code, and Cycle-by-Cycle Execution:

Cycle	SUBD (IMM)			SUBD (DIR)			SUBD (EXT)			SUBD (IND, X)			SUBD (IND, Y)		
	Addr	Data	R/W̄	Addr	Data	R/W̄	Addr	Data	R/W̄	Addr	Data	R/W̄	Addr	Data	R/W̄
1	OP	83	1	OP	93	1	OP	B3	1	OP	A3	1	OP	18	1
2	OP + 1	jj	1	OP + 1	dd	1	OP + 1	hh	1	OP + 1	ff	1	OP + 1	A3	1
3	OP + 2	kk	1	00dd	(00dd)	1	OP + 2	ll	1	FFFF	—	1	OP + 2	ff	1
4	FFFF	—	1	00dd + 1	(00dd + 1)	1	hhll	(hhll)	1	X + ff	(X + ff)	1	FFFF	—	1
5				FFFF	—	1	hhll + 1	(hhll + 1)	1	X + ff + 1	(X + ff + 1)	1	Y + ff	(Y + ff)	1
6							FFFF	—	1	FFFF	—	1	Y + ff + 1	(Y + ff + 1)	1
7													FFFF	—	1

SWI
Software Interrupt
SWI

Operation: PC ⬦ (PC) + $0001
⬦(PCL), SP ⬦ (SP) − $0001
⬦(PCH), SP ⬦ (SP) − $0001
⬦(IYL), SP ⬦ (SP) − $0001
⬦(IYH), SP ⬦ (SP) − $0001
⬦(IXL), SP ⬦ (SP) − $0001
⬦(IXH), SP ⬦ (SP) − $0001
⬦(ACCA), SP ⬦ (SP) − $0001
⬦(ACCB), SP ⬦ (SP) − $0001
⬦(CCR), SP ⬦ (SP) − $0001
I ⬦ 1, PC ⬦ (SWI vector)

Description: The program counter is incremented by one. The program counter, index registers Y and X, and accumulators A and B are pushed onto the stack. The CCR is then pushed onto the stack. The stack pointer is decremented by one after each byte of data is stored on the stack. The I bit in the CCR is then set. The program counter is loaded with the address stored at the SWI vector, and instruction execution resumes at this location. This instruction is not maskable by the I bit.

Condition Codes and Boolean Formulae:

S	X	H	I	N	Z	V	C
—	—	—	1	—	—	—	—

I   1
    Set

Source Form: SWI

Addressing Modes, Machine Code, and Cycle-by-Cycle Execution:

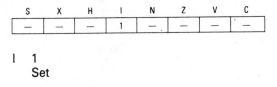

Cycle	SWI (INH)		
	Addr	Data	R/W̄
1	OP	3F	1
2	OP + 1	—	1
3	SP	Rtn lo	0
4	SP − 1	Rtn hi	0
5	SP − 2	(IYL)	0
6	SP − 3	(IYH)	0
7	SP − 4	(IXL)	0
8	SP − 5	(IXH)	0
9	SP − 6	(A)	0
10	SP − 7	(B)	0
11	SP − 8	(CCR)	0
12	SP − 8	(CCR)	1
13	Vec hi	Svc hi	1
14	Vec lo	Svc lo	1

TAB   **Transfer from Accumulator A to Accumulator B**   **TAB**

Operation: ACCB ◄ (ACCA)

Description: Moves the contents of ACCA to ACCB. The former contents of ACCB are lost; the contents of ACCA are not affected.

Condition Codes and Boolean Formulae:

S	X	H	I	N	Z	V	C
—	—	—	—	↕	↕	0	—

N R7
   Set if MSB of result is set; cleared otherwise.

Z $\overline{R7} \cdot \overline{R6} \cdot \overline{R5} \cdot \overline{R4} \cdot \overline{R3} \cdot \overline{R2} \cdot \overline{R1} \cdot \overline{R0}$
   Set if result is $00; cleared otherwise

V 0
   Cleared

Source Form: TAB

Addressing Modes, Machine Code, and Cycle-by-Cycle Execution:

Cycle	TAB (INH)		
	Addr	Data	R/$\overline{W}$
1	OP	16	1
2	OP+1	—	1

TAP

**Transfer from Accumulator A to
Condition Code Register**

TAP

Operation: CCR ◀ (ACCA)

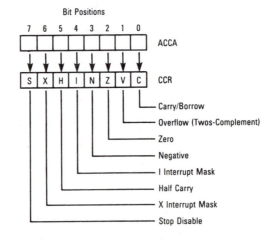

Description: Transfers the contents of bit positions 7–0 of accumulator A to the corresponding bit positions of the CCR. The contents of accumulator A remain unchanged. The X bit in the CCR may be cleared as a result of a TAP instruction but may not be set if it was clear prior to execution of the TAP instruction.

Condition Codes and Boolean Formulae:

S	X	H	I	N	Z	V	C
↕	←	↕	↕	↕	↕	↕	↕

Condition code bits take on the value of the corresponding bit of accumulator A except that the X bit may not change from a zero to a one. Software can leave X set, leave X clear or change X from one to zero. The $\overline{XIRQ}$ interrupt mask can only become set as a result of a $\overline{RESET}$ or recognition of an $\overline{XIRQ}$ interrupt.

Source Form: TAP

Addressing Modes, Machine Code, and Cycle-by-Cycle Execution:

Cycle	TAP (INH)		
	Addr	Data	R/$\overline{W}$
1	OP	06	1
2	OP+1	—	1

TBA   Transfer from Accumulator B to Accumulator A   TBA

Operation: ACCA ◀ (ACCB)

Description: Moves the contents of ACCB to ACCA. The former contents of ACCA are lost; the contents of ACCB are not affected.

Condition Codes and Boolean Formulae:

S	X	H	I	N	Z	V	C
—	—	—	—	↕	↕	0	—

N R7
   Set if MSB of result is set; cleared otherwise.

Z $\overline{R7} \cdot \overline{R6} \cdot \overline{R5} \cdot \overline{R4} \cdot \overline{R3} \cdot \overline{R2} \cdot \overline{R1} \cdot \overline{R0}$
   Set if result is $00; cleared otherwise.

V 0
   Cleared

Source Form: TBA

Addressing Modes, Machine Code, and Cycle-by-Cycle Execution:

Cycle	TBA (INH)		
	Addr	Data	R/W̅
1	OP	17	1
2	OP+1	—	1

TEST

**Test Operation
(Test Mode Only)**

TEST

Description: This is a single-byte instruction that causes the program counter to be continuously incremented. It can only be executed while in the test mode. The MPU must be reset to exit this instruction. Code execution is suspended during this instruction. This is an illegal opcode when not in test mode.

Condition Codes and Boolean Formulae:

S	X	H	I	N	Z	V	C
—	—	—	—	—	—	—	—

None affected

Source Form: TEST

Addressing Modes, Machine Code, and Cycle-by-Cycle Execution:

Cycle	TEST (INH)		
	Addr	Data	R/W̄
1	OP	00	1
2	OP + 1	—	1
3	OP + 2	—	1
4	OP + 3	—	1
5 – n	PREV – 1	(PREV – 1)	1

TPA

Transfer from Condition Code Register to Accumulator A

TPA

Operation: (ACCA) ◄ (CCR)

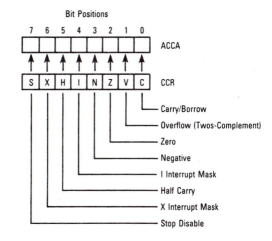

Description: Transfers the contents of the CCR to corresponding bit positions of accumulator A. The CCR remains unchanged.

Condition Codes and Boolean Formulae:

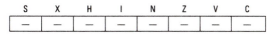

S	X	H	I	N	Z	V	C
—	—	—	—	—	—	—	—

None affected

Source Form: TPA

Addressing Modes, Machine Code, and Cycle-by-Cycle Execution:

Cycle	TPA (INH)		
	Addr	Data	R/W̄
1	OP	07	1
2	OP + 1	—	1

TST

Test

TST

Operation: (ACCX) – $00   **or:**   (M) – $00

Description: Subtracts $00 from the contents of ACCX or M and sets the condition codes accordingly.

The subtraction is accomplished internally without modifying either ACCX or M.

The TST instruction provides only minimum information when testing unsigned values. Since no unsigned value is less than zero, BLO and BLS have no utility. While BHI could be used after TST, it provides exactly the same control as BNE, which is preferred. After testing signed values, all signed branches are available.

Condition Codes and Boolean Formulae:

S	X	H	I	N	Z	V	C
—	—	—	—	↕	↕	0	0

N M7
Set if MSB of result is set; cleared otherwise.

Z $\overline{M7} \cdot \overline{M6} \cdot \overline{M5} \cdot \overline{M4} \cdot \overline{M3} \cdot \overline{M2} \cdot \overline{M1} \cdot \overline{M0}$
Set if result is $00; cleared otherwise

V 0
Cleared

C 0
Cleared

Source Forms: TSTA; TSTB; TST (opr)

Addressing Modes, Machine Code, and Cycle-by-Cycle Execution:

Cycle	TSTA (INH) Addr	Data	R/W̄	TSTB (INH) Addr	Data	R/W̄	TST (EXT) Addr	Data	R/W̄	TST (IND, X) Addr	Data	R/W̄	TST (IND, Y) Addr	Data	R/W̄
1	OP	4D	1	OP	5D	1	OP	7D	1	OP	6D	1	OP	18	1
2	OP+1	—	1	OP+1	—	1	OP+1	hh	1	OP+1	ff	1	OP+1	6D	1
3							OP+2	ll	1	FFFF	—	1	OP+2	ff	1
4							hhll	(hhll)	1	X+ff	(X+ff)	1	FFFF	—	1
5							FFFF	—	1	FFFF	—	1	Y+ff	(Y+ff)	1
6							FFFF	—	1	FFFF	—	1	FFFF	—	1
7													FFFF	—	1

TSX

Transfer from Stack Pointer to Index Register X

TSX

Operation: IX ⬅ (SP) + $0001

Description: Loads the index register X with one plus the contents of the stack pointer. The contents of the stack pointer remain unchanged. After a TSX instruction the index register X points at the last value that was stored on the stack.

Condition Codes and Boolean Formulae:

S	X	H	I	N	Z	V	C
—	—	—	—	—	—	—	—

None affected

Source Form: TSX

Addressing Modes, Machine Code, and Cycle-by-Cycle Execution:

Cycle	TSX (INH)		
	Addr	Data	R/W̄
1	OP	30	1
2	OP + 1	—	1
3	SP	—	1

TSY   Transfer from Stack Pointer to Index Register Y   TSY

Operation: IY ◀ (SP) + $0001

Description: Loads the index register Y with one plus the contents of the stack pointer. The contents of the stack pointer remain unchanged. After a TSY instruction the index register Y points at the last value that was stored on the stack.

Condition Codes and Boolean Formulae:

S	X	H	I	N	Z	V	C
—	—	—	—	—	—	—	—

None affected

Source Form: TSY

Addressing Modes, Machine Code, and Cycle-by-Cycle Execution:

Cycle	TSY (INH)		
	Addr	Data	R/W̄
1	OP	18	1
2	OP + 1	30	1
3	OP + 2	—	1
4	SP	—	1

TXS   Transfer from Index Register X to Stack Pointer   TXS

Operation: SP ◄ (IX) − $0001

Description: Loads the stack pointer with the contents of the index register X minus one. The contents of the index register X remain unchanged.

Condition Codes and Boolean Formulae:

S	X	H	I	N	Z	V	C
—	—	—	—	—	—	—	—

None affected

Source Form: TXS

Addressing Modes, Machine Code, and Cycle-by-Cycle Execution:

Cycle	TXS (INH)		
	Addr	Data	R/W̄
1	OP	35	1
2	OP + 1	—	1
3	FFFF	—	1

TYS   Transfer from Index Register Y to Stack Pointer   TYS

Operation;   SP ◀ (IY) – $0001

Description: Loads the stack pointer with the contents of the index register Y minus one. The contents of the index register Y remain unchanged.

Condition Codes and Boolean Formulae:

S	X	H	I	N	Z	V	C
—	—	—	—	—	—	—	—

None affected

Source Form: TYS

Addressing Modes, Machine Code, and Cycle-by-Cycle Execution:

Cycle	TYS (INH)		
	Addr	Data	R/W̄
1	OP	18	1
2	OP+1	35	1
3	OP+2	—	1
4	FFFF	—	1

WAI

Wait for Interrupt

WAI

Operation;   PC ← (PC) + $0001
→ (PCL), SP ← (SP) − $0001
→ (PCH), SP ← (SP) − $0001
→ (IYL), SP ← (SP) − $0001
→ (IYH), SP ← (SP) − $0001
→ (IXL), SP ← (SP) − $0001
→ (IXH), SP ← (SP) − $0001
→ (ACCA), SP ← (SP) − $0001
→ (ACCB), SP ← (SP) − $0001
→ (CCR), SP ← (SP) − $0001

Description: The program counter is incremented by one. The program counter, index registers Y and X, and accumulators A and B are pushed onto the stack. The CCR is then pushed onto the stack. The stack pointer is decremented by one after each byte of data is stored on the stack.

The MPU then enters a wait state for an integer number of MPU E-clock cycles. While in the wait state, the address/data bus repeatedly runs read bus cycles to the address where the CCR contents were stacked. The MPU leaves the wait state when it senses any interrupt that has not been masked.

Upon leaving the wait state, the MPU sets the I bit in the CCR, fetches the vector (address) corresponding to the interrupt sensed, and instruction execution is resumed at this location.

Condition Codes and Boolean Formulae:

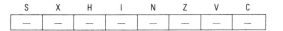

S	X	H	I	N	Z	V	C
—	—	—	—	—	—	—	—

Although the WAI instruction itself does not alter the condition code bits, the interrupt which causes the MCU to resume processing causes the I bit (and the X bit if the interrupt was XIRQ) to be set as the interrupt vector is being fetched.

WAI   **Wait for Interrupt**   **WAI**

Source Form: WAI

Addressing Modes, Machine Code, and Cycle-by-Cycle Execution:

Cycle	WAI (INH)		
	Addr	**Data**	**R/W̄**
1	OP	3E	1
2	OP + 1	—	1
3	SP	Rtn lo	0
4	SP − 1	Rtn hi	0
5	SP − 2	(IYL)	0
6	SP − 3	(IYH)	0
7	SP − 4	(IXL)	0
8	SP − 5	(IXH)	0
9	SP − 6	(A)	0
10	SP − 7	(B)	0
11	SP − 8	(CCR)	0
12 to 12 + n	SP − 8	(CCR)	1
13 + n	Vec hi	Svc hi	1
14 + n	Vec lo	Svc lo	1

XGDX

**Exchange Double Accumulator and
Index Register X**

XGDX

Operation: (iX) ⬥ (ACCD)

Description: Exchanges the contents of double accumulator ACCD and the contents of index register X. A common use for XGDX is to move an index value into the double accumulator to allow 16-bit arithmetic calculations on the index value before exchanging the updated index value back into the X index register.

Condition Codes and Boolean Formulae:

S	X	H	I	N	Z	V	C
—	—	—	—	—	—	—	—

None affected

Source Form: XGDX

Addressing Modes, Machine Code, and Cycle-by-Cycle Execution:

Cycle	XGDX (INH)		
	Addr	Data	R/W̄
1	OP	8F	1
2	OP + 1	—	1
3	FFFF	—	1

XGDY

**Exchange Double Accumulator and
Index Register Y**

XGDY

Operation: (IY) ◆◆ (ACCD)

Description: Exchanges the contents of double accumulator ACCD and the contents of index register Y. A common use for XGDY is to move an index value into the double accumulator to allow 16-bit arithmetic calculations on the index value before exchanging the updated index value back into the Y index register.

Condition Codes and Boolean Formulae:

S	X	H	I	N	Z	V	C
—	—	—	—	—	—	—	—

     None affected

Source Form: XGDY

Addressing Modes, Machine Code, and Cycle-by-Cycle Execution:

Cycle	XGDY (INH)		
	Addr	Data	R/W̄
1	OP	18	1
2	OP+1	8F	1
3	OP+2	—	1
4	FFFF	—	1

B.2

The MC145000 and MC145001 LCD Drivers

MOTOROLA
SEMICONDUCTOR
TECHNICAL DATA

MC145000
MC145001

SERIAL INPUT
MULTIPLEXED LCD DRIVERS
(MASTER AND SLAVE)

The MC145000 (Master) LCD Driver and the MC145001 (Slave) LCD Driver are CMOS devices designed to drive liquid crystal displays in a multiplexed-by-four configuration. The Master unit generates both frontplane and backplane waveforms, and is capable of independent operation. The Slave unit generates only frontplane waveforms, and is synchronized with the backplanes from the Master unit. Several Slave units may be cascaded from the Master unit to increase the number of LCD segments driven in the system. The maximum number of frontplanes is dependent upon the capacitive loading on the backplane drivers and the drive frequency. The devices use data from a microprocessor or other serial data and clock source to drive one LCD segment per bit.

- Direct Interface to CMOS Microprocessors
- Serial Data Port, Externally Clocked
- Multiplexing-By-Four
- Net dc Drive Component Less Than 50 mV
- Master Drives 48 LCD Segments
- Slave Provides Frontplane Drive for 44 LCD Segments
- Drives Large Segments —Up to one Square Centimeter
- Supply Voltage Range = 3 V to 6 V
- Latch Storage of Input Data
- Low Power Dissipation
- Logic Input Voltage Can Exceed V_{DD}
- Accomodates External Temperature Compensation
- Chip Complexities: MC145000 — 1723 FETs or 431 Equivalent Gates
  MC145001 — 1495 FETs or 374 Equivalent Gates

CMOS LSI
(LOW-POWER COMPLEMENTARY MOS)

SERIAL INPUT
MULTIPLEXED LCD DRIVERS
(MASTER AND SLAVE)

P SUFFIX
PLASTIC DIP
CASE 709

P SUFFIX
PLASTIC DIP
CASE 707

FN SUFFIX
PLCC
CASE 776

ORDERING INFORMATION

MC14500xP   Plastic DIP
MC14500xFN   PLCC

MC145000•MC145001

PIN ASSIGNMENTS

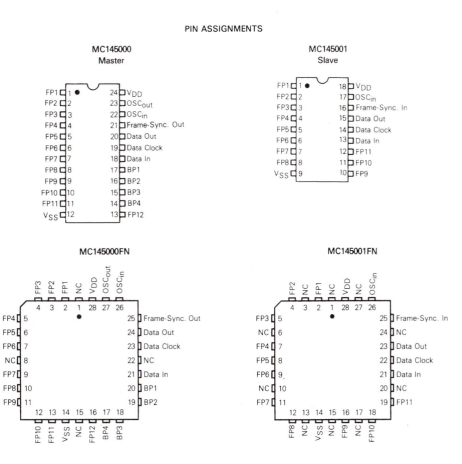

Mux-by-4 LCD Manufacturers

NOTE: Motorola cannot recommend one supplier over
another and in no way suggests that this is a
complete listing of LCD suppliers.

Supplier	Contact Information
Polytronix, Inc.	Phone: (214) 238-7045 FAX: (214) 644-0805
LXD, Inc.	Phone: (216) 292-3300 FAX: (216) 292-4727
UCE, Inc.	Phone: (203) 838-7500 FAX: (203) 838-2566
Hamlin, Inc.	Phone: (414) 648-1000 FAX: (414) 648-1001

MC145000•MC145001

MAXIMUM RATINGS (Voltages referenced to V_{SS})

Characteristic	Symbol	Value	Unit
DC Supply Voltage	V_{DD}	−0.5 to +6.5	V
Input Voltage, Data In and Data Clock	V_{in}	−0.5 to 15	V
Input Voltage, OSC$_{in}$ of Master	$V_{in\ osc}$	−0.5 to V_{DD}+0.5	V
DC Input Current, per Pin	I_{in}	±10	mA
Operating Temperature Range	T_A	−40 to +85	°C
Storage Temperature Range	T_{stg}	−65 to +150	°C

ELECTRICAL CHARACTERISTICS (Voltages referenced to V_{SS})

Characteristic	Symbol	V_{DD} V	−40°C Min	−40°C Max	25°C Min	25°C Typ#	25°C Max	85°C Min	85°C Max	Unit
RMS Voltage Across a Segment "ON" Segment (BPi-FPj)	V_{ON}	3.0	—	—	—	1.73	—	—	—	V
		6.0	—	—	—	3.46	—	—	—	
"OFF" Segment	V_{OFF}	3.0	—	—	—	1.00	—	—	—	V
		6.0	—	—	—	2.00	—	—	—	
Average DC Offset Voltage	V_{dc}	3.0	—	30	—	10	30	—	30	mV
		6.0	—	50	—	20	50	—	50	
Input Voltage "0" Level	V_{IL}	3.0	—	0.90	—	1.35	0.90	—	0.90	V
		6.0	—	1.80	—	2.70	1.80	—	1.80	
"1" Level	V_{IH}	3.0	2.10	—	2.10	1.65	—	2.10	—	V
		6.0	4.20	—	4.20	3.30	—	4.20	—	
Output Drive Current — Backplanes High-Current State* V_O = 2.85 V	I_{BH}	3.0	150	—	75	190	—	35	—	µA
V_O = 1.85 V			220	—	110	200	—	55	—	
V_O = 1.15 V			160	—	80	200	—	40	—	
V_O = 0.15 V			400	—	200	300	—	100	—	
V_O = 5.85 V	I_{BH}	6.0	500	—	250	300	—	125	—	µA
V_O = 3.85 V			1000	—	500	600	—	250	—	
V_O = 2.15 V			800	—	400	500	—	200	—	
V_O = 0.15 V			500	—	250	300	—	125	—	
Low-Current State* V_O = 2.85 V	I_{BL}	3.0	140	—	70	80	—	35	—	µA
V_O = 1.85 V			2.4	—	1.2	2.8	—	0.6	—	
V_O = 1.15 V			2.2	—	1.1	2.5	—	0.5	—	
V_O = 0.15 V			400	—	200	330	—	100	—	
V_O = 5.85 V	I_{BL}	6.0	190	—	95	105	—	45	—	µA
V_O = 3.85 V			15	—	7.5	10	—	3.7	—	
V_O = 2.15 V			13	—	6.5	9	—	3.2	—	
V_O = 0.15 V			850	—	425	570	—	210	—	
Output Drive Current — Frontplanes High-Current State* V_O = 2.85 V	I_{FH}	3.0	80	—	40	60	—	20	—	µA
V_O = 1.85 V			140	—	70	120	—	35	—	
V_O = 1.15 V			180	—	60	100	—	30	—	
V_O = 0.15 V			100	—	50	95	—	25	—	
V_O = 5.85 V	I_{FH}	6.0	140	—	70	90	—	35	—	µA
V_O = 3.85 V			360	—	180	250	—	90	—	
V_O = 2.15 V			400	—	200	240	—	100	—	
V_O = 0.15 V			100	—	50	120	—	25	—	
Low-Current State* V_O = 2.85 V	I_{FL}	3.0	60	—	30	40	—	15	—	µA
V_O = 1.85 V			2.8	—	1.4	2.8	—	0.7	—	
V_O = 1.15 V			2.2	—	1.1	2.5	—	0.5	—	
V_O = 0.15 V			100	—	50	100	—	25	—	
V_O = 5.85 V	I_{FL}	6.0	100	—	50	60	—	25	—	µA
V_O = 3.85 V			15	—	8.0	10	—	4.0	—	
V_O = 2.15 V			13	—	6.5	9	—	3.2	—	
V_O = 0.15 V			200	—	100	175	—	50	—	
Input Current	I_{in}	6.0	—	±0.1	—	±0.00001	±0.1	—	±1.0	µA
Input Capacitance	C_{in}	—	—	—	—	5.0	7.5	—	—	pF
Quiescent Current (Per Package) V_{in} = 0 or V_{DD}, I_{out} = 0 µA	I_{DD}	3.0	—	—	—	2.5	15	—	20	µA
		6.0	—	185	—	50	175	—	130	

*For a time (t ≅ 2.56/osc. freq.) after the backplane or frontplane waveform changes to a new voltage level, the circuit is maintained in the high-current state to allow the load capacitances to charge quickly. Then the circuit is returned to the low-current state until the next voltage level change occurs.

#Data labelled "Typ" is not to be used for design purposes but is intended as an indication of the IC's potential performance.

MC145000•MC145001

SWITCHING CHARACTERISTICS ($C_L = 50$ pF, $T_A = 25°C$)

Characteristic	Symbol	V_{DD}	Min	Typ#	Max	Unit
Data Clock Frequency	f_{cl}	3.0	–	12.5	7.5	MHz
		6.0	–	24	12.5	
Rise and Fall Times — Data clock	t_r, t_f	3.0	–	–	125	μs
		6.0	–	–	10	
Setup Time Data In to Data Clock	t_{su}	3.0	48	–	–	ns
		6.0	16	–	–	
Hold Time Data In to Data Clock	t_h	3.0	– 5	–	–	ns
		6.0	0	–	–	
Pulse Width Data Clock	t_w	3.0	65	–	–	ns
		6.0	40	–	–	

#Data labelled "Typ" is not to be used for design purposes but is intended as an indication of the IC's potential performance.

SWITCHING WAVEFORMS

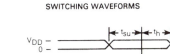
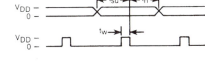

Data In

Data Clock

DEVICE OPERATION

Figure 1 shows a block diagram of the Master unit. The unit is composed of two independent circuits: the data input circuit with its associated data clock, and the LCD drive circuit with its associated system clock.

Forty-eight bits of data are serially clocked into the shift register on the falling edges of the external data clock. Data in the shift register is latched into the 48-bit latch at the beginning of each frame period. (As shown in Figure 3, the frame period, t_{frame}, is the time during which all the LCD segments are set to the desired "ON" or "OFF" states.)

The binary data present in the latch determines the appropriate waveform signal to be generated by the frontplane drive circuits, whereas the backplane waveforms are invariant. The frontplane and backplane waveforms, FPn and BPn, are generated using the system clock (which is the oscillator divided by 256) and voltages from the V/3 generator circuit (which divides V_{DD} into one-third increments). As shown in Figure 3, the frontplane and backplane waveforms and the "ON" and "OFF" segment waveforms have periods equal to t_{frame} and frequencies equal to the system clock divided by four.

Twelve frontplane and four backplane drivers are available from the Master unit. The latching of the data at the beginning of each frame period and the carefully balanced voltage-generation circuitry minimize the generation of a net dc component across any LCD segment.

The Slave unit (Figure 2) consists of the same circuitry as the Master unit, with two exceptions: it has no backplane drive circuitry, and its shift register and latch hold 44 bits. Eleven frontplane and no backplane drivers are available from the Slave unit.

LCD DRIVER SYSTEM CONFIGURATIONS

Figure 4 shows a basic LCD Driver system configuration, with one Master and several Slave units. The maximum number of slave units in a system is dictated by the maximum backplane drive capability of the device and by the system data update rate. Data is serially shifted first into the Master unit and then into the following Slave units on the falling edge of the common data clock. The oscillator is common to the Master unit and each of the Slave units. At the beginning of each frame period, t_{frame}, the Master unit generates a frame-sync pulse (Figure 3) which is received by the Slave units. The pulse is to ensure that all Slave unit frontplane drive circuits are synchronized to the Master unit's backplane drive circuits.

A single multiplexed-by-four, 7-segment (plus decimal point) LCD and possible frontplane and backplane connections are shown in Figure 5. When several such displays are used in a system, the four backplanes generated by the Master unit are common to all the LCD digits in the system. The twelve frontplanes of the Master unit are capable of controlling forty-eight LCD segments (6 LCD digits), and the eleven frontplanes of each Slave unit are capable of controlling forty-four LCD segments (5½ LCD digits).

MC145000•MC145001

FIGURE 1 — BLOCK DIAGRAM OF THE MC145000 (MASTER) LCD DRIVER

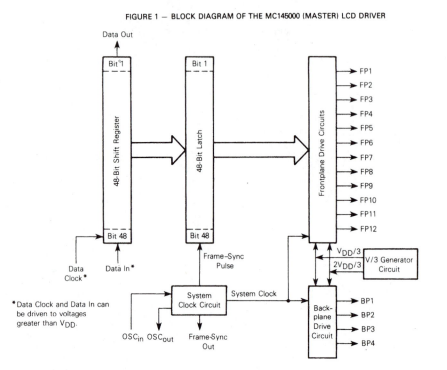

FIGURE 2 — BLOCK DIAGRAM OF THE MC145001 (SLAVE) LCD DRIVER

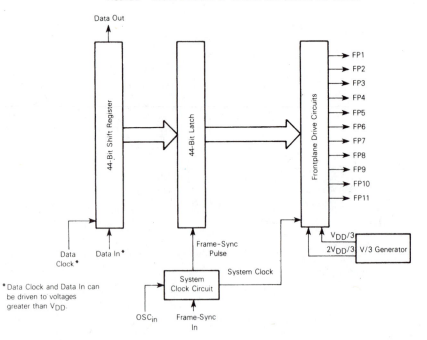

MC145000•MC145001

FIGURE 3 — VOLTAGE WAVEFORMS

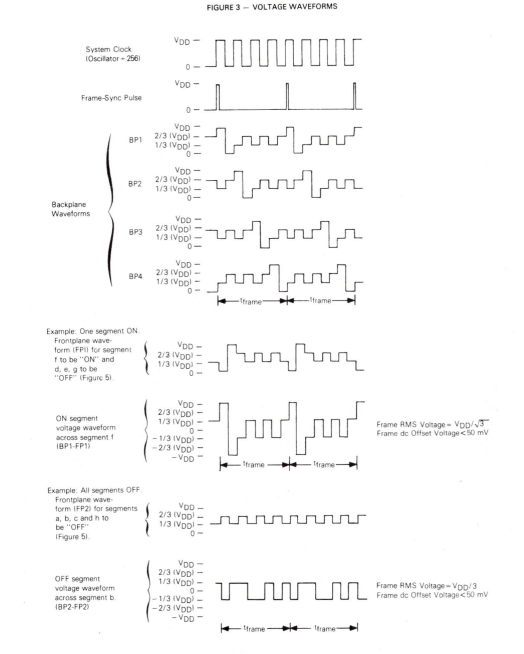

MC145000•MC145001

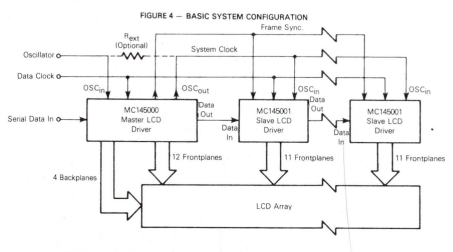

FIGURE 4 — BASIC SYSTEM CONFIGURATION

FIGURE 5 — FRONTPLANE AND BACKPLANE CONNECTIONS TO A MULTIPLEXED-BY-FOUR
7-SEGMENT (PLUS DECIMAL POINT) LCD

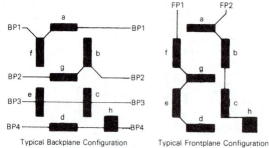

Typical Backplane Configuration

Typical Frontplane Configuration

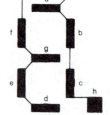

SEGMENT TRUTH TABLE*

	FP1	FP2
BP1	f	a
BP2	g	b
BP3	e	c
BP4	d	h

*Because there is no standard for back-
plane and frontplane connections on
multiplexed displays, this truth table may
be used only for this example.

PIN DESCRIPTIONS

FRONTPLANE DRIVE OUTPUTS
(Master: FP1-FP12)
(Slave: FP1-FP11)

The frontplane drive waveforms for the LCDs.

BACKPLANE DRIVE OUTPUTS
(Master: BP1-BP4)

The backplane drive waveforms for the LCDs.

DATA IN (Master and Slave)

The serial data input pin. Data is clocked into the shift
register on the falling edge of the data clock. A high logic
level will cause the corresponding LCD segment to be turned on, and a low logic level will cause the segment to be
turned off. This pin can be driven to 15 volts regardless of
the value of V_{DD}, thus permitting optimum display drive
voltage.

DATA CLOCK (Master and Slave)

The input pin for the external data clock, which controls
the shift registers. This pin can be driven to 15 volts
regardless of the value of V_{DD}.

DATA OUT (Master and Slave)

The serial data output pin.

FRAME-SYNC OUT (Master)

The output pin for the frame-sync pulse, which is
generated by the Master unit at the beginning of each frame
period, t_{frame}. From Figure 1, the 48-bit latch is loaded during the positive Frame-Sync Out pulse. Therefore, if the Data
Clock is active during this load interval, the display will
flicker.

FRAME-SYNC IN (Slave)

The input pin for the frame-sync pulse from the Master
unit. The frame-sync pulse synchronizes the Slave frontplane drive waveforms to the Master backplane drive
waveforms.

MC145000•MC145001

OSC_{in} (Master and Slave)

The input pin to the system clock circuit. The oscillator frequency is either obtained from an external oscillator or generated in the Master unit by connecting an external resistor between the OSC_{in} pin and the OSC_{out} pin. Figure 6 shows the relationship between resistor value and frequency.

OSC_{out} (Master)

The output pin of the system clock circuit. This pin is connected to the OSC_{in} input of each Slave unit.

V_{DD} (Master and Slave)

The positive supply voltage.

V_{SS} (Master and Slave)

The negative supply (or ground) voltage.

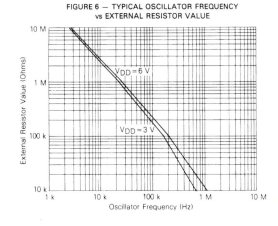

FIGURE 6 — TYPICAL OSCILLATOR FREQUENCY
vs EXTERNAL RESISTOR VALUE

APPLICATIONS

The following examples are presented to give the user further insight into the operation and organization of the Master and Slave LCD Drivers.

An LCD segment is turned either on or off depending upon the RMS value of the voltage across it. This voltage is equal to the backplane voltage waveform minus the frontplane voltage waveform. As previously stated, the backplane waveforms are invariant (see Figure 3). Figure 10 shows one period of every possible frontplane waveform.

For a detailed explanation of the operation of liquid crystal materials and multiplexed displays, refer to a brochure entitled "Multiplexed Liquid Crystal Displays," by Gregory A. Zaker, General Electric Company, Liquid XTAL Displays Operation, 24500 Highpoint Road, Cleveland, Ohio 44122.

Example 1: Many applications (e.g., meters, gasoline pumps, pinball machines, and automobile dashboard displays) require that, for each display update, an entirely new set of data must be shifted into the Master and cascaded Slave units. The correspondence between the frontplane-backplane intersections at the LCD segments and the data bit locations in the 48-bit latch of the Master (or 44-bit latch of the Slave) is necessary information to the system designer. In Figure 1, it is shown that data is serially shifted first into the 48th-bit location of the shift register of the Master. Thus, after 48 data bits have been shifted in, the first bit to be entered has been shifted into bit-location one, the second bit into bit-location two, and so on. Table 1 shows the bit location in the latch that controls the corresponding frontplane-backplane intersection. For example, the information stored in the 26th-bit location of the latch controls the LCD segment at the intersection of FP7 and BP3. The voltage waveform across that segment is equal to (BP3 minus FP7). The same table, but with the column for FP12 deleted, describes the operation of the Slave unit.

In applications of this type, all the necessary data to completely update the display are serially shifted into the Master and succeeding Slave units within a frame period. Typically, a microprocessor is used to accomplish this.

Example 2: Many keyboard-entry applications, such as calculators, require that the most significant digit be entered and displayed first. Then as each succeeding digit is entered, the previously entered digits must shift to the left. It is, therefore, neither necessary nor desirable to enter a completely new set of data for each display change. Figure 7 shows a representation of a system consisting of one Master and three Slave units and displaying 20 LCD digits. If each digit has the frontplane-backplane configuration shown in Figure 5, the relationship between frontplanes, backplanes, and LCD segments in the display is shown in Table 2.

MC145000•MC145001

Digits (or alphanumeric characters) are entered, most-significant digit first, by using a keyboard and a decoder external to the MC145000. Data is entered into the Master and cascaded Slave units according to the following format:

1) Initially, all registers and latches must be cleared by entering 160 zero data bits. This turns off all 160 segments of the display.

2) Entering the most-significant digit from the keyboard causes the appropriate eight bits to be serially shifted into the Master unit. These eight bits control LCD segments a through h of digit 1, and cause the desired digit to be displayed in the least-significant digit location.

3) Entering the second-most-significant digit from the keyboard causes eight more bits to be serially shifted into the Master unit. These eight bits now control LCD segments a

through h of digit 1, and the previously entered eight bits now control segments a through h of digit 2. Thus the two digits are displayed in the proper locations.

4) Entering the remaining 18 digits from the keyboard fills the 20-digit display. Entering an extra digit will cause the first digit entered to be shifted off the display.

Example 3: In addition to controlling 7-segment (plus decimal point) digital displays, the MC145000 and MC145001 may be used to control displays using 5×7 dot matrices. A Master and three Slave units can drive 180 LCD segments, and therefore are capable of controlling five 5×7 dot matrices (175 segments). Two control schemes are presented in Figures 8 and 9; one using a single Master unit, and one using two Master units.

TABLE 1 — THE BIT LOCATIONS, IN THE LATCH, THAT CONTROL THE
LCD SEGMENTS LOCATED AT EACH FRONTPLANE-BACKPLANE INTERSECTION

FRONTPLANES

BACKPLANES	FP1	FP2	FP3	FP4	FP5	FP6	FP7	FP8	FP9	FP10	FP11	FP12
BP1	4	8	12	16	20	24	28	32	36	40	44	48
BP2	3	7	11	15	19	23	27	31	35	39	43	47
BP3	2	6	10	14	18	22	26	30	34	38	42	46
BP4	1	5	9	13	17	21	25	29	33	37	41	45

FIGURE 7 — A 20-DIGIT DISPLAY
(EQUIVALENT TO A 4×40 ARRAY)

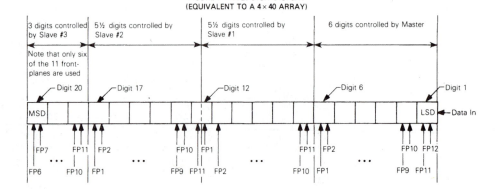

TABLE 2 — THE RELATIONSHIP BETWEEN FRONTPLANE-BACKPLANE INTERSECTIONS
AND LCD SEGMENTS FOR THE SYSTEM CONFIGURATION OF FIGURE 7

	Master							Slave #1				Slave #2						Slave #3				
	FP12	FP11	FP10	FP9	•••	FP2	FP1	FP11	FP10	•••	FP1	FP11	FP10	FP9	•••	FP2	FP1	FP11	FP10	•••	FP7	FP6
BP1	a1	f1	a2	f2	•••	a6	f6	a7	f7	•••	a12	f12	a13	f13	•••	a17	f17	a18	f18	•••	a20	f20
BP2	b1	g1	b2	g2	•••	b6	g6	b7	g7	•••	b12	g12	b13	g13	•••	b17	g17	b18	g18	•••	b20	g20
BP3	c1	e1	c2	e2	•••	c6	e6	c7	e7	•••	c12	e12	c13	e13	•••	c17	e17	c18	e18	•••	c20	e20
BP4	h1	d1	h2	d2	•••	h6	d6	h7	d7	•••	h12	d12	h13	d13	•••	h17	d17	h18	d18	•••	h20	d20
	digit 1				digit 2		•••	digit 6			digit 7		•••	digit 12			digit 13		•••	digit 17	digit 18	••• digit 20

MC145000•MC145001

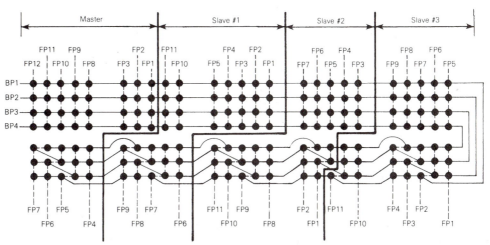

FIGURE 8 — EXAMPLE OF A 5×7 DOT MATRIX DISPLAY SYSTEM CONTROLLED
BY ONE MASTER AND THREE SLAVE UNITS

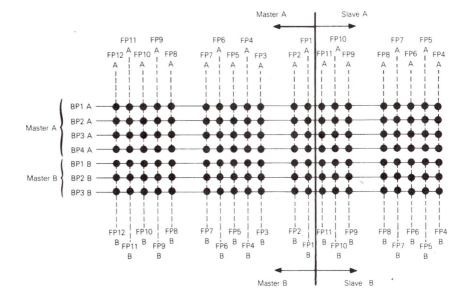

FIGURE 9 — EXAMPLE OF A 5×7 DOT MATRIX DISPLAY SYSTEM
CONTROLLED BY TWO MASTER AND TWO SLAVE UNITS

MC145000•MC145001

FIGURE 10 — POSSIBLE FRONTPLANE WAVEFORMS

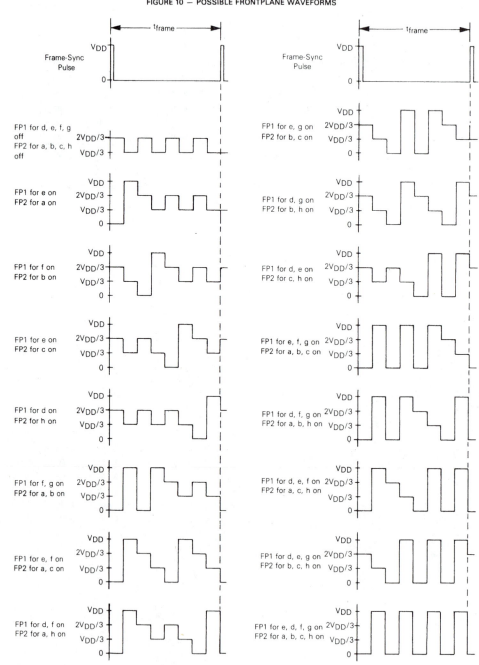

B.3

The MC68HC68T1

Real-Time Clock

with Serial

Interface

MOTOROLA
■■ **SEMICONDUCTOR** ■■■■■■■■■■■■■■■■■■■■■■■
TECHNICAL DATA

MC68HC68T1

P SUFFIX
PLASTIC DIP
CASE 648

DW SUFFIX
SOG
CASE 751G

ORDERING INFORMATION

MC68HC68T1P   Plastic DIP
MC68HC68T1DW SOG Package

Advance Information
Real-Time Clock plus RAM with Serial Interface
CMOS

The MC68HC68T1 HCMOS Clock/RAM peripheral contains a real-time clock/calendar, a 32 × 8 static RAM, and a synchronous, serial, three-wire interface for communication with a microcomputer. Operating in a burst mode, successive Clock/RAM locations can be read or written using only a single starting address. An on-chip oscillator allows acceptance of a selectable crystal frequency or the device can be programmed to accept a 50/60 Hz line input frequency.

The LINE and V_{SYS} pins give the MC68HC68T1 the capability for sensing power-up/power-down conditions, a capability useful for battery-backup systems. The device has an interrupt output capable of signaling the microcomputer of an alarm, periodic interrupt, or power sense condition. An alarm can be set for comparison with the seconds, minutes, and hours registers. This alarm can be used in conjunction with the power supply enable (PSE) output to initiate a system power-up sequence if the V_{SYS} pin is powered to the proper level.

A software power-down sequence can be initiated by setting a bit in the interrupt control register. This applies a reset to the CPU via the $\overline{CPUR}$ pin, sets the clock out (CLK OUT) and power supply enable (PSE) pins low, and disables the serial interface. This condition is held until a rising edge is sensed on the system voltage (V_{SYS}) input pin, signaling system power coming on, or by activation of a previously enabled interrupt if the V_{SYS} pin is powered up.

A watch-dog circuit can be enabled that requires the microcomputer to toggle the slave select (SS) pin of the MC68HC68T1 periodically without performing a serial transfer. If this condition is not sensed, the $\overline{CPUR}$ line resets the CPU.

- Full Clock Features — Seconds, Minutes, Hours (AM/PM), Day-of-Week, Date, Month, Year (0-99), Auto Leap Year
- 32 Byte General-Purpose RAM
- Direct Interface to Motorola SPI and National MICROWIRE™ Serial Data Ports
- Minimum Timekeeping Voltage: 2.2 V
- Burst Mode for Reading/Writing Successive Addresses in Clock/RAM
- Selectable Crystal or 50/60 Hz Line Input Frequency
- Clock Registers Utilize BCD Data
- Buffered Clock Output for Driving CPU Clock, Timer, Colon, or LCD Backplane
- Power-On Reset with First-Time-Up Bit
- Freeze Circuit Eliminates Software Overhead During a Clock Read
- Three Independent Interrupt Modes — Alarm, Periodic, or Power-Down
- CPU Reset Output — Provides Orderly Power Up/Power Down
- Watch-Dog Circuit
- Pin-for-Pin Replacement for CDP68HC68T1
- Chip Complexity: 8500 FETs or 2125 Equivalent Gates
- Also See Application Note ANE425

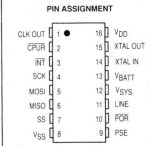

PIN ASSIGNMENT

CLK OUT	1 ●	16 V_{DD}
$\overline{CPUR}$	2	15 XTAL OUT
$\overline{INT}$	3	14 XTAL IN
SCK	4	13 V_{BATT}
MOSI	5	12 V_{SYS}
MISO	6	11 LINE
SS	7	10 $\overline{POR}$
V_{SS}	8	9 PSE

MC68HC68T1

BLOCK DIAGRAM

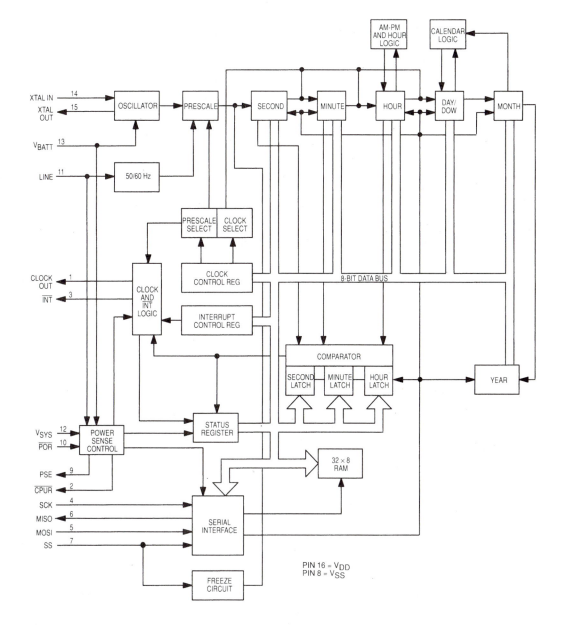

MC68HC68T1

MAXIMUM RATINGS* (Voltages Referenced to V_{SS})

Symbol	Parameter	Value	Unit
V_{DD}	DC Supply Voltage	− 0.5 to + 7.0	V
V_{in}	DC Input Voltage (except Line Input**)	− 0.5 to V_{DD} + 0.5	V
V_{out}	DC Output Voltage	−0.5 to V_{DD} + 0.5	V
I_{in}	DC Input Current, per Pin	± 10	mA
I_{out}	DC Output Current, per Pin	± 10	mA
I_{DD}	DC Supply Current, V_{DD} and V_{SS} Pins	± 30	mA
P_D	Power Dissipation, per Package***	500	mW
T_{stg}	Storage Temperature	− 65 to + 150	°C
T_L	Lead Temperature (10-Second Soldering)	260	°C

*Maximum Ratings are those values beyond which damage to the device may occur.
**See Electrical Characteristics Table
***Power Dissipation Temperature Derating: − 12 mW/°C from 65°C to 85°C.

This device contains protection circuitry to guard against damage due to high static voltages or electrical fields. However, precautions must be taken to avoid applications of any voltage higher than maximum rated voltages to this high-impedance circuit. For proper operation, V_{in} and V_{out} should be constrained to the range $V_{SS} \leq (V_{in}$ or $V_{out}) \leq V_{DD}$.

Unused inputs must always to tied to an appropriate logic voltage level (e.g., either V_{SS} or V_{DD}). Unused outputs must be left open.

ELECTRICAL CHARACTERISTICS (T_A = − 40 to + 85°C, Voltages Referenced to V_{SS})

Symbol	Parameter	Test Condition	V_{DD} V	Guaranteed Limit	Unit
V_{DD}	Power Supply Voltage Range		—	3.0 to 6.0	V
$V_{(stdby)}$	Minimum Standby (Timekeeping) Voltage*		—	2.2	V
V_{IL}	Maximum Low-Level Input Voltage		3.0 4.5 6.0	0.9 1.35 1.8	V
V_{IH}	Minimum High-Level Input Voltage		3.0 4.5 6.0	2.1 3.15 4.2	V
V_{in}	Maximum Input Voltage, Line Input	Power-Sense Mode	5.0	12	V p-p
V_{OL}	Maximum Low-Level Output Voltage	I_{out} = 0 μA I_{out} = 1.6 mA	4.5	0.1 0.4	V
V_{OH}	Minimum High-Level Output Voltage	I_{out} = 0 μA I_{out} = 1.6 mA	4.5	4.4 3.7	V
I_{in}	Maximum Input Current, Except SS	$V_{in} = V_{DD}$ or V_{SS}	6.0	± 1	μA
I_{IL}	Maximum Low-Level Input Current, SS	$V_{in} = V_{SS}$	6.0	− 1.0	μA
I_{IH}	Maximum Pull-Down Current, SS	$V_{in} = V_{DD}$	6.0	100	μA
I_{OZ}	Maximum 3-State Leakage Current	$V_{out} = V_{DD}$ or V_{SS}	6.0	± 10	μA
I_{DD}	Maximum Quiescent Supply Current	$V_{in} = V_{DD}$ or V_{SS}, All Inputs I_{out} = 0 μA	6.0	50	μA
I_{DD}	Maximum RMS Operating Supply Current Crystal Operation	I_{out} = 0 μA, $V_{in} = V_{DD}$ or V_{SS}, all inputs except XTAL IN, Clock Out Disabled, No Serial Access Cycles $f_{XTAL IN}$ = 32 kHz $f_{XTAL IN}$ = 1 MHz $f_{XTAL IN}$ = 2 MHz $f_{XTAL IN}$ = 4 MHz	5.0	0.1 0.6 0.84 1.2	mA
	Maximum RMS Operating Supply Current External Frequency Source Driving XTAL IN, XTAL OUT Open	I_{out} = 0 μA, $V_{in} = V_{DD}$ or V_{SS}, Clock Out Disabled, No Serial Access Cycles $f_{XTAL IN}$ = 32 kHz $f_{XTAL IN}$ = 1 MHz $f_{XTAL IN}$ = 2 MHz $f_{XTAL IN}$ = 4 MHz	5.0	0.024 0.12 0.24 0.5	

*Timekeeping function only, no read/write accesses. Data in the registers and RAM retained.

(Continued)

MC68HC68T1

ELECTRICAL CHARACTERISTICS (Continued)

Symbol	Parameter	Test Condition	V_{DD} V	Guaranteed Limit	Unit
I_{batt}	Maximum RMS Standby Current Crystal Operation	V_{batt} = 3.0 V, $f_{XTAL\ IN}$ = 32 kHz V_{sys} = 0.0 V, $f_{XTAL\ IN}$ = 1 MHz V_{DD} = 0.0 V, $f_{XTAL\ IN}$ = 2 MHz I_{out} = 0 μA, $f_{XTAL\ IN}$ = 4 MHz V_{in} = Don't Care, all inputs except XTAL IN, Clock Out Disabled, No Serial Access Cycles	0.0	25 250 360 600	μA

AC ELECTRICAL CHARACTERISTICS (T_A = –40 to + 85°C, C_L = 200 pF, Input t_r = t_f = 6 ns, Voltages Referenced to V_{SS})

Symbol	Parameter	V_{DD} V	Guaranteed Limit	Unit
f_{SCK}	Maximum Clock Frequency (Refer to SCK t_w, below) (Figures 1, 2, and 3)	3.0 4.5 6.0	TBD 2.1 2.1	MHz
t_{PLH}, t_{PHL}	Maximum Propagation Delay, SCK to MISO (Figures 2 and 3)	3.0 4.5 6.0	200 100 100	ns
t_{PLZ}, t_{PHZ}	Maximum Propagation Delay, SS to MISO (Figures 2 and 4)	3.0 4.5 6.0	200 100 100	ns
t_{PZL}, t_{PZH}	Maximum Propagation Delay, SCK to MISO (Figures 2 and 4)	3.0 4.5 6.0	200 100 100	ns
t_{TLH}, t_{THL}	Maximum Output Transition Time, Any Output (Figures 2 and 3) (Measured Between 70% V_{DD} and 20% V_{DD})	3.0 4.5 6.0	200 100 100	ns
C_{in}	Maximum Input Capacitance	—	10	pF

TIMING REQUIREMENTS (T_A = – 40 to + 85°C, Input t_r = t_f = 6 ns, Voltages Referenced to V_{SS})

Symbol	Parameter	V_{DD} V	Guaranteed Limit	Unit
t_{su}	Minimum Setup Time, SS to SCK (Figures 1 and 2)	3.0 4.5 6.0	200 100 100	ns
t_{su}	Minimum Setup Time, MOSI to SCK (Figures 1 and 2)	3.0 4.5 6.0	200 100 100	ns
t_h	Minimum Hold Time, SCK to SS (Figures 1 and 2)	3.0 4.5 6.0	200 125 125	ns
t_h	Minimum Hold Time, SCK to MOSI (Figures 1 and 2)	3.0 4.5 6.0	200 100 100	ns
t_{rec}	Minimum Recovery Time, SCK (Figures 1 and 2)	3.0 4.5 6.0	200 200 200	ns
$t_{w(H)}$, $t_{w(L)}$	Minimum Pulse Width, SCK (Figures 1 and 2)	3.0 4.5 6.0	400 200 200	ns

(Continued)

MC68HC68T1

TIMING REQUIREMENTS (Continued)

Symbol	Parameter	VDD V	Guaranteed Limit	Unit
t_W	Minimum Pulse Width, $\overline{POR}$	3.0 4.5 6.0	TBD 100 100	ns
t_r, t_f	Maximum Input Rise and Fall Times (Except XTAL IN and $\overline{POR}$) (Figures 1 and 2) (Measured Between 70% V_{DD} and 20% V_{DD})	3.0 4.5 6.0	TBD 2 2	μs

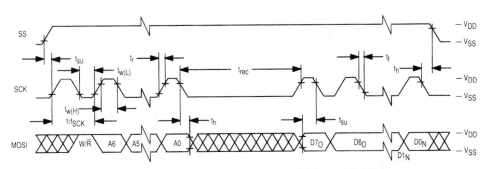

NOTE: Measurement points are V_{IL} and V_{IH} unless otherwise noted on the ac electrical characteristics table.

Figure 1. Write Cycle

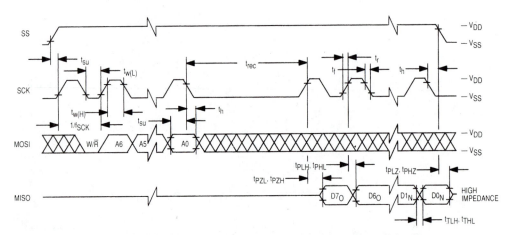

NOTE: Measurement points are V_{OL}, V_{OH}, V_{IL}, and V_{IH} unless otherwise noted on the ac electrical characteristics table.

Figure 2. Read Cycle

MC68HC68T1

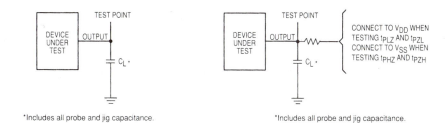

*Includes all probe and jig capacitance.

Figure 3. Test Circuit

*Includes all probe and jig capacitance.

Figure 4. Test Circuit

OPERATING CHARACTERISTICS

The real-time clock consists of a clock/calendar and a 32 x 8 RAM. (See Figure 5). Communication with the device may be established via a serial peripheral interface (SPI) or MICROWIRE bus. In addition to the clock/calendar data from seconds to years, and systems flexibility provided by the 32-byte RAM, the clock features computer handshaking with an interrupt output and a separate square-wave clock output that can be one of seven different frequencies. An alarm circuit is available that compares the alarm latches with the seconds, minutes, and hours time counters and activates the interrupt output when they are equal. The clock is specifically designed to aid in power up/power down applications and offers several pins to aid the designer of battery-backup systems.

CLOCK/CALENDAR

The clock/calendar portion of this device consists of a long string of counters that is toggled by a 1 Hz input. The 1 Hz input is derived from the on-chip oscillator that utilizes one of four possible external crystals or that can be driven by an external frequency source. The 1 Hz trigger to the counters can also be supplied by a 50 or 60 Hz source that is connected to the LINE input pin.

The time counters offer seconds, minutes, and hours data in 12- or 24-hour format. An AM/PM indicator is available that once set, toggles at 12:00 AM and 12:00 PM. The calendar counters consist of day of week, date of month, month, and year information. Data in the counters is in BCD format. The hours counter utilizes BCD for hours data plus bits for 12/24 hour and AM/PM modes. The seven time counters are read serially at addresses $20 through $26. The time counters are written to at addresses $A0 through $A6. (See Table 1 and Figure 6.)

32 x 8 GENERAL-PURPOSE RAM

The real-time clock also has a static 32 x 8 RAM. The RAM is read at addresses $00 through $1F and written to at addresses $80 through $9F. (See Figure 5.)

ALARM

The alarm is set by accessing the three alarm latches and loading the desired data. (See **SERIAL PERIPHERAL INTERFACE**.) The alarm latches consist of seconds, minutes, and hours registers. When their outputs equal the values of the seconds, minutes, and hours time counters, an interrupt is

generated. The interrupt output goes low if the alarm bit in the status register is set and the interrupt output is activated after an alarm time is sensed (see **PIN DESCRIPTIONS, INT** pin). To preclude a false interrupt when loading the time counters, the alarm interrupt bit in the interrupt control register should be reset. This procedure is not required when the alarm time is being loaded.

WATCH-DOG FUNCTION

When Watch Dog (bit 7) in the interrupt control register is set high, the clock's slave select pin must be toggled at regular intervals without a serial data transfer. If SS is not toggled, the MC68HC68T1 supplies a CPU reset pulse at pin 2 and Watch-Dog (bit 6) in the status register is set. (See Figure 7.) Typical service and reset times are shown in Table 2.

CLOCK OUT

The value in the three least significant bits of the clock control register selects one of seven possible output frequencies. (See **CLOCK CONTROL REGISTER**.) This square-wave signal is available at the CLK OUT pin. When the power-down operation is initialized, the output is reset low.

CONTROL REGISTER AND STATUS REGISTER

The operation of the real-time clock is controlled by the clock control and interrupt control registers, which are read/write registers. Another register, the status register, is available to indicate the operating conditions. The status register is a read-only register.

MODE SELECT

The voltage level that is present at the V_{SYS} input pin at the end of power-on reset selects the device to be in the single-supply mode or battery-backup mode.

Single-Supply Mode

If V_{SYS} is powered up when power-on reset is completed, CLK OUT, PSE, and $\overline{CPUR}$ are enabled high and the device is completely operational. $\overline{CPUR}$ is placed low if the voltage level at the V_{SYS} pin subsequently goes 0.7 V below V_{DD}. If CLK OUT, PSE, and $\overline{CPUR}$ are reset low due to a power-down instruction, V_{SYS} brought low and then powered high enables these outputs.

An example of the single-supply mode is where only one supply is available and V_{DD}, V_{BATT}, and V_{SYS} are tied together to the supply.

MC68HC68T1

Battery-Backup Mode

If V_{SYS} is not powered up at the end of power-on reset, CLK OUT, PSE, $\overline{CPUR}$, and SS are disabled (CLK OUT, PSE, and $\overline{CPUR}$ low). This condition is held until V_{SYS} rises to a threshold (approximately 0.7 V) above V_{BATT}. CLK OUT, PSE, and $\overline{CPUR}$ are then enabled and the device is operational. If V_{SYS} falls below a threshold above V_{BATT}, the outputs CLK OUT, PSE, and $\overline{CPUR}$ are reset low.

An example of battery-backup operation occurs if V_{SYS} is tied to V_{DD} and V_{DD} is not receiving voltage from a supply. A rechargeable battery is connected to the V_{BATT} pin. The device retains data and keeps time down to a minimum V_{BATT} voltage of 2.2 V.

POWER CONTROL

Power control is composed of two operations, power sense and power down/power up. Two pins are involved in power sensing, the LINE input pin and the $\overline{INT}$ output pin. Two additional pins, PSE and V_{SYS}, are utilized during power down/up operation.

FREEZE FUNCTION

The freeze function prevents an increment of the time counters, if any of the time registers are being read. Also, alarm operation is delayed if the time registers are being read.

POWER SENSING

When power sensing is enabled (Power Sense Bit in the interrupt control register), ac/dc transitions are sensed at the LINE input pin. Threshold detectors determine when transitions cease. After a delay of 2.68 to 4.64 ms plus the external input RC circuit time constant, an interrupt true bit is set high in the status register. This bit can then be sampled to see if system power has turned back on. (See Figure 8.)

The power-sense circuitry operates by sensing the level of the voltage present at the LINE input pin. This voltage is centered around V_{DD}, and as long as the voltage is either plus or minus a threshold (approximately 0.7 V) from V_{DD}, a power

sense failure is not indicated. With an ac signal present, remaining in this V_{DD} window longer than a maximum of 4.64 ms activates the power-sense circuit. The larger the amplitude of the signal, the less likely a power failure would be detected. A 50 or 60 Hz, 10 V p-p sine-wave voltage is an acceptable signal to present at the LINE input pin to set up the power-sense function.

Power Down

Power down is a processor-directed operation. The power down bit is set in the interrupt control register to initiate power down operation. During power down, the power supply enable (PSE) output, normally high, is placed low. The CLK OUT pin is placed low. The $\overline{CPUR}$ output, connected to the processor reset input pin, is also placed low. In addition, the serial interface (MOSI and MISO) is disabled (see Figure 9).

Power Up

There are four methods that can initiate the power-up mode. Two of the methods require an interrupt to the microcomputer by programming the interrupt control register. The interrupts can be generated by the alarm circuit by setting the alarm bit and the appropriate alarm registers. Also, an interrupt can be generated by programming the periodic interrupt bits in the interrupt control register.

The third method is by initiating the power sense circuit with the power sense bit in the interrupt control register to sense power loss along with the V_{SYS} pin to sense subsequent power-up condition. (See Figure 10.) (Reference Figure 19 for application circuit for third method.)

The fourth method that initiates power-up occurs when the level on the V_{SYS} pin rises 0.7 V above the level of the V_{BATT} pin, after previously falling to the level of V_{BATT} while in the battery-backup mode. An interrupt is not generated when the fourth method is utilized.

While in the single-supply mode, power-up is initiated when the V_{SYS} pin loses power and then returns high. There is no interrupt generated when using this method (see Figure 11).

MC68HC68T1

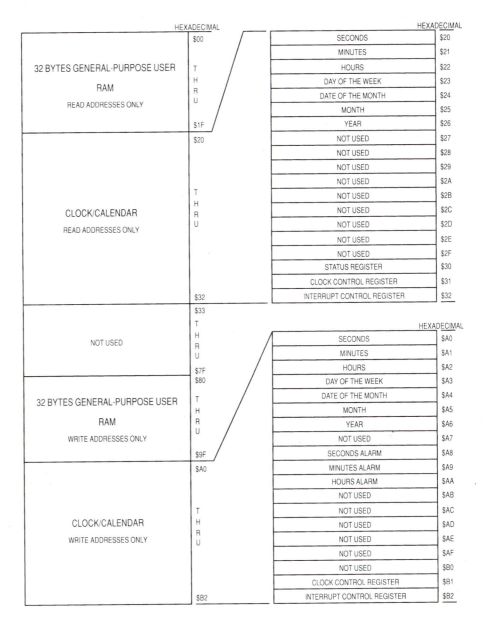

Figure 5. Address Map

MC68HC68T1

Table 1. Clock/Calendar and Alarm Data Modes

Read	Write	Function	Decimal Range	BCD Data Range	BCD Date* Example
Address Location					
$20	$A0	Seconds	0–59	00–59	21
$21	$A1	Minutes	0–59	00–59	40
$22	$A2	Hours** (12 Hour Mode)	1–12	81–92 (AM) A1–B2 (PM)	90
		Hours (24 Hour Mode)	0–23	00–23	10
$23	$A3	Day of Week (Sunday = 1)	1–7	01–07	03
$24	$A4	Date of Month	1–31	01–31	16
$25	$A5	Month (Jan = 1)	1–12	01–12	06
$26	$A6	Year	0–99	00–99	87
N/A	$A8	Seonds Alarm	0–59	00–59	21
N/A	$A9	Minutes Alarm	0–59	00–59	40
N/A	$AA	Hours Alarm*** (12 Hour Mode)	1–12	01–12 (AM) 21–32 (PM)	10
		Hours Alarm (24 Hour Mode)	0–23	00–23	10

N/A = Not Applicable

*Example: 10:40:21 A.M., Tuesday, June 16, 1987.

**Most-Significant data bit, D7, is "0" for 24 hour mode and "1" for 12 hour mode. Data bit D5 is "1" for P.M. and "0" for A.M. in 12 hour mode.

***Data bit D5 is "1" for P.M. and "0" for A.M. in 12 hour mode. Data bits D7 and D6 are Don't Cares.

Table 2. Watchdog Service and Reset Times

	50 Hz		60 Hz		XTAL	
	Min	Max	Min	Max	Min	Max
Service Time	—	10 ms	—	8.3 ms	—	7.8 ms
Reset Time	20 ms	40 ms	16.7 ms	33.3 ms	15.6 ms	31.3 ms

MC68HC68T1

WRITE/READ REGISTERS

HEX ADDRESS READ	HEX ADDRESS WRITE	WRITE/READ REGISTERS (DB7 ... DB0)	FUNCTION
$20	$A0	TENS 0–5 / UNITS 0–9	SECONDS (00–59)
$21	$A1	TENS 0–5 / UNITS 0–9	MINUTES (00–59)
$22	$A2	12 HR 24 / X / PM/AM TENS 0–2 / UNITS 0–9	DB7, 1 = 12 HR, 0 = 24 HR DB5, 1 = PM, 0 = AM HOURS (01–12 OR 00–23)
$23	$A3	X X X X / X UNITS 1–7	DAY OF WEEK (01–07) SUNDAY = 1
$24	$A4	TENS 0–3 / UNITS 0–9	DATE OF MONTH (01–31)
$25	$A5	TENS 0–1 / UNITS 0–9	MONTH (01–12) JAN = 1
$26	$A6	TENS 0–9 / UNITS 0–9	YEAR (00–99)
$31	$B1	7 6 5 4 / 3 2 1 0	CLOCK CONTROL REGISTER
$32	$B2	7 6 5 4 / 3 2 1 0	INTERRUPT CONTROL REGISTER

WRITE-ONLY REGISTERS

READ	WRITE	WRITE-ONLY REGISTERS	FUNCTION
N/A	$A8	TENS 0–5 / UNITS 0–9	SECONDS ALARM (00–59)
N/A	$A9	TENS 0–5 / UNITS 0–9	MINUTES ALARM (00–59)
N/A	$AA	X X / PM/AM TENS 0–2 / UNITS 0–9	HOURS ALARM (01–21 OR 00–23) DB5, 1 = PM, 0 = AM IN 12 HR MODE

READ-ONLY REGISTER

READ	WRITE	READ-ONLY REGISTER	FUNCTION
$B0	N/A	7 6 5 4 3 2 1 0	STATUS REGISTER

RAM DATA BYTE

READ	WRITE	RAM DATA BYTE	FUNCTION
$00 TO $1F	$80 TO $9F	D7 D6 D5 D4 D3 D2 D1 D0	DATA

NOTE: X = Don't Care for write
X = 0 for read
N/A = Not Applicable

Figure 6. Clock/RAM Registers

MC68HC68T1

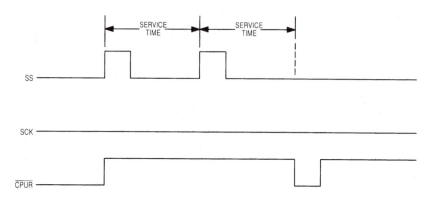

Figure 7. Watch-Dog Operation Waveforms

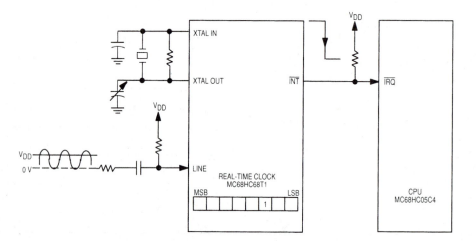

NOTE: A 60 Hz 10 V p-p sine-wave voltage is an acceptable signal to present at the LINE input pin.

Figure 8. Power Sensing Functional Diagram

MC68HC68T1

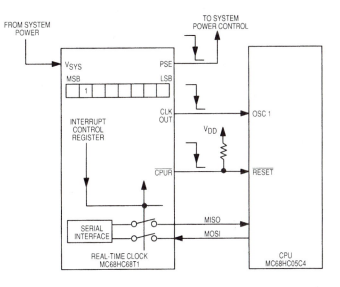

Figure 9. Power Down Functional Diagram

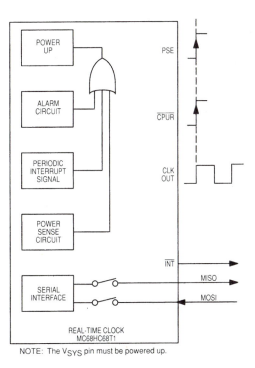

NOTE: The V_{SYS} pin must be powered up.

Figure 10. Power Up Functional Diagram
(Initiated by Interrupt Signal)

Reprinted with permission of Motorola

MC68HC68T1

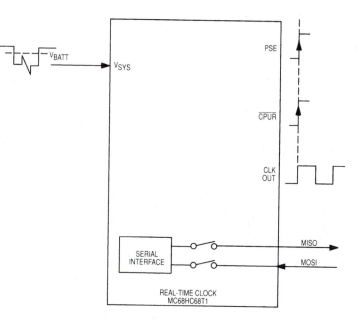

Figure 11. Power Up Functional Diagram
(Initiated by a Rise in Voltage on the V$_{SYS}$ Pin)

PIN DESCRIPTIONS

CLK OUT (PIN 1) — CLOCK OUTPUT

This signal is the buffered clock output which can provide one of the seven selectable frequencies (or this output can be reset low). The contents of the three least-significant bit positions in the clock control register determine the output frequency (50% duty cycle, except 2 Hz in the 50 Hz time-base mode). During power-down operation (Power Down bit in the interrupt control register set high), the clock output is reset low.

$\overline{\text{CPUR}}$ (PIN 2) — CPU RESET

This pin provides an N-channel, open-drain output and requires an external pullup resistor. This active low output can be used to drive the reset pin of a microprocessor to permit orderly power up/power down. The $\overline{\text{CPUR}}$ output is low from 15 to 40 ms when the watch-dog function detects a CPU failure (see Table 2). The low level time is determined by the input frequency source selected as the time standard. $\overline{\text{CPUR}}$ is reset low when power down is initiated.

This output has no ESD protection diode tied to V$_{DD}$ which allows this pin's voltage to rise above V$_{DD}$. Care must be taken in handling this device.

$\overline{\text{INT}}$ (PIN 3) — INTERRUPT

This active-low output is driven from a single N-channel transistor and must be tied to an external pullup resistor.

Interrupt is activated to a low level when any one of the following takes place:

1. Power sense operation is selected (Power Sense Bit in the interrupt control register is set high) and a power failure occurs.
2. A previously set alarm time occurs. The alarm bit in the status register and the interrupt signal are delayed 30.5 ms when 32 kHz or 1 MHz operation is selected, 15.3 ms for 2 MHz operation, and 7.6 ms for 4 MHz operation.
3. A previously selected periodic interrupt signal activates.

The status register must be read to disable the interrupt output after the selected periodic interval occurs. This is also true when conditions 1 and 2 activate the interrupt. If power down has been previously selected, the interrupt also sets the power-up function only if power is supplied to the V$_{SYS}$ pin to the proper threshold level above V$_{BATT}$.

This output has no ESD protection diode tied to V$_{DD}$ which allows this pin's voltage to rise above V$_{DD}$. Care must be taken in the handling of this device.

SCK (PIN 4) — SERIAL CLOCK

This serial clock input is used to shift data into and out of the on-chip interface logic. SCK retains its previous state if the line driving it goes into a high-impedance state. In other words, if the source driving SCK goes to the high-impedance state, the previous low or high level is retained by on-chip control circuity.

Reprinted with permission of Motorola

MC68HC68T1

MOSI (PIN 5) — MASTER OUT SLAVE IN

The serial data present at this port is latched into the interface logic by SCK if the logic is enabled. Data is shifted in, either on the rising or falling edges of SCK, with the most-significant bit (MSB) first.

In Motorola's microcomputers with SPI, the state of the CPOL bit determines which is the active edge of SCK. If SCK is high when SS goes high, the state of the CPOL bit is high. Likewise, if a rising edge of SS occurs while SCK is low (see Figure 13), then the CPOL bit in the microcomputer is low.

MOSI retains its previous state if the line driving it goes into high-impedance state. In other words, if the source driving MOSI goes to the high-impedance state, the previous low or high level is retained by on-chip control circuitry.

MISO (PIN 6) — MASTER IN SLAVE OUT

The serial data present at this port is shifted out of the interface logic by SCK if the logic is enabled. Data is shifted out, either on the rising or falling edge of SCK, with the most-significant bit (MSB) first. The state of the CPOL bit in the microcomputer determines which is the active edge of SCK. (See Figure 13.) ·

SS (PIN 7) — SLAVE SELECT

When high, the slave select input activates the interface logic, otherwise the logic is in a reset state and the MISO pin is in the high-impedance state. The watch-dog circuit is toggled at this pin. SS has an internal pulldown device. Therefore, if SS is in a low state before going to high impedance, SS can be left in a high-impedance state. That is, if the source driving SS goes to the high-impedance state, the previous low level is retained by on-chip control circuitry.

V_{SS} (PIN 8) — NEGATIVE POWER SUPPLY

This negative power supply reference pin is connected to ground.

PSE (PIN 9) — POWER SUPPLY ENABLE

The power supply enable output is used to control system power and is enabled high under any one of the following conditions:

1. V_{SYS} rises above the V_{batt} voltage after V_{SYS} is reset low by a system failure.
2. An interrupt occurs (if the V_{SYS} pin is powered up 0.7 V above V_{BATT}).
3. A power-on reset occurs (if the V_{SYS} pin is powered up 0.7 V above V_{BATT}).

PSE is reset low by writing a high into the power-down bit of the interrupt control register.

POR (PIN 10) — POWER-ON RESET

This active-low Schmitt-trigger input generates an internal power-on reset signal using an external RC network. (See Figures 18 through 21). Both control registers and frequency dividers for the oscillator and line inputs are reset. The status register is reset except for the first time-up bit (bit 4), which is set high. At the end of the power-on reset, single-supply or battery-backup mode is selected.

LINE (PIN 11) — LINE SENSE

The LINE sense input can be used to drive one of two functions. The first function utilizes the input signal as the frequency source for the timekeeping counters. This function is selected by setting the line/$\overline{XTAL}$ bit high in the clock control register. The second function enables the LINE input to detect a power failure. Threshold detectors operating above and below V_{DD} sense an ac voltage loss. The Power Sense bit in the interrupt control register must be set high and crystal or external clock source operation is required. The line/$\overline{XTAL}$ bit in the clock control register must be low to select crystal operation.

V_{SYS} (PIN 12) — SYSTEM VOLTAGE

This input is connected to system voltage. The level on this pin initiates power up if it rises 0.7 V above the level at the V_{BATT} input pin after previously falling below 0.7 V below V_{BATT}. When power-up is initiated, the PSE pin returns high and the CLK OUT pin is enabled. The $\overline{CPUR}$ output pin is also set high. Conversely, if the level of the V_{SYS} pin falls below V_{BATT} + 0.7 V, the PSE, CLK OUT, and $\overline{CPUR}$ pins are placed low. The voltage level present at this pin at the end of $\overline{POR}$ determines the device's operating mode.

V_{BATT} (PIN 13) — BATTERY VOLTAGE

This pin is the *only* oscillator power source and should be connected to the positive terminal of a rechargeable battery. **The V_{BATT} pin always supplies power to the MC68HC68T1, even when the device is not in the battery back-up mode.** To maintain timekeeping, the V_{BATT} pin must be at least 2.2 V. When the level on the V_{SYS} pin falls below V_{BATT} + 0.7 V, **V_{BATT} is internally connected to the V_{DD} pin.**

When the LINE input is used as the frequency source, the unused V_{BATT} and XTAL pins may be tied to V_{SS}. Alternatively, if V_{BATT} is connected to V_{DD}, XTAL IN can be tied to either V_{SS} or V_{DD}.

XTAL IN (PIN 14), XTAL OUT (PIN 15) — CRYSTAL INPUT/OUTPUT

For crystal operation, these two pins are connected to a 32.768 kHz, 1.048576 MHz, 2.097152 MHz, or 4.194304 MHz crystal. If crystal operation is not desired and Line Sense is used as frequency source, connect XTAL IN to V_{DD} or V_{SS} (caution: see V_{BATT} pin description) and leave XTAL OUT open. If an external clock is used, connect the external clock to XTAL IN and leave XTAL OUT open. The external clock must swing from at least 30% to 70% of (V_{DD}-V_{SS}). Preferably, this input should swing from V_{SS} to V_{DD}.

V_{DD} (PIN 16) — POSITIVE POWER SUPPLY

For full functionality, the positive power supply pin may range from 3.0 to 6.0 V with respect to V_{SS}. To maintain timekeeping, the minimum standby voltage is 2.2 V with respect to V_{SS}. **(Caution: Data transfer to/from the MC68HC68T1 must not be attempted if the supply voltage falls below 3.0 V.)**

MC68HC68T1

REGISTERS

CLOCK CONTROL REGISTER (READ/WRITE) — READ ADDRESS $31/WRITE ADDRESS $B1

MSB D7	D6	D5	D4	D3	D2	D1	LSB D0
START	LINE	XTAL SELECT 1	XTAL SELECT 0	50 Hz	CLK OUT 2	CLK OUT 1	CLK OUT 0
STOP	XTAL			60 Hz			

Start-Stop

A high written into this bit enables the counter stages of clock circuitry. A low holds all bits reset in the divider chain from 32 Hz to 1 Hz. The clock out signal selected by bits D0, D1, and D2 is not affected by the stop function except the 1- and 2-Hz outputs.

Line-XTAL

When this bit is high, clock operation uses the 50 or 60 cycle input present at the LINE input pin. When the bit is low, the XTAL IN pin is the source of the time update.

XTAL Select

Accommodation of one of four possible crystals are selected by the value in bits D4 and D5.

0 = 4.194304 MHz	2 = 1.048576 MHz
1 = 2.097152 MHz	3 = 32.768 kHz

The MC68HC68T1 has an on-chip 150 K resistor that is switched in series with the internal inverter when 32 kHz is selected via the clock control register. This eliminates the usual external series requirement present in 32 kHz circuits. At power up, the device sets up for a 4 MHz oscillator and the series resistor is not part of the oscillator circuit. Until this resistor is switched in, oscillations may be unstable with the 32 kHz crystal. (See Figure 12.)

Resistor R1 is recommended to be 22 MΩ for 32 kHz operation. Consult crystal manufacturer for R1 value for other frequencies. Resistor R2 is used for 32 kHz operation only. Use a 100 kΩ to 300 kΩ range as specified by the crystal manufacturer.

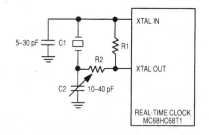

**Figure 12. Recommended Oscillator Circuit
(C1, C2 Values Depend Upon the Crystal Frequency)**

50 Hz-60 Hz

50 Hz may be used as the input frequency at the LINE input when this bit is set high; a low accommodates 60 Hz. The power sense bit in the interrupt control register must be reset low for line frequency operation.

Clock Out

Three bits specify one of the seven frequencies to be used as the square-wave clock output (CLK OUT).

0 = XTAL	4 = Disable (low output)
1 = XTAL/2	5 = 1 Hz
2 = XTAL/4	6 = 2 Hz
3 = XTAL/8	7 = 50/60 Hz for LINE operation
	7 = 64 Hz for XTAL operation

All bits in the clock control register are reset by a power-on reset. Therefore, XTAL is selected as the clock output at this time.

INTERRUPT CONTROL REGISTER (READ/WRITE) — READ ADDRESS $32/WRITE ADDRESS $B2

MSB D7	D6	D5	D4	D3	D2	D1	LSB D0
WATCH-DOG	POWER DOWN	POWER SENSE	ALARM	PERIODIC SELECT			

All bits are reset low by power-on reset.

Watch Dog

When this bit is set high, the watch-dog operation is enabled. This function requires the CPU to toggle the SS pin periodically without a serial transfer requirement. In event this does not occur, a CPU reset is issued at the CPUR pin. The status register must be read before re-enabling the watch-dog function.

Power Down

A high in this location initiates a power down. A CPU reset occurs via the CPUR output, the CLK OUT and PSE output pins are reset low, and the serial interface is disabled.

Power Sense

When set high, this bit is used to enable the LINE input pin to sense a power failure. When power sense is selected, the input to the 50/60 Hz prescaler is disconnected; therefore, crystal operation is required. An interrupt is generated when a power failure is sensed and the power sense and interrupt true bit in the status register are set. When power sense is activated, a logic low must be written to this location followed by a high to re-enable power sense.

Alarm

The output of the alarm comparator is enabled when this bit is set high. When an equal comparison occurs between the seconds, minutes, and hours time counters and alarm latches, the interrupt output is activated. When loading the time counters, this bit should be reset low to avoid a false interrupt. This is not required when loading the alarm latches. See INT pin description for explanation of alarm delay.

Periodic Select

The value in these four bits (D0, D1, D2, and D3) selects the frequency of the periodic output. (See Table 3.)

MC68HC68T1

Table 3. Periodic Interrupt Output Frequencies (at INT Pin)

D3–D0 Value (Hex)	Periodic Interrupt Output Frequency	Frequency Timebase	
		XTAL	Line
0	Disable		
1	2048 Hz	X	
2	1024 Hz	X	
3	512 Hz	X	
4	256 Hz	X	
5	128 Hz	X	
6	64 Hz	X	
	50 or 60 Hz		X
7	32 Hz	X	
8	16 Hz	X	
9	8 Hz	X	
A	4 Hz	X	
B	2 Hz	X	X
C	1 Hz	X	X
D	1 Cycle per Minute	X	X
E	1 Cycle per Hour	X	X
F	1 Cycle per Day	X	X

STATUS REGISTER (READ ONLY) — ADDRESS $30

MSB D7	D6	D5	D4	D3	D2	D1	LSB D0
0	WATCH-DOG	0	FIRST TIME UP	INTER-RUPT TRUE	POWER SENSE INT	ALARM INT	CLOCK INT

Watch Dog

If this bit is set high, the watch-dog circuit has detected a CPU failure.

First Time Up

Power-on reset sets this bit high. This signifies the data in the RAM and Clock is not valid and should be initialized. After the status register is read, the first time-up bit is set low if the POR pin is high. Conversely, if the POR pin is held low, the first time-up bit remains set high.

Interrupt True

A high in this bit signifies that one of the three interrupts (power sense, alarm, or clock) is valid.

Power-Sense Interrupt

This bit set high signifies that the power sense circuit has generated an interrupt. This bit is not reset after a read of this register.

Alarm Interrupt

When the contents of the seconds, minutes, and hours time counters and alarm latches are equal, this bit is set high. The status register must be read before loading the interrupt control register for valid alarm indication after the alarm activates.

Clock Interrupt

A periodic interrupt sets this bit high. (See Table 3.)

All bits are reset low by a power-on reset except the first time-up bit which is set high. All bits except the power sense bit are reset after a read of the status register.

SERIAL PERIPHERAL INTERFACE (SPI)

The serial peripheral interface (SPI) utilized by the MC68HC68T1 is a serial synchronous bus for address and data transfers. The shift clock (SCK), which is generated by the microcomputer, is active only during address and data transfer. In systems using the MC68HC05C4 or MC68HC11A8, the inactive clock polarity is determined by the clock polarity (CPOL) bit in the microcomputer's control register.

A unique feature of the MC68HC68T1 is that the level of the inactive clock is determined by sampling SCK when SS becomes active. Therefore, either SCK polarity is accommodated. Input data (MOSI) is latched internally on the internal strobe edge and output data (MISO) is shifted out on the shift edge (see Table 4 and Figure 13). There is one clock for each bit transferred. Address as well as data bits are transferred in groups of eight.

Table 4. Function Table

Mode	Signal			
	SS	SCK	MOSI	MISO
Disabled Reset	L	Input Disabled	Input Disabled	High Z
Write	H	CPOL = 1 ⊿ / CPOL = 0 ⊽	Data Bit Latch	High Z
Read	H	CPOL = 1 ⌐ / CPOL = 0 _⌐	X	Next Data Bit Shifted Out*

*MISO remains at a High Z until eight bits of data are ready to be shifted out during a read. MISO remains at a High Z during the entire write cycle.

MC68HC68T1

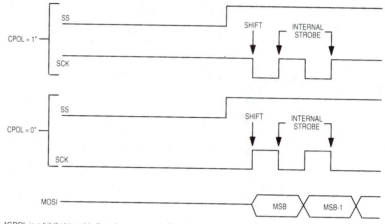

*CPOL is a bit that is set in the microcomputer's Control Register.

Figure 13. Serial Clock (SCK) as a Function of MCU Clock Polarity (CPOL)

ADDRESS AND DATA FORMAT

There are three types of serial transfers:
1. Read or write address
2. Read or write data
3. Watch-dog reset (actually a non-transfer)

The address and data bytes are shifted MSB first, into the serial data input (MOSI) and out of the serial data output (MISO). Any transfer of data requires the address of the byte to specify a write or read Clock or RAM location, followed by one or more bytes of data. Data is transferred out of MISO for a read operation and into MOSI for a write operation. (See Figures 14 and 15.)

Address Byte

The address byte is always the first byte entered after SS goes true. To transmit a new address, SS must first be brought low and then taken high again.

MSB							LSB
A7	A6	A5	A4	A3	A2	A1	A0

A7 — High initiates one or more write cycles. Low initiates one or more read cycles.

A6 — Must be low (zero) for normal operation

A5 — High signifies a clock/calendar location. Low signifies a RAM location

A0-A4 — Remaining address bits. (See Figure 5.)

Address and Data

Data transfers can occur one byte at a time or in multi-byte burst mode. (See Figures 16 and 17.) After the MC68HC68T1 is enabled (SS = high), an address byte selects either a read or a write of the Clock/Calendar or RAM. For a single-byte read or write, one byte is transferred to or from the Clock/Calendar register or RAM location specified by an address. Additional reading or writing requires re-enabling the device and providing a new address byte. If the MC68HC68T1 is not disabled, additional bytes can be read or written in a burst mode. Each read or write cycle causes the Clock/Calendar register or RAM address to automatically increment. Incrementing continues after each byte transfer until the device is disabled. After incrementing to $1F or $9F, the address wraps to $00 and continues if the RAM is selected. When the Clock/Calendar is selected, the address wraps to $20 after incrementing to $32 or $B2.

MC68HC68T1

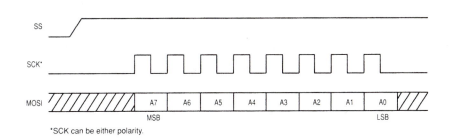

*SCK can be either polarity.

Figure 14. Address Byte Transfer Waveforms

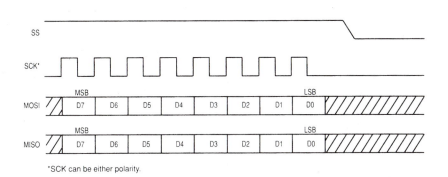

*SCK can be either polarity.

Figure 15. Read/Write Data Transfer Waveforms

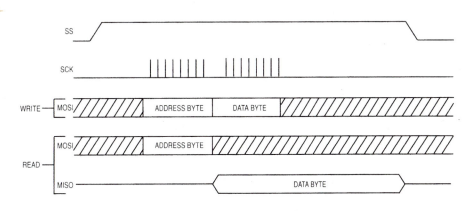

Figure 16. Single-Byte Transfer Waveforms

MC68HC68T1

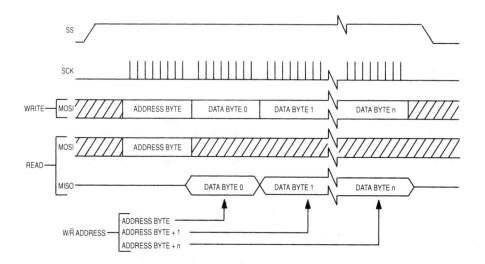

Figure 17. Multiple-Byte Transfer Waveforms

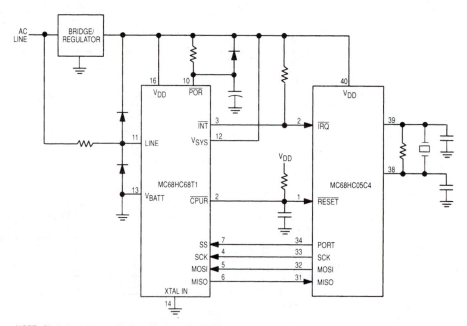

NOTE: Clock circuit driven by line input frequency. Power-on reset circuit included to detect power failure.

Figure 18. Power-Always-On System

MC68HC68T1

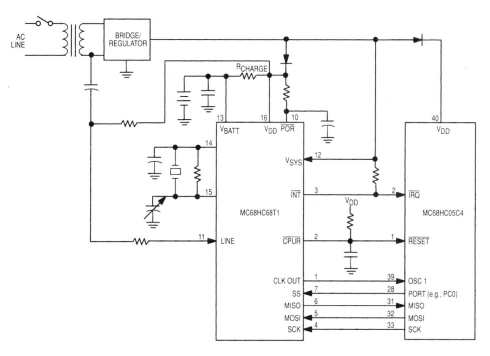

NOTE: The LINE input pin can sense when the switch opens by use of the power sense interrupt. The MC68HC68T1 crystal drives the clock input to the CPU using the CLK OUT pin. On power down when $V_{SYS} < V_{BATT} + 0.7$ V, V_{BATT} powers the clock. A threshold detect activates an on-chip p-channel switch, connecting V_{BATT} to V_{DD}. V_{BATT} always supplies power to the oscillator, keeping voltage frequency variation to a minimum.

POWER-DOWN PROCEDURE

A procedure for power-down operation consists of the following:

1. Set power sense operation by writing bit 5 high in the interrupt control register.
2. When an interrupt occurs, the CPU reads the status register to determine the interrupt source.
3. Sensing a power failure, the CPU does the necessary housekeeping to prepare for shutdown.
4. The CPU reads the status register again after several milliseconds to determine validity of power failure.
5. The CPU sets power down (bit 6) and disables all interrupts in the interrupt control register when power down is verified. This causes the CPU reset and Clock Out pins to be held low and disconnects the serial interface.
6. When power returns and V_{SYS} rises above $V_{BATT} + 0.7$ V, power up is initiated. The CPU reset is released and serial communication is established.

MC68HC68T1

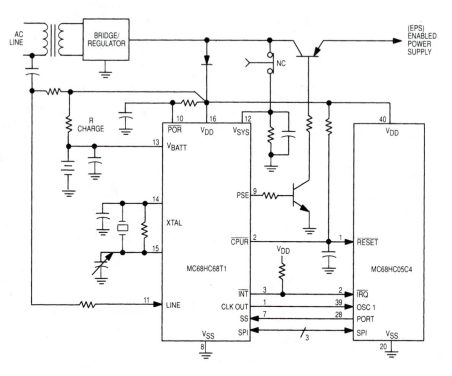

Figure 20. Battery Back-Up System

MC68HC68T1

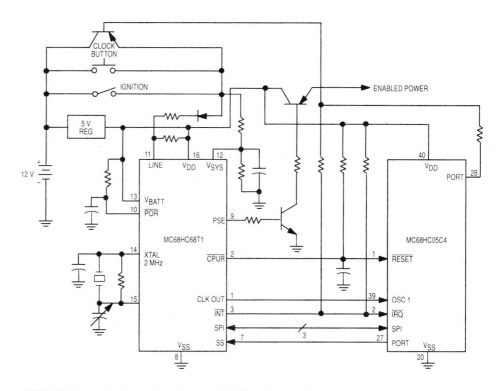

NOTE: The V_{SYS} and Line inputs can be used to sense the ignition turning on and off. An external switch is included to activate the sytem without turning on the ignition. Also, the CMOS CPU is not powered down with the system V_{DD}, but is held in a low power reset mode during power down. When restoring power, the MC68HC68T1 enables the CLK OUT pin and sets the PSE and $\overline{CPUR}$ pins high.

Figure 21. Automotive System

B.4

The MC145050

10-bit A/D

Converter with

Serial Interface

MOTOROLA
■ SEMICONDUCTOR
TECHNICAL DATA

Advance Information

10-Bit A/D Converters
with Serial Interface
CMOS

These ratiometric 10-bit ADCs have serial interface ports to provide communication with MCUs and MPUs. *Either a 10- or 16-bit format can be used.* The 16-bit format can be one continuous 16-bit stream or two intermittent 8-bit streams. The converters operate from a single power supply with no external trimming required. Reference voltages down to 2.5 V are accommodated.

The MC145050 has the same pin out as the 8-bit MC145040 which allows an external clock (ADCLK) to operate the dynamic A/D conversion sequence. The MC145051 has the same pin out as the 8-bit MC145041 which has an internal clock oscillator and an end-of-conversion (EOC) output.

- 11 Analog Input Channels with Internal Sample-and-Hold
- Operating Temperature Range: −40° to 125°C
- Successive Approximation Conversion Time:
    MC145050—21 μs (with 2.1 MHz ADCLK)
    MC145051—88 μs Maximum
- Maximum Sample Rate:
    MC145050—38 ks/s
    MC145051—10.7 ks/s
- Analog Input Range with 5-Volt Supply: 0 to 5 V
- Monotonic with No Missing Codes
- Direct Interface to Motorola SPI and National MICROWIRE Serial Data Ports
- Digital Inputs/Outputs are TTL, NMOS, and CMOS Compatible
- Low Power Consumption: 14 mW
- Chip Complexity: 1630 Elements (FETs, Capacitors, etc.)

MC145050
MC145051

P SUFFIX
PLASTIC
CASE 738

DW SUFFIX
SOG
CASE 751D

ORDERING INFORMATION

MC14505xP	Plastic DIP
MC14505xDW	SOG Package

PIN ASSIGNMENT

AN0	1 ●	20	V_{DD}
AN1	2	19	★
AN2	3	18	SCLK
AN3	4	17	D_{in}
AN4	5	16	D_{out}
AN5	6	15	$\overline{CS}$
AN6	7	14	V_{ref}
AN7	8	13	V_{AG}
AN8	9	12	AN10
V_{SS}	10	11	AN9

★ADCLK (MC145050); EOC (MC145051)

BLOCK DIAGRAM

PIN 20 = V_{DD}
PIN 10 = V_{SS}

MICROWIRE is a trademark of National Semiconductor Corp.
This document contains information on a new product. Specifications and information herein are subject to change without notice.

MC145050•MC145051

MAXIMUM RATINGS*

Symbol	Parameter	Value	Unit
V_{DD}	DC Supply Voltage (Referenced to V_{SS})	-0.5 to $+6.0$	V
V_{ref}	DC Reference Voltage	V_{AG} to $V_{DD} + 0.1$	V
V_{AG}	Analog Ground	$V_{SS} - 0.1$ to V_{ref}	V
V_{in}	DC Input Voltage, Any Analog or Digital Input	$V_{SS} - 0.5$ to $V_{DD} + 0.5$	V
V_{out}	DC Output Voltage	$V_{SS} - 0.5$ to $V_{DD} + 0.5$	V
I_{in}	DC Input Current, per Pin	± 20	mA
I_{out}	DC Output Current, per Pin	± 25	mA
I_{DD}, I_{SS}	DC Supply Current, V_{DD} and V_{SS} Pins	± 50	mA
T_{stg}	Storage Temperature	-65 to 150	°C
T_L	Lead Temperature, 1 mm from Case for 10 Seconds	260	°C

*Maximum Ratings are those values beyond which damage to the device may occur. Functional operation should be restricted to the Operation Ranges below.

This device contains protection circuitry to guard against damage due to high static voltages or electric fields. However, precautions must be taken to avoid applications of any voltage higher than maximum rated voltages to this high-impedance circuit. For proper operation, V_{in} and V_{out} should be constrained to the range $V_{SS} \le (V_{in}$ or $V_{out}) \le V_{DD}$.

Unused inputs must always be tied to an appropriate logic voltage level (e.g., either V_{SS} or V_{DD}). Unused outputs must be left open.

OPERATION RANGES (Applicable to Guaranteed Limits)

Symbol	Parameter	Value	Unit
V_{DD}	DC Supply Voltage, Referenced to V_{SS}	4.5 to 5.5	V
V_{ref}	DC Reference Voltage (Note 1)	$V_{AG} + 2.5$ to $V_{DD} + 0.1$	V
V_{AG}	Analog Ground (Note 1)	$V_{SS} - 0.1$ to $V_{ref} - 2.5$	V
V_{AI}	Analog Input Voltage (Note 2)	V_{AG} to V_{ref}	V
V_{in}, V_{out}	Digital Input Voltage, Output Voltage	V_{SS} to V_{DD}	V
T_A	Ambient Operating Temperature	-40 to 125	°C

NOTES:
1. Reference voltages down to 1.0 V ($V_{ref} - V_{AG} = 1.0$ V) are functional, but the A/D converter electrical characteristics are not guaranteed.
2. Analog input voltages greater than V_{ref} convert to full scale. Input voltages less than V_{AG} convert to zero. See V_{ref} and V_{AG} pin descriptions.

DC ELECTRICAL CHARACTERISTICS

(Voltages Referenced to V_{SS}, Full Temperature and Voltage Ranges per Operation Ranges table, unless otherwise indicated)

Symbol	Parameter	Test Conditions	Guaranteed Limit	Unit
V_{IH}	Minimum High-Level Input Voltage (D_{in}, SCLK, $\overline{CS}$, ADCLK)		2.0	V
V_{IL}	Maximum Low-Level Input Voltage (D_{in}, SCLK, $\overline{CS}$, ADCLK)		0.8	V
V_{OH}	Minimum High-Level Output Voltage (D_{out}, EOC)	$I_{out} = -1.6$ mA $I_{out} = -20$ μA	2.4 $V_{DD} - 0.1$	V
V_{OL}	Maximum Low-Level Output Voltage (D_{out}, EOC)	$I_{out} = +1.6$ mA $I_{out} = 20$ μA	0.4 0.1	V
I_{in}	Maximum Input Leakage Current (D_{in}, SCLK, $\overline{CS}$, ADCLK)	$V_{in} = V_{SS}$ or V_{DD}	± 2.5	μA
I_{OZ}	Maximum Three-State Leakage Current (D_{out})	$V_{out} = V_{SS}$ or V_{DD}	± 10	μA
I_{DD}	Maximum Power Supply Current	$V_{in} = V_{SS}$ or V_{DD}, All Outputs Open	2.5	mA
I_{ref}	Maximum Static Analog Reference Current (V_{ref})	$V_{ref} = V_{DD}$, $V_{AG} = V_{SS}$	100	μA
I_{AI}	Maximum Analog Mux Input Leakage Current between all deselected inputs and any selected input (AN0–AN10)	$V_{AI} = V_{SS}$ to V_{DD}	± 1	μA

MC145050●MC145051

A/D CONVERTER ELECTRICAL CHARACTERISTICS

(Full Temperature and Voltage Ranges per Operation Ranges table; MC145050: 500 kHz ≤ ADCLK ≤ 2.1 MHz unless otherwise noted.)

Characteristic	Definition and Test Conditions		Guaranteed Limit	Unit
Resolution	Number of bits resolved by the A/D converter		10	Bits
Maximum Nonlinearity	Maximum difference between an ideal and an actual ADC transfer function		± 1	LSB
Maximum Zero Error	Difference between the maximum input voltage of an ideal and an actual ADC for zero output code		± 1	LSB
Maximum Full-Scale Error	Difference between the minimum input voltage of an ideal and an actual ADC for full-scale output code		± 1	LSB
Maximum Total Unadjusted Error	Maximum sum of nonlinearity, zero error, and full-scale error		± 1	LSB
Maximum Quantization Error	Uncertainty due to converter resolution		± 1/2	LSB
Absolute Accuracy	Difference between the actual input voltage and the full-scale weighted equivalent of the binary output code, all error sources included		± 1-1/2	LSB
Maximum Conversion Time	Total time to perform a single analog-to-digital conversion	MC145050	44	ADCLK cycles
		MC145051	88	μs
Data Transfer Time	Total time to transfer digital serial data into and out of the device		10 to 16	SCLK cycles
Sample Acquisition Time	Analog input acquisition time window		6	SCLK cycles

Minimum Total Cycle Time	Total time to transfer serial data, sample the analog input, and perform the conversion
	  MC145050: ADCLK = 2.1 MHz, SCLK = 2.1 MHz
	  MC145051: SCLK = 2.1 MHz
			
			26
			93
				μs

Maximum Sample Rate	Rate at which analog inputs may be sampled
	  MC145050: ADCLK = 2.1 MHz, SCLK = 2.1 MHz
	  MC145051: SCLK = 2.1 MHz
			
			38
			10.7
				ks/s

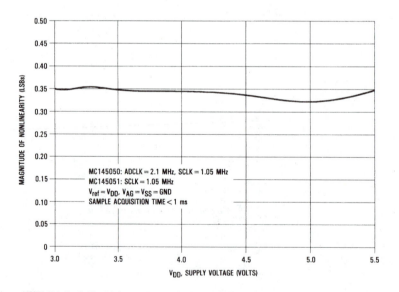

NOTE: This "typical" graph is not to be used for design purposes but is intended as an indication of the IC's potential performance.

Graph 1. Typical Nonlinearity vs Supply Voltage

MC145050•MC145051

AC ELECTRICAL CHARACTERISTICS
(Full Temperature and Voltage Ranges per Operation Ranges table)

Figure	Symbol	Parameter		Guaranteed Limit	Unit
1	f	Clock Frequency, SCLK Note: Refer to t_{wH}, t_{wL} below	(10-bit xfer) Min (11- to 16-bit xfer) Min (10- to 16-bit xfer) Max	0 Note 1 2.1	MHz
1	f	Clock Frequency, ADCLK Note: Refer to t_{wH}, t_{wL} below	Minimum Maximum	500 2.1	kHz MHz
1	t_{wH}	Minimum Clock High Time	ADCLK SCLK	190 190	ns
1	t_{wL}	Minimum Clock Low Time	ADCLK SCLK	190 190	ns
1, 7	t_{PLH}, t_{PHL}	Maximum Propagation Delay, SCLK to D_{out}		240	ns
1, 7	t_h	Minimum Hold Time, SCLK to D_{out}		10	ns
2, 7	t_{PLZ}, t_{PHZ}	Maximum Propagation Delay, $\overline{CS}$ to D_{out} High-Z		150	ns
2, 7	t_{PZL}, t_{PZH}	Maximum Propagation Delay, $\overline{CS}$ to D_{out} Driven	MC145050 MC145051	2 ADCLK cycles + 300 4.3	ns μs
3	t_{su}	Minimum Setup Time, D_{in} to SCLK		100	ns
3	t_h	Minimum Hold Time, SCLK to D_{in}		0	ns
4, 7, 8	t_d	Maximum Delay Time, EOC to D_{out} (MSB)	MC145051	100	ns
5	t_{su}	Minimum Setup Time, $\overline{CS}$ to SCLK	MC145050 MC145051	2 ADCLK cycles + 425 4.425	ns μs
—	t_{CSd}	Minimum Time Required Between 10th SCLK Falling Edge ($\leq$0.8 V) and $\overline{CS}$ to Allow a Conversion	MC145050 MC145051	44 Note 2	ADCLK cycles
—	t_{CAs}	Maximum Delay Between 10th SCLK Falling Edge ($\leq$2 V) and $\overline{CS}$ to Abort a Conversion	MC145050 MC145051	36 9	ADCLK cycles μs
5	t_h	Minimum Hold Time, Last SCLK to $\overline{CS}$		0	ns
6, 8	t_{PHL}	Maximum Propagation Delay, 10th SCLK to EOC	MC145051	4.35	μs
1	t_r, t_f	Maximum Input Rise and Fall Times	SCLK ADCLK D_{in}, $\overline{CS}$	1 250 10	ms ns μs
1, 4, 6-8	t_{TLH}, t_{THL}	Maximum Output Transition Time, Any Output		300	ns
—	C_{in}	Maximum Input Capacitance	AN0-AN10 ADCLK, SCLK, $\overline{CS}$, D_{in}	55 15	pF
—	C_{out}	Maximum Three-State Output Capacitance	D_{out}	15	pF

NOTES:
1. After the 10th SCLK falling edge ($\leq$2 V), at least 1 SCLK rising edge ($\geq$2 V) must occur within 38 ADCLKs (MC145050) or 18.5 μs (MC145051).
2. On the MC145051, a $\overline{CS}$ edge may be received immediately after an active transition on the EOC pin.

MC145050•MC145051

SWITCHING WAVEFORMS

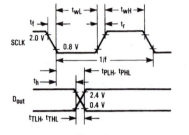

Figure 1

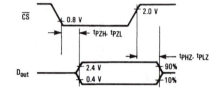

Figure 2

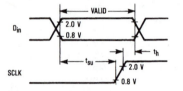

Figure 3

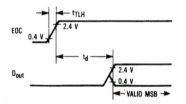

NOTE: D$_{out}$ is driven only when $\overline{CS}$ is active (low).

Figure 4

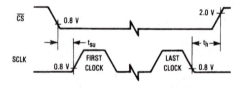

Figure 5

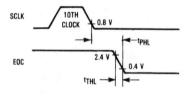

Figure 6

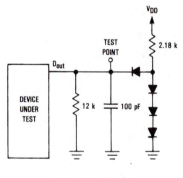

Figure 7. Test Circuit

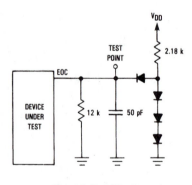

Figure 8. Test Circuit

MC145050•MC145051

PIN DESCRIPTIONS

DIGITAL INPUTS AND OUTPUTS

The various serial bit-stream formats for the MC145050/51 are illustrated in the timing diagrams of Figures 9 through 14. Table 1 assists selection of the appropriate diagram. Note that the ADCs accept 16 clocks which makes them SPI (serial peripheral interface) compatible.

Table 1. Timing Diagram Selection

No. of Clocks in Serial Transfer	Using $\overline{CS}$	Serial Transfer Interval	Figure No.
10	Yes	Don't Care	9
10	No	Don't Care	10
11 to 16	Yes	Shorter than Conversion	11
16	No	Shorter than Conversion	12
11 to 16	Yes	Longer than Conversion	13
16	No	Longer than Conversion	14

$\overline{CS}$ (Pin 15)

Active-Low Chip Select Input. Chip select initializes the chip to perform conversions and provides 3-state control of the data output pin (D_{out}). While inactive high, $\overline{CS}$ forces D_{out} to the high-impedance state and disables the data input (D_{in}) and serial clock (SCLK) pins. A high-to-low transition on $\overline{CS}$ resets the serial data port and synchronizes it to the MPU data stream. $\overline{CS}$ can remain active during the conversion cycle and can stay in the active low state for multiple serial transfers or $\overline{CS}$ can be inactive high after each transfer. If $\overline{CS}$ is kept active low between transfers, the length of each transfer is limited to either 10 or 16 SCLK cycles. If $\overline{CS}$ is in the inactive high state between transfers, each transfer can be anywhere from 10 to 16 SCLK cycles long. See the SCLK pin description for a more detailed discussion of these requirements.

Spurious chip selects caused by system noise are minimized by the internal circuitry. Any transitions on the $\overline{CS}$ pin are recognized as valid only if the level is maintained for a setup time plus two falling edges of ADCLK after the transition.

NOTE

If $\overline{CS}$ is inactive high after the 10th SCLK cycle and then goes active low before the A/D conversion is complete, the conversion is aborted and the chip enters the initial state, ready for another serial transfer/conversion sequence. At this point, the output data register contains the result from the conversion before the aborted conversion. Note that the last step of the A/D conversion sequence is to update the output data register with the result. Therefore, if $\overline{CS}$ goes active low in an attempt to abort the conversion too close to the end of the conversion sequence, the result register may be corrupted and the chip could be thrown out of sync with the processor until $\overline{CS}$ is toggled again (refer to the AC Electrical Characteristics in the spec tables).

D_{out} (Pin 16)

Serial Data Output of the A/D Conversion Result. This output is in the high-impedance state when $\overline{CS}$ is inactive high. When the chip recognizes a valid active low on $\overline{CS}$, D_{out} is taken out of the high-impedance state and is driven with the MSB of the previous conversion result. (For the first transfer after power-up, data on D_{out} is undefined for the entire transfer.) The value on D_{out} changes to the second most significant result bit upon the first falling edge of SCLK. The remaining result bits are shifted out in order, with the LSB appearing on D_{out} upon the ninth falling edge of SCLK. Note that the order of the transfer is MSB to LSB. Upon the 10th falling edge of SCLK, D_{out} is immediately driven low (if allowed by $\overline{CS}$) so that transfers of more than 10 SCLKs read zeroes as the unused LSBs.

When $\overline{CS}$ is held active low between transfers, D_{out} is driven from a low level to the MSB of the conversion result for three cases: Case 1—upon the 16th SCLK falling edge if the transfer is longer than the conversion time (Figure 14); Case 2—upon completion of a conversion for a 16-bit transfer interval shorter than the conversion (Figure 12); Case 3—upon completion of a conversion for a 10-bit transfer (Figure 10).

D_{in} (Pin 17)

Serial Data Input. The four-bit serial input stream begins with the MSB of the analog mux address (or the user test mode) that is to be converted next. The address is shifted in on the first four rising edges of SCLK. After the four mux address bits have been received, the data on D_{in} is ignored for the remainder of the present serial transfer. See Table 2 in **Applications Information**.

SCLK (Pin 18)

Serial Data Clock. This clock input drives the internal I/O state machine to perform three major functions: (1) drives the data shift registers to simultaneously shift in the next mux address from the D_{in} pin and shift out the previous conversion result on the D_{out} pin, (2) begins sampling the analog voltage onto the RC DAC as soon as the new mux address is available, and (3) transfers control to the A/D conversion state machine (driven by ADCLK) after the last bit of the previous conversion result has been shifted out on the D_{out} pin.

The serial data shift registers are completely static, allowing SCLK rates down to dc in a continuous or intermittent mode. There are some cases, however, that require a minimum SCLK frequency as discussed later in this section. SCLK need not be synchronous to ADCLK. At least ten SCLK cycles are required for each simultaneous data transfer. After the serial port has been initiated to perform a serial transfer*, the new

---

*The serial port can be initiated in three ways: (1) a recognized $\overline{CS}$ falling edge, (2) the end of an A/D conversion if the port is performing either a 10-bit or a 16-bit "shorter-than-conversion" transfer with $\overline{CS}$ active low between transfers, and (3) the 16th falling edge of SCLK if the port is performing 16-bit "longer-than-conversion" transfers with $\overline{CS}$ active low between transfers.

MC145050•MC145051

mux address is shifted in on the first four rising edges of SCLK, and the previous 10-bit conversion result is shifted out on the first nine falling edges of SCLK. After the fourth rising edge of SCLK, the new mux address is available; therefore, on the next edge of SCLK (the fourth falling edge), the analog input voltage on the selected mux input begins charging the RC DAC and continues to do so until the tenth falling edge of SCLK. After this tenth SCLK edge, the analog input voltage is disabled from the RC DAC and the RC DAC begins the "hold" portion of the A/D conversion sequence. Also upon this tenth SCLK edge, control of the internal circuitry is transferred to ADCLK which drives the successive approximation logic to complete the conversion. If 16 SCLK cycles are used during each transfer, then there is a constraint on the minimum SCLK frequency. Specifically, there must be at least one rising edge on SCLK before the A/D conversion is complete. If the SCLK frequency is too low and a rising edge does not occur during the conversion, the chip is thrown out of sync with the processor and $\overline{CS}$ needs to be toggled in order to restore proper operation. If 10 SCLKs are used per transfer, then there is no lower frequency limit on SCLK. Also note that if the ADC is operated such that $\overline{CS}$ is inactive high between transfers, then the number of SCLK cycles per transfer can be anything between 10 and 16 cycles, but the "rising edge" constraint is still in effect if more than 10 SCLKs are used. (If $\overline{CS}$ stays active low for multiple transfers, the number of SCLK cycles must be either 10 or 16.)

ADCLK (Pin 19, MC145050 Only)

This pin clocks the dynamic A/D conversion sequence, and may be asynchronous to SCLK. Control of the chip passes to ADCLK after the tenth falling edge of SCLK. Control of the chip is passed back to SCLK after the successive approximation conversion sequence is complete (44 ADCLK cycles), or after a valid chip select is recognized. ADCLK also drives the $\overline{CS}$ recognition logic. The chip ignores transitions on $\overline{CS}$ unless the state remains for a setup time plus two falling edges of ADCLK. The source driving ADCLK must be free running.

EOC (Pin 19, MC145051 Only)

End-of-Conversion Output. EOC goes low on the tenth falling edge of SCLK. A low-to-high transition on EOC occurs when the A/D conversion is complete and the data is ready for transfer.

ANALOG INPUTS AND TEST MODE

AN0 through AN10 (Pins 1-9, 11, 12)

Analog Multiplexer Inputs. The input AN0 is addressed by loading $0 into the mux address register. AN1 is addressed by $1, AN2 by $2, ..., AN10 by $A. Table 2 shows the input format for a 16-bit stream. The mux features a break-before-make switching structure to minimize noise injection into the analog inputs. The source resistance driving these inputs must be ≤ 10 kΩ.

There are three tests available that verify the functionality of all the control logic as well as the successive approximation comparator. These tests are performed by addressing $B, $C, or $D and they convert a voltage of $(V_{ref} + V_{AG})/2$, V_{AG}, or V_{ref}, respectively. The voltages are obtained internally by sampling V_{ref} or V_{AG} onto the appropriate elements of the RC DAC during the sample phase. Addressing $B, $C, or $D produces an output of $200 (half scale), $000, or $3FF (full scale), respectively, if the converter is functioning properly. However, deviation from these values occurs in the presence of sufficient system noise (external to the chip) on V_{DD}, V_{SS}, V_{ref}, or V_{AG}.

POWER AND REFERENCE PINS

V_{SS} and V_{DD} (Pins 10 and 20)

Device Supply Pins. V_{SS} is normally connected to digital ground; V_{DD} is connected to a positive digital supply voltage. Low frequency $(V_{DD} - V_{SS})$ variations over the range of 4.5 to 5.5 volts do not affect the A/D accuracy. (See the Operations Ranges table for restrictions on V_{ref} and V_{AG} relative to V_{DD} and V_{SS}.) Excessive inductance in the V_{DD} or V_{SS} lines, as on automatic test equipment, may cause A/D offsets $> \pm 1$ LSB. Use of a 0.1 μF bypass capacitor across these pins is recommended.

V_{AG} and V_{ref} (Pins 13 and 14)

Analog reference voltage pins which determine the lower and upper boundary of the A/D conversion. Analog input voltages $\geq V_{ref}$ produce a full scale output and input voltages $\leq V_{AG}$ produce an output of zero. CAUTION: The analog input voltage must be $\geq V_{SS}$ and $\leq V_{DD}$. The A/D conversion result is ratiometric to $V_{ref} - V_{AG}$. V_{ref} and V_{AG} must be as noise-free as possible to avoid degradation of the A/D conversion. Ideally, V_{ref} and V_{AG} should be single-point connected to the voltage supply driving the system's transducers. Use of a 0.2 μF bypass capacitor across these pins is strongly urged.

MC145050•MC145051

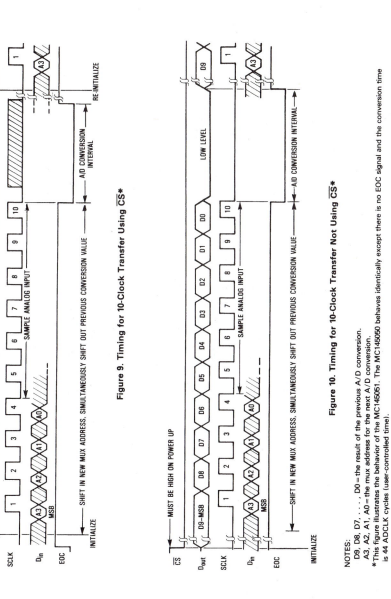

Figure 9. Timing for 10-Clock Transfer Using $\overline{CS}$*

Figure 10. Timing for 10-Clock Transfer Not Using $\overline{CS}$*

NOTES:
D9, D8, D7, . . . , D0 = the result of the previous A/D conversion.
A3, A2, A1, A0 = the mux address for the next A/D conversion.
*This figure illustrates the behavior of the MC145051. The MC145050 behaves identically except there is no EOC signal and the conversion time is 44 ADCLK cycles (user-controlled time).

MC145050•MC145051

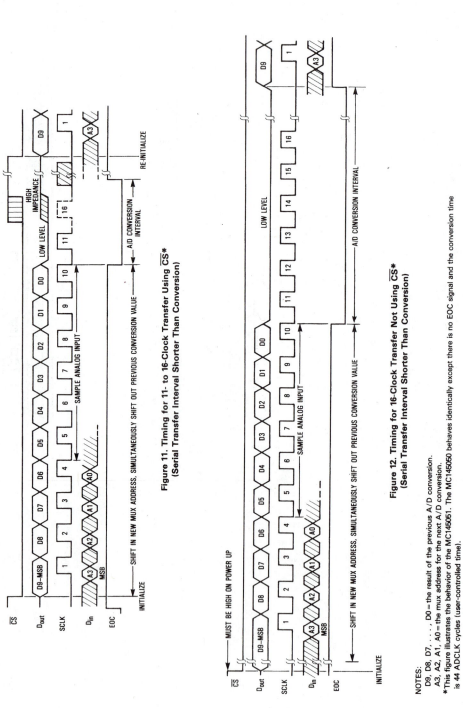

Figure 11. Timing for 11- to 16-Clock Transfer Using CS*
(Serial Transfer Interval Shorter Than Conversion)

Figure 12. Timing for 16-Clock Transfer Not Using CS*
(Serial Transfer Interval Shorter Than Conversion)

NOTES:
D9, D8, D7, . . . , D0 = the result of the previous A/D conversion.
A3, A2, A1, A0 = the mux address for the next A/D conversion.
*This figure illustrates the behavior of the MC145051. The MC145050 behaves identically except there is no EOC signal and the conversion time is 44 ADCLK cycles (user-controlled time).

MC145050•MC145051

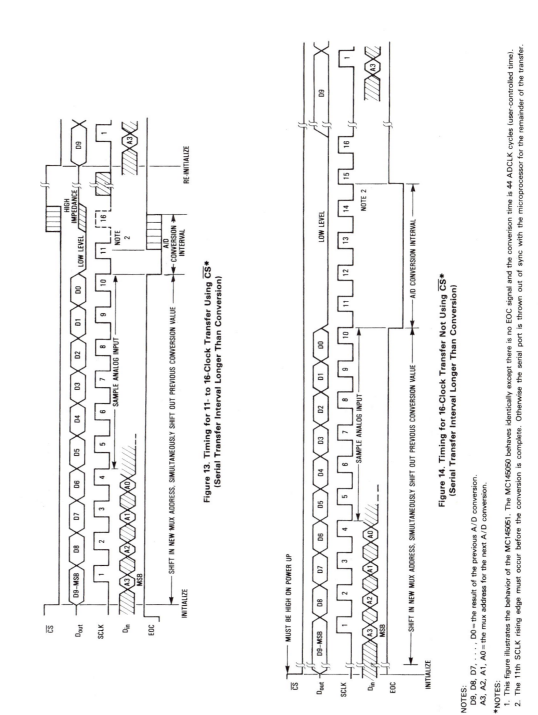

Figure 13. Timing for 11- to 16-Clock Transfer Using CS*
(Serial Transfer Interval Longer Than Conversion)

Figure 14. Timing for 16-Clock Transfer Not Using CS*
(Serial Transfer Interval Longer Than Conversion)

NOTES:
D9, D8, D7, ... , D0 = the result of the previous A/D conversion.
A3, A2, A1, A0 = the mux address for the next A/D conversion.

*NOTES:
1. This figure illustrates the behavior of the MC145051. The MC145050 behaves identically except there is no EOC signal and the conversion time is 44 ADCLK cycles (user-controlled time).
2. The 11th SCLK rising edge must occur before the conversion is complete. Otherwise the serial port is thrown out of sync with the microprocessor for the remainder of the transfer.

MC145050•MC145051

APPLICATIONS INFORMATION

DESCRIPTION

This example application of the MC145050/MC145051 ADCs interfaces three controllers to a microprocessor and processes data in real-time for a video game. The standard joystick X-axis (left/right) and Y-axis (up/down) controls as well as engine thrust controls are accommodated.

Figure 15 illustrates how the MC145050/MC145051 is used as a cost-effective means to simplify this type of circuit design. Utilizing one ADC, three controllers are interfaced to a CMOS or NMOS microprocessor with a serial peripheral interface (SPI) port. Processors with National Semiconductor's MICROWIRE serial port may also be used. Full duplex operation optimizes throughput for this system.

DIGITAL DESIGN CONSIDERATIONS

Motorola's MC68HC05C4 CMOS MCU may be chosen to reduce power supply size and cost. The NMOS MCUs may be used if power consumption is not critical. A V_{DD} to V_{SS} 0.1 μF bypass capacitor should be closely mounted to the ADC.

Both the MC145050 and MC145051 accommodate all the analog system inputs. The MC145050, when used with a 2 MHz MCU, takes 27 μs to sample the analog input, perform the conversion, and transfer the serial data at 2 MHz. Forty-four ADCLK cycles (2 MHz at input pin 19) must be provided and counted by the MCU before reading the ADC results. The MC145051 has the end-of-conversion (EOC) signal (at output pin 19) to define when data is ready, but has a slower 93 μs cycle time. However, the 93 μs is constant for serial data rates of 2 MHz independent of the MCU clock frequency. Therefore, the MC145051 may be used with the CMOS MCU operating at reduced clock rates to minimize power consumption without severely sacrificing ADC cycle times, with EOC being used to generate an interrupt. (The MC145051 may also be used with MCUs which do not provide a system clock.)

ANALOG DESIGN CONSIDERATIONS

Controllers with output impedances of less than 10 kilohms may be directly interfaced to these ADCs, eliminating the need for buffer amplifiers. Separate lines connect the V_{ref} and V_{AG} pins on the ADC with the controllers to provide isolation from system noise.

Although not indicated in Figure 15, the V_{ref} and controller output lines may need to be shielded, depending on their length and electrical environment. This should be verified during prototyping with an oscilloscope. If shielding is required, a twisted pair or foil-shielded wire (not coax) is appropriate for this low frequency application. One wire of the pair or the shield must be V_{AG}.

A reference circuit voltage of 5 volts is used for this application. The reference circuitry may be as simple as tying V_{AG} to system ground and V_{ref} to the system's positive supply. (See Figure 16.) However, the system power supply noise may require that a separate supply be used for the voltage reference. This supply must provide source current for V_{ref} as well as current for the controller potentiometers.

A bypass capacitor of approximately 0.22 μF across the V_{ref} and V_{AG} pins is recommended. These pins are adjacent on the ADC package which facilitates mounting the capacitor very close to the ADC.

SOFTWARE CONSIDERATIONS

The software flow for acquisition is straightforward. The nine analog inputs, AN0 through AN8, are scanned by reading the analog value of the previously addressed channel into the MCU and sending the address of the next channel to be read to the ADC, simultaneously.

If the design is realized using the MC145050, 44 ADCLK cycles (at pin 19) must be counted by the MCU to allow time for A/D conversion. The designer utilizing the MC145051 has the end-of-conversion signal (at pin 19) to define the conversion interval. EOC may be used to generate an interrupt, which is serviced by reading the serial data from the ADC. The software flow should then process and format the data, and transfer the information to the video circuitry for updating the display.

When these ADCs are used with a 16-bit (2-byte) transfer, there are two types of offsets involved. In the first type of offset, the channel information sent to the ADCs is offset by 12 bits. That is, in the 16-bit stream, only the first 4 bits (4 MSBs) contain the channel information. The balance of the bits are don't cares. This results in 3 don't-care nibbles, as shown in Table 2. The second type of offset is in the conversion result returned from the ADCs; this is offset by 6 bits. In the 16-bit stream, the first 10 bits (10 MSBs) contain the conversion results. The last 6 bits are zeroes. The hexadecimal result is shown in the first column of Table 3. The second column shows the result after the offset is removed by a microprocessor routine.

MC145050•MC145051

Table 2. Programmer's Guide for 16-Bit Transfers: Input Code

Input Address in Hex	Channel To Be Converted Next	Comment
$0XXX	AN0	Pin 1
$1XXX	AN1	Pin 2
$2XXX	AN2	Pin 3
$3XXX	AN3	Pin 4
$4XXX	AN4	Pin 5
$5XXX	AN5	Pin 6
$6XXX	AN6	Pin 7
$7XXX	AN7	Pin 8
$8XXX	AN8	Pin 9
$9XXX	AN9	Pin 11
$AXXX	AN10	Pin 12
$BXXX	AN11	Half Scale Test: Output = $8000
$CXXX	AN12	Zero Test: Output = $0000
$DXXX	AN13	Full Scale Test: Output = $FFC0
$EXXX	None	Not Allowed
$FXXX	None	Not Allowed

Table 3. Programmer's Guide for 16-Bit Transfers: Output Code

Conversion Result Without Offset Removed	Conversion Result With Offset Removed	Value
$0000	$0000	Zero
$0040	$0001	Zero + 1 LSB
$0080	$0002	Zero + 2 LSBs
$00C0	$0003	Zero + 3 LSBs
$0100	$0004	Zero + 4 LSBs
$0140	$0005	Zero + 5 LSBs
$0180	$0006	Zero + 6 LSBs
$01C0	$0007	Zero + 7 LSBs
$0200	$0008	Zero + 8 LSBs
$0240	$0009	Zero + 9 LSBs
$0280	$000A	Zero + 10 LSBs
$02C0	$000B	Zero + 11 LSBs
•	•	•
•	•	•
•	•	•
$FF40	$03FD	Full Scale–2 LSB
$FF80	$03FE	Full Scale–1 LSB
$FFC0	$03FF	Full Scale

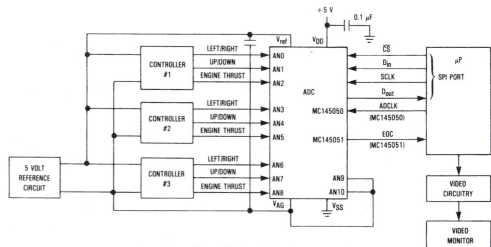

Figure 15. Joystick Interface

MC145050•MC145051

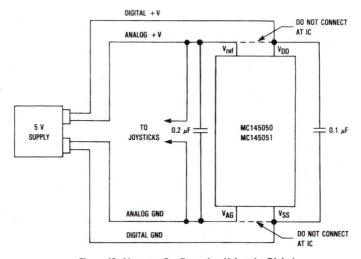

**Figure 16. Alternate Configuration Using the Digital
Supply for the Reference Voltage**

Compatible Motorola MCUs/MPUs

This is not a complete listing of Motorola's MCUs/MPUs.
Contact your Motorola representative if you need additional
information.

Instruction Set	Memory (Bytes)		SPI [1] SCI [2]	Device Number
	ROM	EEPROM		
M6805	2096	—	—	MC68HC05C2
	2096	—	Yes	MC68HC05C3
	4160	—	Yes	MC68HC05C4
	4160 [3]	—	Yes	MC68HSC05C4
	8K [3]	—	Yes	MC68HSC05C8
	4160 [4]	—	Yes	MC68HCL05C4
	8K [4]	—	Yes	MC68HCL05C8
	7700	—	Yes	MC68HC05C8
	—	4160	—	MC68HC805C4
M68000	—	—	—	MC68HC000

[1] SPI = Serial Peripheral Interface.

[2] SCI = Serial Communications Interface.

[3] High speed.

[4] Low power.

References

Uyless D. Black. *Data Communications and Distributed Networks.* 2d ed. Englewood Cliffs, N.J.: Prentice-Hall, 1987.

Uyless D. Black. *Data Networks, Concepts, Theory, and Practice.* Englewood Cliffs, N.J.: Prentice-Hall, 1989.

Frederick F. Driscoll, Robert F. Coughlin, and Robert S. Villangucci. *Data Acquisition and Process Control with the M68HC11 Microcontroller.* Westerville, Ohio, Merrill, 1994.

John Fulcher. *An Introduction to Microcomputer System: Architecture and Interfacing.* Reading, Mass.: Addison Wesley, 1989.

Joseph D. Greenfield. *The 68HC11 Microcontroller.* Philadelphia: W. B. Saunders, 1992.

G. J. Lipovski, *Single- and Multiple-Chip Microcomputer Interfacing.* Englewood Cliffs, N.J.: Prentice-Hall, 1988.

Gene H. Miller, *Microcomputer Engineering.* Englewood Cliffs, N.J.: Prentice-Hall, 1993.

Motorola, *CMOS Application-Specific Standard ICs.* Phoenix, AZ: Motorola, 1990.

Motorola, *M68HC11 Reference Manual.* Phoenix, AZ: Motorola, 1991.

Motorola, *M68HC11A8 Technical Data.* Phoenix, AZ: Motorola, 1991.

Motorola, *Industrial Control Applications.* Phoenix, AZ: Motorola, 1991.

Motorola, *8-bit MCU Applications Manual.* Phoenix, AZ: Motorola, 1992.

John B. Peatman, *Design with Microcontrollers.* New York: McGraw-Hill, 1988.

Peter Spasov. *Microcontroller Technology: The 68HC11.* Englewood Cliffs, N.J.: Prentice-Hall, 1993.

William Stallings. *Handbook of Computer Communications Standards, Vol. I.* Carmel, Ind.: Howard W. Sams & Company, 1987.

INDEX